1993

# Global Warming and Biological Diversity

# Global

# Warming

EDITED BY
ROBERT L. PETERS &
THOMAS E. LOVEJOY

# and

# Biological

# Diversity

YALE UNIVERSITY PRESS
NEW HAVEN & LONDON

Published with assistance from the Louis Stern Memorial Fund.

Set in Joanna type by Keystone Typesetting, Inc., Orwigsburg, Pennsylvania. Printed in the United States of America by Hamilton Printing Company, Castleton, New York.

*Library of Congress*
*Cataloging-in-Publication Data*
Global warming and biological diversity / edited by Robert L. Peters and Thomas E. Lovejoy.
    p.  cm.
  Includes bibliographical references and index.
  ISBN 0-300-05056-9
    1. Global warming—Congresses.
2. Bioclimatology—Congresses.   3. Biological diversity conservation—Congresses.
4. Greenhouse effect, Atmospheric—Congresses.
5. Biological diversity—Congresses. 6. Climatic changes—Congresses.   I. Peters, Robert L., 1951–  .  II. Lovejoy, Thomas E.
QH543.G56   1992
574.5'222—dc20              91-33532

A catalogue record for this book is available from the British Library.

The paper in this book meets the guidelines for permanence and durability of the Committee on Production Guidelines for Book Longevity of the Council on Library Resources.

10   9   8   7   6   5   4   3   2   1

This book is dedicated in gratitude and admiration to
 H. John Heinz III
 G. Evelyn Hutchinson
 Walter E. Westman
for all they did, each in his own way, on behalf of the
kindly fruits of the earth.

*T.E.L. and R.L.P.*

And to my parents, Jean
Brookhart and Bob Peters,
whose love and faith made
this book possible.

*R.L.P.*

As I did stand my watch upon the hill,
I look'd toward Birnam, and anon,
   methought,
The wood began to move.

Shakespeare, Macbeth

# Contents

# Foreword: The Wrong Time For Climate Change

MICHAEL E. SOULÉ

Rapid global climate change could not be happening at a worse time. Massive changes such as those elaborated in this volume would have been less disastrous for biological diversity and integrity if they had occurred in the past. A century ago—in a world less transformed by humans—organisms would have had a better chance of adapting to rapidly increasing temperatures, rising sea levels, and unpredictable changes in precipitation.

The situation then was much more forgiving, because in most regions the biosphere was still seamless; the number of humans on the planet a century ago was much lower than today—about one person was living in the 1890s for every five alive in the 1990s. Another reason for the difference is technology. A hundred years ago, bulldozers, chain saws, power plants, automobiles, and automatic weapons were unimagined. Though parts of the world had already been denatured, including the Mediterranean region and much of Europe, the Far East, and eastern North America, the tropics had yet to be pillaged, either by the hunter with his AK-47 or by the logger with bulldozer and chain saw. The rending of continental habitats into fragments is a largely modern phenomenon. Highway construction, logging, damming, overhunting, overfishing, overgrazing, acid rain, ozone depletion, and eutrophication are modern phenomena, something suddenly, something terribly new.

In the 1890s the forests were not so full of hunters, and people did not have much fire power. Nor was there an international market for tropical wildlife, such as primates, to sell to zoos or laboratories. The demand for reptile skin, furs, ivory, and rhinoceros horn was far smaller than today, and there was almost no international trade in tropical hardwoods, tropical beef, sugarcane, cocoa, cocaine, pineapple, and coffee.

In the temperate zones, forests stretched

north to south in blankets rather than in crazy quilts. Rivers flowed in undammed innocence. Deserts and prairies were unskinned of their topsoil. In North America, grizzly bears and wolves, chestnuts and elms, passenger pigeons and condors, wetlands and old-growth forests flourished.

Now, imagine that the predictions for climate alteration had been borne out a century ago or more, before the explosion of human numbers and technology. It is likely that the expected increases in temperature and severe and unpredictable changes in precipitation would have led to ecological transformation on a vast scale (as did the glaciations some thousands of years before). Nevertheless, the situation would not be nearly as bleak for the organisms or for the humans trying to save them.

For one thing, engineered physical barriers were much less ubiquitous. A century ago, few dams would have prevented the movement of aquatic organisms up and down rivers. Humans had cleared few low and tropical areas, so animals could move freely over coastal plains and between mountain ranges. Such places as Central America, eastern Brazil, the foothills of the Andes, central and southern Africa, most of Madagascar, the Philippines, Indonesia, and Southeast Asia—areas with significant topographic relief—were sparsely settled. Today humans have usurped these arable and loggable lands. A century ago, therefore, the distributions of species were much larger and more continuous, particularly in the lowlands and in coastal areas.

If rapid climate change had occurred under these relatively benign conditions, many species could have adjusted, especially those in regions with considerable topographic relief. Much of the lowland biota could still have found safety in the cooler and more diverse climates of mountainous areas. Now, however, many wild areas have disappeared, and much of the lowland flora and fauna, if they still persist, are prevented from reaching the mountains or find the lower slopes already occupied by reservoirs, highways,

farms, and plantations, not to mention armed humans.

Now, consider the consequences of climate change in the future. For the sake of argument, assume that it is the year 2100 but that the human population and environmental conditions are the same as today. In addition, assume that science has continued to develop normally. Would the management expertise and biological technology of 2100 be sufficiently advanced to deal with a holocaust of the kind anticipated by most of the authors in this volume? In some ways, it would. There is little doubt that advances in biotechnology and conservation biology will bring more effective and economical ways of protecting species and ecosystems from massive extinction and loss of function.

There is little doubt, for example, that completely new technologies will facilitate the storing of DNA from the entire range of taxa. Our descendants will be able to reconstitute species and, yes, entire species assemblages from such genetic archives. This is not to say that a more advanced civilization would choose to convert its biosphere to a homosphere completely controlled by humans, or would consider it desirable to put hundreds of millions of years of evolution in cold storage. Wiser humans would, we hope, live more gently on their planet and would control their runaway natality.

In 2100 conservation biologists will be able to anticipate and treat many of the problems that are only now being recognized and studied. Today's journals are full of articles with important lessons about the control of aggressive alien weeds, about predators and diseases, about subtle but dangerous edge effects that nibble away at nature reserves, about population viability and extinction processes, about the role of so-called keystone species, large and small, plant and animal, about ways to restore fertility to abused landscapes, about ways to link reserves and remnants, and about strategies to obtain and maintain the support of local people in conservation programs. But progress is impeded by lack of funding. Many biologists are frustrated by their inability to fund the training and research needed to understand fragmented systems and to maintain and restore them, particularly in the tropics. We do have one thing in surplus, though: talent.

The situation is similar in the critical fields of animal and plant reproduction. The technologies for storing and using germ plasm, such as artificial insemination and embryo storage and transfer, exist for only a dozen or so species of vertebrates. Seeds from most tropical plants cannot be stored at all, and those plants that are today being kept in germ plasm collections are not being replaced and regenerated fast enough, mostly because of lack of funds.

The point is that our knowledge today is insufficient to construct and manage a comprehensive biodiversity management system. In addition, corruption, social injustice, political immaturity, and instability greatly exacerbate the current crisis. Disparities of wealth between and within nations tend to destabilize the conservation infrastructure, but they make compelling demands on any funds that become available as a result of the thawing of the Cold War. Besides, competition for peace dividends from the biomedical, space, and physics communities will be intense. The only hope for funding the necessary research on ecological processes, biogeography, genetics, behavior, physiology, systematics, restoration, sociology, and economics may rest on initiatives like the proposed National Institutes for the Environment.

It is painfully apparent that we still lack the knowledge or the funds to confront massive global climate change and to prevent a major setback for biological diversity. In spite of all the rhetoric, there is no national or international commitment to save biological diversity. This could change if climate deteriorates fast enough and public opinion is galvanized. Conservation scientists and managers in the United States ask for a mere $100 million a year,[1] while the president of the United States wants to spend $10 billion a year to plant a flag on planet Mars. The irony probably escapes him.

The good news is that books like this one are exactly the right medicine for a society too easily lulled into apathy by simplistic solutions like backyard tree planting. Citizens and scientists are finally becoming invested in the greatest challenge ever faced by any sentient species—the rescue of its planet.

1. This is an estimate of the cost of implementing the recommendation in *Research Priorities for Conservation Biology*, edited by M. E. Soulé and K. Kohm and published by Island Press (1989).

# Preface

THOMAS E. LOVEJOY

Exobiologists are scientists who study the possibility of life elsewhere in the universe. Like oceanographers, but on a galactic scale, they are forced to study what goes on at a great distance and by taking tiny samples. Had there been exobiologists on some other planet studying the earth over the past several centuries they would have had to do so principally by examining the composition of the atmosphere and the changing reflectivity of the planet. The presence of certain gases like methane, which are the product of life processes and hence termed biogenic, would have suggested to them the presence of life on earth.

Such scientists would also have detected the relatively rapid change in atmospheric composition over the last couple of centuries, including the 30% increase in $CO_2$ and the arrival of previously unknown compounds, the chlorofluorocarbons. By planetary standards, this is a very rapid change, differing in rate from the past except for cataclysmic events such as volcanic eruptions or the posited meteoric impact at the end of the Cretaceous Period. Those exobiologists would have concluded that something highly unusual is going on.

In an important sense, the unnatural heating of the planet is not a problem unless it causes biological dislocations. In the end, the best measure of this is biological diversity, locally, regionally, and globally—for the fauna and flora of this planet are not only fundamental resources but also the most sensitive indicators of environmental change. Ruth Patrick has built a career around the way the numbers and kinds of species (especially diatoms) reveal the chemistry and health of aquatic systems (in particular, rivers). This volume examines that principle on a global scale.

It has become apparent that biological systems and biological diversity are not only vulnerable to severe consequences from climate change but also significantly involved in the generation of greenhouse gases. Alberto Setzer's report from Brazil's National Institute for Space Studies (Instituto Nacional de Pesquisas Espaciais) on the burning of Amazon forests in 1987 hit the front pages of the world's newspapers in 1988: 178,000 individual fires were detected by weather satellite, and first estimates were that $20 \times 10^6$ km$^2$ of Brazilian Amazonia had been burned, of which $8 \times 10^6$ km$^2$ was forest, equal to 2% of the Brazilian Amazon forest. It is now understood that the remote sensing techniques of the Advanced Very High Resolution Radiometer (AVHRR) tend to overestimate the area of burning. Our estimates of the extent of biomass burning in the tropics generally are still imprecise, as are our estimates of the amount of biological diversity destroyed in the process. Happily, there has also been, for many reasons, a distinct decrease in burning in Amazonian Brazil. Still, the numbers are big, and the rough estimate is that burning of biomass throughout the world contributes between one and three billion tons of carbon as $CO_2$ to the atmosphere annually.

The chapters that follow are laid out to lead the reader to understand the complex yet largely unknown consequences of global climatic change for biological diversity. The first section provides the necessary insights into global warming and projected climate change so that biological impacts can be considered. It is followed by a retrospective look at climate change and response by vegetation and mammalian fauna—essentially harnessing knowledge of the past to help envision the possible future. Possible biological responses to climate change are the focus of two large sections. One looks at responses from different perspectives: the marine environment, vegetation, the soil, forests. Also included are responses at various levels—physiological, behavioral, ecological, and special cases, such as those of migratory animals and epidemiological systems. The subsequent section provides site-specific and regional perspectives on the consequences for biological diversity.

There are two major differences from the past tens of million of years as the specter of climate change confronts the biota. One is the likelihood that rates of change will be faster than the flora and fauna have ever experienced. Put most simplistically, will species be able to move as fast as their preferred climate, or will they languish behind under increasingly unfavorable climatic regimes until they vanish? The second difference is that they must respond in a highly modified landscape. Increasingly, the terrestrial biota is confined to isolated parks and reserves that are essentially locked in by human populations. Even if species are able to move quickly enough to track their preferred climate, they will have to do so within a major obstacle course set by society's massive conversion of the landscape. As my colleague Robert Peters put it, a species may be impelled to move, but Los Angeles will be in the way.

The final section centers on the conservation implications of climate change. Each reader will draw his or her own conclusions from the chapters that follow, but in my view, unless there is major change, the artificial heating of the planet is likely to generate a tsunami of extinctions equal to that currently playing itself out in the tropical forest regions of the world. And as biological diversity goes up in smoke with tropical deforestation, the $CO_2$ generated is augmenting climate change, which will cause a second major reduction in biological diversity.

Among the repercussions the greenhouse effect is likely to have, the hardest to mitigate is the loss of biological diversity. The impact on agriculture can, in theory at least, be handled by the development of new strains and agricultural extension. The Netherlands can build higher dikes, and Maldivians can be relocated. But landscapes are already so modified that there are limited opportunities for augmenting dispersal by designing corridors. And again one would have to assume rates of dispersal beyond any hitherto demonstrated.

It appears that under normal climate oscillations in the past species did not migrate together, but in the end, new communities resulted composed of different combinations of species. Certainly, species of obvious importance such as known relatives of current crop species can be (and are) maintained in seed banks. But it defies the imagination to protect biological diversity solely through artificial propagation when science is currently unable to estimate the number of species on earth to within an order of magnitude. What is likely to happen is something akin to the great biotic crises of the past. Essentially, it will be like dropping the hand of biotic playing cards, retrieving a random subset, and then reshuffling them—hardly stable ecological ground for a society growing by a billion people each decade.

As dark as the biological preview of climate change is, in another sense, it is biology that gives us hope for avoiding climate change. The power of combined biological activity is enormous: each annual downswing in the overall climb of $CO_2$ levels in the atmosphere amounts to six billion tons of carbon removed from the atmosphere by the photosynthetic activity of the northern hemisphere growing season. Elimination or drastic reduction of forest burning in the tropics plus a massive reforestation project worldwide could easily eliminate two billion tons of $CO_2$ from the average annual net increase of three and a half billion tons. That represents only a thirty-year purchase of time, because as the reforestation matures its rate of carbon storage tapers off. That nonetheless gives us time to work on new energy scenarios, to cope with the legitimate aspirations of the developing countries, and to use the power of energy conservation. In the end, the power

of biological systems is the only hope that we may move to avoid climate change without grave economic dislocation.

It is understood on an increasingly wider basis that biological diversity is important as a set of fundamental resources, as the ultimate library for the development of the life sciences, and as the best indicators of environmental stress and change. Even a single species can make major transformations in society, whether through medical breakthroughs, such as the cowpox virus that generated the concept of vaccination, or whether through a sense of beauty and wonder. Isn't that what a gift of flowers—an act reenacted billions of times—is really all about?

Today biological diversity is signaling that the sheer numbers of people combined with our effects on the environment have almost reached the point of no return. Like a giant hermit crab, the collective effect of humanity has outgrown its home. Unlike a hermit crab, we have no larger shell to seek. Instead, we must look for ways to readjust our lives so as not to despoil our home.

I am not one of those who feel we must retreat into a thoughtless Luddite existence. Yet it is clear that tinkering with the biosphere is not the solution. It is, after all, a highly complex system with ten million to thirty million different kinds of working parts and myriad feedback systems positive and negative, each with its own lag time, of which we have little ken. In the end, the biosphere is too intricate, too complex, and too unpredictable to manage under any circumstances other than an atmosphere with a stable composition. That is the challenge to society at the end of the twentieth century. Otherwise, "antediluvian" will take on an additional and unfortunate twenty-first-century meaning.

# Acknowledgments

Both this book and the conference that stimulated it were made possible by farsighted individuals and organizations who realized the importance of the potential consequences of global warming on biological diversity. Foremost is World Wildlife Fund–U.S., which was the primary convener of the conference. The conference was also supported by the Joyce Mertz-Gilmore Foundation, the U.S. Environmental Protection Agency, the National Science Foundation, the U.S. Department of Energy, and the U.S. National Park Service. World Wildlife Fund–U.S., the W. Alton Jones Foundation, and the Joyce Mertz-Gilmore Foundation provided substantial support necessary for preparation of the manuscript.

We particularly want to thank Bob Crane and Larry Condon (Joyce Mertz-Gilmore), J. P. Myers and Patricia Edgerton (W. Alton Jones), Roger Dahlman (DOE), Steve Seidel, Joel Smith, and William Reilly (EPA), Patrick Webber and Frank Harris (NSF), Bill Greg and Boyd Evison (NPS), Russell Train, Catherine Fuller, and Paige McDonald (WWF), Russ Mittermeier (Conservation International), and Congresswoman Claudine Schneider for their confidence in this project and their help in its execution. We would like to thank our Yale University Press editors Edward Tripp and Jean Thomson Black for their faith in the book and gentle prodding, and Stacy C. Roberts who did a superb job of handling logistics for the conference. Two outside readers, Robert D. Holt (University of Kansas) and an anonymous scholar, provided excellent, detailed reviews of the entire manuscript. Finally, we would like to thank the contributing authors for their willingness to lead the way in thinking about this new issue and for their helpfulness and patience during the editing process.

# PART I

# Overview: *What Is Global Warming?*

CHAPTER ONE

# Introduction

ROBERT L. PETERS

## I. OVERVIEW

The chapters in this book are based on presentations given at the World Wildlife Fund's Conference on Consequences of the Greenhouse Effect for Biological Diversity, which was held October 4–6, 1988, at the National Zoological Park in Washington, D.C. It was the first meeting to focus on the effects of global warming on the conservation of natural ecosystems. The primary goal was to ensure that future scientific and policy discussions of global warming would pay adequate attention to natural ecosystems, an area that had previously been overlooked, in contrast to the attention focused on agriculture and the possible effects of rising sea levels on coastal habitations. This conference therefore was designed to pull together existing information about ecological responses to climate, to stimulate biologists into focusing their efforts on global warming, to draw general conclusions about conservation consequences, and to communicate these conclusions to the scientific, policy, funding, and conservation management communities.

In preparing for the conference, we identified experts in a variety of fields, including animal and plant physiology, ecology, animal behavior, and epidemiology. Some workers were knowledgeable about specific ecosystems, including tropical forests, eastern North American deciduous forests, arctic tundra, and arctic marine systems. Others were synthesists who focused on interactions between various environmental components, such as precipitation and soil chemistry, and on synergisms between climate change and other human activities, such as deforestation. Stephen H. Schneider of the National Center for Atmospheric Research provided the scientists with a generic global warming scenario (chapter 4), based on computer models of future climate, and we challenged the scientists to respond to this scenario by projecting how their particular ecological systems would be likely to respond. In some cases the projections were based on ecological computer models that were tightly linked to the climate projections, whereas in other cases the projections were based on deduction and knowledge of how ecological systems have responded to past climate changes. We asked each specialist to identify not only what is currently known about how a particular system would respond but also what is not known. In pulling together current knowledge, the conference thus identified gaps and suggested future research.

## II. THE CLIMATE SCENARIO

The climate scenario that the biologists' projections were based on (chapter 4) was originally developed by Stephen Schneider, Linda Mearns, and Peter Gleick for a report to the American Association for the Advancement of Science. The scenario presents plausible forecasts of change for a variety of climate variables, including temperature, precipitation, and sea level rise (see table 4.1), which were used by the biologists as the basis for assessing impacts. The forecasts are based on state-of-the-art modeling results from several general circulation models (GCMs), which are complex computer simulations of climate (chapter 4 discusses the nature and limitations of GCMs).

The scenario reflects the view of most climatologists that a doubling of preindustrial $CO_2$ concentrations, or the equivalent warming caused by a combination of $CO_2$ and other greenhouse gases, would cause a warming of $3° \pm 1.5°C$ (NRC 1983, WMO 1982). Some recent estimates of future warming have suggested that the warming could be even larger, including an estimate by D. A. Lashof (1989), who suggests that given positive feedbacks, even $8°-10°C$ is possible. How fast will this warming be reached? The Intergovernmental Panel on Climate Change (IPCC 1990) predicts that the most likely rate of warming, given present knowledge, is "$1°C$

above the present value (about $2°C$ above that in the preindustrial period) by 2025 and $3°C$ above today's (about $4°C$ above preindustrial) before the end of the next century." At this rate, given that even a $1°C$ rise can have a large effect on ecological systems, they are likely to experience substantial stress within the next fifty years. Warming could be faster or slower that this estimate—the exact rate will depend heavily on whether people are successful in curtailing production of greenhouse gases and on the significance of what are now poorly understood feedback processes.

Two points with important conservation implications are stressed by Schneider et al. First, the projected rate of warming would be fast compared with past normal warmings, perhaps fifty times as fast. Therefore, even if climatologists are underestimating the rate of warming by a factor of two, the warming will still occur many times faster than usual. Second, along with heating would come changes—often increases—in the frequency of such extreme events as fires, hurricanes, and droughts, and those events may be more important than temperature change itself in changing patterns of biological diversity.

Climatologists are often asked how confident they are in the climate predictions, given the many areas of uncertainty. Most climatologists (and the NRC, IPCC, and WMO) assign high probability to warming for several reasons. First, there is no doubt about the basic greenhouse theory. We know that a planet's temperature is strongly affected by its atmospheric composition—it is the earth's present gas mix that keeps the atmosphere warm enough for life; without it the earth's average temperature would be about $-18°C$ (Gribben 1988). Similarly, it is an excess of carbon dioxide that makes Venus too hot and a lack of it that makes Mars too cold. Given that the greenhouse effect is accepted, where does the uncertainty lie? Scientists are cautious because many aspects of the climate system are poorly understood, and a number of them have the potential to dampen warming, including certain types of increased cloud cover, ocean currents, and anthropogenic aerosol production. Although research and debate continue, at present none has been convincingly demonstrated as likely to substantially offset warming. Alternatively, it is equally plausible that positive feedbacks could make the warming even greater than projected; one possibility is that atmospheric warming will melt permafrost, thawing frozen stores of peat, which will then decompose to release additional $CO_2$ and methane (see Post et al. 1990 for a review of what is known and not known about the global carbon cycle). So, on balance, we are left with a strong, identified force for warming—the greenhouse effect—and several possible forces for either warming or cooling, the magnitude and certainty of which are in doubt.

An additional source of confidence is that the climate models do a reasonable job of predicting both present-day seasonal temperatures and ancient paleoclimates (see chapter 5 for an example of how fossil plant data are used to test a GCM). These results lead Schneider et al. to conclude that "taken together, these verifications provide a strong circumstantial case that the modeling of sensitivity of temperature to greenhouse gases is probably valid within the oft cited three-fold range" (1.5°–4.5°C).

A final piece of suggestive evidence is that global mean temperature has shown a large increase of 0.5°C during the past century (see Schneider 1989:84, Jones and Wigley 1990) at the same time greenhouse concentrations have been rising. Although most climatologists believe we cannot yet determine statistically whether this warming is or is not due to the greenhouse effect, the trend is in the expected direction.

Given that evidence, we may not be certain about future climate, but we can use our best guess in making conservation decisions. We are in the position of a homeowner who has bought a house in the Santa Monica mountains of California, only to find that the best guess of fire-control experts is that a continuing buildup of shrubby vegetation will fuel a large, home-destroying fire during the next

thirty years. The experts admit there are uncertainties—it is possible that some unknown factor will stunt the vegetation before the fuel load reaches a critical level, or it is possible that unpredictable changes in weather patterns will decrease the risk—but their best guess is that the fire will happen. I think most homeowners in this situation would sell the house and move, or they would buy insurance, collectively manage the vegetation to decrease fire probability, and put a fire-resistant roof on the house. In dealing with the greenhouse effect, we do not have the option of moving, which leaves us with preventing the warming or mitigating the effects.

Whether or not this particular unnatural warming occurs, conservation will never again be thought of as being carried out in a stable environment. We now realize that climate fluctuations, whether they be warmer or colder, will change and stress natural ecosystems, providing a challenge to the flexibility of our conservation strategies.

## III. GENERAL CONCLUSIONS ABOUT BIODIVERSITY

In the analogy above, the worst-case outcome is clear—the house will burn down. Understanding the effects of global warming on natural ecosystems, however, is more complicated. The most general conclusion to be drawn from the conference is that many ecological systems will be dramatically changed by warming. Among the projections are large shifts in the ranges of species, hundreds of kilometers toward the north in the north temperate zone, causing ecological communities to break up and reassort. Also, many species extinctions are likely, given that for some species climate will become unsuitable in much or all of their present range. In some cases, entire food webs may be disturbed, as suggested by Vera Alexander (chapter 17) for the Arctic marine ecosystem.

Unfortunately for accurate prediction, the nature of these effects is complicated and poorly understood (chapter 24 gives a good overview of the degree of complexity). We have already mentioned the many uncertainties surrounding the climate projections. Although there is reasonable confidence in projections of an increase in global average temperature—3°C or more during the next century—there is poor resolution at the regional level, let alone at the local level where conservation managers must act. Projections for precipitation changes associated with warming are even poorer. In most cases it is unknown at the regional level whether precipitation will increase or decrease. Provisional projections have been made at the continental level, suggesting for example that continental interiors, notably in North America, may become significantly drier. To those uncertainties must be added effects on and interactions among soil, water, atmospheric chemistry, sea level rise, and storm and fire frequencies. George Woodwell (chapter 3) outlines some of the ways in which biospheric feedbacks may act to affect the rate of warming. For example, a "1°C increase in temperature is widely recognized as increasing the rate of respiration 10%–30% while having little effect on photosynthesis," thereby stimulating the release of both carbon dioxide and methane, adding to the greenhouse effect.

Synergistic interactions with other human-caused disturbances, including acid rain and other pervasive pollutants, stratospheric ozone depletion, and habitat destruction, must all be accounted for in anticipating the future state of biological diversity. Added to those physical changes must be the indirect effects of changes in the biota caused by warming, including the migration into protected areas of new pathogens, competitors, and predators, including those introduced by people.

## IV. SPECIFIC PROJECTIONS OF POTENTIAL BIOLOGICAL RESPONSES

Because vegetation type is a primary determinant of ecosystem type, playing a major role in determining the associated fauna and soil microbiota (chapter 9), it is important that we predict changes in the distributions and compositions of plant associations. Ian Woodward (chapter 8) uses a GCM coupled to vegetation models to project where vegetation types would be located throughout the world if there were an equivalent doubling of carbon dioxide. Other workers have focused on vegetation changes in specific locations, notably North American temperate forests. Margaret Davis and Catherine Zabinski (chapter 22) present future range maps for four important eastern trees: sugar maple, beech, yellow birch, and hemlock. They project that in response to 3°C of global warming these species would die out in the southern parts of their ranges, withdrawing hundreds or a thousand kilometers or more toward Canada. These four species may be thought of as representing the eastern temperate deciduous forest, and die-offs of the many forest species sharing the same ecological requirements as these trees would change the face of eastern North America. Reserves in this area would be likely to lose most of the species within them. Davis and Zabinski note that many understory plants may be even more susceptible to warming effects than the dominant trees.

Daniel Botkin and Robert Nisbet (chapter 21) complement the work of Davis and Zabinski by modeling climate-induced changes in temperate eastern forest at specific geographic sites. They project that a site in the Boundary Waters Canoe Area of Minnesota could change over from the present balsam fir–dominated softwood forest to an eastern hardwood forest, possibly within as little as thirty years. Near Grayling, Michigan, the jack pine that is essential habitat for the endangered Kirtland's warbler could die out and be replaced by a sugar maple–dominated forest, also within approximately thirty years.

Kirtland's warbler provides just one example of how animal species would be affected by changes in vegetation. Hank Shugart and Tom Smith (chapter 11) describe how computer models of vegetation response to a variety of environmental variables, including climate variables, can be used to predict the carrying capacity of habitat for a variety of bird species. Dan Rubenstein (chapter 14) observes that the social structure of many mammals, including topi antelope, elephants, and feral horses, changes as vegetation structure and quality change in response to precipitation. Such changes in social structure in turn can affect survivorship or the outcome of sexual competition.

Climate will also have direct effects on animals: William Dawson (chapter 12) points out that some reptiles, birds, and many mammals cope with thermal stress by actively evaporating water (panting and sweating), so rising temperatures coupled with decreased water availability may limit species ranges. He believes that reproductive biology may be particularly susceptible to thermal stress, citing, among other research, studies on domesticated mammals showing decreased fertility and fetal survival in response to increases in ambient temperature. Ectothermic animals, such as insects, whose body temperatures vary widely with temperature, may show more dramatic changes in basic physiology and thereby geographic range due to direct effects of climate variables. For example, Rubenstein models how changes in temperature or humidity can alter substantially the rates of metabolism, fecundity, and feeding in insects, in turn determining the range of species. Richard Tracy (chapter 13) discusses how changing thermal regimes might change the composition of animal communities, favoring those species that are relatively adapted to high temperatures. He uses the example of a community of darkling beetle species, each of which inhabits a slightly different thermal niche. He also presents data from power plant cooling ponds that suggest warming may simplify animal communities by decreasing the number of species present.

In addition to latitudinal changes in species ranges, a local increase in temperature would also cause vegetation and associated animals to shift their altitude upward approximately 500 meters for a 3°C warming (Peters and Darling 1985). Because mountain tops are smaller than bottoms, a species shifting upward will generally experience a decrease in the size of its range. Dennis Murphy and Stuart Weiss (chapter 26) focus their attention on butterfly and mammal populations in mountain ranges of the U.S. Great Basin, projecting, that based on species-area relationships, that many local extinctions would occur as ranges decrease, leading to an estimated 23% loss of butterfly species per mountain range. Not surprisingly, sedentary species would be hardest hit, with a 30% loss. They project that mammals would lose approximately 44% of their species. For both groups, small mountainous areas would lose more species than large ones. (Also see Leverenz and Lev 1987 for modeling results projecting upward shifting of coniferous tree species in the western United States, with resultant range decreases for some commercially important tree species.)

The ability of a species to shift its range will depend on its dispersal mechanisms. Birds are good dispersers because they can fly, mammals can walk, and plants can send out seeds to create new colonies where the climate becomes suitable. Their ability to disperse, however, depends not only on their intrinsic dispersal rate—whether they are highly mobile dispersers—but also on whether they face barriers to dispersal. I note in chapter 2 that even under past natural rates of climate change, barriers such as mountains have caused extinctions where species were squeezed up against them and could not follow the shifting climate. In the future environment, most species will be isolated in habitat islands surrounded by roads, cities, and fields. Norman Myers (chapter 25) reemphasizes this, pointing out that interaction between habitat destruction and climate change is synergistic, in that the combination of the two forces would threaten more species than the sum of their individual efforts.

If the rate of climate change is so fast that the preferred climate travels north faster than a species can follow, extinction is possible. Thompson Webb (chapter 5) calls this mismatch between climate change and the response of a species (or vegetation type) "disequilibrium." Davis and Zabinski (chapter 22) provide estimates of average migration rates for North American trees, between 20 and 40 kilometers per century, which are too low by an order of magnitude to track shifting climate. In the projected warming, with warming rates as much as fifty times higher than normal, extinction is much more likely than during past natural warmings.

Because different species have different dispersal abilities and respond individually to various ecological forces, communities tend to fragment as species shift their ranges in different directions. Both Webb (chapter 5) and Russell Graham (chapter 6) present extensive evidence showing the breakup and reassortment of plant and animal associations during past climate changes. For instance, Webb describes how 18,000 years ago spruce and sedges grew together in open woodland associations, but by 10,000 years ago the spruce forest had closed and sedges were no longer associated with spruce. This differential shifting means that climate change will indirectly stress species by forcing them to cope with new predators, competitors, and diseases.

Changes in the distribution of pests, disease vectors, and diseases will affect natural resources, species of conservation concern, and human health. The forest industry, for example, is concerned about possible range expansion of pests destructive to valuable timber trees (Winget 1988). Andrew Dobson (chapter 16), an epidemiologist, focuses on the African tsetse fly, carrier of sleeping sickness. Areas in which sleeping sickness is prevalent cannot be used for cattle production and are therefore de facto wildlife refuges. Dobson projects that warming would cause a substantial shifting of the sleeping-sickness belt, opening wildlife areas to human settlement while making new areas unsuitable for cattle. He further projects that the United States could expect a northward expansion of important tropical disease vectors that are not now major problems, including the ma-larial mosquito. With respect to isolated populations of rare species within nature reserves, I point out in chapter 2 that dispersal rates for pests and diseases tend to be high, often more than 100 kilometers per year, but rates for other species may be much slower, such as the 40 or so kilometers per century that can be covered by tree dispersal. This means that isolated populations, while unable to track shifting climate themselves, may nonetheless be easily found by threatening diseases and pests.

Because warming will be much greater as one travels away from the equator, arctic ecosystems will experience even greater warming and ecological change than the temperate ecosystems. Several authors describe dramatic changes in tundra systems: Ian Woodward (chapter 8) suggests that tundra vegetation could be pushed as much as 4° of latitude toward the north. This echoes projections made elsewhere (Emanuel et al. 1985) that as much as 37% of tundra vegetation could be replaced by forest if climate warms an average of 3°C. Dwight Billings and Kim Peterson (chapter 18) focus on coastal wet tundra and suggest that warming would melt permafrost with subsequent thermokarst erosion and loss of peat and sediments. Vera Alexander (chapter 17) describes the critical importance of continued sea ice to the arctic marine food web. Without sea ice, which disappears from much of the Arctic Ocean under some warming scenarios, marine mammals would lack ice floes on which to rest, travel, and pup, and the effective growing season for phytoplankton, which the entire food chain depends on, would be significantly shortened. Although speculative, the result could be, she believes, the collapse of large marine mammal populations.

One group of animals that depends on tundra and thus could be negatively affected are migratory shorebirds. Pete Myers (chapter 15) describes the potential plight of shorebirds that need to synchronize their migrations with food availability. If shorebirds arrive in the arctic at their usual time, only to find that warming has caused insect abundance to peak early, there may not be enough food left to fledge their young. In general, migratory animals throughout the world will be at great risk because of the need for precise synchrony of movement and resources, a synchrony often mediated by climate.

Jumping to the tropics, where temperature rise is expected to be relatively small, Gary Hartshorn (chapter 10) stresses that the projected changes in rainfall patterns could cause substantial disturbance in tropical forests. Timing of fruiting and flowering are determined in large measure by the temporal distribution of droughts and rainy periods, and if these change, ecosystem effects can be severe. Hartshorn presents case studies for Barro Colorado Island in Panama and La Selva Biological Station in Costa Rica. On Barro Colorado, when rain continued during the normal dry season, normal flowering and fruiting failed and there was mass starvation and emigration among fruit-eating birds and mammals. But unusual *droughts* can also have severe consequences, including loss of a large proportion of adult trees. Another major source of disturbance in some tropical forests is hurricanes, which are projected to increase in frequency with warming. Massive blowdowns caused by hurricanes can change local ecologies substantially, and the storms themselves can directly threaten animal species such as the endangered Puerto Rican parrot (Snyder et al. 1987) and the whooping crane, both of which experienced significant population declines as the result of hurricanes during this century. More recently, in 1987 Hurricane Hugo destroyed 70% of the nesting trees of the largest population of the endangered red-cockaded woodpecker, found in the Francis Marion National Forest in South Carolina.

Disturbance regimes in general play major roles in determining suitability of habitat for species, and will play major roles in facilitating turnover from one species or vegetation type to another in response to climate change. Jerry Franklin et al. (chapter 19) report that adult coniferous trees in northwestern U.S. forests are relatively resilient to climate change and that the forest collectively ameliorates local climate. The result is that even with substantial warming, the forest, in the absence of major disturbance, would be able to survive for decades, at least until the mature trees die of old age. Franklin emphasizes, however, that turnover would actually be fairly rapid given an increased frequency of fires, caused by hotter, drier conditions. The same would be true of forest loss due to blowdowns or cutting. When the old forest is removed, in many areas conditions would not be suitable for its reestablishment. Walter Westman and George Malanson (chapter 20) make the same point for California chaparral, namely that increased disturbance by fire would play a major role in the turnover of vegetation types.

The estimate of sea level rise presented in chapter 4 is a range of 10–100 cm caused by an equivalent doubling of $CO_2$. That wide range reflects a high degree of uncertainty, in particular about the relative contribution of Antarctic melting, and a wide range of estimates by different studies (see IPCC 1990 for a table summarizing recent studies). The IPCC (1990) attempted to estimate how rapid this rise would be. It projects that for business as usual—that is, no significant reduction in greenhouse gas production—by 2030 sea level is likely to have risen 8–29 cm, with a best estimate of 18 cm. By 2070 it expects 21–71 cm, with a best estimate of 44 cm. Lest this amount of rise seem insignificant, it is important to realize that a relatively small increase in local sea level can translate into greatly increased rates of coastal erosion and height of storm surges. For example, Steve Leatherman (unpublished manuscript) has recounted the fate of several Chesapeake Bay islands and associated towns that were washed away over the past 100 years as local sea level rose only 30 cm. Further, some regional or local increases will be substantially larger than the global average—for example, the IPCC cites Mikolajewicz et al. (1990), whose "dynamic ocean model showed regional differences of up to a factor of two relative to the global-mean value."

Several authors, including Pete Myers (chapter 15), Larry Harris and Wendell Cropper (chapter 23), and Carleton Ray and coauthors (chapter 7), describe losses of coastal wetlands as a major concern. The EPA has projected that between 40% and 73% of all U.S. coastal marsh will be lost if the sea rises one meter (Titus et al. 1991). Harris describes effects on the southern Florida ecosystem, including "loss of the Everglades" to extensive saltwater intrusion. Among other things, climate-induced habitat change would further threaten the already endangered Florida panther (*Felis concolor*) and snail kite (*Rostrhamus sociabilis*).

## V. SLOWING THE RISE OF GREENHOUSE GASES

The scale of the possible disruption to natural ecosystems makes the slowdown of greenhouse gas production a top priority. This will be difficult, not only because fossil fuel use and other sources of greenhouse gases will probably increase as the world's population and economies grow, as pointed out by Woodwell (chapter 3), but also because effective action will demand a high degree of international cooperation. Nonetheless, many actions could slow the rise of greenhouse gas concentrations substantially.

To deal with carbon dioxide, which is the single greatest contributor to global warming (at present, roughly half the total), Woodwell argues the need for a phased but dramatic decrease in fossil fuel use, moving toward ultimate "abandonment of fossil fuels as the primary source of energy for industrial societies." Keystones of this policy would be energy conservation and technological shifts to renewable resources. A second important step is to halt destruction of forests, particularly high biomass forests, such as climax tropical rainforest and old-growth conifer forest in the American Northwest. Forest destruction may currently contribute nearly a fifth as much carbon to the atmosphere as fossil fuel use, and deforestation rates are rising rapidly (this contribution will ultimately

decline, either because no forests remain or—the preferable option—because they are protected). If carried out on a large scale, reforestation of areas presently without forest could soak up substantial amounts of carbon (chapter 8), although in the near term, reforestation cannot begin to make up for forest loss overall. (The argument that young forests store carbon more rapidly than mature ones and that, therefore, we can combat the greenhouse effect by cutting mature forests and replacing them with new forests is false. The huge trees of an old forest contain so much carbon that it would take centuries for a new forest to store the same amount.)

Methane, although not yet as important a greenhouse gas as carbon dioxide, is increasing in concentration twice as rapidly and, molecule for molecule, is twenty times more efficient at trapping heat. It may prove even more difficult to control than $CO_2$ because it is created by a wide variety of diffuse sources including rice paddies, swamps, cattle digestion, landfills, oil wells, coal mines, and burning tropical forests. As is the case with $CO_2$, we can diminish methane release by protecting forests and reducing dependency on fossil fuels, although many of the activities producing methane are too intimately tied to agricultural production for easy control.

The next-largest contributor to warming after carbon dioxide is the group of chlorofluorocarbons (CFCs). CFCs not only add significantly to global warming but also are the primary agents involved in thinning the ozone layer, which protects both plants and animals from ultraviolet radiation. CFCs offer the greatest opportunity for near-term reduction, by replacement with other compounds. Although the 1987 Montreal Protocol placed international limits on their use, it is generally accepted that more and faster reduction is necessary (see Gribben and Kelly 1989). There are lesser greenhouse gases, notably ozone and nitrous oxide, that have specific sources and possible control methods, but they will not be discussed here in detail.

Fortunately, we have the technological ability to make many of the necessary changes, provided we have the social will. One recent study concludes, with respect to carbon dioxide alone, that sensible steps to increase energy efficiency, reduce commitment to fossil fuels, and slow deforestation could decrease carbon dioxide's contribution by nearly two thirds, compared with doing business as usual (Mintzer 1987). (Obviously, there are large uncertainties in such projections, dependent as they are on assumptions about future patterns of population growth and resource utilization.) Further, the models in I. M. Mintzer's study predicted that the total contribution of greenhouse gas to warming could vary by a factor of four, depending on whether strong measures to control sources are implemented. Therefore, the good news is that we still have the ability to decrease future warming substantially, provided that changes begin soon (for other discussions of strategies see Oppenheimer and Boyle 1990, Lyman et al. 1990, Gribben and Kelly 1989, Schneider 1989, Ogden and Williams 1989).

## VI. RESEARCH NEEDS

The bad news is that even if all greenhouse gas production stopped immediately, concentrations already in the atmosphere would cause enough warming (approximately 1°C) for substantial ecological disruption. Moreover, given the reluctance of governments to undertake economically painful reforms in energy, industry, forest, and agriculture policy to deal with a perceptually distant threat, we can expect substantial emissions to continue for some time. Therefore, conservationists and scientists concerned with natural ecosystems must begin research to understand and mitigate negative effects on biological diversity.

Panel members of a discussion group on research needs at World Wildlife Fund's conference suggested a number of areas in which research should be concentrated:

• Because species ranges will shift and communities will experience invasion by new species, we need to understand dispersal mechanisms: At what speed can we expect particular species to move? Why are some species good dispersers, while others are not? What characteristics of a species, and of the community it is invading, make invasion easy or difficult?

• Another important area is the relation between ecosystem structure and stability. Because many ecosystems will be losing species (as well as gaining some new ones), we should understand how loss of species will affect, for example, an ecosystem's productivity and resistance to further destabilization. Could the biological systems supporting human life be threatened by the high loss of species predicted under conditions of global warming?

• Species that are likely to be particularly sensitive to climate effects should be identified, with ecological study and conservation plans designed as appropriate.

• We need monitoring programs that can first gather baseline ecological information on, for example, community composition, species ranges, and reproductive success, and then monitor subsequent changes. Such programs might be established in protected areas, such as the U.S. national parks, and in other sites of ecological importance, such as altitudinal or latitudinal ecotones.

• We need to improve our links between different scales. Ideally, a vegetation modeler should be able to obtain physiological data taken on the leaf or plant scale and use them to build up to large-scale vegetation responses on the plot, watershed, or larger scales, integrating with ecological and climate information.

• Biologists need to work with climatologists to help them develop climate models that provide the output necessary to answer biological questions. One of the critical needs is for GCMs with greater accuracy on the regional level. Also, climate models need to generate appropriate data on the frequency and spacing of climate events. For example, as Margaret Torn pointed out, knowing what the probability is of a 20-day gap in rainfall during fire season is much more useful for estimating fire probability than simply knowing the average summer rainfall.

- Many of the panelists and contributors to this book noted the profound lack of basic biological information about how species and communities are affected by climate. Dan Botkin said, "There is not sufficient information to make any kind of statement about the response of the major vegetation types to be able to enter into models." Specifically, the panelists called for greatly increased research on plant and animal physiology, genetics, ecology, and behavior. To take one example, plant physiology, the majority of experiments to date have been limited to testing the effects of a single atmospheric or climate variable, often $CO_2$ concentration, on a single plant species. Panelists called for multivariable experiments that will illuminate how organisms will respond to different combinations of $CO_2$, temperature, precipitation, and relative humidity. The next step, already being taken in a few cases (e.g., open-field $CO_2$ experiments), is to observe the effects of changes in climate and atmospheric variables on competition and herbivory.

- Several panelists emphasized the need to coordinate the creation of biological models with field or laboratory data collection. In this way quantitative, standardized data can be collected that will give realistic parameters for the models. Panelists said they need standardization so they "can compare one person's studies with another" (Dan Botkin).

- Many of the contributors argued that research on climate effects needs to be multidisciplinary to a degree unusual in ecological research, as exemplified by the merging of climate modeling, research on fossil plant distributions, and the resulting projections of future distributions. Bill Dawson said, "We have all plowed our own individual furrows, and I think that everybody is guilty of not looking up over the walls of that furrow." Dan Botkin observed that present academic structures provide inadequate rewards for either long-term

studies or interdisciplinary research. Steve Schneider spoke for the climatologists, saying, "I would hope ecologists would drop in frequently and find out what we are doing, give us midcourse direction, and tell us also why it is that these variables are important."

- Another common note at the conference was concern that more long-term studies are not being done. For example, Bill Dawson, an animal physiologist, observed that most studies in his field have been "acute" and there has been inadequate study of long-term physiological responses. One example of the type of sustained, climate-related ecological research considered necessary is the two decades of work by P. R. Ehrlich and his coworkers on population dynamics in checkerspot butterflies (Ehrlich 1965, Ehrlich et al. 1980).

- The needed research will require substantial increases in sustained funding. One difficulty in obtaining this funding has been identified by Steve Schneider, who notes that Congress, which funds most scientific research, works on two-, four-, and six-year cycles, while climate-change research needs funding on the scale of decades. He held out hope that funding agencies, particularly the National Science Foundation, are becoming more committed to long-term research, but, he said, "we ought to push it faster." Another difficulty recognized by the panel is that environmental scientists (in the broad sense) are less than effective at working together for common goals, unlike the physicists who have jointly persuaded the federal government to spend billions of dollars on mammoth projects. Charles Hall (Syracuse University) put the need for support succinctly: "I think we need to have a concerted effort to go to Congress and say, We need the kind of support you have been giving physicists. . . . You had better help us train the people who are going to understand these issues into the next century, because . . . it's a hell of a lot more important than atom smashers or your next aircraft carrier."

One intriguing effort toward developing a well-coordinated, well-funded scientific effort in climate change has been the recent formation of the Committee for a National Institute for the Environment (NIE), which proposes the creation of a government-funded NIE. The committee envisions the NIE as analogous to the National Institutes of Health, which have been tremendously successful in targeting and solving health problems while building a superb biomedical infrastructure in this country. Similarly, an NIE would provide a high-profile, coordinated way to address specific environmental problems by assessing need and sponsoring problem-solving research. Although an NIE is not necessarily the only answer to the current need for better funding, some such coordinated effort is necessary to raise the priority of climate-ecosystem research.

## VII. MITIGATING THREATS TO BIOLOGICAL DIVERSITY

In addition to basic research, the apparent inevitability of significant warming means the need for carefully planned conservation management will increase (chapter 2). Adequate numbers of large reserves will become more important than ever, and conservationists will finally have to get serious about dealing with the issue of nonreserve habitat. That is, since reserves alone will be inadequate to deal with the extent of the conservation crisis, we will have to find ways of increasing the conservation contribution of nonreserve habitat. Further, conservation will require the development of better techniques to control species abundances, to introduce missing species, and to rebuild communities. Therefore, restoration ecology will increase in importance, including ex situ and translocation techniques (see Jordan et al. 1988).

Because of the high level of uncertainty about regional climate changes, conservation managers will need to take a risk-management approach, preparing contingency plans for a variety of possible climatic futures. Be-

cause of the rapidity of the projected warming, and because of the poor state of knowledge about both ecosystem response and ecosystem management, it is time to begin the necessary research and planning.

## REFERENCES

Ehrlich, P. R. 1965. The population biology of the butterfly *Euphydryas edithia*: II. The structure of the Jasper Ridge Colony. *Evolution* 19:322.

Ehrlich, P. R., D. D. Murphy, M. C. Singer, C. B. Sherwood, R. R. White, and I. L. Brown. 1980. Extinction, reduction, stability, and increase: The responses of checkerspot butterfly populations to the California drought. *Oecologia* 46:101.

Emanuel, W. R., H. H. Shugart, and M. Stevenson. 1985. Response to comment: Climatic change and the broad-scale distribution of terrestrial ecosystem complexes. *Clim. Change* 7:457.

Gribbin, J. 1988. The greenhouse effect. *New Scientist, Inside Science* 13 (Oct. 22, 1988).

Gribbin, J., and M. Kelly. 1989. *Winds of Change: Living with the Greenhouse Effect*. London: Hodder and Stoughton, Headway.

IPCC (Intergovernmental Panel on Climate Change). 1990. *Climate Change: The IPCC Scientific Assessment*. J. T. Houghton, G. J. Jenkins, and J. J. Ephraums, eds. New York: Cambridge University Press.

Jones, P. D., and T. M. L. Wigley. 1990. Global warming trends. *Sci. Am.* 263(2):84.

Jordan, William R. III, R. L. Peters, and E. B. Allen. 1988. Ecological restoration as a strategy for conserving biological diversity. *Envir. Manag.* 12(1):55.

Lashof, D. A. 1989. The dynamic greenhouse: Feedback processes that may influence future concentrations of atmospheric trace gases and climatic change. *J. Climate Change* 14:213.

Leverenz, J. W., and D. J. Lev. 1987. Effects of $CO_2$-induced climate changes on the natural ranges of six major commercial tree species in the western U.S. Paper presented at the National Forest Products Assoc. Conference, "Rising $CO_2$ and Changing Climate: Forest Risks and Opportunities," Boulder, Colo., June 25–27, 1984.

Lyman, F., I. Mintzer, K. Courrier, and J. Mackenzie. 1990. *The Greenhouse Trap*. Boston: Beacon Press.

Mikolajewicz, U., B. Santer, and E. Maier-Reimer. 1990. Ocean response to greenhouse warming. Max-Planck-Institut für Meteorologie, Report 49.

Mintzer, I. M. 1987. *A Matter of Degrees: The Potential for Controlling the Greenhouse Effect*. Research Report 5.

Washington, D.C.: World Resources Institute.

NRC (National Research Council). 1983. *Changing Climate*. Washington, D.C.: National Academy Press.

Ogden, J. M., and R. H. Williams. 1989. *Solar Hydrogen: Moving beyond Fossil Fuels*. Washington, D.C.: World Resources Institute.

Oppenheimer, M., and R. H. Boyle. 1990. *Dead Head: The Race against the Greenhouse Effect*. New York: Basic Books.

Peters, R. L., and J. D. Darling. 1985. The greenhouse effect and nature reserves. *Bioscience* 35(11):707.

Post, W. M., T. Peng, W. R. Emanuel, A. W. King, V. H. Dale, and D. L. DeAngelis. 1990. The global carbon cycle. *Am. Scient.* 78:310.

Schneider, S. H. 1989. *Global Warming: Are We Entering the Greenhouse Century?* San Francisco: Sierra Club Books.

Snyder, N. F. R., J. W. Wiley, and C. B. Kepler. 1987. *The Parrots of Luquillo: Natural History and Conservation of the Puerto Rican Parrot*. Los Angeles: Western Foundation of Vertebrate Zoology.

Titus, J. G., R. Park, S. P. Leatherman, R. Weggel, M. Green, P. Mausel, S. Brown, C. Gaunt, M. Treehan, and G. Yohe. 1991. Greenhouse effect and sea level rise: Loss of land and the cost of holding back the sea. *Coastal Manag.* 19(2).

Winget, C. H. 1988. Forest management strategies to address climate change. In *Preparing for Climate Change*, Proceedings of the first North American conference on preparing for climate change: A cooperative approach, pp. 328–333. Rockville, Md.: Government Institutes.

WMO (World Meteorological Organization). 1982. *Report of the JSC/CAS: A Meeting of Experts on Detection of Possible Climate Change*, Moscow, October 1982. Rep. WCP29. Geneva: World Meteorological Organization.

# CHAPTER TWO

# Conservation of Biological Diversity in the Face of Climate Change

ROBERT L. PETERS

## I. INTRODUCTION

We can infer how the biota might respond to climate change by observing present and past distributions of plants and animals, which are largely determined by temperature and moisture patterns. For example, one race of the dwarf birch (*Betula nana*) can grow only where the temperature never exceeds 22°C (Ford 1982), suggesting that it would disappear from those areas where global warming causes temperatures to exceed 22°C. Recent historical observations of changes in range or species dominance, such as the gradual replacement of spruce (*Picea rubens*) by deciduous species during the past 180 years in the eastern United States, can also suggest future responses (Hamburg and Cogbill 1988). Insight into long-term responses to large climatic changes can be gleaned from studies of fossil distributions of, particularly, pollen (Davis 1983; see also chapters 5 and 22) and small mammals (Graham 1986; chapter 6). Such observations tell us that plants and animals are very sensitive to climate. Their ranges move when the climate patterns change—species die out in areas where they were once found and colonize new areas where the climate becomes newly suitable.

We can expect similar responses to projected global warming during the next 50 to 100 years, including disruption of natural communities and extinction of populations and species. Even many species that are today widespread will experience large range reductions. Efficient dispersers may be able to shift their ranges to take advantage of newly suitable habitat, but most species will at best experience a time lag before extensive colonization is possible and hence in the short term will show range diminishment (see chapter 5 for a discussion of vegetation-climate disequilibrium). At worst, many species will never be able to recover without human intervention since migration routes are cut off by development or other habitat loss caused by humans.

Although this chapter will focus on the terrestrial biota, ocean systems may show similar shifts in species ranges and community compositions if ocean waters warm or the patterns of water circulation change. For example, recent El Niño events demonstrate the vulnerability of primary productivity and species abundances to changes in ocean currents and local temperatures (Duffy 1983, Glynn 1984; see also chapters 7 and 17).

## II. THE NATURE OF ECOLOGICALLY SIGNIFICANT CHANGES

Although the exact rate and magnitude of future climate change is uncertain, given imperfect knowledge about the behavior of clouds, oceans, and biotic feedbacks, there is widespread consensus among climatologists that ecologically significant warming will occur during the next century. For example, the National Academy of Sciences has concluded that both global mean surface warming and an associated increase in global mean precipitation are "very probable" (NAS 1987).

It should be emphasized that although projections can be made about global averages, regional projections are much less certain (Schneider 1988). It is known that warming will not be even over the earth, with the high latitudes, for example, likely to be warmer than the low latitudes (Hansen et al. 1988). Regional and local peculiarities of topography and circulation will play a strong role in determining local climates.

For the purposes of discussion in this chapter, I will take average global warming to be 3°C, since that figure is a commonly used benchmark, but it must be recognized that additional warming well beyond 3°C may be reached during the next century if the production of anthropogenic greenhouse gases continues. I will also make the conservative assumption that 3°C warming is not reached until A.D. 2070. Additional warming or faster warming would cause additional biological disruption beyond that laid out here.

The threats to natural systems are serious for the following reasons. First, 3°C of warming would present natural systems with a warmer world than has been experienced in the past 100,000 years (Schneider and Londer 1984); 4°C would make the earth its warmest since the Eocene epoch, 40 million years ago (Barron 1985; see also chapter 5). This warming would not only be large compared with recent natural fluctuations, but it would also be fast, perhaps fifteen to forty times faster than past natural changes (chapter 4). Such a rate of change may exceed the ability of many species to adapt. Even widespread species are likely to have drastically curtailed ranges, at least in the short term. Moreover, human encroachment and habitat destruction will make wild populations of many species small and vulnerable to local climate changes.

Second, ecological stress would not be caused by temperature rise alone. Changes in global temperature patterns would trigger widespread alterations in rainfall patterns (Hansen et al. 1981, Kellogg and Schware 1981, Manabe et al. 1981), and we know that for many species precipitation is a more important determinant of survival than temperature per se. Indeed, except at treeline, rainfall is the primary determinant of vegetation structure, trees occurring only where annual precipitation exceeds 300 mm (chapter 8). Because of global warming, some regions would see dramatic increases in rainfall, and others would lose their present vegetation because of drought. For example, the U.S. Environmental Protection Agency (1989) concluded, based on several studies, that a long-term drying trend is likely in the midlatitude, interior continental regions during the summer. Specifically, W. W. Kellogg and R. Schware (1981), based on rainfall patterns during past warming periods, projected that substantial decreases in rainfall in North America's Great Plains are possible—perhaps as much as 40% by the early decades of the next century.

Other environmental factors important in determining vegetation type and health would change because of global warming:

Soil chemistry would change as, for example, changes in storm patterns alter leaching and erosion rates (Kellison and Weir 1987; chapter 24). Increased carbon dioxide concentrations may accelerate the growth of some plants at the expense of others, possibly destabilizing natural ecosystems (NRC 1983, Strain and Bazzaz 1983; see discussion in Woodward chapter 8). And rises in sea level may inundate coastal biological communities (NRC 1983, Hansen et al. 1981, Hoffman et al. 1983, Titus et al. 1984).

As mentioned, a variety of computer projections conclude that warming will be greater at higher latitudes (Hansen et al. 1987). This suggests that although tropical systems may be more diverse and are currently threatened by habitat destruction, temperate zone and arctic species may ultimately be in greater jeopardy from climate change, at least from temperature per se (see chapter 10 for discussion of precipitation effects on tropical forest). Arctic vegetation would experience widespread changes (chapters 8 and 18; Edlund 1987). A recent attempt to map climate-induced changes in world biotic communities projects that high-latitude communities would be particularly stressed, and boreal forest, for example, was projected to decrease by 37% in response to global warming of 3°C (Emanuel et al. 1985).

A final point, important in understanding species response to climate change, is that weather is variable, and extreme events, like droughts, floods, blizzards, and hot or cold spells, may have more effect on species distributions than average climate (e.g., Knopf and Sedgwick 1987). For example, in northwestern forests, global warming is expected to increase fire frequency, leading to rapid alteration of forest character (chapter 19).

## III. THE SHIFT OF SPECIES' RANGES IN RESPONSE TO CLIMATE CHANGE

We know that when temperature and rainfall patterns change, species' ranges change. Not surprisingly, species tend to track their climatic optima, retracting their ranges where conditions become unsuitable while expanding them where conditions improve (Peters and Darling 1985, Ford 1982). Even small temperature changes of less than one degree within this century have been observed to cause substantial range changes. For example, the white admiral butterfly (Ladoga camilla) and the comma butterfly (Polygonia c-album) greatly expanded their ranges in the British Isles during the past century as the climate warmed approximately 0.5°C (Ford 1982). The birch (Betula pubescens) responded rapidly to warming during the first half of this century by expanding its range north into the Swedish tundra (Kullman 1983).

On a larger ecological and temporal scale, entire vegetation types have shifted in response to past temperature changes no larger than those that may occur during the next 100 years or less (Baker 1983, Bernabo and Webb 1977, Butzer 1980, Flohn 1979, Muller 1979, Van Devender and Spaulding 1979). As the earth warms, species tend to shift to higher latitudes and altitudes. From a simplified point of view, rising temperatures have caused species to colonize new habitats toward the poles, often while their ranges contracted away from the equator as conditions there became unsuitable.

During several Pleistocene interglacials, the temperature in North America was apparently 2° to 3°C higher than now. Sweet gum trees (Liquidambar) grew in southern Ontario (Wright 1971); Osage oranges (Maclura) and pawpaws (Asimina) grew near Toronto, several hundred kilometers north of their present distributions; manatees swam in New Jersey; and tapirs and peccaries foraged in North Carolina (Dorf 1976). During the last of those interglacials, which ended more than 100,000 years ago, vegetation in northwestern Europe, which is now boreal, was predominantly temperate (Critchfield 1980). Other significant changes in species' ranges have been caused by altered precipitation accompanying past global warming, including expansion of prairie in the American Midwest during a global warming episode approximately 7000 years ago (Bernabo and Webb 1977).

It should not be imagined that because species tend to shift in the same general direction, existing biological communities move in synchrony. Conversely, because species shift at different rates in response to climate change, communities often dissociate into their component species (fig. 2.1). Recent studies of fossil packrat (*Neotoma* spp.) middens in the southwestern United States show that during the wetter, moderate climate of 22,000–12,000 years ago, there was not a concerted shift of plant communities. Instead, species responded individually to climatic change, forming stable but, by present-day standards, unusual assemblages of plants and animals (Van Devender and Spaulding 1979). In eastern North America, too, postglacial communities were often ephemeral associations of species, changing as individual ranges changed (Davis 1983, Graham 1986; chapter 6).

A final aspect of species response is that species may shift altitudinally as well as latitudinally. When climate warms, species shift upward. Generally, a short climb in altitude corresponds to a major shift in latitude: a 3°C cooling of 500 meters in elevation equals roughly 250 kilometers in latitude (MacArthur 1972). Thus, during the middle Holocene, when temperatures in eastern North America were 2°C warmer than at present, hemlock (*Tsuga canadensis*) and white pine (*Pinus strobus*) were found 350 meters higher on mountains than they are today (Davis 1983).

Because mountain peaks are smaller than bases, species that shift upward in response to warming typically occupy smaller and smaller areas, have smaller populations, and may thus become more vulnerable to genetic and environmental pressures (chapter 26). Species originally situated near mountaintops might have no habitat to move up to and may be entirely replaced by the relatively thermophilous species moving up from below (fig. 2.2). Examples of past extinctions attributed to upward shifting include alpine plants once living on mountains in Central and South America, where vegetation zones have

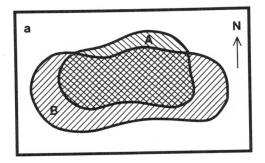

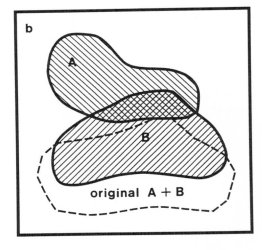

*Figure* 2.1. (a) Initial distribution of two species, A and B, whose ranges largely overlap. (b) In response to climate change, latitudinal shifting occurs at species-specific rates and the ranges dissociate.

shifted upward by 1000–1500 m since the last glacial maximum (Flenley 1979, Heusser 1974). See chapter 26 for projections of some local extinction rates for butterflies, birds, and mammals.

## A. Magnitude of Projected Latitudinal Shifts

If the proposed $CO_2$-induced warming occurs, species shifts similar to those in the Pleistocene epoch would occur, and vegetation belts would move hundreds of kilometers toward the poles (chapter 22; Frye 1983, Peters and Darling 1985). A 300-km shift in the temperate zone is a reasonable minimum estimate for a 3°C warming, based on the positions of vegetation zones during analogous warming periods in the past (Dorf 1976, Furley et al. 1983).

Additional confirmation that shifts of this magnitude or greater may occur comes from attempts to project future range shifts for some species by looking at their ecological requirements. For example, the forest industry is concerned about the future of commercially valuable North American species, like the loblolly pine (*Pinus taeda*). This species is limited on its southern border by moisture stress on seedlings. Based on its physiological requirements for temperature and moisture, W. F. Miller et al. (1987) projected that the southern range limit of the species would shift approximately 350 kilometers northward in response to a global warming of 3°C. Davis and Zabinski (chapter 22) have projected possible northward range withdrawals among several North American tree species, including sugar maple (*Acer saccharum*) and

beech (*Fagus grandifolia*), from 600 kilometers to as much as 2000 kilometers in response to the warming caused by a doubled $CO_2$ concentration. Beech would be most responsive, withdrawing from its present southern extent along the Gulf Coast and retreating into Canada.

## B. Mechanisms Underlying Range Shifts

The range shifts described above are the sum of many local processes of extinction and colonization that occur as climate changes the suitability of habitats. Those changes in habitat suitability are determined by both direct climate effects on physiology, including temperature and precipitation, and indirect effects secondarily caused by other species, themselves affected by temperature.

There are numerous examples of climate's direct influence on survival and consequently on distribution. In animals, the direct range-limiting effects of excessive warmth include lethality, as in corals (Glynn 1984), and interference with reproduction, as in the large blue butterfly *Maculinea arion* (Ford 1982). For insects, Daniel Rubenstein (chapter 14) shows that the way a critical trait, such as basic metabolic rate or fecundity, responds to temperature—whether the response is linear or dome-shaped—may determine the way in which a species' range alters in response to climate change.

In plants, excessive heat and associated decreases in soil moisture may decrease survival and reproduction (chapter 8). Coniferous seedlings, for example, are injured by soil temperatures above 45°C, although other types of plants can tolerate much higher temperatures (Daubenmire 1962). The northern limits of many plants are determined by minimum temperature isotherms below which some key physiological process does not occur. For instance, the gray hair grass (*Corynephorus canescens*) is largely unsuccessful at germinating seeds below 15°C and is bounded to the north by the 15°C July mean isotherm (Marshall 1978). Moisture extremes exceeding physiological tolerances also determine species' distributions. Thus, the European

*Figure 2.2.* (a) Present altitudinal distribution of three species, A, B, and C. (b) Species distribution after a 500-meter shift in altitude in response to a 3°C rise in temperature (based on Hopkin's bioclimatic law, MacArthur 1972). Species A becomes locally extinct. Species B shifts upward and the total area it occupies decreases. Species C becomes fragmented and restricted to a smaller area, while species D successfully colonizes the lowest habitats.

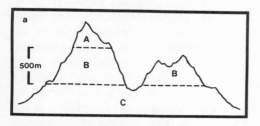

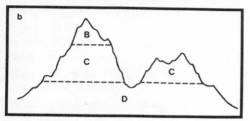

range of the beech tree (*Fagus sylvatica*) ends to the south where rainfall is less than 600 mm annually (Seddon 1971), and dog's mercury (*Mercurialis perennis*), an herb restricted to well-drained sites in Britain, cannot survive in soil where the water table reaches as high as 10 cm below the soil surface (Ford 1982).

The physiological adaptations of most species to climate are conservative, and it is unlikely that most species could evolve significantly new tolerances in the time allotted to them by the coming warming trend. Indeed, the evolutionary conservatism in thermal tolerance of many plant and animal species—beetles, for example (Coope 1977)—is the underlying assumption that allows us to infer past climates from faunal and plant assemblages.

Interspecific interactions altered by climate change will have a major role in determining new species distributions. Temperature can influence predation rates (chapter 14; Rand 1964), parasitism (Aho et al. 1976), and competitive interactions (Beauchamp and Ullyott 1932). C. Richard Tracy (chapter 13) describes how groups of species, in particular a genus of darkling beetles (*Eleodes*), may partition the thermal habitat by time of day and season. Climate-induced changes in the ranges of tree pathogens and parasites may be important in determining future tree distributions (Winget 1988). Soil moisture is a critical factor in mediating competitive interactions among plants, as is the case where dog's mercury (*Mercurialis perennis*) excludes oxlip (*Primula elatior*) from dry sites (Ford 1982).

Given the new associations of species that occur as climate changes, many species will face "exotic" competitors for the first time. Local extinctions may occur as climate change increases the frequency of droughts and fires, favoring invading species. One species that might spread, given such conditions, is *Melaleuca quinquenervia*, a bamboolike Australian eucalypt. This species has already invaded the Florida Everglades, forming dense monotypic stands where drainage and frequent fires have dried the natural marsh community (Courtenay 1978, Myers 1983).

The preceding effects, both direct and indirect, may act in synergy, as when drought makes a tree more vulnerable to attack by insect pests.

## C. Dispersal Rates and Barriers

The ability of species to adapt to changing conditions will depend heavily on their ability to track shifting climatic optima by dispersing colonists. In the case of warming, a North American species, for example, would most likely need to establish colonies to the north or at higher elevations. Survival of plant and animal species would therefore depend either on long-distance dispersal of colonists, such as seeds or migrating animals, or on rapid iterative colonization of nearby habitat until long-distance shifting results. A plant's intrinsic ability to colonize will depend on its ecological characteristics, including fecundity, viability and growth characteristics of seeds, nature of the dispersal mechanism, and ability to tolerate selfing and inbreeding upon colonization. If a species' intrinsic colonization ability is low, or if barriers to dispersal are present, extinction may result if all of its present habitat becomes unsuitable.

Many complete or local extinctions have occurred because species were unable to disperse rapidly enough when climate changed. For example, a large, diverse group of plant genera, including water shield (*Brasenia*), sweet gum (*Liquidambar*), tulip tree (*Liriodendron*), magnolia (*Magnolia*), moonseed (*Menispermum*), hemlock (*Tsuga*), arborvitae (*Thuja*), and white cedar (*Chamaecyparis*), had a circumpolar distribution in the Tertiary period (Tralau 1973). But during the Pleistocene ice ages, all went extinct in Europe while surviving in North America. Presumably, the east-west orientation of such barriers as the Pyrenees, the Alps, and the Mediterranean, which blocked southward migration, was partly responsible for their extinction (Tralau 1973). Other species of plants and animals thrived in Europe during the cold periods but could not survive conditions in postglacial forests. One such previously widespread dung beetle, *Aphodius hodereri*, is now extinct throughout the world except in the high Tibetan plateau where conditions remain cold enough for its survival (Cox and Moore 1985). Other species, like the Norwegian mugwort (*Artemisia norvegica*) and the springtail *Tetracanthella arctica*, now live primarily in the boreal zone but also survive in a few cold, mountaintop refugia in temperate Europe (Cox and Moore 1985).

Those natural changes were slow compared with predicted changes in the near future. The change to warmer conditions at the end of the last ice age spanned several thousand years yet is considered rapid by geologic standards (Davis 1983). We can deduce that if such a slow change was too fast for many species to adapt, the projected warming—possibly forty times faster—will have more severe consequences. For widespread, abundant species, like the loblolly pine (modeled by Miller et al. 1987), even substantial range retraction might not threaten extinction; but rare, localized species, whose entire ranges might become unsuitable, would be threatened unless dispersal and colonization were successful. Even for widespread species, major loss of important ecotypes and associated germplasm is likely (chapter 22).

A key question is whether the dispersal capabilities of most species prepare them to cope with the coming rapid warming. If the climatic optima of temperate-zone species do shift hundreds of kilometers toward the poles within the next 100 years, then those species would have to colonize new areas rapidly. To survive, a localized species whose entire present range becomes unsuitable might have to shift poleward at several hundred kilometers or more per century. Although some species, such as plants propagated by spores or "dust" seeds, may be able to match those rates, many species could not disperse fast enough to compensate for the expected climatic change without human assistance, particularly given the presence of dispersal barriers (Perring 1965, Rapoport 1982). Even wind-assisted dispersal may fall short of the mark for many species. In the case of the Engelmann spruce (*Picea engelman-*

*nii*), a tree with light, wind-dispersed seeds, fewer than 5% of seeds travel even 200 m downwind, leading to an estimated migration rate of 1–20 km per century (Seddon 1971); this reconciles well with rates derived from fossil evidence for North American trees of between 10 km and 45 km per century (chapter 22; also Roberts 1989). As described in the next section, many migration routes will likely be blocked by the cities, roads, and fields replacing natural habitat.

Although many animals may be physically capable of great mobility, the distribution of some is limited by the distributions of particular plants, that is, suitable habitat; their dispersal rates therefore may be largely determined by those of co-occurring plants. Behavior may also restrict dispersal even of animals physically capable of moving great distances. Dispersal rates below 2.0 km per year have been measured for several species of deer (Rapoport 1982), and many tropical deep-forest birds simply do not cross even small unforested areas (Diamond 1975). On the other hand, some highly mobile animals may shift rapidly, as have some European birds (Edgell 1984).

Even if animals can disperse efficiently, suitable habitat may be reduced under changing climatic conditions. For example, it has been suggested that tundra nesting habitat for migratory shorebirds might be reduced by high-arctic warming (chapter 15).

## IV. SYNERGY OF HABITAT DESTRUCTION AND CLIMATE CHANGE

We know that even slow, natural climate change caused species to become extinct. What is likely to happen given the environmental conditions of the coming century? Some clear implications for conservation follow from the preceding discussion of dispersal rates. Any factor that would decrease the probability that a species could successfully colonize new habitat would increase the probability of extinction. Thus, as previously described, species are more likely to become extinct if there are physical barriers to colonization, such as oceans, mountains, and cities. Further, species are more likely to become extinct if their remaining populations are small. Smaller populations mean fewer colonists can be sent out, reducing the probability of successful colonization.

Species are more likely to become extinct if they occupy a small geographic range. It is less likely that some part will remain suitable when the climate changes than it would be if the ranges were larger. Also, if a species has lost much of its range because of some other factor, like clearing of the richer and moister soils for agriculture, it is possible that remaining populations will be located in poor habitat and therefore be more susceptible to new stresses.

For many species, all of those conditions will be met by human destruction of habitat, which increasingly confines the natural biota to small patches of original habitat, patches isolated by vast areas of human-dominated urban or agricultural lands. Habitat destruction in conjunction with climate change sets the stage for an even larger wave of extinction than previously imagined, based on consideration of human encroachment alone. Small, remnant populations of most species, surrounded by cities, roads, reservoirs, and farm land, would have little chance of reaching new habitat if climate change makes the old unsuitable. Few animals or plants would be able to cross Los Angeles on the way to new habitat. Figure 2.3 illustrates the combined effects of habitat loss and warming on a hypothetical reserve (see chapter 23 for discussion of how direct human effects and climate change will affect Florida's endangered fauna).

## V. AMELIORATION AND MITIGATION

Because of the difficulty of predicting regional and local changes, conservationists and reserve managers must deal with increased uncertainty in making long-range plans. However, even given imprecise regional projections, informed guesses can be made at least about the general direction of

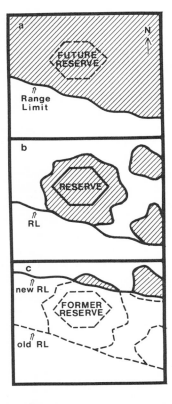

Figure 2.3. Climatic warming may cause species within biological reserves to disappear. Hatching indicates: (a) species distribution before either human habitation or climate change, (b) fragmented species distribution after human habitation but before climate change, (c) species distribution after habitation and climate change. (RL indicates southern limit of species range.)

change, specifically that most areas will tend to be hotter and that continental interiors in particular are likely to experience decreased soil moisture.

How might the threats posed by climatic change to natural communities be mitigated? One basic truth is that the less populations are reduced by development now, the more resilient they will be in the face of climate change. Thus, sound conservation now, in which we try to conserve more than just the minimum number of individuals of a species necessary for present survival, would be an excellent way to start planning for climate change.

In terms of responses specifically directed at the effects of climate change, the most environmentally conservative response would be to halt or slow global warming by cutting back on production of fossil fuels, methane, and chlorofluorocarbons. Extensive planting of trees to capture carbon dioxide could help slow the rise in carbon dioxide concentrations (Sedjo 1989; chapter 3). Nonetheless, even if the production of all greenhouse gases stopped today, it is likely that their concentrations in the air are now high enough to cause ecologically significant warming anyway (Rind 1989). Therefore, those concerned with the conservation of biological diversity must begin to plan mitigation activities now.

To make intelligent plans for siting and managing reserves, we need more knowledge. We must refine our ability to predict future conditions in reserves. We also need to know more about how temperature, precipitation, $CO_2$ concentrations, and interspecific interactions determine range limits (e.g., Picton 1984, Randall 1982) and, most important, how they can cause local extinctions.

Reserves that suffer from the stresses of altered climatic regimes will require carefully planned and increasingly intensive management to minimize species loss. For example, modifying conditions within reserves may be necessary to preserve some species; depending on new moisture patterns, irrigation or drainage may be needed. Because of changes in interspecific interactions, competitors and predators may need to be controlled and invading species weeded out. The goal would be to maintain suitable conditions for desired species or species assemblages, much as the habitat of Kirtland's warbler is periodically burned to maintain pine woods (Leopold 1978; see also chapter 21). On the other hand, if native species die out because of physiological intolerance to climate change, despite management efforts, then some invading species might actually be encouraged as ecological replacements for those that have disappeared.

In attempting to understand how climatically stressed communities may respond, and how they might be managed to prevent the gradual depauperation of their constituents, restoration studies or, more properly, community creation experiments can help. Communities may be created outside their normal climatic ranges to mimic the effects of climate change. One such out-of-place community is the Leopold Pines, at the University of Wisconsin Arboretum in Madison, where there is periodically less rainfall than in the normal pine range several hundred kilometers to the north (Jordan 1988). Researchers have found that although the pines themselves do fairly well once established at the Madison site, many of the other species that would normally occur in a pine forest (especially the small shrubs and herbs, such as *Trientalis borealis*, the northern star flower) have not flourished, despite several attempts to introduce them.

If management measures are unsuccessful and old reserves do not retain necessary thermal or moisture characteristics, individuals of disappearing species might be transferred to new reserves. For example, warmth-intolerant ecotypes or subspecies might be transplanted to reserves nearer the poles. Other species may have to be periodically reintroduced in reserves that experience occasional climate extremes severe enough to cause extinction, but where the climate would ordinarily allow the species to survive with minimal management. Such transplantations and reintroductions, particularly involving complexes of species, will be difficult. In many plants, for example, flowering times are determined by photoperiod, and in such species southern strains flower later in the year than northern ones (McMillan, 1959). A southern strain transplanted to the north therefore might wait to flower too late in the season, when it is too cold for successful reproduction. Despite such difficulties, applicable restoration technologies are being developed for many species (Botkin 1977, Jordan et al. 1988, Lovejoy 1985).

To the extent that we can still establish reserves, pertinent information about changing climate and subsequent ecological response should be used in deciding how to design and locate them to minimize the effects of changing temperature and moisture. One implication is that more reserves may be needed. The existence of multiple reserves for a given species or community type increases the probability that if one reserve becomes unsuitable for climatic reasons, the organisms may still be represented in another reserve.

Reserves should be heterogeneous with respect to topography and soil types, so that even given climatic change, remnant populations may be able to survive in suitable microclimates. Species may survive better in reserves with wide variations in altitude, since, from a climatic point of view, a small altitudinal shift corresponds to a large latitudinal one. Thus, to compensate for a 2°C rise in temperature, a species in the northern hemisphere can achieve almost the same result by increasing its altitude only 500 meters that it would by moving 300 km to the north (MacArthur 1972). Corridors between reserves, important for other conservation reasons, would allow some natural migration of species to track climate shifting. Corridors along altitudinal gradients are likely to be most practical because they can be relatively short compared with the longer distances necessary to accommodate latitudinal shifting.

As climatic models become more refined, reserves may be positioned to minimize the effects of temperature and moisture changes. In the northern hemisphere, for example, where a northward shift in climatic zones is likely, it makes sense to locate reserves as near the northern limit of a species' or community's range as possible, rather than farther south, where conditions are likely to become unsuitable more rapidly. Maximizing the size of reserves will increase long-term persistence of species by increasing the availability of suitable microclimates, by increasing the probability of altitudinal variation, and by increasing the latitudinal distance available to shifting populations. Flexible zoning around reserves may allow us to move reserves in the future to track climatic optima, for example,

by trading present range land for reserve land. The success of this strategy, however, would depend on a highly developed restoration technology, capable of guaranteeing, in effect, the portability of species and whole communities.

## VI. PREPARING FOR THE FUTURE

What concrete steps should be taken now by agencies and organizations responsible for the management of reserves and natural resources? How can they act to preserve species that may soon be dying out over large portions of their ranges? From the conservation manager's point of view, the changes will present many difficult practical and philosophical questions. Should the manager strive to preserve all the species within a reserve, given that climate change is causing some to disappear? Should management be used to conserve examples of community types, given that, on the time scale of climate change, communities are temporary assemblages of species likely to break up as the earth warms? How should the recent evolution of a "let nature take its course" management philosophy within, for example, the U.S. National Park Service be reconciled with the increasingly intensive management that will be necessary to conserve many species in a warming world?

Not only is this problem complex, but management response will also be difficult because, although the changes will be rapid from an ecological point of view, they will be slow in relation to management's traditionally short-term planning horizon. Thus, to ensure rapid response, continuity, and adequate resources, the high-level authorities within conservation organizations, management agencies, and funding bodies must give this issue high priority. The following are some areas that conservation management should emphasize.

### A. Begin Monitoring

One of the most important steps is to begin the collection of baseline data on how species and communities respond to climate.

Within a reserve, for example, abundances, ranges, and reproductive success of important species can be measured and analyzed in terms of ongoing climate measurements. Baseline information is necessary to identify the beginning of warming effects, to distinguish short-term from long-term changes, to help identify susceptible species and communities, to identify the nature of potential changes (such as the direction of climate-driven plant succession), and to provide the basis for identifying the relations between changes in climatic variables and resultant changes in the biota.

Changes might be expected to show up first at high latitudes, in low-lying marine coastal environments, and generally at ecotones between vegetation types determined by both temperature and precipitation. Changeover from one vegetation type to another might first be identified where disturbance events create succession. Monitoring for climate change may be done at different locations from those used for other sorts of monitoring. For example, at Merritt Island National Wildlife Refuge in Florida, transects for studying the effects of burning on vegetation are typically laid out in the center of a plant community, but a climate transect would more likely be at the interface between two communities.

Some programs already exist that could be focused on climate effects. For example, the National Park Service has ongoing ecological and species monitoring programs at several parks, including monitoring and research into the declines of Fraser fir (*Abies fraseri*) and red spruce (*Picea rubens*) of Great Smoky Mountains National Park. Similarly, the integrated studies of small watersheds initiated at several parks as part of the National Acid Precipitation Assessment Program will provide useful information for assessing long-term trends in ecosystem processes associated with climate change. Coordination of monitoring at a variety of sites, along with development of remote sensing and geographical information systems, can yield information on important trends at the landscape level. Such

efforts can provide the scientific basis for the flexible regionwide planning that will be needed to develop effective management responses.

## B. Undertake Ecological Research

Monitoring should be backed up by specific experiments on species and community responses to climate variables. One of the common themes in this book is the lack of good basic knowledge about how species react to climate. To take a single example, Walter Whitford (chapter 9) stresses the paucity of information about climate effects on soil biota, yet these biota will play a key role not only in determining the nature of future soils but also in the influence of soils on carbon cycling, which will help determine the greenhouse effect itself. From the point of view of a reserve manager, autecological studies can demonstrate which species have their ranges within a park determined by climate. Species of particular interest, such as endangered species, could receive special attention as to the effect of climate on, for example, food supply. Paleoecological studies and dendrochronology within protected areas or other important sites can shed additional light on past climate change and biotic response within reserves.

## C. Identify Sensitive Communities, Species, and Populations

The results of monitoring, research, and analysis based on present information should allow identification of species or communities of special concern, including those that are nationally or globally rare or endangered. Climate-sensitive species could be targeted for additional monitoring, research, and the development of management techniques.

## D. Develop Contingency Plans

Long-term plans for endangered species, for species likely to become rare, and for protected areas should have provisions for climate change. Even though precise local or regional climate projections are not available, contingency plans could be developed, particularly for sensitive biota. For example, contingency plans could be made based on assumptions of local average warmings of 2°, 4°, 6°, and 8°C, or on assumptions of various rainfall increases and decreases. Given that in many areas increased temperature will add to water stress, it would be reasonable to make long-term plans for dealing with lower water availability. They might include plans for mitigating effects on sensitive species or political or legal maneuvers to ensure that biological resources receive adequate water in the face of future competition from other users, such as agriculture and urban development.

## E. Develop Regional Plans for Nonreserve Habitat

As the location and abundance of habitat and critical resources change with climate, management of specific wildlife species will need to incorporate populations and resources lying outside protected areas. Therefore, nature reserves and other management units will increasingly be forced to become partners in planning and management that transcend the scale of protected areas. There are precedents: endangered species are already managed as outlined in multi-agency, multi-institution endangered species recovery plans. The current efforts to provide a basin-wide conservation plan for the grizzly bear in the Greater Yellowstone area might provide a model for such regional planning.

## F. Develop Management Techniques

The increased disturbance likely to result from climate change demands a large increase in resources for the development and implementation of new management techniques, particularly those of restoration ecology. Restoration and transplantation techniques are poorly developed at present and require extensive investment in research (see Jordan et al. 1988).

## G. Develop Philosophical Approaches to Management

Conservationists need to begin the process of deciding philosophical questions that will af-

fect management. Given the likelihood of community breakups, should efforts be expended on maintaining existing community types? As conditions become unsuitable for species existing within reserves today, should herculean efforts be expended to maintain them? Should the parks be used as transplantation sites for southern species in need of new habitat? At what point should efforts to maintain a particular species within a reserve be stopped and resources used elsewhere? How can reserves become integrated components of regional, national, and global strategies for conservation of species? Given stresses on the natural world, will the role of parks as refuges increase relative to their recreation role? At the moment, it is easier to ask these questions than to answer them.

## H. Dedicate Additional Reserve Lands

Global warming is a strong argument for the enlargement or creation of additional parks and other reserved lands. As mentioned above, multiple refuges provide additional chances that some protected habitat will remain suitable for a particular species as climate changes. Moreover, as climate change reduces adequate habitat and other resources for species within a reserve, enlargement may be necessary, as it was when Redwoods National Park was expanded during the 1970s to prevent external logging from threatening the park's ecosystems.

## VII. SUMMARY

In the geologic past, natural climate changes have caused large-scale geographic shifts in species' ranges, changes in the species composition of biological communities, and species extinctions. If the widely predicted greenhouse effect occurs, natural ecosystems will respond in ways similar to responses in the past, but the effects will be more severe because of the rapid rate of the projected change. Moreover, population reduction and habitat destruction due to human activities will prevent many species from colonizing new habitat when their old becomes unsuitable. The synergy between climate change and habitat destruction would threaten many more species than either factor alone.

These effects would be pronounced in temperate and arctic regions, where temperature increases are projected to be relatively large. It is unclear how the tropical biota would be affected by the smaller temperature increases projected for the lower latitudes, because little is known about the physiological tolerances of tropical species, but precipitation changes may cause substantial disruptions. Throughout the world, geographically restricted species might face extinction, while widespread species are likely to survive in some parts of their range. In the northern mid and high latitudes, new northward habitat will become suitable even as die-offs occur to the south. However, it may be difficult for many species to take advantage of this new habitat because their dispersal rates are slow relative to the rate of warming, and therefore ranges of even many widespread species are likely to show a net decrease during the next century. Range retractions will be proximally caused by temperature and precipitation changes, increases in fires, changes in the ranges and severity of pests and pathogens, changes in competitive interactions, and additional effects of nonclimatic stresses like acid rain and low-level ozone.

The best solutions to the ecological upheaval resulting from climatic change are not yet clear. In fact, little attention has been paid to the problem. What is clear, however, is that these climatological changes would have tremendous impact on communities and populations isolated by development and by the middle of the next century may dwarf any other consideration in planning for reserve management. The problem may seem overwhelming. One thing, however, is worth keeping in mind: if populations are fragmented and small, they are more vulnerable to the new stresses brought about by climate change. Thus, one of the best things that can be done in the short term is to minimize further encroachment of development on existing natural ecosystems. Further, we must refine climatological predictions and increase our understanding of how climate affects species, both individually and in their interactions with each other. Such studies may allow us to identify those areas where communities will be most stressed, as well as alternate areas where they might best be saved. Meanwhile, efforts to improve techniques for managing communities and ecosystems under stress, and also for restoring them when necessary, must be carried forward energetically.

## REFERENCES

Aho, J. M., J. W. Gibbons, and G. W. Esch. 1976. Relationship between thermal loading and parasitism in the mosquitofish. In *Thermal Ecology II*, G. W. Esch and R. W. McFarlane, eds., pp. 213–218. Springfield, Va.: Technical Information Center, Energy Research and Development Administration.

Baker, R. G. 1983. Holocene vegetational history of the western United States. In *Late Quaternary Environments of the United States. Vol. 2: The Holocene*, H. E. Wright, Jr., ed., pp. 109–125. Minneapolis: University of Minnesota Press.

Barron, E. J. 1985. Explanations of the Tertiary global cooling trend. *Palaeogeo. P.* 50:17.

Beauchamp, R.S.A., and P. Ullyott. 1932. Competitive relationships between certain species of fresh-water triclads. *J. Ecol.* 20:200.

Bernabo, J. C., and T. Webb III. 1977. Changing patterns in the Holocene pollen record of northeastern North America: A mapped summary. *Quatern. Res.* 8:64.

Botkin, D. B. 1977. Strategies for the reintroduction of species into damaged ecosystems. In *Recovery and Restoration of Damaged Ecosystems*, J. Cairns, Jr., K. L. Dickson, and E. E. Herricks, eds., pp. 241–260. Charlottesville: University Press of Virginia.

Butzer, K. W. 1980. Adaptation to global environmental change. *Prof. Geogr.* 32(3):269.

Coope, G. R. 1977. Fossil coleopteran assemblages as sensitive indicators of climatic changes during the Devensian (Last) cold stage. *Philos. Trans. Roy. B* 280:313.

Courtenay, W. R., Jr. 1978. The introduction of exotic organisms. In *Wildlife and America*, H. P. Brokaw, ed., pp. 237–252. Washington, D.C.: Council on Environmental Quality, U.S. Government Printing Office.

Cox, B. C., and P. D. Moore. 1985. *Biogeography: An Ecological and Evolutionary Approach*. Oxford: Blackwell Scientific Publications.

Critchfield, W. B. 1980. Origins of the eastern deciduous forest. In *Proceedings, Dendrology in the Eastern Deciduous Forest Biome*, September 11–13, 1979, pp. 1–14. Virginia Polytech. Inst. and State Univ. School of Forestry and Wildlife Resources publication FWS-2-80.

Daubenmire, R. F. 1962. *Plants and Environment: A Textbook of Plant Autecology.* New York: John Wiley and Sons.

Davis, M. B. 1983. Holocene vegetational history of the eastern United States. In *Late Quaternary Environments of the United States.* Vol. 2: *The Holocene,* H. E. Wright, Jr., ed., pp. 166–181. Minneapolis: University of Minnesota Press.

Diamond, J. M. 1975. The island dilemma: Lessons of modern biogeographic studies for the design of natural preserves. *Biol. Conser.* 7:129.

Dorf, E. 1976. Climatic changes of the past and present. In *Paleobiogeography: Benchmark Papers in Geology 31,* C. A. Ross, ed., pp. 384–412. Stroudsburg, Pa.: Dowden, Hutchinson, and Ross.

Duffy, D. C. 1983. Environmental uncertainty and commercial fishing: Effects on Peruvian guano birds. *Biol. Conser.* 26:227.

Edgell, M.C.R. 1984. Trans-hemispheric movements of Holarctic Anatidae: The Eurasian wigeon (*Anas penelope* L.) in North America. *J. Biogeogr.* 11:27.

Edlund, S. A. 1987. Effects of climate change on diversity of vegetation in arctic Canada. In *Preparing for Climate Change,* Proceedings of the first North American conference on preparing for climate change: A cooperative approach, J. C. Topping, Jr., ed., pp. 186–193. Washington, D.C.: Government Institutes.

Emanuel, W. R., H. H. Shugart, and M. P. Stevenson. 1985. Response to comment: "Climatic change and the broad-scale distribution of terrestrial ecosystem complexes." *Clim. Change* 7:457.

Environmental Protection Agency (EPA). 1989. *The Potential Effects of Global Climate Change on the United States.* Washington, D.C.: EPA.

Flenley, J. R. 1979. *The Equatorial Rain Forest.* London: Butterworths.

Flohn, H. 1979. Can climate history repeat itself? Possible climatic warming and the case of paleoclimatic warm phases. In *Man's Impact on Climate,* W. Bach, J. Pankrath, and W. W. Kellogg, eds., pp. 15–28. Amsterdam: Elsevier.

Ford, M. J. 1982. *The Changing Climate.* London: George Allen and Unwin.

Frye, R. 1983. Climatic change and fisheries management. *Nat. Resources J.* 23:77.

Furley, P. A., W. W. Newey, R. P. Kirby, and J. M. Hotson. 1983. *Geography of the Biosphere.* London: Butterworths.

Glynn, P. 1984. Widespread coral mortality and the 1982–83 El Nino warming event. *Envir. Conser.* 11(2):133.

Graham, R. W. 1986. Plant-animal interactions and Pleistocene extinctions. In *Dynamics of Extinction,* D. K. Elliott, ed., pp. 131–154. Somerset, N.J.: Wiley & Sons.

Hamburg, S. P., and C. V. Cogbill. 1988. Historical decline of red spruce populations and climatic warming. *Nature* 331:428.

Hansen, J., D. Johnson, A. Lacis, S. Lebedeff, P. Lee, D. Rind, and G. Russell. 1981. Climate impact of increasing atmospheric carbon dioxide. *Science* 213:957.

Hansen, J., A. Lacis, D. Rind, G. Russell, I. Fung, and S. Lebedeff. 1987. Evidence for future warming: How large and when. In *The Greenhouse Effect, Climate Change, and U. S. Forests,* W. E. Shands and J. S. Hoffman, eds. Washington, D.C.: Conservation Foundation.

Hansen, J., I. Fung, A. Lacis, S. Lebedeff, D. Rind, R. Ruedy, G. Russell, and P. Stone. 1988. Prediction of near-term climate evolution: What can we tell decision-makers now? In *Preparing for Climate Change,* Proceedings of the first North American conference on preparing for climate change: A cooperative approach, J. C. Topping, Jr., ed., pp. 35–47. Washington, D.C.: Government Institutes.

Heusser, C. J. 1974. Vegetation and climate of the southern Chilean lake district during and since the last interglaciation. *Quatern. Res.* 4:290.

Hoffman, J. S., D. Keyes, and J. G. Titus. 1983. *Projecting Future Sea Level Rise.* Washington, D.C.: U.S. Environmental Protection Agency.

Jordan, W. R., III. 1988. Ecological restoration: Reflections on a half-century of experience at the University of Wisconsin–Madison Arboretum. In *Biodiversity,* E. O. Wilson, ed., pp. 311–316. Washington, D.C.: National Academy Press.

Jordan, W. R., III, R. L. Peters, E. B. Allen. 1988. Ecological restoration as a strategy for conserving biological diversity. *Envir. Manag.* 12(1):55.

Kellison, R. C., and R. J. Weir. 1987. Selection and breeding strategies in tree improvement programs for elevated atmospheric carbon dioxide levels. In *The Greenhouse Effect, Climate Change, and U.S. Forests,* W. E. Shands and J. S. Hoffman, eds., pp. 285–293. Washington, D.C.: Conservation Foundation.

Kellogg, W. W., and R. Schware. 1981. *Climate Change and Society: Consequences of Increasing Atmospheric Carbon Dioxide.* Boulder, Colo.: Westview Press.

Knopf, F. L., and J. A. Sedgwick. 1987. Latent population responses of summer birds to a catastrophic, climatological event. *Condor* 89:869.

Kullman, L. 1983. Past and present tree lines of different species in the Handolan Valley, Central Sweden. In *Tree Line Ecology*, P. Morisset and S. Payette, eds., pp. 25–42. Quebec: Centre d'études nordiques de l'Université Laval.

Leopold, A. S. 1978. Wildlife and forest practice. In *Wildlife and America*, H. P. Brokaw, ed., pp. 108–120. Washington, D.C.: Council on Environmental Quality, U.S. Government Printing Office.

Lovejoy, T. E. 1985. Rehabilitation of degraded tropical rainforest lands. Commission on Ecology Occasional Paper 5. Gland, Switzerland: International Union for the Conservation of Nature and Natural Resources.

MacArthur, R. H. 1972. *Geographical Ecology*. New York: Harper and Row.

Manabe, S., R. T. Wetherald, and R. J. Stouffer. 1981. Summer dryness due to an increase of atmospheric $CO_2$ concentration. *Clim. Change* 3:347.

Marshall, J. K. 1978. Factors limiting the survival of *Corynephorus canescens* (L) Beauv. in Great Britain at the northern edge of its distribution. *Oikos* 19:206.

McMillan, C. 1959. The role of ecotypic variation in the distribution of the central grassland of North America. *Ecol. Monogr.* 29: 285.

Miller, W. F., P. M. Dougherty, and G. L. Switzer. 1987. Rising $CO_2$ and changing climate: Major southern forest management implications. In *The Greenhouse Effect, Climate Change, and U. S. Forests*, W. E. Shands and J. S. Hoffman, eds., pp. 157–187. Washington, D.C.: Conservation Foundation.

Muller, H. 1979. Climatic changes during the last three interglacials. In *Man's Impact on Climate*, W. Bach, J. Pankrath, and W. W. Kellogg, eds., pp. 29–41. Amsterdam: Elsevier Scientific Publishing.

Myers, R. L. 1983. Site susceptibility to invasion by the exotic tree *Melaleuca quinquenervia* in southern Florida. *J. Appl. Ecol.* 20(2):645.

NAS (National Academy of Sciences). 1987. *Current Issues in Atmospheric Change*. Washington, D.C.: National Academy Press.

NRC (National Research Council). 1983. *Changing Climate*. Washington, D.C.: National Academy Press.

Perring, F. H. 1965. The advance and retreat of the British flora. In *The Biological Significance of Climatic Changes in Britain*, C. J. Johnson and L. P. Smith, eds., pp. 51–59. London: Academic Press.

Peters, R. L., and J. D. Darling. 1985. The greenhouse effect and nature reserves. *Bioscience* 35(11):707.

Picton, H. D. 1984. Climate and the prediction of reproduction of three ungulate species. *J. Appl. Ecol.* 21:869.

Rand, A. S. 1964. Inverse relationship between temperature and shyness in the lizard *Anolis lineatopus*. *Ecology* 45:863.

Randall, M.G.M. 1982. The dynamics of an insect population throughout its altitudinal distribution: *Coleophora alticolella* (Lepidoptera) in northern England. *J. Anim. Ecol.* 51:993.

Rapoport, E. H. 1982. *Areography: Geographical Strategies of Species*. New York: Pergamon Press.

Rind, D. 1989. A character sketch of greenhouse. *EPA J.* 15(1):4.

Roberts, L. 1989. How fast can trees migrate? *Science* 243:735.

Schneider, S. H. 1988. The greenhouse effect: What we can or should do about it. In *Preparing for Climate Change*, Proceedings of the first North American conference on preparing for climate change: A cooperative approach, J. C. Topping, Jr., ed., pp. 19–34. Washington, D.C.: Government Institutes.

Schneider, S. H., and R. Londer. 1984. *The Coevolution of Climate and Life*. San Francisco: Sierra Club Books.

Seddon, Brian. 1971. *Introduction to Biogeography*. New York: Barnes and Noble.

Sedjo, R. A. 1989. Forests: A tool to moderate global warming? *Environment* 31(1):14.

Strain, B. R., and F. A. Bazzaz. 1983. Terrestrial plant communities. In *$CO_2$ and Plants*, E. R. Lemon, ed., pp. 177–222. Boulder, Colo.: Westview Press.

Titus, J. G., T. R. Henderson, and J. M. Teal. 1984. Sea level rise and wetlands loss in the United States. *National Wetlands Newsletter* 6(5):3.

Tralau, H. 1973. Some quaternary plants. In *Atlas of Palaeobiogeography*, A. Hallam, ed., pp. 499–503. Amsterdam: Elsevier Scientific Publishing.

Van Devender, T. R., and W. G. Spaulding. 1979. Development of vegetation and climate in the southwestern United States. *Science* 204:701.

Winget, C. H. 1988. Forest management strategies to address climate change. In *Preparing for Climate Change*, Proceedings of the first North American conference on preparing for climate change: A cooperative approach, J. C. Topping, Jr., ed., pp. 328–333. Washington, D.C.: Government Institutes.

WMO (World Meteorological Organization). 1982. Report of the JSC/CAS: A meeting of experts on detection of possible climate change (Moscow,

October 1982). Geneva, Switzerland: Rep. WCP29.

Wright, H. E., Jr. 1971. Late Quaternary vegetational history of North America. In *The Late Cenozoic Glacial Ages*, K. K. Turekian, ed., pp. 425–464. New Haven: Yale University Press.

Much of this chapter was previously published as a chapter in *Global Climate Change and Life on Earth*, edited by R. Wyman, published by Chapman and Hall. Much of the text on global warming was also previously published as an article in *Forest Ecology and Management*, in a special 1989 volume containing the proceedings of the symposium Conservation of Diversity in Forest Ecosystems, University of California at Davis, July 25, 1988. It draws heavily on other previously published versions, including those in *Endangered Species Update* 5(7):1–8 and *Preparing for Climate Change*, Proceedings of the first North American conference on preparing for climate change: A cooperative approach, J. C. Topping, ed. Many of the ideas and all the figures derive from a paper by R. L. Peters and J. D. Darling published in *Bioscience*, December 1985. See Peters and Darling 1985 for a complete list of acknowledgments for help with this work.

CHAPTER THREE

# How Does the World Work? Great Issues of Life and Government Hinge on the Answer

GEORGE M. WOODWELL

## I. LAWS OF MAN AND NATURE

Paul Sears, always fresh in thought, once referred to ecology as "subversive" in an editorial in *Bioscience* published in 1964. The editorial is as pertinent today as it was more than 25 years ago. He pointed to the fact that the laws of human society and the laws of nature are at odds, at least in this civilization, and that a reconciliation will require substantial revisions of government that some might consider downright revolutionary.

The discrepancy elaborated by Sears mattered little as long as the earth was large and the demands placed on it were small: we could proceed with the development of a civilization that assumed that the world would take care of itself. In that dream of the world, nations could pursue what they saw as their self-interest, even to the point of wars that brought devastation not only to people and cities but also to large areas of the earth otherwise little affected by human activities. And industries could proceed with the general and profitable process, articulated so brilliantly by Garrett Hardin a few years ago (1968, 1972), of gathering profits and spreading costs among the public at large. This scurrilous process has become so thoroughly commonplace and acceptable that the public now regularly accepts the burden of proving again and again that poison released into the environment or introduced into food is bad for people. And scientists have been infused with an unbecoming industrial bias to the point where they often demand a special higher standard of objectivity and proof in actions that might reduce industrial profits but improve public welfare than for their everyday business of developing knowledge. The shift has been introduced into law in the case of the hazards of reactors in the United States: the Price-Anderson Act protects the reactor industry by limiting its liability and that of the federal government in a reactor accident. The costs above the limits fall on the public affected, who are expected to have confidence in, and welcome as a neighbor, an intrinsically dangerous technology that neither the industry nor the government trusts. The assumption that the world is large and resilient has taken a strange and dangerous turn.

Nonetheless, the concept of a large and resilient world open to infinite compromise persists. To the extent that any concept of environment whatsoever enters into most of the day-to-day negotiations of the global economy, it is a concept of a limitless world, a cornucopia of goods and services in support of the human enterprise. Such a world allows growth: infinite growth in human numbers, growth in technology, growth in business. It does not require planning, nor does it require limits. The idea of a limitless world is beguiling, and its popularity is understandable. Such a world can be run on economic principles; governments merely supply some order in the market. This is the currently dominant model of how the world works. Tacit acceptance of this model colors decisions in government, economics, business, and international affairs. It is the model that the current perversion of political conservatism is hawking globally. In this model the natural world is virtually infinitely resilient, and the resolution of the inevitable conflicts over resources, even air, water, and land, lies in compromise: the people share their air, their health and vigor, the length of their lives, their welfare and that of their children, with others and with commercial and industrial interests that have established a transcendent right to use, even destroy the air, water, and land for profit. And scholars become evaluators of "risks," comparing degrees of human misery, always ephemeral and diffuse, with profits, by definition well defined.

Abundant experience challenges the underlying assumption of resilience and provides ample basis for recognizing the incremental destruction of environment and human health. We were surprised to learn that radioactivity from bombs detonated in the atmosphere in the 1950s and early 1960s circled the hemisphere within three to four weeks and accumulated to high levels in living systems. In a few more weeks it was transferred to the southern hemisphere to be deposited there as well by rain on land and sea. Persistent pesticides and other toxins such as PCBs follow a similar route and may accumulate in living systems with devastating effects on fish and birds and people. It is a safe assumption that every organism on earth now carries a burden of diverse toxins produced by human activities. A careful review of the potential effects of nuclear war revealed only as recently as 1983 that not only would the radioactivity move throughout the world rapidly but the dust and smoke from the explosions would also cause a severe global cooling at the surface of the earth (Peterson 1983). The cooling might last for several months to a year or more and kill through cold and starvation many more people than were killed directly in the blast or by the radiation (SCOPE 1986). Now, most recently, the general public has become aware and appropriately alarmed that the accumulation of carbon dioxide and other heat-trapping gases in the atmosphere is opening a period of continuous warming globally. The rate of warming may be sufficient to move climatic zones out from under forests and agriculture and to cause a wave of biotic impoverishment globally (WMO/UNEP 1988; see also Abrahamson 1989; Smith and Tirpak 1989; Woodwell 1989, 1990).

The failures of current dreams are many, and it is clear that "if present trends continue" (CEQ 1980), the future of this civilization seems limited. The human enterprise is in trouble. Rapid global changes are under way that are beyond the control of individuals or even individual governments. The potential of climatic change alone for biotic impoverishment and human disruption has been equated to that of nuclear war. The problem is the more acute because the biotic, chemical, and physical processes that keep the world a suitable place for people are not well known and are commonly ignored by the public. A reappraisal of how the world works is appropriate. If the basic tenets concerning environment have been wrong, what tenets might replace them? What would work?

## II. WHAT WORKS IN A SHRINKING WORLD?

The most revealing evidence of the details of the problem is the global environmental changes already entrained. Each offers insight as to how the world works and how we might modify our activities to accommodate some of the more obvious laws of nature. The evidence comes from several sources: the wave of biotic impoverishment that has followed the sweep of civilization across the planet and is now more rapid than ever previously, and accelerating (Woodwell 1990); the hazards of the persistent toxins, including radioactivity and pesticides and various industrial wastes that move globally and are concentrated in biotic systems; the reality of the depression of the ozone concentration in the high atmosphere and of the warming of the earth in response to changes in the composition of the atmosphere through human activities. Two of these global hazards have already led to international action: the Test-Ban Treaty of 1962, combined with international pressure on the nations that did not sign it, has driven most of the testing of nuclear weapons underground, and radioactivity from bomb tests is no longer a serious new source of biotic hazards; the depletion of the ozone layer has brought rapid action among the nations through the Vienna Convention and the Montreal Protocol to reduce emissions of the chlorofluorocarbons (CFCs). This problem has not yet been resolved, but an international mechanism appears to be working. The climatic changes resulting from the accumulation of heat-trapping gases in the atmosphere have long been predicted and now appear to be upon us. This issue is more complicated and will require far more serious changes in the human enterprise than any issue yet addressed. It also offers more insights into how the world works and de-

mands for its amelioration a series of steps that will help resolve other problems, long ignored (WMO/UNEP 1988; see also Abrahamson 1989, Woodwell 1989).

## III. THE BIOTA AS CAUSE AND EFFECT OF CLIMATIC CHANGE

The warming of the earth is commonly seen as a physical problem: the heat-trapping gases accumulate in the atmosphere and establish a new, higher temperature for the equilibrium between the incoming solar radiation and the reradiation from the earth into the blackness of space. The higher temperature means that more energy is available on the surface of the earth for the evaporation of water. Greater evaporation of water in the tropics means that more energy is transported to the higher latitudes, which are warmed differentially to temperatures twice as high as the average warming of the earth. Temperatures in the tropics remain about the same because there is so much water to be evaporated there, stabilizing the temperature of the entire region. The higher-latitude continental centers warm and, because they are warmer, become drier. Climatic zones migrate and the rate of migration affects the survival of species. No one knows how rapidly the earth can be expected to warm, but the most reasonable assumption is that it will warm in proportion to the changes in the atmosphere and that the full warming might occur as rapidly as the heat-trapping gases accumulate. Such an assumption means that in the middle and higher latitudes the warming may proceed at rates of tenths to one degree per decade over the next decades. A one-degree change in mean temperature is encountered by moving latitudinally 60–100 miles. Such rates of change are to be expected over the next decades unless steps are taken to slow or stop the accumulation of heat-trapping gases (WMO/UNEP 1988, Houghton et al. 1990).

These are the physical changes in outline. They obscure the biotic changes, which are not only cause and effect but also potential cure. They are, through the destruction of forests, for example, one of the major causes of the increase in carbon dioxide and methane in the atmosphere. The biotic exchanges are complex, however. A listing of the stocks and flows of carbon emphasizes the importance of the biota in determining the composition of the atmosphere and therefore the temperature of the earth.

*Stocks:*

The atmosphere contains about 750 billion tons of carbon ($750 \times 10^9$ t C).

The vegetation and soils globally contain an estimated 2000 billion tons of carbon as organic matter in various forms, about three times the amount in the atmosphere.

*Flows:*

The annual release into the atmosphere currently through combustion of fossil fuels is about 5.6 billion tons of carbon as carbon dioxide.

The annual release of carbon into the atmosphere from deforestation in 1980 has been estimated as 1–3 billion tons (Houghton et al. 1985); the rate is higher now as deforestation proceeds (Houghton 1989, Woodwell 1988).

Photosynthesis globally on land absorbs into plants from the atmosphere about 100 billion tons of carbon annually, nearly one-seventh of the atmospheric burden (Reiners 1973; see also Houghton and Woodwell 1989, Mooney et al. 1987).

Respiration, including the respiration of plants, animals, and the organisms of decay, releases under normal circumstances an amount of carbon that approximately equals the amount absorbed in photosynthesis, and the atmospheric composition is maintained at equilibrium (Reiners 1973; see also Houghton and Woodwell 1989, Mooney et al. 1987).

The annual accumulation of carbon in the atmosphere is currently about 3 billion tons, which is the excess of the current releases above the transfer into the oceans and the absorption into the terrestrial biota.

The stock of carbon controlled by biotic processes is about three times the amount in the atmosphere, and a change in the flows between the pool of carbon held in plants and soils and the pool in the atmosphere has the potential for affecting the atmosphere significantly. The potential changes extend well beyond the current release from deforestation, now probably in excess of 3 billion tons of carbon annually. Any shift in the magnitude of the photosynthetic or respiratory flows could be significant. A warming of the earth, especially a rapid warming, can be expected to cause such a shift. The pattern is far from certain, but experience suggests that a warming (and almost any other severe disturbance) will stimulate respiration more than it will affect photosynthesis. A 1°C increase in temperature is widely recognized as increasing the rate of respiration 10%–30% while having little effect on photosynthesis (Woodwell 1983). The stimulation of respiration increases the rate of release of the products of respiration, including both carbon dioxide and methane. The process constitutes a positive feedback, a biotic influence that adds heat-trapping gases to the atmosphere, both products of respiration, and speeds the warming.

Evidence from the climatic changes that have occurred during 160,000 years of recent glacial time is consistent with the assumption of a positive biotic feedback but does not prove it correct: as temperatures rose in glacial times, the carbon dioxide content of the atmosphere also rose; as temperatures declined, the carbon dioxide content declined. The warming of the earth that has occurred over the past century has probably stimulated the decay of organic matter in plants and soils of the middle and high latitudes and may be contributing an additional increment of carbon as carbon dioxide and methane to the atmosphere (Houghton and Woodwell 1989). The magnitude of this additional release is not known; it could be one to several billion tons of carbon annually.

Biotic influences on the composition of the atmosphere are direct and large. They are large enough to change the amount of carbon dioxide in the atmosphere by several percent within a few weeks. The continued destruction of the biota globally is not only contributing directly to biotic impoverishment but is also speeding the changes in the atmosphere responsible for the warming of the earth. The safest assumption is that the warming will proceed rapidly over the next decades. Rapid change is the enemy of life. If uncontrolled and allowed to follow the current course the climatic changes will:

- Be continuous and accelerating. The earth is not simply moving to another climatic equilibrium. It is moving from a period of slowly changing climates to one in which the warming will accelerate and be continuous.
- Be rapid by comparison with the capacity of forests to migrate. The most rapid changes will be in the middle and high latitudes of both hemispheres where the rates of warming may approach, possibly exceed, 1° per decade, especially as the biotic feedback begins to be felt.
- Be rapid by comparison with the capacity of agriculture to migrate or shift to new crops. The experience of the summer of 1988 is exemplary. An especially hot, dry summer lowered grain yields 20%–30% in North America. Despite the versatility of agriculture, there was no recourse in this instance.

Such changes in climate are open ended, rapid, and accelerating; they are devastating to agriculture, forests, forestry, parks, and biotic reserves globally but especially serious in the middle and high latitudes; they are replete with surprises such as the ozone hole and the possibility of unpredictable changes in climate as the Arctic Ocean becomes ice-free and oceanic currents change. These transitions will be joined by the complications of an increase in sea level as oceanic waters expand in response to the warming and as glaciers melt. The rate of rise over a century is variously estimated at 0.5 m to 2.0 m or more, depending on the extent of the warming and the behavior of glaciers as the warming progresses (Smith and Tirpak 1989). It could be higher. The floods in Bangladesh come not only from deforestation in the mountains of the Himalayas but also from storm surges in the Bay of Bengal. As sea level rises such surges reach farther inland.

## IV. WHAT CAN BE DONE?

These changes in the human circumstance are the product of a continuous flow of small decisions by individuals, families, villages, businesses, industries, and governments over decades. They are a special product of the industrial revolution that has brought the great surge in human population and the spread of technology globally. They are also the product of growth, including the growth of the human population, now headed toward the third doubling in a little more than a century as we reach toward 10 billion people by 2030 or sooner. Each of the decisions was rational at the time and seems rational now in the limited context of most human activities, but together they have brought crisis to the world. The crisis for conservation is not different from the crisis for all of humanity: how to restabilize the human habitat and preserve opportunities for our children to live comfortably in it.

We can resolve the immediate crisis of climatic change by slowing, then stopping, the accumulation of heat-trapping gases in the atmosphere. We have the potential for controlling emissions of carbon dioxide and methane from fossil fuels and from deforestation. We can also control emissions of CFCs, which share the capacity for trapping heat. But we have little power to control biotic sources of carbon dioxide or methane. As the earth warms those sources will increase in size and could exceed current releases from fossil fuels. If we are to act effectively to slow or deflect a continuous warming, early action is clearly appropriate. Action will require special attention to sources of energy and for the first time in human history to biotic resources globally:

- A 50% reduction within a decade in the use of fossil fuels below the use in the mid-1980s. This reduction will have to be followed immediately by further reductions toward the final abandonment of fossil fuels as the primary source of energy for industrial societies.
- Cessation of deforestation.
- A major program of reforestation. Reforestation of approximately two million square kilometers will store in the soils and plants of the forest about one billion tons of carbon annually for 40 to 50 years as the forest develops (Woodwell 1989).

Such steps are clearly possible, but they will require a new sensitivity in dealing with the human future and a new recognition of the importance of biotic resources, not as a potential source of private gain but as the basis of human existence. The idea is hardly new, but the reality of it seems novel in an age when we have come to think of energy and foreign exchange as the world's most versatile and important resources.

## V. ELEMENTS OF A NEW MODEL

What replaces laissez-faire? The tendency is to think only of top-down, ad hoc solutions, governmental edicts and incentives, financial and otherwise, that will end the problem. Top-down solutions are required, but more than edicts and money will be needed. A new concept of how the world works will be the new touchstone for success in human affairs. What are the dreams of the moment that have sufficient reality to offer patterns of behavior that will work and sufficient appeal to catch the imaginations of millions? A real solution will have both bottom-up and top-down potential.

If there is an answer that can dominate over human depravity and greed, it lies in knowledge now available, in the fact that the earth is a living system that is not only vulnerable but failing, not for some of the earth's inhabitants, but for all. That is the central point in the analyses of climatic change. There is no possibility of a solution to climatic change without a shift away from fossil fuels toward reliance on enduring sources of energy, one of which is solar energy captured by plants, without a cessation of deforestation globally, without specific efforts at reforestation globally. But these steps are the mere beginning in the effort to move the laws of man into consonance with the laws of nature.

We know some of the elements required for sustained use of the planet, but the transition will be a major intellectual challenge whose roots lie in conservation: The key element is life and the context is global. The starting point is natural communities or ecosystems, self-maintaining biotic units that form the matrix within which we have built civilization. The principles are those of ecology, only partially elaborated, but firm of foundation. The objective is the continuity of life, all life.

The laws that count are natural laws that govern the structure, function, and development of ecosystems in a world that is a biotic system first and a human-made system second. We know much about this world and how to live in it. But it is not a world of infinite growth, of open systems, of limitless assimilative capacities for all wastes from the human enterprise. It is a world of life and growth and death and primary production and biotic diversity and biotic impoverishment, often irreversible. But it is also a benign and comfortable world of predictable causes and effects.

The key point is that safety for humans and the human enterprise in toto lies in rigorous protection of the biota and its processes globally. The magnitude of the human enterprise globally is in question; it is in question regionally and locally as well. Although the absolute number of people that the earth can support may remain uncertain, we have clearly exceeded the limit as to the regional intensity and global magnitude of human activities. The intensity and magnitude are not simply a function of the number of people but the product of the density of population and the energy they use. The regional and local limitations appear in the toxification of water, land, and people, and in biotic impoverishment. The global limits appear in the global changes now under way. Just as with the management of poisons, so with the management of the earth as a whole: if living systems apart from people are protected, human interests will be secure. A narrow focus on human health alone as commonly defined, or on human interests as defined by commerce globally, is permissive and self-defeating.

How we interpret the human crises of this difficult time, their seriousness and urgency, the degree of response that is appropriate, and the probable effects of any course of action depends on how we think the world works. If the world is large in proportion to the demands on it, we can continue without significant change. If the signals we have now remind us that the world is life itself, then we have a new challenge in making an accommodation to a finite and living world that will require a stewardship we have to this moment been unwilling and unable to provide. The details of that stewardship are bringing a revolution in government in the United States and will bring a revolution internationally—or lead to a crippling, global biotic impoverishment.

## REFERENCES

Abrahamson, D., ed. 1989. *The Challenge of Global Warming*. Washington, D.C.: Island Press.

CEQ (Council on Environmental Quality). 1980. *Global 2000: Report of the President's Council on Environmental Quality*. Washington, D.C.: Executive Office of the President.

Hardin, G. 1968. The tragedy of the commons. *Science* 162:1243.

Hardin, G. 1972. *Exploring New Ethics for Survival: The Voyage of the Spaceship Beagle*. New York: Viking Press.

Houghton, R. A. 1989. Emissions of greenhouse gases. In *Deforestation Rates in Tropical Forests and Their Climatic Implications*, N. Myers, ed. London: Friends of the Earth.

Houghton, R. A., and G. M. Woodwell. 1989. Global climatic change. *Sci. Am.* April:36.

Houghton, R. A., W. H. Schlesinger, S. Brown, and J. F. Richards. 1985. Carbon dioxide exchange between the atmosphere and terrestrial ecosystems. In *Atmospheric Carbon Dioxide and the Global Carbon Cycle*, J. E. Trabalka, ed. Washington, D.C.: Department of Energy, Technical Information Service.

Houghton, J. T., G. J. Jenkins, and J. J. Ephraums. 1990. *Climate Change: The IPCC Scientific Assessment*. Geneva and Nairobi: WMO/UNEP.

Mooney, H. A., P. Vitousek, and P. Matson. 1987. Exchange of materials between terrestrial ecosystems and the atmosphere. *Science* 238:926.

Peterson, J., ed. 1983. *The Aftermath: The Human and Ecological Consequences of Nuclear War*. New York: Pantheon Books.

Reiners, W. A. 1973. A summary of the world carbon cycle and recommendations for critical research. In *Carbon and the Biosphere*, Proceedings of the 24th Brookhaven Symposium in Biology, Upton, N.T., May 1972, G. M. Woodwell and E. V. Pecan, eds. Technical Information Center, U.S. Atomic Energy Commission.

SCOPE 28. 1986. *Environmental Consequences of Nuclear War*, vols. 1 and 2. New York: John Wiley and Sons. These two volumes review a topic first opened by the editor of *Ambio*, who devoted a 1983 issue of the journal to the topic, and by a conference in Washington, D.C., "The long-term worldwide biological consequences of nuclear war," in October 1983.

Sears, P. B. 1964. Ecology: A subversive subject. *Bioscience* 7:11.

Smith, J. B., and D. Tirpak. 1989. *The Potential Effect of Global Climate Change on the United States*. Washington, D.C.: Environmental Protection Agency.

WMO/UNEP. 1988. Developing policies for responding to climatic change. WMO/TID 225. Geneva: WMO. The biotic implications have been summarized by G. M. Woodwell.

Woodwell, G. M. 1983. Biotic effects on the concentration of atmospheric carbon dioxide: A review and projection. In *Changing Climate*, pp. 216–241. Washington, D.C.: NAS Press.

Woodwell, G. M. 1988. Letter to *Science*. *Science* 241:1736.

Woodwell, G. M. 1989. The warming of the industrialized middle latitudes, 1985–2050: Causes and consequences. *Clim. Change* 15:31.

Woodwell, G. M., ed. 1990. *The Earth in Transition: Patterns and Processes of Biotic Impoverishment*. New York: Cambridge University Press.

CHAPTER FOUR

# Climate-Change Scenarios for Impact Assessment

STEPHEN H. SCHNEIDER,
LINDA MEARNS, AND
PETER H. GLEICK

## I. INTRODUCTION

Many scientists are uncomfortable with public discussion of critical scientific issues involving significant uncertainties. This is especially true if important policy implications attend speculative scientific subjects. The increasing concentrations of so-called greenhouse gases (e.g., $CO_2$, $CH_4$, $N_2O$, chlorofluorocarbons) provide one of the best current examples of a problem in which the public need for reliable scientific knowledge exceeds the ability to provide it. Indeed, potential societal and environmental effects have been assessed and reassessed by national and international groups for decades (Bach et al. 1979, Carbon Dioxide Assessment Committee 1983, Bolin et al. 1986, Williams 1978, Climate Research Board 1978, Farrell 1987, Seidel and Keyes 1983, Pearman 1987, NRC 1987, IPCC 1990, NAS 1991). As our understanding of the scope of those effects has improved, the need for regional and temporal details about future climatic conditions has become increasingly urgent.

The desire to predict climatic changes arises from concern over the rapidity and magnitude of those changes and from the need in some cases to plan how to respond to the changes far in advance. Society may be affected by rises in sea level, intensification of tropical cyclones, declines in the quantity and quality of freshwater resources, alterations in agricultural productivity, increases in direct threats to health, and impacts on unmanaged ecosystems (e.g., Smith and Tirpak 1988). Yet we cannot easily or precisely determine how the complex climatic system will react to anthropogenic pollutants.

One response to the need for information on future climatic changes has been the analysis of large climatic changes in the geologic past (Schneider 1987, Budyko et al. 1987, Barron and Hecht 1985, Berger et al. 1984).

Unfortunately, although such paleoclimatic metaphors are relevant for estimating future climatic sensitivity to large changes in radiation that force climate to change, they are not exact analogies to the rate and character of present greenhouse-gas increases (e.g., see the discussion in section 5.5.3 of IPCC 1990). Looking at more recent climatic records—the so-called historical analogue method—can also provide insights into climatic behavior and societal vulnerabilities (Lough et al. 1983, Jager and Kellogg 1983, Pittock and Salinger 1982). But these methods are also based on climatic cause-and-effect processes that could be different from future greenhouse-gas radiative effects (Schneider 1984). Therefore, scientists estimating future climatic changes have focused on large-scale models of the climate—general circulation models (GCMs)—that attempt to represent mathematically the complex physical interactions among the atmosphere, oceans, ice, biota, and land. As these models have evolved, more and more information has become available and more comprehensive simulations have been performed. Nevertheless, the complexities of the real climate system still vastly exceed the comprehensiveness of today's GCMs and the capabilities of today's computers (see IPCC 1990 for a state-of-the-art review). Many important uncertainties are unlikely to be resolved before some significant climatic changes are felt, and certainly not before we are committed to some long-term environmental and societal effects.

Society is thus faced with a classic example of the need to make decisions based on imperfect information. Some projected climatic effects appear severe, but perhaps they can be mitigated if we know what to expect and if we choose to respond. At the same time, there is a risk of investing resources to prevent a problem that may not appear, or that may appear where least expected. The need to know details about the timing and distribution of future climate changes has been stated in many scientific and political forums, and detailed climate impact studies have been commissioned (Kates et al. 1985, Parry 1985, Senate Committee on Governmental Affairs 1979, CEQ 1980, working group II of IPCC 1990). The American Association for the Advancement of Science conducted one such study on climatic variability, climatic change, and United States water resources (Waggoner 1990). Our contribution to that study involved formulating scenarios of plausible climatic changes that might be used by hydrologists and water-resource planners for more detailed studies of regional hydrologic effects. Those scenarios are summarized (and updated by Stephen H. Schneider) here. Hydrologists simply cannot make informed statements on the water-resource implications of future climatic changes without such scenarios. The same is true for biologists concerned with species movements, habitat transformation, and forest yields.

## II. FORECASTING CHANGES IN METEOROLOGICAL VARIABLES

To shed some light on these questions, we offer here a set of forecasts on changes in some important meteorological variables, over a range of temporal, spatial, and statistical scales. We believe that carefully qualified, explicit scenarios of plausible future climatic changes are preferable to impact speculations based on implicit or casually formulated forecasts. Therefore we have prepared table 4.1 to provide our impact-assessment colleagues with ranges of climate changes that reflect state-of-the-art modeling results. Also, we hope that the table will spark discussions among climatologists over its plausibility and that improved and expanded projections will thus evolve.

These projections are based on our analysis of available results and provide what we believe are plausible estimates of the direction and magnitude of some important anthropogenic climatic changes over the next 50 years or so—a typical estimate for an equivalent doubling of carbon dioxide—together with a simple high, medium, or low level of confidence for each variable. (By

Table 4.1.

| Phenomena | Projection of probable global annual average change (1) | Distribution of change | | | | Confidence of projection | | Estimated time for research that leads to consensus (years) |
|---|---|---|---|---|---|---|---|---|
| | | Regional average | Change in seasonality | Interannual* variability | Significant transients | Global average | Regional average | |
| Temperature ** | +2° to +5° C | −3° to +10° C | yes | down? | yes | high | medium | 0 to 10 |
| Sea level | 0 to 80 cm*** | (2) | no | ??? | yes*** | high | medium | 5 to 20 |
| Precipitation | +7% to +15% | −20% to +20% | yes | up? | yes | high | low | 10 to 40 |
| Direct solar radiation | −10% to +10% | −30% to +30% | yes | ??? | possible | low | low | 10 to 40 |
| Evapotranspiration | +5% to +10% | −10% to +10% | yes | ??? | possible | high | low | 10 to 40 |
| Soil moisture | ??? | −50% to +50% | yes | ??? | yes | ??? | medium | 10 to 40 |
| Runoff | increase | −50% to +50% | yes | ??? | yes | medium | low | 10 to 40 |
| Severe storms | ??? | ??? | (3) | ??? | yes | ??? | ??? | 10 to 40 |

(1) For an equivalent doubling of atmospheric $CO_2$ from the preindustrial level.

(2) Sea level will increase locally at approximately the global rate, except where local geological activity prevails or if ocean currents change.

(3) Some suggestions of longer season and increased intensity of tropical cyclones as a result of warmer sea surface temperatures.

??? No basis for quantitative or qualitative forecast.

* Inferences based on preliminary results for the U.S. from Rind et al. (1990).

** Based on three-dimensional model results. If only trace-gas increases were responsible for the twentieth-century warming trend of about 0.5°C, then this range should be reduced by perhaps 1°C.

*** Assumes only small changes in Greenland or West Antarctic ice sheets in twenty-first century. For equilibrium, hundreds of years would be needed; up to several meters of additional sea-level rise could be accompanied by centuries of ice-sheet melting from an equilibrium warming of 3 or more degrees C.

SOURCE: Modified (by Schneider) from Schneider et al. 1990.

equivalent doubling, we mean that carbon dioxide together with other, trace greenhouse gases has a radiative effect equivalent to doubling the preindustrial value of carbon dioxide from about 280 ppm to 560 ppm.) As another measure of the nature of the uncertainties, we include a rough estimate of the time that may be necessary to achieve a widespread scientific consensus on the direction and magnitude of the changes. In some cases—such as the magnitude of changes in sea level and global annual average temperature and precipitation—such a consensus has virtually been reached (e.g., NRC 1987, IPCC 1990). In other cases—such as changes in the extent of cloud cover, patterns of regional precipitation changes evolving over decades, potential for intensifying storms, and the daily, monthly, or interannual variance of many climatic variables—the large uncertainties surrounding present projections will be reduced only with considerably more research, probably measured in decades (see figure 11.4 in in IPCC 1990). We emphasize the importance of the high degree of uncertainty in climate-change forecasts. This uncertainty may be underrepresented by some who develop or use detailed regional climate-change scenarios for the purpose of studying impacts or advocating policy responses.

Consider, for example, the first row on table 4.1: temperature change. The global average change of $+2°$ to $+5°C$ is the current range for general circulation models (chapter 5 in IPCC 1990). This range is typical of that in most national and international assessments (Dickinson 1986) for an equivalent doubling of greenhouse gases, neglecting transient delays. Based on curve fits to current global warming trends, many scientists prefer to revise that range downward somewhat to $1.5°$ to $4.5°C$ or perhaps even $1°$ to $4°C$ (e.g., Wigley and Raper 1990a). The neglect of transients means that the range given is based on the assumption that trace-gas concentrations have been elevated over a long enough period for the climate to come into equilibrium with the increased concentration of green-

house gases. In reality, the large heat capacity of the oceans will delay realization of full equilibrium warming by perhaps many decades (Hoffert et al. 1980, Schneider and Thompson 1981, Bryan et al. 1982). This implies that at any specific time when we reach an equivalent $CO_2$ doubling (by, say, 2030), the actual global temperature increase may be considerably less than the $+2°$ to $+5°C$ listed in the table. However, this "unrealized warming" (Hansen et al. 1985) will eventually occur when the climate system's thermal response catches up to the greenhouse-gas forcing.

On a finer scale, forecasts of regional or watershed changes in temperature, evaporation, and precipitation are most germane to estimating hydrological or other regional consequences of greenhouse warming. But, as table 4.1 suggests, such regional forecasts are more uncertain. Regional temperature ranges given in the table are much larger than global changes and even allow for some regions of negative change (Schlesinger and Mitchell 1987, IPCC 1990). For example, surface temperature increases projected for the higher northern latitudes are up to several times larger than the projected global average response, at least in equilibrium. Because of the importance of regional impact information, other techniques are being developed to evaluate smaller-scale hydrologic effects of large-scale climatic changes (Gleick 1986). P. H. Gleick employed a regional hydrologic model driven by large-scale climate change scenarios from various GCM inputs (Gleick 1987). Other techniques embed a mesoscale atmospheric model into a limited region of the global-scale GCM (e.g., Giorgi 1990).

Even more uncertain than regional details, but perhaps most important, are estimates for such measures of climatic variability as the frequency and magnitude of severe storms, enhanced heat waves, or reduced frost probabilities (Parry and Carter 1985; Mearns et al. 1984, 1990; Wigley 1985). For example, some modeling evidence suggests that hurricane intensities will increase with climatic changes (Emanuel 1987). Such issues are just now beginning to be considered and evaluated from equilibrium climate-model results and will, of course, have to be studied for realistic transient cases to be of maximum value to impact assessors.

Other uncertainties raised by the transient nature of the actual trace-gas forcing are the emission and removal rates of $CO_2$, $CH_4$, and other greenhouse gases. Figure 4.1 shows three plausible scenarios based on high, medium, and low emission rates (WMO). These uncertainties have been added to those associated with estimates of climate sensitivity and the delay associated with oceanic heat capacity. In any case, since the earth has apparently not experienced global average temperatures more than 1° to 2°C higher than at present over the past glacial cycle of 150,000 years (Barnola et al. 1987), all but the slowest scenario represent a rapid, large climatic change to which the environment and society will have to adapt.

The principal technical advance needed to build consensus on the reliability of time-evolving regional forecasts of hydrological

Figure 4.1. Three scenarios for global temperature change to 2100 derived from combining uncertainties in future trace greenhouse gas projections with those of modeling the climatic response to those projections. Sustained global temperature changes beyond 2°C (3.6°F) would be unprecedented during the era of human civilization. The middle to upper range represents climatic change at a pace 10 to 100 times faster than long-term natural average rates of change. (Source: Jager 1988.)

variables is the development, testing, and verification of coupled atmosphere, ocean, and land-surface models for realistic transient scenarios (Thompson and Schneider 1982). Fortunately, there has been recent progress in the development of such models (Stouffer et al. 1989, Washington and Meehl 1989), although they reaffirm Schneider and Thompson's (1981) contention that reliable regional climatic projections evolving over decades will require coupling of dynamical atmosphere and ocean models. Therefore, we suggest in table 4.1 that it will be a decade or more before the scientific community reaches such a consensus, because it will be at least a decade before high-resolution models of atmospheric, oceanic, biospheric, and hydrospheric subsystems can be run in coupled mode and data can be obtained to validate these simulations (figure 11.4 in IPCC 1990).

Some scientists may object in principle to the approach taken in table 4.1, arguing that since the confidence levels cited are intuitive, they may be incorrect, or even that some predicted change could be in the opposite direction of that listed. Indeed, predictions about something as complex as global climate will always be somewhat uncertain. Nevertheless, many policymakers are likely to have the opposite reaction—the information in the table, even if certain, may still not contain enough detail to justify major policy decisions. Policy analysts typically want regional details even finer than those available from, say, the maps in the survey article by Schlesinger and Mitchell (1987), a few of which are reproduced here as figures 4.2, 4.3, 4.4, and 4.5. Note especially the large warming in high latitudes (especially over sea ice margins) in winter associated with simulated reductions in snow cover or sea ice. Warming of oceans could cause thermal expansion of oceans, melting of some ice on land, and possibly increased buildup of snow on high, cold ice sheets. Taken together, these factors contribute to the rise in sea level of 0 m to about 1 m typically estimated for the twenty-first century (chapter 9 of IPCC 1990).

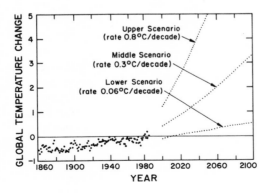

TEMPERATURE DIFFERENCES FOR DJF

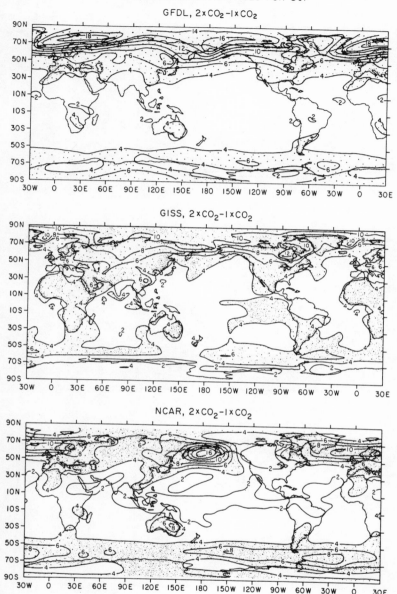

*Figure 4.2.* Geographical distribution of the surface air temperature change (in degrees Celsius), (2 × CO₂) − (1 × CO₂), for December, January, and February, simulated with a version of the Goddard Fluid Dynamics Laboratory GCM (GFDL), the Goddard Institute for Space Studies GCM (GISS), and the National Center for Atmospheric Research GCM (NCAR). Stippling indicates temperature increases larger than 4°C. For reference to particular versions of the GCMs used, see Schlesinger and Mitchell 1987.

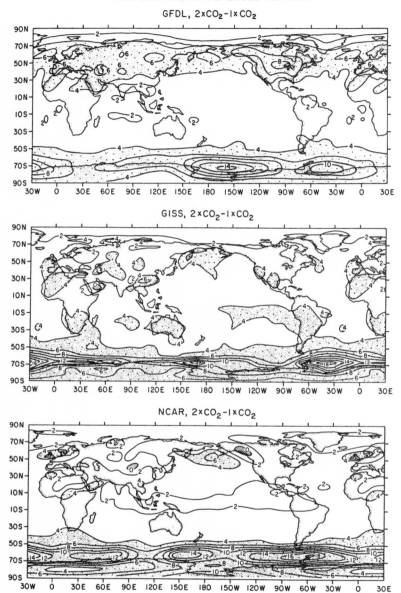

Figure 4.3. Same as figure 4.2, but for June, July, and August.

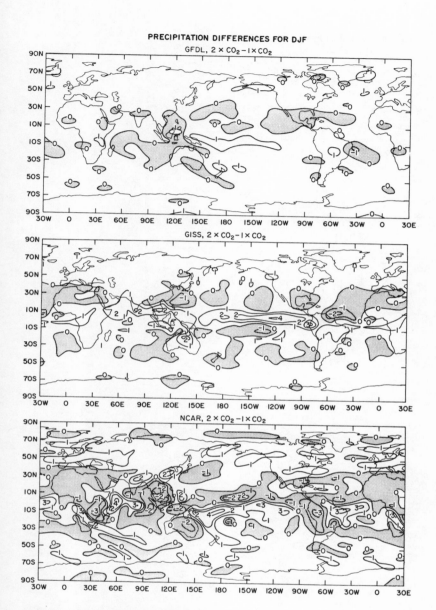

**PRECIPITATION DIFFERENCES FOR DJF**

Figure 4.4. Geographical distribution of the precipitation rate change (in millimeters per day), (2 × $CO_2$) − (1 × $CO_2$), for December, January, and February, simulated with a version of the GFDL GCM, the GISS GCM, and the NCAR GCM. Stippling indicates a decrease in precipitation rate. For details of the GCM versions chosen, see Schlesinger and Mitchell 1987.

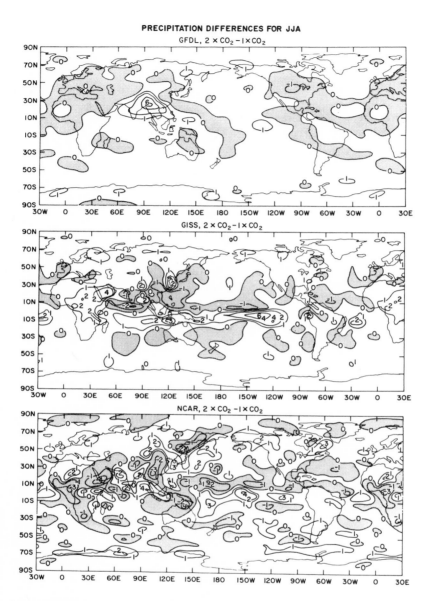

PRECIPITATION DIFFERENCES FOR JJA

GFDL, $2 \times CO_2 - 1 \times CO_2$

GISS, $2 \times CO_2 - 1 \times CO_2$

NCAR, $2 \times CO_2 - 1 \times CO_2$

Figure 4.5. Same as figure 4.4, but for June, July, and August.

Perhaps more important for biological assessment is the prediction of soil moisture, for which one GCM example is given in figure 4.6. Note the large decrease in summertime soil moisture (especially in midlatitudes), a result common to many models (e.g., Rind et al. 1990). The schism between the reticence of some scientists to make any forecasts and the insistence of some policy analysts on high levels of regional detail cannot be resolved simply by having the latter fashion their own implicit scenarios for impact assessment. Rather, we believe, it is better to have knowledgeable scientists putting forth forecasts based on the best available information.

Finally, within the foreseeable future even the highest-resolution three-dimensional GCMs suitable for integrations over fifty or more years will not have a grid much less than 100 km; individual clouds and most biological field research, for example, are on scales far smaller than that. GCMs will not, therefore, be able to resolve individual thunderstorms or the important local and mesoscale effects of hills, coastlines, lakes, vegetation boundaries, and heterogeneous soil. For regions that have relatively uniform surface characteristics, such as a thousand-kilometer savannah or a tropical forest with little eleva-

tion change, parametric representations of surface albedo, soil type, and evapotranspiration could be used to estimate local changes. Alterations in climate predicted within a box would likely apply fairly uniformly across such nicely behaved, homogeneous areas. On the other hand, steep topography or lakes smaller than GCM grids can mediate climate. Therefore, even if GCM predictions were accurate at grid scale, they would not necessarily be appropriate to local conditions.

Large-scale observed climatic anomalies are translated to local variations in figure 4.7. This analysis (Gates 1985) of the local climatic variability for the state of Oregon was based on several years of data using a technique known as empirical orthogonal functions. The north-south Cascade Mountains translate a simple change in the frequency or intensity of westerly winds into a characteristic climatic signature of either wetter on the west slope and drier on the east or vice versa. In

*Figure 4.6.* $CO_2$-induced change in soil moisture expressed as a percentage of soil moisture obtained from a computer model with quadrupled $CO_2$ compared to a control run with normal $CO_2$ amounts. Note the nonuniform response of this ecologically important variable to the uniform change in $CO_2$. (Source: Manabe and Wetherald 1986.)

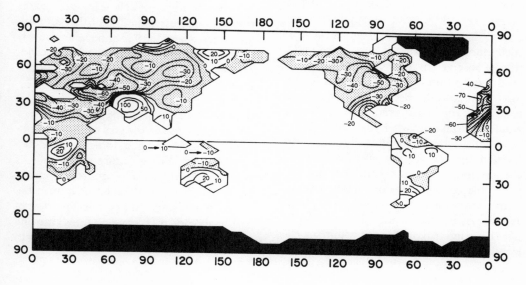

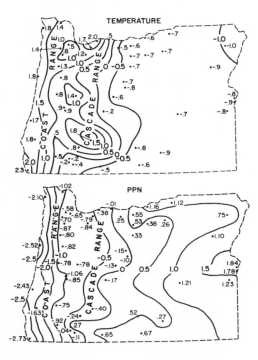

*Figure 4.7.* The distribution of the relationship between large-scale (area-averaged) and local variations of the monthly mean surface air temperature (above) and precipitation (below), as given by the first empirical orthogonal function determined from thirty years of observational monthly means at 49 stations in Oregon in comparison with the statewide average. (Source: Gates 1985.)

other words, a general circulation model of altered westerlies could be applied to the map to determine the effect on a local area. Such a model, constructed from variations of climate observed over several years, seems an ideal way to translate the GCM grid to the local or mesoscale. Because empirical data have been used, however, such a relation would be valid only where the causes of recent climatic variations or oscillations carry forward and include the effect of climatic changes forced by trace gases. It is not obvious that the signature of climatic change from increases in trace gases will be the same as past vacillations, many of which could have been internal oscillations within the climate system, not the result of external forc-

ing such as changes in trace gases. Thus other translations of scale need to be considered.

One might embed a high-resolution mesoscale model within one box of a GCM, using as boundary conditions for the mesoscale model the wind, temperature, and so forth predicted by the GCM at the grid boundaries (Dickinson et al. 1989). The mesoscale model could then account for local topography, soil type, and vegetation cover and translate GCM forecasts to local topography. Figure 4.8 is an example for the western United States. For such a method to have any reasonable hope of success, however, the GCM must produce accurate climatic statistics for the special grid box. To return to the Oregon case in figure 4.7, if the climatic average of the GCM's winds in the unperturbed case (the control case) has the wrong westerly component, the climate change will probably be misrepresented in a region where topography amplifies any such error in the wind direction. A prerequisite for that kind of scale transition, therefore, is a sufficiently accurate control climate for the important variables; only then does it make sense to take the next step of imposing a scenario of trace-gas increase on the GCM to estimate how the local-scale climate might change.

Practically, while testing scale transitions in steep topography and other rapidly varying local features, modelers should examine the behavior of their models using grid boxes that are much less pathological; that is, examine boxes where local features are relatively homogeneous and where translation of local-to-grid scales should prove a less serious obstacle (e.g., Mearns et al. 1990).

Uncertainty about parametric representations of feedback mechanisms like clouds or sea ice is one reason the goal of climate modeling—reliable, verified forecasting of key variables such as temperature and rainfall—is not possible yet. Another source of uncertainty external to the models is human behavior. Forecasting, for example, the effect of carbon dioxide on climate requires knowing how much carbon dioxide or $CH_4$ is going to

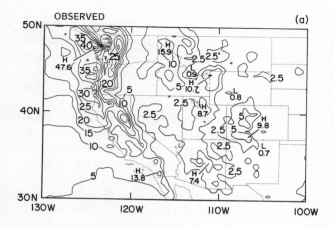

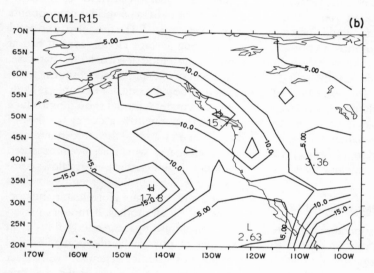

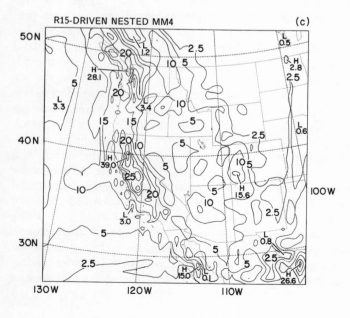

Figure 4.8. Average January total precipitation (centimeters): (a) observations, (b) R15 general circulation model (i.e., 4.5° latitude × 7.5° longitude), (c) mesoscale model driven by output of R15 model. (Source: Giorgi 1990.)

be emitted (Nordhaus and Yohe 1983, Edmonds and Reilly 1984, Ausubel et al. 1988) and how that emission will be distributed or removed by the physical, chemical, and biological processes of the carbon cycle.

What the climate models can do well is analyze the sensitivity of the climate to uncertain or even unpredictable variables. In the case of carbon dioxide, one could construct plausible scenarios of economic, technological, and population growth to project growth of $CO_2$ emission and model the climatic consequences. Such uncertain climatic factors as cloud feedback could be varied over a plausible range. The calculations would indicate which uncertain factors are most important in making the climate sensitive to carbon dioxide increases. One could then concentrate research on those factors. The results would also suggest the range of climatic futures that ecosystems and societies may be forced to adapt to and at what potential rates. How to respond to such information, of course, is a political value issue (e.g., Schneider 1989a, b).

### III. VERIFICATION OF CLIMATE MODELS

The most perplexing question about climate models is whether they can be trusted to provide grounds for altering social policies, such as those governing carbon dioxide emissions. How can models so fraught with uncertainties be verified? There are actually several methods. Although none is sufficient alone, together they can provide significant, albeit circumstantial, evidence of a model's credibility.

The first verification method is checking the model's ability to simulate today's climate. The seasonal cycle (see figure 4.9) is one good test because the temperature changes are several times larger, on a hemispheric average, than the temperature change from an ice age to an interglacial period. GCMs map the seasonal cycle well, which suggests they are on the right track as far as fast physics, such as cloud feedback, are concerned. The seasonal test, however, does not indicate how well a model simulates such

slow processes as changes in deep ocean circulation or ice cover, which may have an important effect on the time scales (decade to century) over which $CO_2$ is expected to double.

A second verification technique is isolating individual physical components of the model, such as its parameterizations, and testing them against reality. For example, one can check whether the model's parameterized cloudiness matches the observed cloudiness of a particular box. But this technique cannot guarantee that the complex interactions of individual model components are properly treated. The model may be good at predicting average cloudiness but bad at representing cloud feedback. In that case, simulation of overall climatic response to increased carbon dioxide, for example, is likely to be inaccurate. A model should reproduce to better than, say, 10% accuracy the flow of thermal energy among the atmosphere, surface, and space, which are well-measured quantities. Together, these energy flows compose the well-established greenhouse effect on earth and constitute a formidable and necessary test for all models. A model's performance in simulating these energy flows is an example of physical verification of model components. Some scientists (Ellsaesser 1984, Lindzen 1990) have argued that critically important hydrological feedback processes are not well simulated by GCMs. Recent satellite analyses (figure 4.10) by A. Raval and V. Ramanathan (1989) suggest that models do a credible job of simulating the so-called water vapor–greenhouse effect so critical to a model's sensitivity to greenhouse-gas forcing (see Schneider 1990 for a discussion).

A third method for determining overall, long-term simulation skill is the model's ability to reproduce the diverse climates of the ancient earth (COHMAP 1988, Barron and Hecht 1985, Imbrie 1987, Schneider 1987) or even of other planets (Kasting et al. 1988). Paleoclimatic simulations of the Mesozoic Era, glacial-interglacial cycles, or other extreme past climates help in understanding the coevolution of the earth's climate with

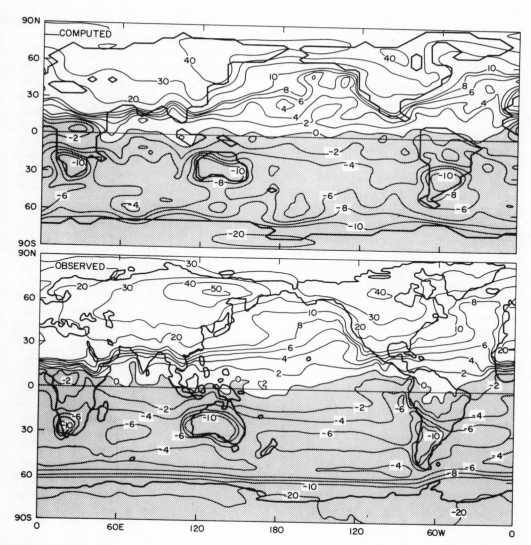

Figure 4.9. A three-dimensional climate model has been used to compute the winter-to-summer temperature extremes all over the globe. The model's performance can be verified against the observed data shown in the lower map. (Source: Manabe and Stouffer 1980.)

Figure 4.10. Comparison of greenhouse effect, heat-trapping factor, and surface temperature, obtained from three sources: ERBE annual values, obtained by averaging April, July, and October 1985 and January 1986 satellite measurements; 3-D climate-model simulations for a perpetual April simulation (NCAR Community Climate Model); line-by-line radiation-model calculations by Dr. A. Arking using $CO_2$, $O_3$, and $CH_4$. The line-by-line model results come close to the CCM and the ERBE values. (Source: Raval and Ramanathan 1989.)

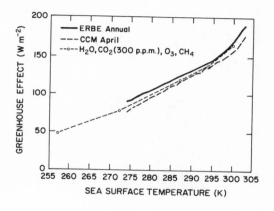

living things. As verifications of climate models, however, they are also crucial to estimating the climatic and biological future.

Overall validation of climatic models thus depends on constant appraisal and reappraisal of performance in the above categories. But those are indirect or surrogate validations for greenhouse-gas increases. Also important are direct validations, such as a model's response to such century-long forcings as the 25% increase in carbon dioxide and large increases of other trace greenhouse gases (Ramanathan et al. 1985) since the Industrial Revolution. Indeed, most climatic models are sensitive enough to predict that a warming of 1°C should have occurred during the past century. The precise forecast of the past 100 years also depends on how the model accounts for such factors as changes in the solar constant or volcanic dust (Schneider and Mass 1975, Gilliland and Schneider 1984, Hansen et al. 1981). Indeed, as figure 4.11 shows, the typical prediction of

a degree or so of warming is broadly consistent but somewhat larger than observed. Possible explanations (see Schneider 1989a) for the discrepancy include (1) the models are too sensitive to increases in trace greenhouse gases by a rough factor of two, (2) modelers have not properly accounted for such competitive external forcings as volcanic dust or changes in solar energy output, (3) modelers have not accounted for other external forcings such as regional tropospheric aerosols from agricultural, biological, and industrial activity (e.g., Wigley 1989), (4) modelers have not properly accounted for internal processes that could lead to stochastic or chaotic behavior (Hasselmann 1976, Dalfes et al. 1983, Lorenz 1968), (5) modelers have not properly accounted for the large heat capacity of the oceans, which could take up some of the heating of the greenhouse effect and delay, but not ultimately reduce, warming of the lower atmosphere, (6) both present models and observed climatic trends could be correct, but models are typically run for equivalent doubling of the $CO_2$ concentration, whereas the world has experienced only a quarter of this increase, and nonlinear processes have been properly modeled and produced a sensitivity appropriate for doubling but not for a 25% increase, and (7) the incomplete and inhomogeneous network of thermometers has underestimated actual global warming this century.

Figure 4.11. Observed global mean temperature changes (1861–1989) compared with predicted values from an upwelling-diffusion climate model. The modeled results show how much temperature change would have occurred if the equilibrium sensitivity to $CO_2$ doubling were as labeled on the figure and if greenhouse-gas forcing were the only cause of climate change. (Source: Wigley and Raper 1990a).

## IV. SUMMARY

Despite the litany of excuses for the disagreement between observed global temperature trends in the past century and those anticipated by most GCMs, the roughly twofold discrepancy is not overwhelming. T.M.L. Wigley and S.C.B. Raper (1990a and 1990b) assert that present models are most consistent with twentieth-century temperature records for equilibrium sensitivity to $CO_2$ doubling between about 1° and 3°C, although higher or lower values would be possible if short-term natural climatic variability were unusually large during the past century. Most climatologists do not yet proclaim that the

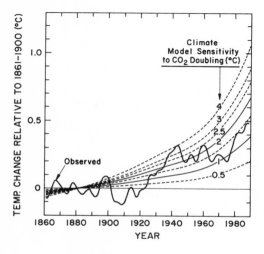

observed twentieth-century temperatures have been caused beyond doubt by the greenhouse effect. Thus, a greenhouse-effect signal cannot yet be said to be unambiguously detected (at, say, the 99% confidence level) in the record. It is still possible that the observed trend and the predicted warming could be chance occurrences. One cannot easily rule out other factors, such as solar constant variations or cooling effects from sulfur dioxide emissions from volcanic, biological, and industrial sources. These possibilities simply have not been adequately accounted for over the past century—except perhaps during the past decade or so, when adequate instruments have been measuring them. Nevertheless, this empirical test of model predictions against a century of observations certainly is consistent to a rough factor of two to three. That test is reinforced by the good simulation by most climatic models of the seasonal cycle for surface air temperature, of diverse ancient paleoclimates (see Lorius et al. 1990), of hot conditions on Venus and cold conditions on Mars, and of the present distribution of climates on earth, and by the agreement between model-generated and satellite-observed energy radiated to space. When taken together, these verifications provide a strong circumstantial case that the modeling of sensitivity of temperature to greenhouse gases is probably valid within the oft cited three-fold range. Another decade or two of observations of trends in the earth's climate, of course, should produce signal-to-noise ratios sufficiently obvious that almost all scientists will agree whether present estimates of climatic sensitivity to increasing trace gases have been predicted well or not.

As the art evolves, such estimates will need to be regularly revised and improved and the implications for the environment and society reassessed. Through this iterative process we hope that a clearer understanding of potential climatic changes will develop. Of course, while scientists study and debate, the world becomes committed to a growing dose of greenhouse gases and their effects. Anthropogenic climate change, on a sustained and global basis, could be evolving (as in figure 4.1) at an extremely rapid rate when compared with most paleoclimatic trends. It is questionable whether natural ecosystems and many human activities can adapt to the mid or upper rates of change without major disruptions, suggesting some urgency in resolving uncertainties in both climatic scenarios and impact assessments and in considering actions to slow the buildup of greenhouse gases (Schneider 1989b).

## REFERENCES

Ausubel, J. H., A. Grubler, and N. Nakicenovic. 1988. Carbon dioxide emissions in a methane economy. *Clim. Change* 12:245.

Bach, W., J. Pankrath, and W. Kellogg, eds. 1979. *Man's Impact on Climate*. Amsterdam: Elsevier.

Barnola, J. M., D. Raynaud, Y. S. Korotkevich, and C. Lorius. 1987. Vostok ice core provides 160,000-year record of armospheric $CO_2$. *Nature* 329:408.

Barron, J., and A. D. Hecht, eds. 1985. *Historical and Paleoclimatic Analysis and Modeling*. New York: John Wiley and Sons.

Berger, A., J. Imbrie, J. Hays, G. Kukla, and B. Saltzman, eds. 1984. In *Milankovitch and Climate*, parts 1 and 2. Dordrecht, The Netherlands: D. Reidel.

Bolin, B., et al., eds. 1986. *The Greenhouse Effect, Climatic Change and Ecosystems*. Chichester: John Wiley and Sons.

Bryan, K., F. G. Komro, S. Manabe, and M. J. Spelman. 1982. Transient climate response to increasing atmospheric carbon dioxide. *Science* 215: 56.

Budyko, M. I., A. B. Ronov, and A. L. Yanshin. 1987. *History of the Earth's Atmosphere*. New York: Springer-Verlag.

Carbon Dioxide Assessment Committee. 1983. *Climate Change*. Washington, D.C.: National Academy Press.

Climate Research Board. 1978. *International Perspectives on the Study of Climate and Society*. Washington, D.C.: National Academy of Sciences.

COHMAP Members (P. M. Anderson et al.). 1988. Climatic changes of the last 18,000 years: Observations and model simulations. *Science* 241:1043.

CEQ (Council on Environmental Quality). 1980. *The Global 2000 Report to the President*. Washington, D.C.: U.S. Government Printing Office.

Dalfes, H., S. H. Schneider, and S. L. Thompson. 1983. Effects of bioturbation on climatic spectra inferred from deep-sea cores. *J. Atmos. Sci.* 40: 1648.

Dickinson, R. E. 1986. The impact of human activities on climate: A framework. In *Sustainable Development of the Biosphere*, W. C. Clark and R. E. Munn, eds., pp. 252–289. Cambridge: Cambridge University Press.

Dickinson, R. E., R. M. Errico, F. Giorgi, and G. T. Bates. 1989. A regional climate model for the western United States. *Clim. Change* 15:383.

Edmonds, J. A., and J. Reilly. 1984. Global energy in $CO_2$ to the year 2050. *Energy J.* 4:21.

Ellsaesser, H. W. 1984. The climatic effect of $CO_2$: A different view. *Atmos. Envir.* 18:431.

Emanuel, K. A. 1987. The dependence of hurricane intensity on climate. *Nature* 326:483.

Farrell, M. P., ed. 1987. *Master Index for the Carbon Dioxide Research State-of-the-Art Report Series*. Washington, D.C.: Department of Energy.

Gates, W. L. 1985. The use of general circulation models in the analysis of the ecosystem impacts of climatic change. *Clim. Change* 7:267.

Gilliland, R. L., and S. H. Schneider. 1984. Volcanic, $CO_2$, and solar forcing of northern and southern hemisphere surface air temperatures. *Nature* 310:38.

Giorgi, F. 1990. Simulation of regional climate using a limited area model nested in a general circulation model. *J. Climate* 3(9):941.

Gleick, P. H. 1986. Methods for evaluating the regional hydrologic impacts of global climatic changes. *J. Hydrol.* 88:97.

Gleick, P. H. 1987. Regional hydrologic consequences of increases in atmospheric $CO_2$ and other trace gases. *Clim. Change* 10:137.

Hansen, J., D. Johnson, A. Lacis, S. Lebedeff, P. Lee, D. Rind, and G. Russell. 1981. Climate impact of increasing atmospheric carbon dioxide. *Science* 213:957.

Hansen, J., G. Russell, A. Lacis, I. Fung, and D. Rind. 1985. Climate response times: Dependence on climate sensitivity and ocean mixing. *Science* 229:857.

Hasselmann, K. 1976. Stochastic climate models, Part I: Theory. *Tellus XXVIII* 6:473.

Hoffert, M. I., A. J. Callegari, and C. T. Hsieh. 1980. The role of deep-sea heat storage in the secular response to climatic forcing. *J. Geo. Res.* 85:667.

Imbrie, J. 1987. Abrupt terminations of Late Pleistocene ice ages: A simple milankovitch explanation. In *Abrupt Climatic Change: Evidence and Implications*, W. H. Berger and L. D. Labeyrie, eds., pp. 365–367. Dordrecht, The Netherlands: D. Reidel.

IPCC (Intergovernmental Panel on Climate Change). 1990. *Climate Change: The IPCC Scientific Assessment*, J. T. Houghton, G. J. Jenkins, and J. J. Ephraums, eds. Cambridge: Cambridge University Press.

Jager, J. 1988. Developing policies for responding to climatic change: A summary of the discussion and recommendations of the workshops held in Villach, 28 September to 2 October 1987. WCIP-1, WMO/TD, no. 225.

Jager, J., and W. W. Kellogg. 1983. Anomalies in temperature and rainfall during warm arctic seasons. *Clim. Change* 5:39.

Kasting, J. F., O. B. Toon, and J. B. Pollack. 1988. How climate evolved on the terrestrial planets. *Sci. Am.* 258 (February):90.

Kates, R. H., J. H. Ausubel, and M. Berberian. 1985. *Climate Impact Assessment*, SCOPE 27. New York: John Wiley and Sons.

Lindzen, R. S. 1990. Some coolness concerning global warming. *Bull. Am. Meteor. Soc.* 77:288.

Lorenz, E. N. 1968. Climate determinism. *Meteor. Monogr.* 8(30):1.

Lorius, C., J. Jouzel, D. Raynaud, J. Hansen, and H. Le Treut. 1990. The ice-core record: Climate sensitivity and future greenhouse warming. *Nature* 347:139.

Lough, J. M., T.M.L. Wigley, and J. P. Palutikof. 1983. Climate and climate impact scenarios for Europe in a warmer world. *J. Clim. Appl. Meteor.* 22:1673.

Manabe, S., and R. J. Stouffer. 1980. Sensitivity of a global climate model to an increase in $CO_2$ concentration in the atmosphere. *J. Geo. Res.* 85:5529.

Manabe, S., and R. Wetherald. 1986. Reduction in summer soil wetness induced by an increase in atmospheric carbon dioxide. *Science* 232:626.

Mearns, L. O., R. W. Katz, and S. H. Schneider. 1984. Extreme high temperature events: Changes in their probabilities and changes in mean temperature. *J. Clim. Appl. Meteor.* 23:1601.

Mearns, L. O., S. H. Schneider, S. L. Thompson, and L. R. McDaniel. 1990. Analysis of climate variability in general circulation models: Comparison with observations and changes in variability in $2 \times CO_2$. *J. Geo. Res.* 95:20, 469.

NAS (National Academy of Sciences). 1991. *Policy Implications of Greenhouse Warming*. Washington, D.C.: National Academy Press.

NRC (National Research Council). 1987. *Current Issues in Atmospheric Change*. Washington, D.C.: National Academy Press.

Nordhaus, W., and G. Yohe. 1983. Future paths of energy and carbon dioxide emissions. In *Changing Climate*, Report of the Carbon Dioxide Assessment Committee, pp. 87–153. Washington, D.C.: National Academy Press.

Parry, M. L. 1985. Estimating the sensitivity of natural ecosystems and agriculture to climatic change. *Clim. Change* 7:1.

Parry, M. L., and T. R. Carter. 1985. The effect of climatic variations on agricultural risk. *Clim. Change* 7:95.

Pearman, G. I., ed. 1987. *Greenhouse: Planning for Climate Change.* Leiden, The Netherlands: E. J. Brill.

Pittock, A. B., and J. Salinger. 1982. Towards regional scenarios for a $CO_2$-warmed earth. *Clim. Change* 4:23.

Ramanathan, V., R. J. Cicerone, H. B. Singh, and J. T. Kiehl. 1985. Trace gas trends and their potential role in climate change. *J. Geo. Res.* 90:5547.

Raval, A., and V. Ramanathan. 1989. Observational determination of the greenhouse effect. *Nature* 342:758.

Rind, D., R. Goldberg, J. Hansen, C. Rosenzweig, and R. Ruedy. 1990. Potential evapotranspiration and the likelihood of future drought. *J. Geo. Res.* 95,D7:9983.

Schlesinger, M. E., and J.F.B. Mitchell. 1987. Climate model simulations of the equilibrium climatic response to increased carbon dioxide. *Rev. Geophys.* 25:760.

Schneider, S. H. 1984. In *Climate Processes and Climate Sensitivity,* J. Hansen and T. Takahashi, eds., pp. 187–201. Washington, D.C.: American Geophysical Union.

Schneider, S. H. 1987. Climate modeling. *Sci. Am.* 256(5):72.

Schneider, S. H. 1989a. The greenhouse effect: Science and policy. *Science* 243:771.

Schneider, S. H. 1989b. *Global Warming: Are We Entering the Greenhouse Century?* San Francisco: Sierra Club Books.

Schneider, S. H. 1990. The global warming debate heats up: An analysis and perspective. *Bull. Am. Meteor. Soc.* 71:1292.

Schneider, S. H., and C. Mass. 1975. Volcanic dust, sunspots, and temperature trends. *Science* 190:741.

Schneider, S. H., and S. L. Thompson. 1981. Atmospheric $CO_2$ and climate: Importance of the transient response. *J. Geo. Res.* 86:3135.

Schneider, S. H., P. Gleick, and L. O. Mearns. 1990. Prospects for climate change. In *Climate Change*

and *U. S. Water Resources,* P. E. Waggoner, ed., pp. 41–73. New York: John Wiley and Sons.

Seidel, S., and D. Keyes. 1983. *Can We Delay a Greenhouse Warming? The Effectiveness and Feasibility of Options to Slow a Build-Up of Carbon Dioxide in the Atmosphere.* Washington, D.C.: Environmental Protection Agency and Strategic Studies Staff, Office of Policy Analysis, Office of Policy and Resources Management.

Senate Committee on Governmental Affairs. 1979. *Carbon Dioxide Accumulation in the Atmosphere: Synthetic Fuels and Energy Policy.* Washington D.C.: U.S. Government Printing Office.

Smith, J. B., and D. Tirpak, eds. 1988. *The Potential Effects of a Global Climate Change on the United States: Draft Report to Congress.* Vol. II, Chap. 9. Washington, D.C.: Environmental Protection Agency.

Stouffer, R. J., S. Manabe, and K. Bryan. 1989. Interhemispheric asymmetry in climate response to a gradual increase of atmospheric $CO_2$. *Nature* 342:660.

Thompson, S. L., and S. H. Schneider. 1982. $CO_2$ and climate: The importance of realistic geography in estimating the transient response. *Science* 217:1031.

Waggoner, P. E., ed. 1990. *Climate Change and U.S. Water Resources.* New York: John Wiley and Sons.

Washington, W. M., and G. A. Meehl. 1989. Climate sensitivity due to increased $CO_2$: Experiments with a coupled atmosphere and ocean general circulation model. *Clim. Dynamics* 4:1.

Wigley, T.M.L. 1985. Impact of extreme events. *Nature* 316:106.

Wigley, T.M.L. 1989. Possible climate change due to $SO_2$-derived cloud nuclei. *Nature* 339:365.

Wigley, T.M.L., and S.C.B. Raper. 1990a. Paper presented at Second World Climate Conference, Geneva.

Wigley, T.M.L., and S.C.B. Raper. 1990b. Natural variability of the climate system and detection of the greenhouse effect. *Nature* 344:324.

Williams, J. 1978. *Carbon Dioxide, Climate and Society.* Oxford: Pergamon Press.

WMO (World Meteorological Organization). 1988. Report of the Workshops Held in Villach (28 September–2 October 1987) and Bellagio (9–13 November 1987). WCIP-1.

# PART II

# Responses to Past Climate:
# The Fossil Record

CHAPTER FIVE

# Past Changes in Vegetation and Climate: Lessons for the Future

THOMPSON WEBB III

## I. INTRODUCTION

Knowledge of past climates helps in judging the severity and uniqueness of potential future changes in climate. We need to know how the predicted changes for global warming measure up. Are they larger or more rapid than past changes? How reliable are the predictions? What impacts have past climatic changes had on the vegetation, and what impacts might we anticipate in the future? These are some of the critical questions environmental planners need to answer before formulating policy for the future. The purpose of my chapter is to review some of what is known about past climate changes and to illustrate how these changes have affected the vegetation across eastern North America during the past 18,000 years. I also show how paleoclimatic data are being used to test the reliability of climate models and list some of the implications of these results for anticipating how future climatic changes may affect the biosphere. My review highlights information from P. J. Bartlein (1988), COHMAP (1988), A. D. Hecht (1985), M. L. Hunter et al. (1988), P. C. McDowell et al. (1991), W. F. Ruddiman and H. E. Wright (1987), and Webb (1986) and documents the utility of response surfaces (Bartlein et al. 1986) for estimating the potential vegetational response to climate change, as explained in section IV.

I will focus my discussion largely on the climates of the past 18,000 years because the rise in global mean temperature of $5° \pm 1°C$ during this interval (see appendix to this chapter) closely approximates the $4.2° \pm 1.2°C$ rise predicted for the near future as the result of a doubling in the effective concentration of greenhouse gases (Schlesinger 1989, chapter 4). Because the magnitude of warming is similar in both cases, we can examine atmospheric and ecological trends associated with the past warming to help understand how future warming might change the biosphere. Further, by studying past climates we have a way to test the accuracy of the climate models used to predict the regional climate patterns of the future; we can use the models to simulate past climates and then use our knowledge of actual past conditions to verify the results.

The past records show that the predicted global mean temperature will exceed any during the past 18,000 years. Although sometime between 6000 and 9000 years ago the global mean temperature may have been 1° ± 1°C higher than today (Webb and Wigley 1985), the last time it was as much as 4°C higher than today—the rise projected by the climate models—was in the Eocene (40 million years ago) or earlier (Barron 1985, Crowley 1989).

Also, the rate at which the global mean temperature will rise under global warming conditions—3°C per century or faster—will be faster than any natural warming during the past 18,000 years. For brief times during that period, local and regional climates may have warmed at rates similar to 3°C per century (Rind et al. 1986), but such a rate of increase for the global mean temperature seems highly unusual. Chapters by Margaret Davis and Catherine Zabinski (chapter 22) and by Daniel Botkin and Robert Nisbet (chapter 21) explore some of the biological consequences of such a rate of temperature increase, and I later note that it is likely many plant species will be unable to move their ranges rapidly enough to keep up with such a rapidly shifting climate, and may thus create a mismatch called a disequilibrium response.

## II. PAST CHANGES IN CLIMATE

We can look at how global mean temperature has changed over time and use this variable to compare past climate changes on different time scales. Figure 5.1 shows the patterns of variability in global mean temperature on five time scales—each scale measured in units ten times larger than the previous. The shortest scale covers a single century (1880–1980), the

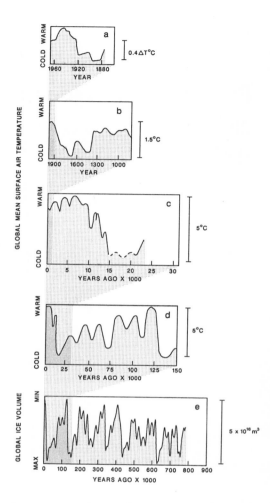

Figure 5.1. General trends in global climate for various time scales ranging from decades to hundreds of millennia. Note that local and regional climate changes may not resemble these global trends. The time series of climatic information are from instrumental data for 1880–1970 (a), from historical information primarily from the North Atlantic region and Europe for the last thousand years (b), from pollen data and alpine glaciers for 30,000 years (c), from marine plankton data covering 150,000 years (d), and from oxygen-isotope fluctuations in foraminifera shells in deep-sea sediments covering roughly one million years (e). Modified from Bernabo 1978.

longest roughly one million years. I have recently evaluated the quality of data on which each of these five time series is based and described the need for more data to improve the two time series for 1000 and 30,000 years; however, the current time series are adequate to support the general conclusions of this chapter (Webb 1991).

Over the past million years, the most noticeable pattern of global variation has been the oscillation between glacial and interglacial climates, with glacial periods occurring roughly every 100,000 years (a quasi-periodicity of 100,000 years; fig. 5.1e). A detailed view of the last 150,000 years (fig. 5.1d) shows three things: First, the last interglacial period, which may have been slightly warmer than today (by 2° ± 1°C at most), lasted from 130,000 to 120,000 years ago. Second, the present interglacial began about 10,000 years ago. Third, the last full-glacial period was from 23,000 to 13,000 years ago (fig. 5.1c). Over the past million years, the earth's climate has varied continuously and has been in interglacial and full-glacial modes for only 20% of the time. The rest of the time, it was somewhere in between.

Over these long time spans, the variability of global temperature has been more than matched by geographic variations in temperature, moisture, and atmospheric circulation (CLIMAP 1981, COHMAP 1988). For example, during the last full-glacial period, sea surface temperatures varied geographically from being more than 10°C lower than those today in the North Atlantic Ocean to being 2°C higher than those today in the tropical Pacific Ocean. These geographic variations and the changes in ice sheets and sea ice are what plant and animal distributions respond to. Rainfall variations in the northern tropics have also been large, with peak rainfall occurring every 20,000 years (Prell and Kutzbach 1987). These variations have caused biomes like the Sahara Desert to vary widely in extent and composition (COHMAP 1988). Though qualitatively different, the variability of past tropical climates has therefore matched that of many mid- to high-latitude climates.

Over short time spans of 100 to 1000 years (fig. 5.1a, b), climate has also varied continuously, but the magnitude of change has been less than that between glacial and interglacial periods. Within the North Atlantic region, historical, archeological, and fossil records show there was a warmer period from A.D. 800 to 1200 when Scandinavians were able to settle Iceland and Greenland (Lamb 1979), but the global extent of those warmer temperatures is as yet unknown. The Little Ice Age followed this period (Grove 1988), and instrumental records indicate that the earth has warmed by 0.5° ± 0.1°C since the end of the Little Ice Age in A.D. 1850 ± 30 years (Jones et al. 1986, 1989). The maximum rate for this warming was about 0.5°C per century. (The end of the Younger Dryas episode about 10,200 years ago, ±300 years, is often cited as a time of rapid warming with rates of 2°C or more per 100 years, but we have yet to demonstrate that the warming at those rates was global.) These short-term changes have been large and rapid enough to have affected human events (Wigley et al. 1981).

We remain uncertain of the causes for the short-term (decade to millennium) changes in climate, but much progress has been made in understanding causes of the long-term changes, which cycle with a periodicity of 20,000 years or longer. Knowledge of the causes of long-term changes has allowed the design of modeling experiments that are key to testing how well climate models predict spatial patterns in climate (COHMAP 1988).

## III. CAUSES OF LONG-TERM CLIMATE VARIATIONS

We understand fairly well the factors that pace the long-term periodic changes in global mean temperature during the past 150,000 years. Such changes seem to be caused indirectly by two types of long-term rhythmic variations in the orientation of the earth's axis within the earth's orbit about the sun: tilt and precession (fig. 5.2). These orbital changes alter the amount of radiation received each season and thus alter seasonality, which is a

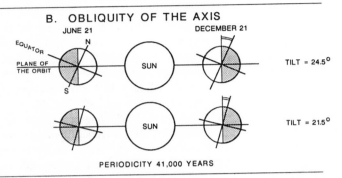

## A. ECCENTRICITY OF THE ORBIT

ALMOST ELLIPTICAL          ALMOST CIRCULAR

PERIODICITY 100,000 YEARS

## B. OBLIQUITY OF THE AXIS

JUNE 21          DECEMBER 21

EQUATOR
N
PLANE OF
THE ORBIT
SUN
S

TILT = 24.5°

SUN

TILT = 21.5°

PERIODICITY 41,000 YEARS

## C. PRECESSION OF THE EQUINOXES

23.5°

MAR. 20
JUNE 21          DEC. 21
SEPT. 22

PERIHELION
TODAY

DEC. 21
MAR. 20          SEPT. 22
JUNE 21

5,500
YEARS AGO

SEPT. 22
DEC. 21          JUNE 21
MAR. 20

11,000
YEARS AGO

AXIAL WOBBLE

PERIODICITY ca. 21,000 YEARS

Figure 5.2 Variations in the earth's orbit that control long-term climatic change (modified from Imbrie and Imbrie 1979 by McDowell et al. 1991). A. Eccentricity: The more elliptical the orbit, the more seasonal the variation in insolation. B. Obliquity: The more tilted the earth's axis, the greater the differences between summer and winter radiation within the same hemisphere, and the greater the differences between simultaneous temperatures in the northern and southern hemispheres. C. Precession: The precession of the earth within its elliptical orbit changes the time of year when the earth is near or far from the sun. This alternately increases and decreases the severity of winters and summers. Today the earth is closest to the sun on January 4 (the perihelion) and farthest on July 4 (the aphelion). Today's winter solstice (shortest day of the year) for the northern hemisphere is December 21, the summer solstice is June 21, and the equinoxes are March 20 and September 22.
Reproduced by permission of Cambridge University Press.

measure of how warm the summers are in contrast to the severity of the winters. The seasonal variations in temperature affect atmospheric and oceanic circulation and lead to increases and decreases in glacial ice volume, atmospheric carbon dioxide concentrations, and sea-surface temperatures. The radiation changes and those in ice volume and other factors also cause changes in global mean temperatures.

The two long-term orbital rhythms do not alter the total annual amount of radiation falling on the earth. Instead, they alter how that radiation is distributed in time and space within and between the northern and southern hemispheres. Working together over thousands of years, the rhythms increase and decrease seasonality, first making winters colder and summers hotter, and then moderating them. The northern and southern hemispheres are affected differently, as described below. The earth is currently in a period of relative moderation, when winters are relatively mild and summers are cool (especially in the northern hemisphere).

The first of the two types of rhythmic change is variation in the tilt of the earth's axis, which has a cycling period of about 41,000 years as it tilts from 24.5° from the vertical to 21.5° and back again (fig. 5.2b). An increase in the angle of tilt has two noticeable effects on climate. First, it increases seasonality within each hemisphere, meaning that summers become hotter and winters colder. The more the axis tilts, the more each hemisphere is tilted toward the sun during its summer, and the farther away it tilts during the winter. The result for both hemispheres is to increase solar radiation during summers and decrease it during winters. Therefore, at times like the present, when the angle of tilt is moderate (23.5° vs. 24.5°), annual seasonality within a hemisphere is moderate. Second, because an increase in tilt makes northern winters colder while making the concurrent southern summer warmer, and vice versa, it also increases the thermal contrast between the hemispheres.

The second source of rhythmicity is variation in how close to the sun the earth is at a given time of year (fig. 5.2c). Because the earth's orbit is elliptical, the earth is closer to the sun during half of the year than it is during the other half. At present, the earth is closest on January 4 and farthest on July 4, but 11,000 years ago the conditions were reversed and the earth was closest in June. This shift in timing is called precession of the equinoxes, and it occurs with a periodicity of 21,000 years. It is caused by a cyclical wobble in the position of the earth's axis. The result of precession is to vary, as the equinoxes shift, the amount of solar radiation falling at a given time of year on the northern and southern hemispheres. For example, if precession puts the northern hemisphere closest to the sun during winter (January 4), northern winters are warmer than if the earth were farther away; the concurrent southern hemisphere summer is hotter. When the cycle is reversed and the northern hemisphere is closest to the sun during summer (July 4), northern summers are hotter, and the concurrent southern winter is warmer, but the southern summers are colder.

These two effects, tilt and precession, work together to pace the cyclical patterns seen in figure 5.1d. Each effect has its own cyclical period (41,000 versus 21,000 years), and depending on how the cycles coincide, they can counteract or reinforce each other to cause seasonal cooling or warming. For example, in the northern hemisphere, seasonality is greatest when tilt is large and precession places the earth farthest from the sun during winter and closest in summer. Today, the earth is tilted at 23.5° (1° less than maximum) and is closest to the sun during winter in the northern hemisphere. Both factors make the northern hemisphere winter warmer and its summer cooler than they would be at maximum tilt and reversed precession. About 11,000 years ago, both tilt and precession acted to make northern winters colder and summers hotter than now: the earth's axial tilt was greater by 1° and the northern hemisphere was closest to the sun during the summer. The result was that the northern hemisphere received 8% more summertime and 8% less wintertime radiation than either today or 18,000 years ago (fig. 5.3).

A third type of orbital variation also affects solar radiation: change in the shape, or eccentricity, of the earth's orbit (fig. 5.2a). When the earth's orbit is more elliptical, the variation in the maximum seasonal contrast is greater and can be as high as 10% versus the 8% variation shown in figure 5.3 (Prell and Kutzbach 1987). The cycle of this orbital change (ca. 100,000 years) is too long to warrant further consideration here (see Berger et al. 1984).

Paleoclimatologists are still searching for the full explanation of how tilt and precession have influenced global climates and caused the ice sheets to wax and wane (Ruddiman et al. 1986), but the research of J. D. Hays et al. (1976) and of J. Imbrie et al. (1984) strongly implicates orbital variation as the pacemaker for the long-term changes in climate (1000–100,000 years). These changes in seasonal distribution of radiation result in overall global increases and decreases in mean temperature. Possible mechanisms by which seasonal changes in solar radiation affect the climate system include land-sea contrasts in heat capacity and resultant effects on the strength of monsoons and hence tropical rainfall, and changes in ocean circulation and their possible role in changing the concentrations of atmospheric carbon dioxide. Many researchers are currently active in trying to identify the full set of mechanisms that translate the orbitally induced changes in radiation into major changes in the global climate system (Berger et al. 1984, Imbrie et al. 1989).

## IV. CLIMATES OF THE PAST EIGHTEEN THOUSAND YEARS AND TESTS OF CLIMATE MODELS

Our understanding of how the orbital changes affect climate is good enough that these effects can be incorporated into current general circulation models (GCMs) that are used to help predict future climates (see chapter 4 for discussion of GCMs). If the modelers plug in appropriate values for seasonal radiation, plus other atmospheric and earth-surface factors that determine climate, the models will simulate past or future climate conditions, including global mean and regional temperatures, that are in equilibrium with the radiation and earth-surface values. These factors that determine the model's predictions are called boundary conditions, and include not only the orbital variations described above but other factors like extent of ice sheets and sea ice.

Besides simulating future climate conditions by running models with twice the current concentration of carbon dioxide (Schlesinger and Mitchell 1985), we can test the accuracy of the models by giving them appropriate boundary conditions for specific times in the past and allowing them to "predict" past climates. The model's prediction can then be compared with the actual climate of that time, as ascertained from fossil and other observational evidence (see section V), and the model's accuracy judged.

Such an experiment was designed and completed by J. E. Kutzbach and P. Guetter (1986) and COHMAP (1988), who modeled a series of climates over the past 18,000 years. These researchers used a GCM (the National Center for Atmospheric Research Community Climate Model—the NCAR CCM) that incorporated solar radiation variations, patterns of sea-surface temperatures, the size of the continental ice sheets, values for surface reflectivity, and concentrations of aerosols and atmospheric carbon dioxide (fig. 5.3). All have varied in the past 18,000 years, with changes in solar radiation, ice sheets, and sea-surface temperatures having the greatest effect on the climate outcome.

The model was used to predict a series of equilibrium climates at selected dates between 18,000 years ago and today. Figure 5.4 shows one series of such snapshots for eastern North America, mapping the changes in temperature and moisture conditions. These changes reflect the sequential changes in atmospheric circulation caused by the direct influence of the retreating Laurentide ice

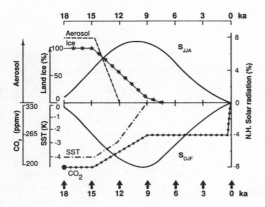

*Figure* 5.3. Changing external and surface boundary conditions during the past 18,000 years. Right: Solar radiation for the northern hemisphere (N.H.) as % of today's values. Left: Land ice as % of the maximum during the past 20,000 years; sea-surface temperatures (SST) in degrees Celsius below current temperature; $CO_2$ in parts per million by volume; and aerosols, or atmospheric particulate matter, no scale. Top and bottom: Thousands of years before today (0). $S_{JJA}$ is the solar radiation for July through August, and $S_{DJF}$ is the solar radiation for December through February. The SST lowering by 4°C is an areal average for both the ice-free and ice-covered oceans from climate-model estimates generated by Kutzbach and Guetter (1986) using data from CLIMAP (1981). Reproduced from COHMAP (1988). Copyright © 1988 by AAAS.

sheet (Kutzbach 1987, Webb et al. 1987), as well as changing patterns of seasonal radiation and sea-surface temperatures. Both January and July temperatures increased between 18,000 and 6000 years ago, especially in the northern part of the continent. After 6000 years ago, the temperatures in the north decreased by a degree or two. Many of the predicted changes are in good to fair agreement with the paleoclimatic data in terms of the changing patterns of the vegetation and lake levels (Bartlein and Webb 1985, COHMAP 1988, Webb et al. 1987).

On a global scale, one of the key results from the model was that orbitally caused changes in solar radiation caused large changes in monsoonal circulation over Africa and Asia (Kutzbach 1981). In the model, the

extra summertime radiation in the northern hemisphere tropics from 12,000 to 6000 years ago warmed the land surface relative to the oceans and thereby enhanced the monsoonal circulation of warm moist air from the oceans over the continents. This additional moist air led to increased continental rainfall in the model, and there is good evidence that rainfall really did increase at that time. Data indicate that before 6000 years ago, water levels were high in lakes that are today dry within the Sahara, western India, and many other tropical regions (Kutzbach and Street-Perrott 1985, Street-Perrott and Harrison 1985). This good agreement between the data and model results is a key verification of the ability of this GCM to simulate spatial patterns of climate correctly. Other GCMs have produced similar results (Kutzbach 1981, Kutzbach and Gallimore 1988, Mitchell et al. 1988). One implication of these results for wildlife is that the increased moisture in the Sahara 9000 years ago allowed crocodiles and hippopotamuses to spread their range far out into what is desert today. The Sahara then was not the biogeographical barrier that it is today for certain taxa (COHMAP 1988).

The model predictions of past climates can also be checked against vegetation patterns. We can take a past climate predicted by the GCM, then predict what geographical distribution of vegetation would have been created by the climate (see chapter 8 for discussion of how climate determines vegetation patterns), and check these simulated vegetation patterns against the actual vegetation patterns of the time. If the simulated climate predicts a vegetation distribution that matches the actual distribution, then we gain confidence that the simulated climate is a close fit for the actual climate of the time.

In one example, Webb et al. (1987) modeled the distributions of spruce (*Picea*) in eastern North America at different times in the past and then compared those predictions with actual distributions. The study modeled distributions of spruce pollen, because actual data on prehistoric distributions of spruce and other trees are based on surveys of pol-

Simulated by CCM

Mean January Temperature (°C)

Mean July Temperature (°C)

Mean Annual Precipitation (mm)

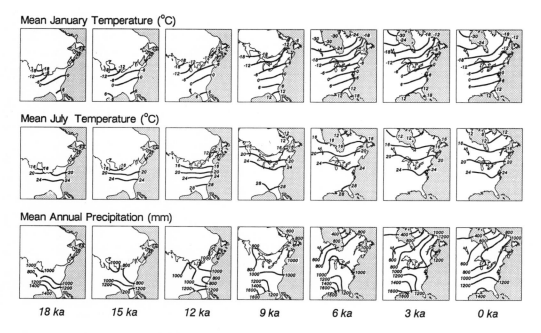

| 18 ka | 15 ka | 12 ka | 9 ka | 6 ka | 3 ka | 0 ka |

*Figure 5.4.* Climate-model simulations of January and July temperatures (in degrees Celsius) and annual precipitation (in millimeters) in eastern North America from 18,000 years ago to present (18 ka to 0 ka). Outlined area in the north from 18,000 to 9000 years ago marks the retreating Laurentide ice sheet. The maps for today are modern observed values, and the maps of simulated conditions for earlier dates were constructed by adding to these observed values the difference between the values simulated for a given date and the values from the model control run, which is the model simulation for the climate today. This procedure minimized the effect of model bias on the simulated values (see Webb et al. 1987).

len distribution (see section V). Technically, the process went as follows: Webb et al. used observed modern temperature and precipitation patterns and the modern distribution of spruce pollen to calculate a "response surface" for spruce (fig. 5.5). Response surfaces are mathematical representations of the abundances of plant taxa or pollen types in terms of nonlinear functions of temperature and precipitation (Bartlein et al. 1986). Next

the response surfaces were used to transform simulated temperature and precipitation values into a time sequence of maps showing simulated distributions of spruce pollen throughout eastern North America. When these maps are compared with observed patterns of actual pollen distribution, the patterns match quite well (fig. 5.6). This good fit further confirms the ability of climate models to simulate the distribution of combinations of climate variables in space.

## V. POLLEN DISTRIBUTIONS AND PAST VEGETATION CHANGE IN EASTERN NORTH AMERICA

Pollen data have long been a key source of information about past vegetation change. They are collected by extracting cores of lake and bog sediments in which the pollen has been preserved through time. The sediments are then radiocarbon dated to determine a time scale for the depths in the cores, and the pollen from various depths or times is identified and counted under a microscope. The

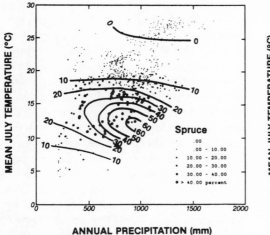

Figure 5.5. Three-dimensional response surface for spruce pollen in which the modern percentage of spruce pollen is modeled as a function of July temperature, January temperature, and annual precipitation. The two graphs portray two-dimensional views of the three-dimensional surface. Modified from Webb et al. 1987.

Figure 5.6. Maps of observed and simulated pollen percentages for spruce. The map of observed values is computer-generated and produced from the same data used in figure 5.8. The simulated values were produced by applying the response surface for spruce (fig. 5.5) to the climate-model-simulated temperature and precipitation estimates in figure 5.4. Modified from Webb et al. 1987 and COHMAP 1988. Copyright © 1988 by AAAS.

temporal changes of pollen data from each site indicate the changing vegetation near each site. Another source of information about past distributions is fossils of whole or partial plants, called macrofossils because they are larger than pollen microfossils. In desert regions where lakes are dry, macrofossils can be collected from packrat middens (Cole 1982, Spaulding et al. 1983, Van Devender et al. 1987). H.J.B. Birks and H. H. Birks (1980), B. E. Berglund (1986), and B. Huntley and Webb (1988) describe the collection and analysis of pollen and plant macrofossil data.

Studies of modern pollen data show that the relative abundance of each major pollen type correlates with the relative abundance of

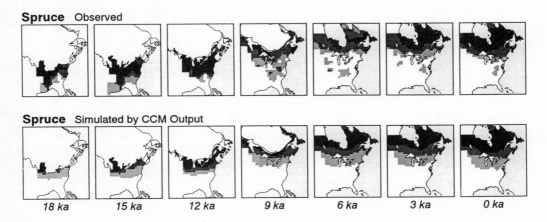

the plant that produced it (Bradshaw and Webb 1985, Prentice 1988, Prentice and Webb 1986). When the current patterns for various pollen types are mapped, the distribution patterns show good correspondence with the modern vegetation patterns (Webb 1987, 1988). For example, in eastern North America, herbaceous pollen types (prairie forbs, Cyperaceae) are most abundant in samples from the prairie and tundra; spruce (*Picea*) pollen, with some birch (*Betula*) and pine (*Pinus*) pollen, dominates in samples from the boreal forest; birch and pine pollen dominate in the mixed forest; oak (*Quercus*) and pollen from other deciduous trees dominate in the deciduous forest; and pollen from southern pines dominates in the southeastern conifer forest (figs. 5.7 and 5.8). The maps of the pollen types have abundance gradients that indicate where one vegetation region merges into another (an ecotone), and the varying abundance of the different pollen types reveals composition changes within each region.

Pollen and macrofossil records from the past 18,000 years show that major changes in

*Figure 5.7.* Modern vegetation regions in eastern North America. Modified from Webb 1988.

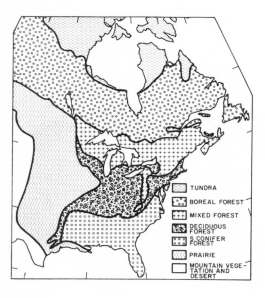

TUNDRA
BOREAL FOREST
MIXED FOREST
DECIDUOUS FOREST
S.CONIFER FOREST
PRAIRIE
MOUNTAIN VEGE-TATION AND DESERT

climate caused large rearrangements in the composition and structure of the major vegetation regions across North America (Barnosky et al. 1987; Webb 1987, 1988; Thompson 1988) and Europe (Huntley and Birks 1983, Huntley 1988). Maps of pollen data from the past 18,000 years show how individual taxa have changed in abundance, location, and association in conjunction with the retreating ice sheet and changing climate conditions. For example, maps of spruce pollen (fig. 5.8) show spruce trees were abundant 18,000 years ago in the midwest, then spread to the east just south of the ice by 12,000 years ago. By 8000 years ago spruce had markedly decreased in abundance, only to increase again in area and abundance in central Canada about 6000 years ago as the modern boreal forest formed.

When spruce trees were growing south of the ice sheet from 18,000 to 12,000 years ago, they grew in a parkland in association with sedges. By 10,000 years ago, the patterns for abundant spruce and sedge had ceased to overlap, and spruce trees grew in closed forests for the first time. Birch populations then increased relative to the spruce populations in forests south of the retreating ice sheet, and it is only after 6000 years ago that the modern boreal forest formed with its composition of spruce and birch along with alder (*Alnus*) and fir (*Abies*) (see Webb 1987).

The climate and vegetational changes between 12,000 and 10,000 years ago were particularly rapid and large (Jacobson et al. 1987, Webb et al. 1987). It is in this period that the spruce parkland disappeared as a large vegetational region, and a few species of large mammals went extinct. Russell Graham (chapter 6) discusses some of the consequences of this and other climate and vegetational changes on animal populations (see also Graham and Grimm 1990).

This history of changing composition and structure for spruce-dominated vegetation, in which each species responded individually, is not unique; taxa from other major vegetation regions experienced similar changes (Davis 1983, Jacobson et al. 1987, Webb 1988). Hunt-

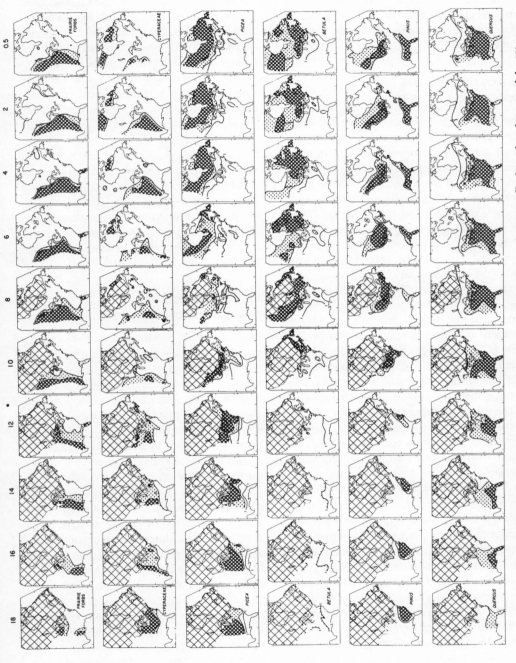

Figure 5.8. Contour (isopoll) maps of the pollen percentages in eastern North America for prairie forbs (Compositae − Ambrosia + Chenopodiaceae/Amaranthaceae), sedge, spruce, birch, pine, and oak in 2000-year intervals from 18,000 to 500 years ago. Dark stippling indicates the areas of highest abundance, and the 1% contour appears on the maps for spruce, birch, and oak pollen. For details see Webb 1988. The large cross-hatching in the north indicates the location of the Laurentide ice sheet. Modified from Webb 1988.

ley (1988, 1990) shows similar individualistic species responses and vegetational rearrangements for Europe, and R. S. Thompson (1988) has used plant macrofossils and pollen data to demonstrate similar changes in the western United States.

## VI. IMPLICATIONS FOR BIOLOGICAL DIVERSITY

The maps of pollen data in figure 5.8 show how plant species responded in the past. Their independent movements led to the appearance and disappearance of vegetation regions (see also Jacobson et al. 1987, Huntley 1990) and were accompanied by local or regional extinctions and colonizations. They show that plants can migrate and thus survive in the face of large climate changes occurring at natural rates (Davis 1976, Huntley and Webb 1989, Johnson and Webb 1989).

The climate record of the past million years (fig. 5.1e) shows that the predicted magnitude of future changes will not be as big a problem as the predicted *rate* of change. My view is that the large changes in Quaternary (the last two million years) climates probably had little impact on global species diversity. Because the fossil record shows that species of marine invertebrates and of higher plants have generally had a longevity of 1 million to 10 million years, such species have had to adjust many times to large environmental changes at the natural rates illustrated in figure 5.1 (Webb 1987). Such species now and in the past must therefore be well adapted to coping with the past magnitude and rates of climate change. Any not so adapted in terms of growth, reproductive, and dispersal rates would never have survived for 100,000 years or more and would be long extinct. My own hypothesis is that during the last interglacial and glacial periods, global species diversity for higher plants and vertebrates was similar to that during the current interglacial. Despite the local and global extinctions of some species (chapter 6; Van der Hammen et al. 1971), the glacial-interglacial swings in climate neither greatly increased the rate of speciation

nor increased the rate of extinction, and no major net loss in species numbers has occurred globally during the past million years. The major recent long-term changes in climate, therefore, have little affected species diversity globally, even though short-term local and regional impacts have been large.

Such observations should not decrease concern about the consequences of the climate changes predicted for the next century or so. Current predictions of an unusually rapid global change (approximately 3°C per century) raises concerns about how well adapted extant species may be to coping with this future rate of change. Records of past data do not show such a rapid change in the global mean annual temperature, which should also be associated with large rapid regional changes in temperature and precipitation.

## VII. EQUILIBRIUM VS. DISEQUILIBRIUM RESPONSES

One of the important factors determining whether species will survive climate change is how rapidly they are able to change their ranges or abundances to adapt to changing climate. Paleoecologists have long argued about whether plant taxa have rapid enough dispersal mechanisms and population-growth responses to remain in equilibrium with climate at all times in the past, especially during periods of relatively rapid change (Prentice 1986). If, for example, regional climate changes slowly, so that a plant species is able to colonize new habitat as soon as or shortly after the climate there becomes suitable, the plant species is said to be in equilibrium with the climate. On the other hand, if climate change is rapid relative to the species' dispersal ability, and the plant species cannot colonize new climatic habitat within a short time after it forms, then disequilibrium occurs. Paleoecologists have wondered to what extent past distributions and abundances of plants have reflected prevailing climate conditions, and to what extent biotic factors (e.g., inherent rates of dispersal or population expansion) have interceded and become rate-limiting, thus leading to taxon abundances and community composition out of equilibrium with climate (Birks 1981).

The vegetational response will always lag somewhat behind the creation of favorable habitats by climatic change. Some lag in a taxon's response, therefore, is not necessarily evidence of disequilibrium conditions. In light of the continuous nature of climatic change, a useful criterion for judging whether a plant taxon (i.e., its abundance distribution) is in equilibrium with climate is to compare its response time (the time it takes to respond significantly to a given climate change by changing its local abundance and/or its geographic range) to the rate of climate change. This comparison can be expressed as a ratio of the taxon's response time to the rate of climate change (technically, the period of climatic forcing; Webb 1986). If the ratio is small (200 yr/20,000 yr), then conditions for dynamic equilibrium prevail; but if the ratio is large (200 yr/200 yr), then disequilibrium exists.

Paleoecologists have not yet agreed on the response times for various tree taxa, but the minimum value that can be assigned seems to be 50 to 200 years for tree composition to change significantly in response to climate change. Those response times are fast enough for the tree taxa to stay in equilibrium with most major past climate changes (Webb 1986) but are similar in length to the predicted time scale for climate change induced by greenhouse gases. This observation implies that disequilibrium conditions will prevail for tree taxa in the future. Their intrinsic ability to respond may be far outstripped, and the predicted fast rates of climate change may lead to community disruption and the extinction of some species. Furthermore, the human control of landscapes and destruction of habitats can only add to the problems that species will have in trying to cope with future changes (Hunter et al. 1988). The evidence from the past therefore shows that plants respond to climate but that many slow-maturing taxa like trees will have trouble with

the future rate and magnitude of change. Their difficulties in responding will affect the structure and composition of the vegetation, which will limit populations of animals that are otherwise mobile and fully capable of tracking the rapid climate changes.

## VIII. SUMMARY

Changes in globally averaged climate during the past 18,000 years have been similar in magnitude (ca. 4° ± 1°C) to those predicted for global warming induced by greenhouse gases. Changes in past climates illustrate some of the types of regional climatic changes that may be associated with future changes in global climates.

Large changes in taxon ranges and abundance were associated with the past climate changes. Taxa exhibited individualistic patterns of change. Changes should therefore be expected in the composition of vegetation regions and ecosystems, if the effects of future climate changes are similar to those of the past (Jacobson et al. 1987, Overpeck and Bartlein 1989).

The predicted rate of global climate change is estimated to be at least an order of magnitude faster than rates in the past. This fast rate will be similar in speed or even faster than the response times for taxa with long-lived individuals like trees. One possible scenario is that trees in current forests may die before better-adapted tree taxa reach the site. See Davis (1986, 1989; and chapter 22) and Botkin and Nisbet (chapter 21) for further discussion.

A global mean temperature that is higher than today's by 3°C or more will lead to taxa growing in areas where they have never been or have not been for millions of years. This could be another potential source for extinction because some northern or alpine taxa may be cut off from old habitats and may be unable to find any suitable new habitats.

If nature reserves are going to serve a key role in preserving biological diversity, their design should be carefully considered (Hunter et al. 1988, Peters and Darling 1985).

The climate changes predicted for the future may cause many taxa to shift their ranges by large distances, and corridors between reserves will be needed to give individuals from many taxa some chance of moving rapidly enough. To allow for future climate changes, reserves should be chosen that contain a diversity of environments.

## APPENDIX: ESTIMATES OF THE PAST GLOBAL MEAN TEMPERATURE

We have no direct measures of the past global mean annual temperature before ca. 1850 B.C.E. Though extensive (e.g., fig. 1 in COHMAP 1988), the coverage provided by paleoclimatic records alone is insufficient to yield a fully accurate estimate by an areal averaging of available data. The nearest an estimate has come to being a global average of temperature estimates for a network of sampling sites is the estimate for the spatial average for sea-surface temperatures in the ice-free oceans 18,000 years ago of about 1.5° ± 0.3°C lower than the temperatures in the same area today (CLIMAP 1981). This estimate is, of course, not global, but is the largest areal average available.

Estimation schemes for Quaternary data (i.e., data from the past 2 million years) must, therefore, involve a combination of data and climate model results. The scheme used to obtain the calibrated time series of global mean temperature estimates in parts d and e of figure 5.1 was to start by using oxygen isotope variations in cores with shells of marine benthic and planktonic foraminifera, which indicate variations in global ice volume, as an uncalibrated proxy time series for the global mean annual temperature (with a temporal imprecision of ± 2000 years) and then to calibrate this time series by estimating the temperature difference between maximum glacial and interglacial conditions. Climate-model simulations of maximum glacial conditions (18,000 years ago) were key to this calibration and required (1) the use of many types of data to estimate the surface boundary conditions (e.g., sea-surface temperatures, sea-ice extent, ice sheet size, and surface re-

flectivity) at the time of glacial maximum for use in general circulation models (GCMs) of the atmosphere and ocean, (2) the simulation by GCMs of the climate conditions that were in equilibrium with the boundary conditions, (3) the averaging of the model-simulated surface temperatures to obtain an estimate of the global mean temperature for each modeling study, and (4) the use of these estimates of global mean temperature to calibrate the oxygen isotope variations. Climate-modeling studies by Broccoli and Manabe (1987), Gates (1976), Manabe and Hahn (1977), Hansen et al. (1984), Kutzbach and Guetter (1986), and Hyde et al. (1989) provided an estimate of 5° ± 1°C for the magnitude of global mean annual temperature variation between extreme glacial and interglacial climates, and this value was used to calibrate the oxygen-isotope variations in parts d and e of figure 5.1. This temperature estimate depends upon the accuracy of the surface boundary conditions and the accuracy of the model.

## REFERENCES

Barnosky, C. W., P. M. Anderson, and P. J. Bartlein. 1987. The northwestern U.S. during deglaciation: Vegetational history and paleoclimatic implications. In *North America and Adjacent Oceans during the Last Deglaciation*, W. F. Ruddiman and H. E. Wright, Jr., eds., The Geology of North America, vol. K-3, pp. 289–321. Boulder, Colo.: Geological Society of America.

Barron, E. J. 1985. Explanations of the Tertiary global cooling trend. *Palaeogeo. P.* 50:17.

Bartlein, P. J. 1988. Late Tertiary and Quaternary palaeoenvironments. In *Vegetation History*, B. Huntley and T. Webb III, eds., pp. 113–152. Dordrecht: Kluwer Academic.

Bartlein, P. J., and T. Webb III. 1985. Mean July temperature at 6000 yr B.P. in eastern North America: Regression equations for estimates from fossil-pollen data. *Syllogeus* 55:301.

Bartlein, P. J., I. C. Prentice, and T. Webb III. 1986. Climate response surfaces based on pollen from some eastern North American taxa. *J. Biogeogr.* 13:35.

Berger, A., J. Imbrie, J. Hays, G. Kukla, and B. Saltzman, eds. 1984. *Milankovitch and Climate*, parts 1 and 2. Dordrecht: D. Reidel.

Berglund, B. E., ed. 1986. *Handbook of Holocene Palaeo-ecology and Palaeohydrology*. New York: John Wiley and Sons.

Bernabo, J. C. 1978. *Proxy Data: Nature's Records of Past Climates*. Washington, D.C.: Environmental Data Service, NOAA, U.S. Dept. of Commerce, pp. 1–8.

Birks, H.J.B. 1981. The use of pollen analysis in the reconstruction of past climates: A review. In *Climate and History: Studies in Past Climate and Their Impact on Man*, T.M.L. Wigley, M. J. Ingram, and G. Farmer, eds., pp. 111–138. Cambridge: Cambridge University Press.

Birks, H.J.B., and H. H. Birks. 1980. *Quaternary Palaeoecology*. London: Edward Arnold.

Bradshaw, R.H.W., and T. Webb III. 1985. Relationships between contemporary pollen and vegetation data from Wisconsin and Michigan. *Ecology* 66:721.

Broccoli, A. J., and S. Manabe. 1987. The influence of continental ice, atmospheric carbon dioxide, and land albedo on the climate of the last glacial maximum. *Clim. Dynamics* 1:87.

CLIMAP Project Members. 1981. Seasonal reconstructions of the earth's surface at the last glacial maximum. Geological Society of America Map and Chart Series, MC-36, pp. 1–18.

COHMAP Members. 1988. Climatic changes of the last 18,000 years: Observations and model simulations. *Science* 241:1043.

Cole, K. L. 1982. Late Quaternary zonation of vegetation in the eastern Grand Canyon. *Science* 212:1142.

Crowley, T. J. 1989. Paleoclimate perspectives on a greenhouse warming. In *Climate and Geosciences*, J. C. Duplessy, A. Berger, and S. H. Schneider, eds., pp. 179–207. Dordrecht: Kluwer Academic.

Davis, M. B. 1976. Pleistocene biogeography of temperate deciduous forests. *Geosci. and Man* 13:13.

Davis, M. B. 1983. Quaternary history of deciduous forests in eastern North America. *Annals of the Missouri Botanical Garden* 70:550.

Davis, M. B. 1986. Climatic instability, time lags, and community disequilibrium. In *Community Ecology*, J. Diamond and T. J. Case, eds., pp. 269–284. New York: Harper and Row.

Davis, M. B. 1989. Lags in vegetation response to greenhouse warming. *Clim. Change* 15:75.

Gates, W. L. 1976. Modeling the ice-age climate. *Science* 191:1138.

Graham, R. W., and E. C. Grimm. 1990. Effects of global climate change on the patterns of terrestrial biological communities. *Trends Ecol. and Evol.* 5:289.

Grove, J. M. 1988. *The Little Ice Age*. London: Methuen.

Hansen, J., A. Lacis, A. Rind, G. Russell, P. Stone, I. Fung, R. Ruedy, and J. Lerner. 1984. Climate sensitivity: Analysis of feedback mechanisms. In *Climate Processes and Climate Sensitivity*, J. E. Hansen and T. Takahashi, eds., pp. 130–163. Washington, D.C.: American Geophysical Union, Maurice Ewing Series 5.

Hays, J. D., J. Imbrie, and N. Shackleton. 1976. Variations in the earth's orbit: Pacemaker of the ice volume cycle. *Science* 194:1121.

Hecht, A. D., ed. 1985. *Paleoclimate Analysis and Modeling*. New York: John Wiley and Sons.

Hunter, M. L., Jr., G. L. Jacobson, Jr., and T. Webb III. 1988. Paleoecology and coarse-filter approach to maintaining biological diversity. *Conser. Biol.* 2:375.

Huntley, B. 1988. Glacial and Holocene vegetation history: Europe. In *Vegetation History*, B. Huntley and T. Webb III, eds., pp. 341–383. Dordrecht: Kluwer Academic.

Huntley, B. 1990. European vegetation history: Paleovegetation maps from pollen data—13,000 yr B.P. to present. *J. Quatern. Sci.* 5:103.

Huntley, B., and H.J.B. Birks. 1983. *An Atlas of Past and Present Pollen Maps for Europe: 0–13,000 Years Ago*. Cambridge: Cambridge University Press.

Huntley, B., and T. Webb III, eds. 1988. *Vegetation History*. Dordrecht: Kluwer Academic.

Huntley, B., and T. Webb III. 1989. Migration: Species' response to climatic variations caused by changes in the earth's orbit. *J. Biogeogr.* 16:5.

Hyde, W. T., T. J. Crowley, K.-Y. Kim, and G. R. North. 1989. Comparison of GCM and energy budget model simulations of seasonal temperature changes over the past 18,000 years. *J. Climate* 2:864.

Imbrie, J., and K. P. Imbrie. 1979. *Ice Ages: Solving the Mystery*. Short Hills, N.J.: Enslow.

Imbrie, J., J. D. Hays, D. G. Martinson, A. McIntyre, A. C. Mix, J. J. Morley, N. J. Pisias, W. L. Prell, and N. J. Shackleton. 1984. The orbital theory of Pleistocene climate: Support from a revised chronology of the marine del-18 O record. In *Milankovitch and Climate*, parts 1 and 2, A. Berger, J. Imbrie, J. Hays, G. Kukla, and B. Saltzman, eds., pp. 269–305. Dordrecht: D. Reidel.

Imbrie, J., A. McIntyre, and A. Mix. 1989. Oceanic response to orbital forcing in the late Quaternary: Observational and experimental strategies.

In *Climate and Geosciences*, J. C. Duplessy, A. Berger, and S. H. Schneider, eds. Dordrecht: Kluwer Academic.

Jacobson, G. L. Jr., T. Webb III, and E. C. Grimm. 1987. Patterns and rates of vegetation change during the deglaciation of eastern North America. In *North America and Adjacent Oceans during the Last Deglaciation*, W. F. Ruddiman and H. E. Wright, Jr., eds., The Geology of North America, vol. K-3, pp. 277–288. Boulder, Colo.: Geological Society of America.

Johnson, W. C., and T. Webb III. 1989. The role of blue jays (*Cyanocitta cristata* L.) in the postglacial dispersal of fagaceous trees in eastern North America. *J. Biogeogr.* 16:561.

Jones, P. D., T.M.L. Wigley, and P. B. Wright. 1986. Global temperature variations between 1861 and 1984. *Nature* 322:430.

Jones, P. D., P. M. Kelly, C. M. Goodess, and T. Karl. 1989. The effect of urban warming on the northern hemisphere temperature average. *J. Climate* 2:285.

Kutzbach, J. E. 1981. Monsoon climate of the early Holocene: Climate experiment with the earth's orbital parameters for 9000 years ago. *Science* 214:59.

Kutzbach, J.E. 1987. Model simulations of the climatic patterns during the deglaciation of North America. In *North America and Adjacent Oceans during the Last Deglaciation*, W. F. Ruddiman and H. E. Wright, Jr., eds., The Geology of North America, vol. K-3, pp. 425–446. Boulder, Colo.: Geological Society of America.

Kutzbach, J. E., and R. Gallimore. 1988. Sensitivity of a coupled atmosphere/mixed layer ocean model to changes in orbital forcing at 9000 years B.P. *J. Geo. Res.* 93:803.

Kutzbach, J. E., and P. Guetter. 1986. The influence of changing orbital parameters and surface boundary conditions on climate simulations for the past 18,000 years. *J. Atmos. Sci.* 43:1726.

Kutzbach, J. E., and F. A. Street-Perrott. 1985. Milankovitch forcing of fluctuations in the level of tropical lakes from 18 to 0 kyr B.P. *Nature* 317:130.

Lamb, H. H. 1979. Climatic variation and changes in the wind and ocean circulation: The little ice age in the northeast Atlantic. *Quatern. Res.* 11:1.

Manabe, S., and D. G. Hahn. 1977. Simulation of the tropical climate of an ice age. *J. Geo. Res.* 82:3889.

McDowell, P. C., T. Webb III, and P. J. Bartlein. 1991. Long-term environmental change. In *Earth as Transformed by Human Action*, B. L. Turner II, W. C. Clark, R. W. Kates, J. T. Mathews, and J. Richards, eds. Cambridge: Cambridge University Press.

Mitchell, J.F.B., N. S. Grahame, and K. J. Needham. 1988. Climate simulations for 9000 years before present: Seasonal variations and effect of the Laurentide ice sheet. *J. Geo. Res.* 93:8283.

Overpeck, J. T., and P. J. Bartlein. 1989. Assessing the response of vegetation to future climate change: Ecological response surfaces and paleoecological model validation. In *The Potential Effects of Global Climate Change on the United States*, J. B. Smith and D. A. Tirpak, eds., EPA-230-05-89-054. Washington, D.C.: Environmental Protection Agency.

Peters, R. L., and J.D.S. Darling. 1985. The greenhouse effect and nature reserves. *Bioscience* 35:707.

Prell, W. F., and J. E. Kutzbach. 1987. Monsoon variability over the past 150,000 years. *J. Geo. Res.* 92:8411.

Prentice, I. C. 1986. Vegetation responses to past climatic changes. *Vegetatio* 67:131.

Prentice, I. C. 1988. Records of vegetation in time and space: The principles of pollen analysis. In *Vegetation History*, B. Huntley and T. Webb III, eds., pp. 17–42. Dordrecht: Kluwer Academic.

Prentice, I. C., and T. Webb III. 1986. Pollen percentages, tree abundances and the Fagerlind effect. *J. Quatern. Sci.* 1:35.

Rind, D., D. Peteet, W. Broecker, A. McIntyre, and W. F. Ruddiman. 1986. The impact of cold North Atlantic sea-surface temperatures on climate: Implications for the Younger Dryas cooling (11–10K). *Clim. Dynamics* 1:3.

Ruddiman, W. F., and H. E. Wright, Jr., eds. 1987. *North America and Adjacent Oceans during the Last Deglaciation*. The Geology of North America, vol. K-3. Boulder, Colo.: Geological Society of America.

Ruddiman, W. F., M. Raymo, and A. McIntyre. 1986. Matuyama 41,000-year cycles: North Atlantic Ocean and northern hemisphere ice sheets. *Earth Plan. Sci.* 80:117.

Schlesinger, M. E. 1989. Model projections of the climatic changes induced by increased atmospheric $CO_2$. In *Climate and Geosciences*, J. C. Duplessy, A. Berger, and S. H. Schneider, eds., pp. 375–415. Dordrecht: Kluwer Academic.

Schlesinger, M. E., and J.F.B. Mitchell. 1985. Model projections of the equilibrium climatic response to increased carbon dioxide. In *Detecting the Climatic Effects of Increasing Carbon Dioxide*, M. C. MacCracken and F. M. Luther, eds., DOE/ER-0237, pp. 83–147. Washington, D.C.: Department of Energy.

Spaulding, W. G., E. B. Leopold, and T. R. Van Devender. 1983. Late Wisconsin paleoecology of the American Southwest. In *The Late Quaternary Environments of the United States*. Vol. 1, *The Late Pleistocene*, S. C. Porter, ed., pp. 259–293. Minneapolis: University of Minnesota Press.

Street-Perrott, F. A., and S. P. Harrison. 1985. Lake levels and climate reconstruction. In *Paleoclimate Analysis and Modeling*, A. D. Hecht, ed., pp. 163–195. New York: John Wiley and Sons.

Thompson, R. S. 1988. Glacial and Holocene vegetation history: Western North America. In *Vegetation History*, B. Huntley and T. Webb III, eds., pp. 415–458. Dordrecht: Kluwer Academic.

Van der Hammen, T., T. A. Wijmstra, and W. H. Zagwijn. 1971. The floral record of the Late Cenozoic of Europe. In *Late Cenozoic Glacial Ages*, K. K. Turekian, ed., pp. 391–424. New Haven: Yale University Press.

Van Devender, T. R., R. S. Thompson, and J. L. Betencourt. 1987. Vegetation history of the deserts of the southwestern North America: The nature and timing of the Late Wisconsin–Holocene transition. In *North America and Adjacent Oceans during the Last Deglaciation*, W. F. Ruddiman and H. E. Wright, Jr., eds., The Geology of North America, vol. K-3, pp. 323–352. Boulder, Colo.: Geological Society of America.

Webb, T. III. 1986. Is vegetation in equilibrium with climate? How to interpret late-Quaternary pollen data. *Vegetatio* 67:75.

Webb, T. III. 1987. The appearance and disappearance of major vegetational assemblages: Long-term vegetational dynamics in eastern North America. *Vegetatio* 69:177.

Webb, T. III. 1988. Glacial and Holocene vegetation history: Eastern North America. In *Vegetation History*, B. Huntley and T. Webb III, eds., pp. 385–414. Dordrecht: Kluwer Academic.

Webb, T. III. 1991. The spectrum of temporal climatic variability: Current estimates and the need for global and regional time series. In *Records of Past Global Change*, R. W. Bradley, ed. Boulder, Colo.: Office of Interdisciplinary Earth Studies, UCAR.

Webb, T. III, and T.M.L. Wigley. 1985. What past climates can indicate about a warmer world. In *Detecting the Climatic Effects of Increasing Carbon Dioxide*, M. C. MacCracken and F. M. Luther, eds., DOE/ER-0237, pp. 239–257. Washington, D.C.: Department of Energy.

Webb, T. III, P. J. Bartlein, and J. E. Kutzbach. 1987. Climatic change in eastern North America during the past 18,000 years: Comparisons of pollen data with model results. In *North America and Adjacent Oceans during the Last Deglaciation*, W. F. Ruddiman and H. E. Wright, Jr., eds., The Geology of North America, vol. K-3, pp. 447–462. Boulder, Colo.: Geological Society of America.

Wigley, T.M.L., M. J. Ingram, and G. Farmer, eds. 1981. *Climate and History: Studies in Past Climates and Their Impact on Man*. Cambridge: Cambridge University Press.

A National Science Foundation grant (ATM87-13981) from the Program for Climate Dynamics to COHMAP (Cooperative Holocene Mapping Project), a Department of Energy grant (FG02-85ER60304) from the Carbon Dioxide Research Division, and a visiting fellowship from CIRES (Cooperative Institute for Research in Environmental Sciences) at the University of Colorado, Boulder, supported this research. I thank K. Anderson, P. Klinkman, and P. Newby for technical assistance; P. J. Bartlein, T. J. Crowley, B. Huntley, and P. McDowell for helpful discussions; and M. B. Davis and R. Peters for comments on the manuscript.

# CHAPTER SIX

# Late Pleistocene Faunal Changes as a Guide to Understanding Effects of Greenhouse Warming on the Mammalian Fauna of North America

RUSSELL W. GRAHAM

## I. INTRODUCTION

Climatic models for the future predict rapid global warming as a consequence of human-induced increases in atmospheric carbon dioxide and other greenhouse gases (Hansen et al. 1988, Schneider 1989). It is difficult to forecast precisely the effects that this global warming will have on the earth's biota. Obviously, it is impractical for ecologists and conservationists to conduct experiments at a scale consistent with the global changes anticipated. However, comparisons can be made with natural warming events in the past. The oxygen isotope record in the shells of a type of fossil plankton called foraminifera, preserved in sediment layers of the deep sea, can be used to document prehistoric climate patterns. This information shows there were at least 63 different alternations between warm and cold climates during the last 1.6 million years (Ruddiman et al. 1986), with the patterns and timing of global warming best known for changes since the last glacial maximum, 20,000 years ago (fig. 6.1).

Responses of plants and animals to that last warming period, as determined by the fossil record, can be used to understand in a general way how the earth's biota will respond to the projected greenhouse warming. This chapter will describe what is known about such past responses, particularly for small mammals, and will draw conclusions about future responses. It must be borne in mind that past warming events were not identical (Graham 1986a, Shackleton 1987), nor will they be exactly the same as future warming events. For example, the middle Holocene warming, or Hypsithermal (ca. 9000–4500 years ago), in North America was not of the same duration or intensity as the terminal Pleistocene warming event (Graham and Mead 1987). Therefore, paleobiological data cannot be used to make specific predictions

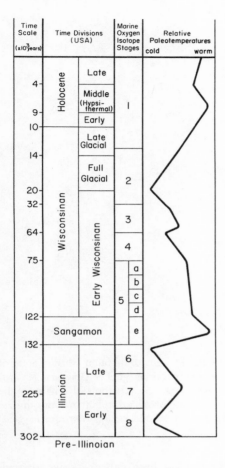

| Time Scale (x10³years) | Time Divisions (USA) | | Marine Oxygen Isotope Stages | Relative Paleotemperatures cold → warm |
|---|---|---|---|---|
| 4 | Holocene | Late | 1 | |
| 9 | Holocene | Middle (Hypsi-thermal) | 1 | |
| 10 | Holocene | Early | 1 | |
| 14 | Wisconsinan | Late Glacial | | |
| 20 | Wisconsinan | Full Glacial | 2 | |
| 32 | Wisconsinan | | 2 | |
| 64 | Wisconsinan | Early Wisconsinan | 3 | |
| 75 | Wisconsinan | Early Wisconsinan | 4 | |
| | Wisconsinan | Early Wisconsinan | 5 a | |
| | Wisconsinan | Early Wisconsinan | 5 b | |
| | Wisconsinan | Early Wisconsinan | 5 c | |
| 122 | Wisconsinan | Early Wisconsinan | 5 d | |
| 132 | Sangamon | | 5 e | |
| | Illinoian | Late | 6 | |
| 225 | Illinoian | Late | 7 | |
| | Illinoian | Early | 8 | |
| 302 | Pre-Illinoian | | | |

*Figure 6.1.* Late Quaternary time chart for North America with marine oxygen isotope stages and relative paleotemperatures. Modified from Bowen et al. 1986.

about the exact composition of future biotas, and generalizations about climatic warming must be tempered with careful attention to the exact nature of the process.

## II. INDIVIDUALISTIC AND STATIC COMMUNITY MODELS

In theory, there are two ways the terrestrial biota of North America could have responded to environmental changes during the late Quaternary (Graham and Grimm 1990). One hypothesis, a static community model, proposes that large groups of species

(i.e., communities) shifted as tightly linked and coevolved assemblages. The other hypothesis, an individualistic model, suggests that individual species responded to the changes in accordance with their own tolerance limits. The biological consequences of these two hypotheses are profoundly different. The static model predicts that communities are stable and remain as units through long periods of geologic time. Furthermore, it suggests that modern communities can be used as direct analogs in paleoenvironmental, as well as future, reconstructions. Conversely, the individualistic model suggests that communities are loosely organized collections of species whose coexistence is dependent on their tolerance limits and subsequent distribution along environmental gradients. Therefore, their association may appear ephemeral in geologic time and modern communities may not serve as direct analogs for either past or future environments.

The evidence from paleoecological studies clearly demonstrates that the individualistic model is closer to the truth. For example, late Quaternary climatic changes did not cause simple latitudinal or altitudinal shifts in life zones. In the eastern United States, paleovegetational analyses show that species responded individually to changing climates (Cushing 1965, Davis 1976, Jacobson et al. 1987, Webb 1987). Likewise, in the western United States, fossil evidence indicates that plant species responded to environmental changes by making individualistic altitudinal shifts and that communities did not shift as tightly linked groups of species (Spaulding et al. 1983).

Animals, especially mammals, have shown similar individualistic responses to late Quaternary climatic changes (Graham 1986b, Graham and Lundelius 1984). Their individual range shifts have acted to break up and reassort mammal communities, with the result that those of today are very different from those identified from fossils of the late Pleistocene. For example, it is common in late Pleistocene faunas of the northeastern and upper midwestern United States to find species within contemporaneous deposits that today inhabit the different habitats of arctic tundra, grassland, and forests of both coniferous and deciduous trees (Graham and Mead 1987, Graham et al. 1987). Such prehistoric communities that differ from contemporary ones are termed nonanalog. These associations are not restricted to mammalian assemblages but are also characteristic of Pleistocene floras as well as insect and terrestrial molluscan assemblages (Graham and Lundelius 1984). They are found not only in North America but in other continents as well, including Australia, Eurasia, and Africa (Graham and Lundelius 1984).

The following are three examples of small mammals that dramatically shifted their ranges in highly individualistic ways during the late Pleistocene. Comparison of their range maps (figs. 6.2, 6.3, and 6.4) will make it obvious that direction and magnitude of shifting varied among species. Around 17,000 years ago the northern pocket gopher (Thomomys talpoides) inhabited southwestern Wisconsin (Foley 1984), more than 600 km east of its modern distribution (fig. 6.2). This species is also found in western Iowa in deposits that date from 23,200 to 14,800 years ago (Rhodes 1984) and in undated or poorly dated sites in northeastern Iowa (Rosenberg 1983) and north central Missouri (Parmalee and Oesch 1972). The eastern limit of the northern pocket gopher's distribution moved farther westward (from southwestern Wisconsin to east central South Dakota) sometime after 14,000 years ago. The least shrew (Cryptotis parva) existed as far west as southwestern New Mexico and the Laramie Range of Wyoming during the late Wisconsinan (Harris 1985). Those fossil localities are more than 700 km and 400 km west of its current distribution (fig. 6.3). The collared lemming (Dicrostonyx spp.) was widespread south of the Laurentide ice sheet more than 1600 km south of its present distribution (fig. 6.4). Again, this species, except for some isolated high-altitude populations, appears to have been restricted farther north by the late glacial (ca. 14,000

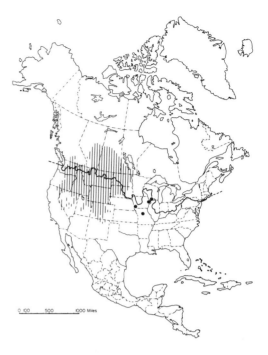

Figure 6.2. Modern distribution (shaded area) of the northern pocket gopher (Thomomys talpoides) and select fossil localities (dots) that illustrate the eastern extension of its distribution during the late Wisconsinan. Solid line marks Wisconsinan glacial maximum.

years ago). All three of these species are small rat-size animals, so individuals do not have the ability to migrate long distances (probably less than a few kilometers in their lifetimes). Therefore, changes in their distributions have taken numerous generations, and the extensive geographic areas involved reflect regional or global environmental changes rather than local ones. These patterns in the shifts of species ranges are not restricted to the few that have been mentioned but are demonstrated by a variety of mammalian species (Lundelius et al. 1983). In fact, the changes illustrated by the small mammals are also replicated by many of the large ones, for example, barren ground muskox (Ovibos moschatus), caribou (Rangifer tarandus), and mountain goat (Oreamnos spp.).

Other vertebrate groups also responded individually to the climatic fluctuations at the

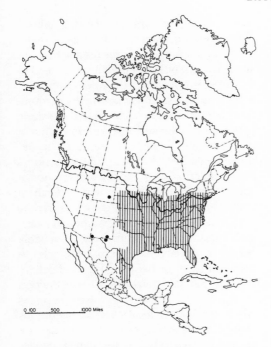

*Figure* 6.3. Modern distribution (shaded area) of the least shrew (*Cryptotis parva*) and select fossil localities (dots) that illustrate the western extension of its distribution during the late Wisconsinan. Solid line marks Wisconsinan glacial maximum.

*Figure* 6.4. Modern distribution (shaded area) of the collared lemming (*Dicrostonyx* spp.) and fossil localities (dots) that illustrate the southern extension of its distribution during the late Wisconsinan. Solid line marks Wisconsinan glacial maximum.

end of the Pleistocene (Lundelius et al. 1983). Changes in the distribution of the gopher tortoise (*Gopherus* spp.) during the late Quaternary are counterintuitive to the standard concept of cold-blooded vertebrates expanding northward with climatic warming. Instead, this tortoise shifted its distribution south with warming during the late Pleistocene. K. B. Moodie and T. R. Van Devender (1979) believe that these changes occurred because seasonal extremes were accentuated during this time and that they were more important in limiting the distribution of this tortoise than increases in the mean annual temperature.

In summary, climatic warming during the late glacial (14,000 to 10,000 years ago) caused individual species distributions to change along environmental gradients in different directions, at different rates, and during different times. The individualistic response of the biota caused the emergence of new community patterns. Changes in the distributions of individual taxa continued during the Holocene (the last 10,000 years), but the magnitude of these changes was not the same as those of the late glacial (Semken 1983). Similarly, individualistic changes can be anticipated in the future. If, as predicted (Schneider 1989), the rate of warming caused by the greenhouse effect is greater than in past events, then the individualistic responses may be even more profound.

### III. SAVING BIOLOGICAL DIVERSITY: SHOULD WE SAVE SPECIES OR COMMUNITIES?

Since climate change causes individualistic movement of species and alteration of community composition, these processes must be considered for any plans in the conservation of biodiversity. Biodiversity is often described as having three levels, including diversity of species, genetic types, and communities, each of which would be affected differently by global warming, and each of which would require different conservation strategies.

A primary strategy designed to conserve all three levels of diversity has been to protect natural areas from development by designating them reserves. Typically, such reserves may be surrounded by developed land, becoming habitat islands, and several authors have pointed out that such islands are vulnerable to a variety of stresses (Boecklen and Simberloff 1986, Peters and Darling 1985). One of the fundamental assumptions that underlies the reserve-creation strategy is that environmental conditions and hence community patterns within the reserves have been stable for long periods of time and that they will continue to be stable in the future. However, paleobiological data do not support either of those assumptions. Environments and community patterns have changed constantly in the past, and they will continue to change naturally in the future, particularly under the influence of global warming. We can expect that reserves will lose some of their species and that some of the community types within reserves will change substantially.

Most conservationists consider the extinction of species unacceptable, and substantial resources have been expended in their conservation. Similar efforts have been made for the conservation of communities. However, the ephemeral nature of communities, on a geologic time scale, particularly due to climate change, suggests that significant resources should not be devoted to trying to conserve specific community types within reserves. Rather, emphasis should be on management strategies that allow maximum survival of species and genetic strains and that allow natural communities to evolve.

For species to survive in the long-term, they must be able to respond to environ-

mental change, in part by tracking shifting climatic environments. In the case of mammals within reserves, this means that reserves should ideally be large enough and located so that species can change their geographic distributions. During this process new community patterns would evolve. To facilitate these processes, biological reserves in different physiographic, biological, and climatic settings should be linked with habitat corridors along differing (e.g., east-west vs. north-south) environmental gradients (Graham 1988, Hunter et al. 1988, Peters and Darling 1985; chapter 2). If species cannot make necessary individualistic shifts because suitable habitat is too far away, because obstacles to dispersal exist, or because the species have weak dispersal adaptations, then extinction may occur. As an example of what happens when regional or local climatic conditions change substantially, consider the worst-case scenario of Holocene environmental changes in the eastern Sahara of Africa. Early Holocene (10,000 to 8000 years ago) climatic conditions in this area supported extensive vegetation, a Sudanic biome with perhaps a parkland megafauna and widespread human occupation (McHugh et al. 1988). Today, the eastern Sahara is one of the most hyperarid regions in the world. Clearly, biological reserves designed as habitat islands without migratory corridors under early Holocene environmental conditions would have been rendered unsuitable by the climatic and environmental changes of the later Holocene (Graham 1988).

Moreover, this example is particularly apt because it demonstrates the effects of changing climate not only on a natural biotic community but also on the people associated with the community. By the late Holocene times, when the eastern Sahara dried, sophisticated human societies had developed along the Nile in Egypt and throughout most of the Middle East. In fact, even during the middle Neolithic (ca. 6200 years ago), the Nubian desert supported organized human villages, which subsisted on hunting and gathering as well as agriculture (Wendorf et al. 1985). Then the later Holocene environmental changes spelled the end of the parkland biotic community and dramatically changed the human community as well.

These changes occurred in response to warming and drying that progressed at a natural rate. But it has been projected that the greenhouse warming may be up to forty times faster than such natural rates of change (chapter 4). This warming may be so rapid that species would not have time to adjust even if migratory corridors exist (chapter 22; Peters and Darling 1985). And even if species were mobile enough to utilize such corridors, there may not be the political will to develop them in time. In that case, other alternatives, including translocation and restoration, must be instituted (Griffith et al. 1989, Jordan et al. 1988, Peters and Darling 1985). Whatever the choice, it is clear from the paleobiological data that environments and species distributions are constantly changing, and this must be considered in conservation plans.

## IV. MARCH OF LIFE ZONES AND ELIMINATION OF TUNDRA

Models for greenhouse warming suggest that continental areas and areas at high latitudes may experience the greatest climate change (Hansen et al. 1988), and tundra ecosystems might therefore experience particular stress. A simple static community model would suggest that warmer climates should force the tundra ecosystem farther north, where it would be blocked by the Arctic Ocean. Decreases in tundra extent have been more methodically projected by S. A. Edlund (1987) and by W. R. Emanuel et al. (1985), the latter projecting a worldwide tundra loss of 32% for a 3°C rise in average global temperature. It is expected that these changes would not consist of simple replacement of tundra vegetation and animals by other community types, but that patterns of biotic change would be complex, resulting in species mixes and patches of vegetation types different from those that currently exist. In part this is

deducible by observing patterns of change in tundra extent and character in response to warming that occurred since the last glacial maximum (ca. 18,000 years ago).

During the full glacial, tundralike environments existed at both arctic and temperate latitudes (fig. 6.5). In each of these areas, the Pleistocene tundralike environments varied significantly from modern ones. To underscore the differences from modern tundra communities, J. V. Matthews (1982) has referred to Pleistocene arctic environments in Alaska and Siberia as types of an extinct biome, the arctic steppe. That term emphasizes the drier nature of the Pleistocene arctic environments in comparison to the muskeg tundra of the region today. J. C. Ritchie (1984) points out that the Russian paleoecologists have used the term "arctic steppe" differently to describe an existing plant community that North American ecologists regard as herb tundra in modern environments, composed of predominantly herbaceous plants including grasses and sedges.

The Pleistocene tundralike communities south of the Laurentide ice sheet (fig. 6.5) were also quite different from modern tundra communities (Jacobson et al. 1987). In the mammalian fauna, cold-adapted species like the collared lemming, which today lives in modern tundra habitats, occurred in areas along the ice front in the northern plains, upper midwest, and northeastern United States during the Pleistocene (fig. 6.4). However, in each of those areas the tundra animals were intermixed with species that today occur primarily in boreal and deciduous forests, as well as grasslands (Graham and Mead 1987). Again, there is no modern analog for these Pleistocene assemblages or, presumably, the environments that supported them.

Even though the Pleistocene arctic and temperate tundralike environments differed from modern tundra, they shared some similarities, especially within the mammalian communities. Both arctic and temperate regions contained collared lemmings, brown lemmings (Lemmus trimucronatus), caribou, and barren ground muskox. They also differed

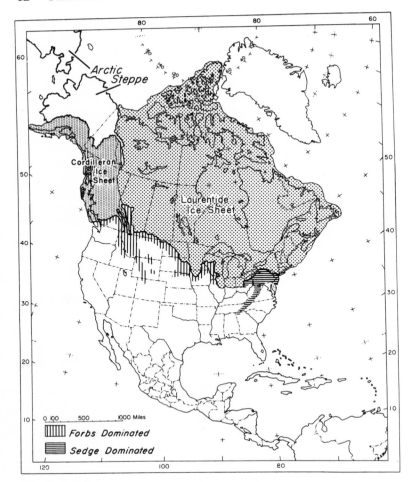

Figure 6.5. Location of arctic steppe and temperate tundralike environments (partially adapted from Jacobson et al. 1987) in relation to the Laurentide and Cordilleran ice sheets during the full glacial (ca. 18,000 years ago). The Des Moines Lobe (Iowa) is shown although its southward advance did not occur until ca. 14,000 years ago.

from one another in important ways. Saiga antelope (*Saiga tatarica*) was restricted to Beringia (Kurten and Anderson 1980), whereas northern pocket gopher, eastern chipmunk (*Tamias striatus*), and thirteen-lined ground squirrel (*Spermophilus tridecemlineatus*) were found with tundralike mammals in the temperate latitudes.

The full glacial mammalian faunas do indicate that an environmental gradient ex-

tended along the ice front from east to west (Graham and Mead 1987). In the northeastern United States, the tundra mammals were intermixed primarily with species that today inhabit deciduous and boreal forests, with a few grassland forms. From the upper midwest to the northern plains, the number of deciduous forest species decreased and the number of grassland species increased significantly. These regional differences in the temperate tundralike environments were also reflected in paleovegetational patterns. G. L. Jacobson et al. (1987) consider the western region of the ice front as forb dominated and the eastern region as sedge dominated.

A comparison of the tundralike biotas of the late Pleistocene with today's tundra suggests that complete elimination of tundra species over large areas might not be caused by future warming. As previously stated, the paleobiological data indicate that future, new communities should evolve because of individual species' responses to environmental changes. The new communities thus formed will probably mix species that inhabit different biomes today, such as tundra and grassland. Vegetation structure may be more important to some animals than species composition of the vegetation, so that even given widespread change, animals will be able to survive by using structurally analogous new plant assemblages. This may partially explain the composition of the present arctic steppe and what may be anticipated in future arctic ecosystems.

I do not mean that the arctic steppe or Pleistocene tundralike environments south of the Laurentide ice sheet will evolve again and that future arctic environments will be the same as those at the ice front during the full glacial. But I do suggest that many of the arctic species might be able to survive in open environments that are quite different in species composition from the modern tundra. Likewise, I do not mean to suggest that the concern for extinctions should be diminished, but it should be tempered with consideration of the habitat variability of individual species. Finally, it should be noted that all arctic species survived the last interglacial environment about 125,000 years ago, even though it was significantly warmer than the present one has been to date.

## V. IMPORTANCE OF SEASONAL EXTREMES

In addition to changes in mean temperature, the greenhouse effect will alter seasonal temperature and moisture patterns. Climatic extremes during different seasons are often more important than mean annual temperature or precipitation in determining species ranges. By seasonal extremes, I mean the greatest differences between summer and winter temperatures as well as the wettest and driest season. Specifically, the coldest winter and the hottest summer temperatures are more critical to populations at the margins of a species distribution than the mean annual temperature. Likewise, the seasonal pattern of precipitation has more effect on marginal populations than mean annual precipitation. In fact, for organisms, the seasonal distribution of effective moisture—the net result of precipitation and evapotranspiration—is probably the most limiting aspect of precipitation.

Obviously, for long-term changes in the distributions of species, changes in seasonal extremes (e.g., colder winters, hotter summers) must be maintained for a substantial period of time, but perhaps as little as decades. One severe winter or summer may reduce or eliminate local populations, but effects of only a single such event are likely to be short-lived because population growth and recolonization can occur when conditions return to normal. However, if extreme events occur often, they could have long-term effects on a species' regional distribution.

As an example, consider the southern boundary of a northerly distributed mammal species such as the collared lemming (fig. 6.4). The most southern occurrence of this species today is above 50°N latitude. This limit is probably controlled by summer high temperatures. Extremely hot summers would eliminate populations along the southern margin of the distribution and therefore tend to move the southern boundary farther north. Conversely, cooler summers would allow populations to colonize new areas farther south, effectively extending the southern limit of the species distribution. Likewise, the northern limit of distribution of a species like the least shrew will be limited by winter extremes (fig. 6.3). The eastern and western limits of a species range (e.g., the northern plains pocket gopher and the least shrew, figs. 6.2 and 6.3) are controlled more by seasonal patterns in effective moisture than strictly by temperature.

As pointed out by J. Hansen et al. (1988), seasonal extremes may be hard to model accurately because "temperature maps for any given month and year represent natural fluctuations [noise] of the climate system as well as long-term trends due to greenhouse forcing, [and consequently the] natural fluctuations are an unpredictable 'sloshing' around of a nonlinear fluid dynamical system." It is clear from the paleobiological record that not all major warming events are the same but that they may differ in the types of seasonal patterns they produce. For example, the warming of previous interglacials, as well as the middle Holocene Hypsithermal, had differing patterns of extremes and therefore different consequences for the North American biota from the warming at the end of the Pleistocene, about 10,000 years ago (Graham 1986a, Graham and Mead 1987).

## VI. TERMINAL PLEISTOCENE EXTINCTIONS

Because of the possibility that greenhouse warming will cause extinctions, it is useful to look to the past for similar extinction events that may have been due at least partly to climate change. One such event occurred at the Pleistocene-Holocene boundary 10,000 years ago, when more than 32 genera of large terrestrial mammals became extinct in North America alone. South America and Australia also experienced massive losses in their mammalian megafauna, with fewer extinctions in Eurasia and Africa. The cause of those extinctions is highly debatable, but overpredation by human populations and environmental changes resulting from climatic fluctuations are the predominant reasons proposed (Martin and Klein 1984).

How might climate change have contributed to the extinctions? Perhaps the most encompassing effect of climate change was habitat destruction (Graham 1986a). With the individualistic reorganization of the biota, new communities and habitats appeared and old ones were eliminated. If the rates and magnitudes of the changes were low, then species could adapt and extinction probabilities would also be low. Rapid rates and

large magnitude changes would produce high probabilities of extinction.

The rate and magnitude of environmental changes at the end of the Pleistocene were substantially different from those of previous glacial-interglacial fluctuations (Graham 1986a). Jacobson et al. (1987) have also demonstrated that the rate of vegetational change at the time of extinction, between 10,000 and 12,000 years ago, was significantly greater than any other changes in the last 18,000 years, except for the last 1000 years. Given that such habitat changes played a causal role in past extinctions, and given that faster and greater changes are more likely to cause extinctions, then the very fast rise in temperature projected for the next 75 or so years (Schneider 1989) may be expected to cause a substantial number of vertebrate extinctions.

However, rate and magnitude are not the only important variables in climatic models for extinction. As discussed above, changes in seasonal extremes, length of growing season, and composition of the vegetational mosaic are also critical factors. R. W. Graham and E. L. Lundelius (1984) as well as R. D. Guthrie (1984) have suggested that increasing seasonal extremes at the end of the Pleistocene caused the evolution of new and less patchy vegetational mosaics, which would have had a variety of ramifications for the megaherbivore guild. First, the changes may have disrupted feeding strategies for many of the large mammal herbivores. With the individualistic reorganization of the vegetational mosaic, mammalian herbivores would have had to select a diet from a new array of plants. For a herbivore guild composed of a few species, this might have been a relatively easy process. In a more complex system like the Pleistocene megafauna, the changes may have heightened competition among the herbivores (Graham and Lundelius 1984). Second, Guthrie (1984) has suggested that the new vegetational mosaics that emerged at the end of the Pleistocene may have had a lower effective nutritional value. This could have been accomplished by increasing the dominance of noxious and toxic plant species and decreasing the abundance of more edible ones. For instance, the more open spruce forest of the late Pleistocene (Jacobson et al. 1987) would have contained a greater biomass of edible understory plants than the modern spruce forests. Reductions in the understory biomass would have decreased the carrying capacity of the Holocene spruce forest environment for both simple and diverse large-mammal herbivore guilds.

Guthrie (1984) also suggested another source of increased stress for large herbivores at the end of the Pleistocene, 10,000 years ago—a shorter growing season. Although global average temperatures were rising as the world came out of the ice age, the growing season could have been shorter if seasonal extremes were accentuated, yielding hotter summers and colder winters, perhaps with more snow. If greenhouse warming makes future Holarctic winters warmer, as projected by climate models (Hansen et al. 1988; chapter 4), we might expect the opposite of what the large herbivores experienced—we might anticipate longer growing seasons and higher productivity in these high-latitude environments (see chapter 8 for discussion of resultant vegetation changes).

In conclusion, it is difficult to isolate a single cause for the late Pleistocene extinction event. However, concomitant changes in herbivore diet, lowering of effective nutritional value of the vegetation, and reduction of growing seasons must have had compounded effects on the megaherbivore guild. Pressures exerted by human predation may have added to the environmental stresses.

## VII. CONCLUSIONS

The paleobiological record of warming events during the late Quaternary is not a crystal ball for the precise prediction of future biotic changes as a result of greenhouse warming. However, this record does provide insights into the processes that might result from the warming, even though it is anticipated to occur at an unprecedented rate. Specifically, the paleobiological record indicates that species will respond to environmental changes individually and not as tightly linked assemblages of species (communities). New community patterns will evolve. The design of biological reserves should, therefore, be more concerned with the conservation of species and genetic types and less with attempts to preserve static community patterns. This can be facilitated by developing corridors between reserves so species can migrate and respond naturally. If that is not feasible, then other means of allowing species to adjust their distributions must be found, including human-assisted dispersal.

The individualistic community model also suggests that climatic warming may not necessarily push the tundra biome into the Arctic Ocean en masse, but this biome may be reorganized in ways similar to, although not identical to, the tundralike environments that existed during the late Pleistocene. Because vegetation structure may be more important to animal survival than vegetation species composition, animal species may be able to survive in some areas where species composition changes dramatically in response to climate change. This does not mean that all species are safe from risk of extinction, but it does suggest that extinction probability may need to be evaluated on a case-by-case basis for individual animal species.

Finally, it is apparent that for biological modeling, climatic models must consider how future warming events will alter seasonal extremes. For instance, will a mean annual increase in temperature result from increases in summer temperatures only, with winters remaining essentially the same, or will summers be unchanged and winters become warmer? Of course, a mean annual increase in temperature could be caused by warming both summer and winter temperatures. The effects on the seasonal distribution of effective moisture will also be of paramount importance for some species. It is changes in these seasonal extremes in climate, and not the mean annual parameters, that will cause geographic shifts in the ranges of species.

## REFERENCES

Boecklen, W. J., and D. Simberloff. 1986. Area-based extinction models in conservation. In *Dynamics of Extinction*, D. K. Elliott, ed., pp. 247–276. New York: John Wiley and Sons.

Bowen, D. Q., G. M. Richmond, D. S. Fullerton, V. Sibrava, R. J. Fulton, and A. A. Velichko. 1986. Chart 1, Correlation of Quaternary glaciations in the northern hemisphere. In *Quaternary Glaciations in the Northern Hemisphere*, V. Sibrava, D. Q. Bowen, and G. M. Richmond, eds. Quaternary Science Reviews 5.

Cushing, E. J. 1965. Problems in the Quaternary phytogeography of the Great Lakes region. In *The Quaternary of the United States*, H. E. Wright, Jr., and D. G. Frey, eds., pp. 403–416. Princeton: Princeton University Press.

Davis, M. B. 1976. Pleistocene biogeography of temperate deciduous forests. *Geosci. and Man* 13: 13.

Edlund, S. A. 1987. Effects of climate change on diversity of vegetation in arctic Canada. In *Preparing for Climate Change*, Proceedings of the first North American conference on preparing for climate change: A cooperative approach, J. C. Topping, Jr., ed., pp. 186–193. Washington, D.C.: Government Institutes.

Emanuel, W. R., H. H. Shugart, and M. P. Stevenson. 1985. Response to comment: "Climatic change and the broad-scale distribution of terrestrial ecosystem complexes." *Clim. Change* 7:457.

Foley, R. L. 1984. Late Pleistocene (Woodfordian) vertebrates from the Driftless Area of southwestern Wisconsin, the Moscow Fissure local fauna. *Illinois State Museum Reports of Investigations* 39:1.

Graham, R. W. 1986a. Plant-animal interactions and Pleistocene extinctions. In *Dynamics of Extinctions*, D. K. Elliott, ed., pp. 131–154. New York: John Wiley and Sons.

Graham, R. W. 1986b. Response of mammalian communities to environmental changes during the late Quaternary. In *Community Ecology*, J. Diamond and T. J. Case, eds., pp. 300–313. New York: Harper and Row.

Graham, R. W. 1988. The role of climatic change in the design of biological reserves: The paleoecological perspective for conservation biology. *Conser. Biol.* 2:391.

Graham, R. W., and E. C. Grimm. 1990. Effects of global climate change on the patterns of terrestrial biological communities. *Trends Ecol. and Evol.* 5(9):289.

Graham, R. W., and E. L. Lundelius, Jr. 1984. Co-evolutionary disequilibrium and Pleistocene extinctions. In *Quaternary Extinctions: A Prehistoric Revolution*, P. S. Martin and R. G. Klein, eds., pp. 223–249. Tucson: University of Arizona Press.

Graham, R. W., and J. I. Mead. 1987. Environmental fluctuations and evolution of mammalian faunas during the last deglaciation in North America. In *North America and Adjacent Oceans during the Last Deglaciation*. W. F. Ruddiman and H. E. Wright, Jr., eds., The Geology of North America, vol. K-3, pp. 371–402. Boulder, Colo.: Geological Society of America.

Graham, R. W., H. A. Semken, and M. A. Graham, eds. 1987. *Late Quaternary Mammalian Biogeography and Environments of the Great Plains and Prairies*. Springfield: Illinois State Museum Scientific Papers, vol. 22.

Griffith, B., J. M. Scott, J. W. Carpenter, and C. Reed. 1989. Translocation as a species conservation tool: Status and strategy. *Science* 245:477.

Guthrie, R. D. 1984. Mosaics, allelochemics, and nutrients. In *Quaternary Extinctions: A Prehistoric Revolution*, P. S. Martin and R. G. Klein, eds., pp. 259–298. Tucson: University of Arizona Press.

Hansen, J., I. Fung, A. Lacis, S. Lebedeff, D. Rind, R. Ruedy, G. Russell, and P. Stone. 1988. Prediction of near-term climate evolution: What can we tell decision-makers now? In *Preparing for Climate Change*, Proceedings of the first North American conference on preparing for climatic change: A cooperative approach, J. C. Topping, Jr., ed. Washington, D.C.: Government Institutes.

Harris, A. H. 1985. *Late Pleistocene Vertebrate Paleoecology of the West*. Austin: University of Texas Press.

Hunter, M. L., G. L. Jacobson, Jr., and T. Webb III. 1988. Paleoecology and the coarse-filter approach to maintaining biological diversity. *Conser. Biol.* 2(4):375.

Jacobson, G. L. Jr., T. Webb III, and E. C. Grimm. 1987. Patterns and rates of vegetation change during the deglaciation of eastern North America. In *North America and Adjacent Oceans during the Last Deglaciation*, W. F. Ruddiman and H. E. Wright, Jr., eds., pp. 277–288. Boulder, Colo.: Geological Society of America.

Jordan, W. R. III, R. L. Peters, and E. B. Allen. 1988. Ecological restoration as a strategy for conserving biological diversity. *Envir. Manag.* 12(1):55.

Kurten, B., and E. Anderson. 1980. *Pleistocene Mammals of North America*. New York: Columbia University Press.

Lundelius, E. L. Jr., R. W. Graham, E. Anderson, J. Guilday, J. A. Holman, D. W. Steadman, and S. D.

Webb. 1983. Terrestrial vertebrate faunas. In *Late Quaternary Environments of the United States*. Vol. 1, *The Late Pleistocene*, S. C. Porter, ed., pp. 311–353. Minneapolis: University of Minnesota Press.

Martin, P. S., and R. G. Klein, eds. 1984. *Quaternary Extinctions: A Prehistoric Revolution*. Tucson: University of Arizona Press.

Matthews, J. V. 1982. East Beringia during late Wisconsin time: A review of the biotic evidence. In *Paleoecology of Beringia*, D. M. Hopkins, J. V. Matthews, C. E. Schweger, and S. B. Young, eds., pp. 127–152. New York: Academic Press.

McHugh, W. P., J. F. McCauley, C. V. Haynes, C. S. Breed, and G. G. Schaber. 1988. Paleorivers and geoarchaeology in the southern Egyptian Sahara. *Geoarchaeology* 3:1.

Moodie, K. B., and T. R. Van Devender. 1979. Extinction and extirpation in the herpetofauna of the southern High Plains with emphasis on *Geochelone wilsoni* (Testudinae). *Herpetologica* 35:198.

Parmalee, P. W., and R. D. Oesch. 1972. Pleistocene and recent faunas from the Brynjulfson caves, Missouri. *Illinois State Museum Reports of Investigations* 25:1.

Peters, R. L., and J. D. S. Darling. 1985. The greenhouse effect and nature reserves. *Bioscience* 35: 707.

Rhodes, R. S. II. 1984. Paleoecology and regional paleoclimatic implications of the Farmdalian Craigmile and Woodfordian Waubonsie mammalian local faunas, southwestern Iowa. *Illinois State Museum Reports of Investigations* 40:1.

Ritchie, J. C. 1984. *Past and Present Vegetation of the Far Northwest of Canada*, p. 163. Toronto: University of Toronto Press.

Rosenberg, R. S. 1983. The paleoecology of the late Wisconsinan Eagle Point local fauna, Clinton County, Iowa. M.A. thesis, University of Iowa.

Ruddiman, W. F., M. Raymo, and A. McIntyre. 1986. Matuyama 41,000-year cycles: North Atlantic Ocean and northern hemisphere ice sheets. *Earth Plan. Sci.* 80:117.

Schneider, S. H. 1989. The greenhouse effect: Science and policy. *Science* 243:771.

Semken, H. A. Jr. 1983. Holocene mammalian biogeography and climatic change in the eastern and central United States. In *Late Quaternary Environments of the United States*. Vol. 2, *The Holocene*, H. E. Wright, Jr., ed., pp. 182–207. Minneapolis: University of Minnesota Press.

Shackleton, N. J. 1987. Oxygen isotopes, ice volume, and sea level. *Quat. Sci. Rev.* 6:183.

Spaulding, W. G., E. B. Leopold, and T. R. Van Devender. 1983. Late Wisconsin paleoecology of the American Southwest. In *Late Quaternary Environments of the United States*. Vol. 1, *The Late Pleistocene*, S. C. Porter, ed., pp. 259–293. Minneapolis: University of Minnesota Press.

Webb, T. III. 1987. The appearance and disappearance of major vegetational assemblages: Long-term vegetational dynamics in eastern North America. *Vegetatio* 69:177.

Wendorf, F., A. E. Close, and R. Schild. 1985. Prehistoric settlements in the Nubian desert. *Am. Scient.* 73:132.

I wish to thank P. Kollen and M. A. Graham for their assistance in manuscript preparation. I appreciate the efforts of J. Snider, who drafted the illustrations. E. C.Grimm provided valuable discussions about the concepts of arctic steppe and the nature of Pleistocene tundralike environments. I thank R. Peters and the World Wildlife Fund for their support and for the opportunity to participate in the enlightening conference on the greenhouse effects and wildlife that initiated the production of this chapter. The manuscript was vastly improved by comments from R. Peters, and I thank him and two anonymous reviewers.

# PART III

# General Ecological and Physiological Responses

CHAPTER SEVEN

# Effects of Global Warming on the Biodiversity of Coastal-Marine Zones

G. CARLETON RAY,
BRUCE P. HAYDEN,
ARTHUR J. BULGER, JR., AND
M. GERALDINE MCCORMICK-RAY

## I. INTRODUCTION

In this chapter we present a preliminary assessment of how future temperature change may affect coastal and marine environments. Coastal zones are particularly worthy of attention because they are the most productive, richest in species, most perturbed by humans, and (probably) most affected by global change of all marine or marine-related systems. To arrive at some predictions of possible effects of global warming, we first briefly describe the nature of biodiversity in coastal-marine environments. Second, we review basic knowledge about how temperature affects marine organisms. Third, we project possible changes that warming could cause in specific ecosystems, using as examples coral reefs, coastal marshes and wetlands, and the sea-ice habitats of marine mammals. Finally, based on the preceding examples, we draw some general conclusions about possible effects on coastal-marine biological diversity.

We recognize that our conclusions are very preliminary for several reasons. First, global warming models themselves require further refinement, especially in the treatment of oceans. In particular, the models do not cope well with the redistribution of heat within three-dimensional ocean space. Nevertheless, while voicing this caveat, we accept for purposes of discussion the scenarios presented by Stephen H. Schneider et al. in chapter 4, in which global temperatures may rise 2°–5°C and sea levels may rise 10–100 cm by about the year 2050. A second major difficulty is the translation of climate information to generalizations about biological and ecological response. We still have only crude ideas about how species' tolerances and ecosystems may respond to temperature change and what the relationships are between species diversity and ecosystem function. Nevertheless, even taking into account the difficulty of the problem and the crude nature of the projections, preliminary attempts at predictions are essential in view of the ongoing and massive alterations of environments by human activities.

## II. THE NATURE OF COASTAL-MARINE DIVERSITY

Globally, the coastal zone includes the continental plains and continental shelves—or about 8% of the earth's surface. Typical coastal environments include tidal marshes, coral reefs, seagrass beds, mangroves, estuaries, banks (for example, Georges Bank off New England), and other diverse and productive environments. Few environments on earth rival coastal zones in productivity or biodiversity. In part, this reflects the generally high taxonomic diversity of marine biota: G. C. Ray (1985, 1988) and Robert M. May (1988) observe that although marine ecosystems may contain only about a fifth of known species, at higher taxonomic levels the oceans may hold much greater biological diversity than the land. For example, every known animal phylum but one exists in marine systems, but only about half of them exist on land.

Given the great area and volume of oceans, it is not surprising that marine biogeographic diversity—the number of biogeographic biomes, realms, and provinces—is also high. B. P. Hayden et al. (1984) describe 21 types of oceanic and coastal-margin realms and 45 coastal provinces, without even considering deep-water realm candidates such as trenches and abyssal zones. Including such zones could at least double the number of realms and yield more than 300 biotic provinces for the total coastal-ocean volume, compared with the 8 realms and 193 provinces described for terrestrial environments by M.D.F. Udvardy (1975). This patterning of biogeographic diversity is determined in part by climate, and if climate change occurs the location of many of these spatial regions can be expected to shift. Also, individualistic and

noncongruent range changes by different species, as demonstrated paleontologically by Russell W. Graham for terrestrial communities, may result in changes of species composition within regions (chapter 6).

As the ranges and species composition of marine-coastal ecosystems change in response to climate, basic ecological processes of the ecosystems (collectively, process diversity) will also change. Such processes include nutrient cycling, release of trace gases, ecological succession, and primary productivity (i.e., fixation of carbon). Given the complexity of ecosystem processes, it is difficult to predict exactly how they might change. For example, even something as basic as how species richness affects or is affected by primary productivity is not well understood. We can only note that high productivity is found in both species-rich coral reefs and relatively species-poor temperate estuaries. Understanding process diversity is more difficult in the oceans than on land because of the extreme temporal and spatial complexity with which species are distributed. For example, more complex food webs may be found in ecosystems with greater three-dimensionality (Briand and Cohen 1987). Further, given that the sea contains a greater size range of species than the land, from microbiota to the great whales, greater trophic complexity is implied.

Given the complexity and present lack of understanding about process diversity, substantial additional research will be necessary before we are able to make comprehensive projections of process response to climate change (see chapter 18, on tundra microcosms, for an example of research on process response to climate in a terrestrial system). The importance of process-level research has been emphasized by the U.S. National Committee for the International Union of Biological Sciences (IUBS), which recently suggested that the IUBS adopt a program "to understand biological diversity in the context of the structure and function of ecosystems" (Simpson 1988). This has been taken up by a workshop with the intent of developing an international research program on the eco-system function of biodiversity (Di Castri and Younes 1989). Reaching this understanding will not be a simple task. We have all become aware of how difficult it is to describe com-munities, patterns, and processes within tropical forests. This will be much more diffi-cult for the seas and coasts, which are so much more extensive and inaccessible.

## III. BIOLOGICAL EFFECTS OF TEMPERATURE

Aquatic systems are fundamentally different from terrestrial ones because of the physical properties of water. Water is denser than air, has a higher specific heat, is a better heat conductor, is an excellent solvent, and is less subject to rapid temperature change. At ele-vated temperatures water becomes less dense and also holds less oxygen; changes in density affect stratification and circulation, and changes in oxygen concentration can pose physiological constraints. All of these factors influence species' adaptations. Below, we consider only the effects of temperature.

### A. Thermal Physiology

The vast majority of marine organisms are thermoconformers; that is, they cannot main-tain body temperatures much different from the water in which they occur. Also, for many species, temperature is a master switch that induces many important behavioral and physiological responses, such as migration and reproduction. Most marine organisms are exquisitely sensitive to temperature, since it is important for the short-term survival of individuals and the long-term survival of spe-cies. Furthermore, thermoconformers typi-cally are most active near their upper, rather than lower, thermal limits; this provides the greatest scope for activity and maximizes power output, e.g., swimming speed (Ben-nett and Rubin 1979). Unfortunately, this also increases vulnerability to temperature rise be-cause relatively small temperature increases near the upper thermal limit can be lethal.

Within the limits of a species' tolerance,

temperature controls rates of metabolism and development, governs distribution, and influences a host of biological processes. Some examples illustrate the diversity of these effects. Temperature affects the sex ratio of fish (Conover and Heins 1987), sea turtles (Mrosovsky and Yntema 1980), and copepods (Tande 1988). It determines em-bryonic development (Igumnova 1985) and size after metamorphosis of some fish (Seikai et al. 1986). Temperature can differentially affect zooplankton competition (Threlkeld 1986). Whether Baltic salmon larvae will emerge from the gravel in the night or in daylight is determined by the interaction of temperature and photoperiod (Braennaes 1987). For marine plants, which generally have a more restricted temperature range than terrestrial plants, oxygen supply may in-fluence heat tolerance: tropical algae are not very heat resistant and most are killed by 12-hour exposure to 32°–35°C (Gessner 1970).

The thermal responses of species have been shaped by evolution. An example is the capacity to adjust thermal tolerance season-ally, by shifting to alternate enzyme systems and to different cell membrane components. Such processes are collectively termed accli-matization and are responsible for the sur-vival of many temperate-zone species. The same animal may be active in summer at a temperature that is lethal in winter and active at a temperature in winter that would pro-duce torpor in summer. Animals that inhabit environments where water temperature is only moderately variable seasonally (the deep sea, the tropics, and polar latitudes) typ-ically have a relatively narrow range of ther-mal tolerance and are called stenothermal to distinguish them from more broadly tolerant species, which are called eurythermal. Even short-term temperature increases of only a few degrees may be lethal to stenothermal species. Antarctic ice fishes of the genus Tre-matomus die of heat stress at 6°C (Schmidt-Nielsen 1983); some corals exhibit symptoms of heat stress when temperature is elevated by as little as 4°C. Many temperate fishes are also relatively stenothermal and avoid seasonal extreme temperatures by migrating (Briggs 1974, Moyle and Cech 1982). The smallest acclimatization capacity usually occurs in early stages of development or gonad maturation, and it is at these stages that temperature change may have the most profound effects.

Acclimatization also occurs for metabolic rate, which is a good index of the rate at which an organism uses oxygen and energy. Commonly, a temperature increase of 10°C multiplies metabolic rate two- to three-fold, a substantial increase in oxygen and energy demands. Following a temperature increase some thermoconformers can actively shift enzyme and other systems by a process called metabolic rate compensation, so that cellular processes are gradually returned to the previous rate, thereby conserving oxygen and energy. Conversely, many thermoconformers have imperfect metabolic rate compensation, and many others have none at all. This is especially true for stenothermal species from thermally stable environments; these species are thus quite vulnerable to even very small temperature increases.

## B. Temperature Synergisms and Ecological Effects

Temperature acts in conjunction with many other variables. D. W. McLeese (1956) has illustrated the lobster's (*Homarus americanus*) response to changes in oxygen, temperature, and salinity and how its tolerance for any one of these variables is influenced by the others (fig. 7.1). Many species exhibit such complex synergisms. In addition, each species' life stage may be differently affected.

How temperature regimes may affect the interdependency of some boreal and temperate fishes and invertebrates is illustrated by figure 7.2. The feathery calanus copepod (*Calanus plumchrus*) and the walleyed pollock (*Theragra chalcogramma*) are adapted to low temperatures; they are also strongly linked in a common food web. Both occur in the Bering Sea's outer-shelf to basin waters (NOAA 1988). The copepod's age at sexual maturity and maximum life span is 1 year; adults thus

have only one opportunity to reproduce. Pollock live about 15 years and reach sexual maturity in about 4 years. This means that an adult pollock can potentially reproduce annually within an 11-year adulthood. Alteration of water temperature could disrupt the ecological relationship between copepods and pollock because survival of each species will be different in response to temperature change. As other variables are considered, the situation clearly becomes more complex.

Two other examples are illustrated by figure 7.2. Two anadromous fishes, the American shad (*Alosa sapidissima*) and the striped bass (*Morone saxatilis*), are coexisting species with different thermal requirements. Altered temperature regimes would probably alter their relative abundances where they coexist, with consequences for local food web patterns, as these two fishes have very different food habits. Similarly, two temperate benthic invertebrates, the northern lobster and the American oyster (*Crassostrea virginica*), provide examples of potential effects on reproduction and distribution. Although adult lobsters and oysters have broad temperature tolerances, spawning and development occur within relatively narrow ranges, which are very different for these two species.

A restricted temperature range for re-

*Figure 7.1.* Relations among oxygen, temperature, and salinity for the American lobster, *Homarus americanus*, and their combined influence on the tolerance zone. From McLeese 1956.

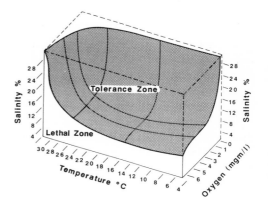

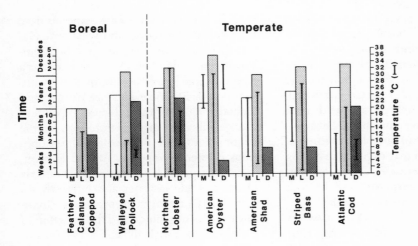

*Figure 7.2.* Temporal and temperature characteristics of the life histories of selected temperate and boreal species, representing geographically and ecologically distinct environments. Species are also representative of holoplankton (feathery calanus copepod), benthic invertebrates (American oyster, American lobster), shelf fishes (walleye pollock, Atlantic cod), and anadromous fishes (American shad, striped bass). For temporal ranges read bars and left vertical axis: M, maximum time to reach sexual maturity; L, longevity; D, development time for eggs and larvae. For temperature values read lines within bars and right vertical axis: M, spawning range; L, adult tolerance range; and D, preferred temperature for hatching and larval growth. From Kinne 1970 and NOAA 1984, 1988.

production is characteristic of most marine animals and acts to stabilize a species to a particular niche (Kinne 1970). Optimum temperature regimes are important in defining centers for essential life-history functions, such as reproduction. From such centers, peripheral areas may be occupied by recruitment. It follows that temperature change may alter the distributions and population dynamics of many species, including valuable commercial species, with serious consequences for coastal fishing economies.

Both biological and ecological effects of temperature are well illustrated by the American shad, a species whose life-history pattern is known to be finely tuned to temperature cues (fig. 7.3). Shad reach sexual maturity in coastal marine waters and spawn in fresh wa-

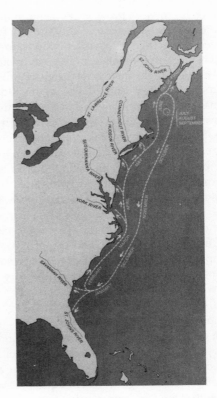

*Figure 7.3.* Migration routes and spawning areas of the American shad, *Alosa sapidissima*, of the eastern United States. See text for explanation. From Leggett 1973; copyright © 1973 by Scientific American, Inc. All rights reserved.

ter. As do many coastal fishes of eastern North America, shad migrate along the continental shelf largely in response to seasonal temperature differences (Leggett 1973, Briggs 1974, Moyle and Cech 1982) and may travel 20,000 km in a 5-year life span (Dadswell et al. 1987). This is an example of a species that has wide temperature tolerance but is behaviorally stenothermal. Shad can tolerate temperatures of 3°–25°C and may occur in spawning areas at temperatures of 5°–23°C. However, the vast majority of prespawning adults track water temperatures of only 13°–18°C. Furthermore, each population seeks to spawn in its native river. In Canada, spawning runs peak in June when rivers warm to 15°C, whereas spawning peaks in Florida when rivers cool to 15°C (Leggett 1973).

Regional warming may have two important consequences for shad. First, the number of repeat spawners is likely to be reduced, especially in southern populations where higher temperatures elevate metabolic rate, thereby increasing the energy cost of spawning. Shad cannot make up this additional cost because they do not feed on the spawning run. In the event of coastal warming, postspawning mortality may therefore increase, changing the life histories and population dynamics of affected populations. Second, coastal warming may disrupt the timing of reproduction. Shad may arrive at their spawning rivers at the appropriate time in their annual migration and spawning cycle but find that the river is at an inappropriate temperature for them. Will they then attempt to spawn at this altered temperature or forgo spawning and continue the offshore migration, to spawn in the wrong river at the right temperature?

In conclusion, almost every response and every process, from the functioning of cells to ecological processes, proceeds within a specific thermal regime within aquatic ecosystems. If sea temperatures rise, impacts on reproduction, life history, and migration are to be expected for the majority of coastal species that migrate or reproduce in response to temperature. Whereas extinction is

unlikely because many of these species have large geographic ranges and wide temperature tolerances, the displacement of assemblages consisting of hundreds of species is a foregone conclusion, as is borne out by observations at El Niño events (see Wooster and Fluharty 1986). Even a moderate temperature change will affect many, if not most, marine communities and the ecological relationships of member species. Altered environmental temperature regimes may act directly on species' life histories and indirectly on community or ecosystem processes. Unfortunately, the magnitude and the direction of change are extremely difficult to predict.

## IV. REGIONAL ECOSYSTEM EFFECTS

Another perspective may be gained through examination of ecosystems or communities of organisms. So diverse are marine and coastal environments and species that we will approach this task for only three examples, one each from the tropics, temperate zones, and polar regions.

### A. Coral Reef Communities

Coral reefs are thought to host the most species-rich communities of the marine environment. Their exuberant complexity makes the sparse life of open ocean waters seem desertlike by comparison (Sheppard 1983). They are comparable to tropical rain forests in that damage to their ecosystems may affect thousands of species adversely. Two-thirds of all marine fish species are associated with tropical reefs (Nelson 1984), and diversity among reef-associated invertebrates is similarly vast. This fact has obvious implications for the many human societies that depend on coral reefs for food, sport, protection of shorelines from storm damage, and tourism.

At the nucleus of the coral reefs are simple colonial animals of the phylum Coelenterata—the reef-building corals—that have photosynthetic plant cells (dinoflagellates) living symbiotically in their tissues. This plant-animal partnership is responsible for an efficient recycling of nutrients. Geologically, reefs are made up of calcium carbonate produced by corals and other organisms, most notably coralline algae, that has accumulated layer upon layer over thousands of generations. Reef communities are dependent on this reef structure for substrate and shelter and also on the reef's primary production, to which the corals and algae make substantial contributions.

Coral reefs are largely limited to tropical waters, but even there, no reefs occur where the waters are too deep, too muddy, too diluted by fresh water, or too hot. No coral reefs occur where temperatures exceed 30°C for extended periods (Sheppard 1983, Levinton 1982). It is important to recognize that many, if not most, coral reefs are already near their upper thermal limit, at least for some months of the year. Thus, even the small increases of 2°–3°C that are predicted for the surface waters of tropical oceans have profound implications for the structure, function, and distribution of reef ecosystems.

At least three major effects of temperature increase on coral reefs may be identified. The

first is physiological. Corals may expel their symbiotic algae in response to heat stress so that they appear bleached (fig. 7.4). Without those cells, corals cannot grow, and unless the algae become reestablished, the corals die within a few months. Warming of tropical eastern Pacific waters contributed to widespread coral bleaching and death in 1982–83, and surface water temperatures above 30°C are thought to have been responsible for the widespread bleaching of corals in the Caribbean Sea in 1987 (Roberts 1987).

Second, mechanical damage to reefs could increase. If corals die and reef growth stops, the reefs will become more vulnerable to erosion (Roberts 1987). Furthermore, warming of tropical oceans may increase the frequency of hurricanes (Wendland 1977). Mechanical damage due to storms is a major source of coral mortality (Levinton 1982). Hurricanes can strip all living coral from long stretches of reef, but the branching corals, such as those of the genus *Acropora*, are especially vulnerable. Estimates for the minimum recovery time following a single storm range from 10 to 20 years.

Third, thermal expansion of ocean water,

*Figure* 7.4. Bleached corals from Looe Key, Florida, December 1987. Photograph by G. C. Ray.

among other factors, causes elevation of sea level. The rate of sea level rise may affect the extent, structure, and functioning of coral reef communities. Two major categories of corals, often present on the same reef, have very different growth rates. Rapidly growing branching corals, such as *Acropora*, have maximum growth rates of 100 mm/yr, whereas some massive corals, such as *Montastrea*, have maximum growth rates of only 7 mm/yr (Levinton 1982). Both types might be able to keep pace with a sea level rise under the low scenario of 3 mm/yr, but the high scenario of 12 mm/yr could favor the more damage-prone, fast-growing, branching species. This change would alter the kinds of substrate and shelter available to reef inhabitants. However, it must be kept in mind that growth rates for individual coral colonies are not the same as accretion rates of entire reefs. Accretion is a result of many processes including coral growth and physical and biological erosion. The maximum vertical accretion rate of many reefs is estimated to be only about 10 mm/yr (Buddemeier and Smith 1988, Grigg and Epp 1989). Coral reefs may be drowned when sea level rises faster than that rate.

In conclusion, should global warming be realized, profound local and regional effects could result for coral reefs. These could include an increase in the frequency, magnitude, and area of coral bleaching; an increase in physical damage resulting from an increase in hurricane frequency; large areas of impoverished, disturbance-dominated reef communities; and drowning of reefs by high rates of sea level rise. On the other side of the coin, widespread extinctions are not expected because of the dispersal capacity and large geographic ranges of most corals and other reef-associated species. Also, the species diversity of cooler-temperature reef communities could be enhanced by warming, as reef communities become established at higher latitudes.

## B. Temperate Barriers, Lagoons, and Marshes

The sea level along the U.S. Atlantic coast has, over the past several decades, slowed its rise (Hicks and Hickman 1988), but we assume here that there will be a substantial sea level rise by A.D. 2050. We also assume that the current rate of relative sea level rise of 2 mm/yr will be increased to 3 mm/yr in the low scenario and to 12 mm/yr in the high scenario (chapter 4). Nonetheless, we voice some skepticism that water would be available to meet the high-scenario rate, given that the rate was only 8.4 mm/yr during post-Pleistocene melting.

Take, for example, the Virginia Coast Reserve on the seaward side of the Delmarva Peninsula. The reserve is owned by the Nature Conservancy and is a Biosphere reserve (Man and the Biosphere Programme of UNESCO) as well as a National Science Foundation long-term ecological research site with an active research program of the University of Virginia. The lagoons of the reserve consist of open waters containing marshes of three types: mainland fringing marshes, barrier island fringing marshes, and mid-lagoon marshes. Between 15,000 and 8000 B.P., sea level rise was about 8.4 mm/yr, largely from the melting of the polar ice caps. Little is known about the status of the lagoons or even if they existed at this early date. Between 6000 and 2000 B.P. the rate of sea level rise was about 3 mm/yr, from land subsidence and eustatic sea level rise, and the lagoons were open water environments with little evidence of mid-lagoon marshes (Harrison et al. 1965). After A.D. 1200, land subsidence slowed, sea level rise declined to the low rate of 1.3 mm/yr, and mid-lagoon marshes flourished from Assateague Island, Virginia, to the entrance of Chesapeake Bay (Harrison et al. 1965). Presently, the rate of sea level rise is about 2 mm/yr.

Mid-lagoon marshes in general lack an inorganic sediment supply so that their upward growth rate approximates only about 1.5 mm/yr (NAS 1987)—less than the current rate of sea level rise. For the Virginia Coast Reserve, there is now an annual net loss of mid-lagoon marsh (fig. 7.5A), and mid-lagoons are rapidly being converted to open waters. Analysis of survey maps reveals that there has been a 16% loss of marsh in the

**Virginia Coast Reserve**

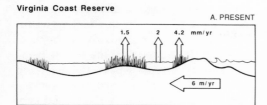

A. PRESENT

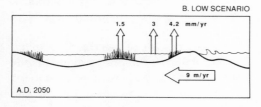

B. LOW SCENARIO

A.D. 2050

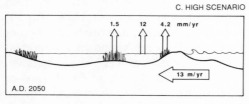

C. HIGH SCENARIO

A.D. 2050

*Figure* 7.5. Rates of change in upward marsh growth for the Virginia Coast Reserve: A. Present conditions. B. Low sea level rise scenario. C. High sea level rise scenario. Horizontal arrows indicate the landward erosion rate of the barrier islands. See text for explanation.

lagoons between 1852 and 1968, most of which was mid-lagoon marsh (Knowlton 1971). In contrast, the barrier-island fringing marshes of the seaward sides of the lagoons now easily keep up with present sea level rise. Estimates for their upward growth rates are around 4.2 mm/yr (NAS 1987). This high rate is possible because beach erosion supplies sand to the marsh surface by overwash across the barrier islands, resulting in a net landward migration of the island mass. The current rate of migration of the barrier islands of the reserve, as indicated by shoreline erosion rates, is 6 m/yr.

For the low scenario, the annual rate of sea level rise is predicted to be 3 mm/yr (fig. 7.5B). This is twice the upward growth rate of mid-lagoon marshes. We therefore expect that the lagoons of the reserve would be free

of mid-lagoon marshes by A.D. 2050. However, this low rate of sea level rise is less than the current 4.2 mm/yr upward growth rate of the fringing marshes. Under this scenario, the expected island migration rate could increase by 2050 to 9 m/yr. The supply of sand to the fringing marshes would also increase so that the presently estimated 4.2 mm/yr upward marsh growth rate might be conservative. For the high scenario, the estimated rate of sea level rise is expected to be 12 mm/yr, indicating an island migration rate of 13 m/yr by 2050 (fig. 7.5C). We believe that sand supply would still be sufficient to support fringing marshes, but there is little prospect for the survival of the mid-lagoon marshes.

In conclusion, for both scenarios sea level rises are in excess of mid-lagoon marsh growth rates. Accordingly, open, marsh-free lagoons are to be expected, and it is clear that marshland habitat will be drastically reduced. The marshes and barriers of the Virginia Coast Reserve have among the highest erosion rates for the entire United States (Dolan and Kimball 1988). Even given lower erosion rates in other areas, water levels are still likely to rise enough that the majority of Atlantic and Gulf Coast barriers, lagoons, and marshes and their dependent fisheries will be altered markedly in the case of either scenario. It would be hard to come to a different conclusion given the history presented at the beginning of this section.

### C. Polar Seas and Marine Mammals

The global warming models considered here predict that the greatest temperature increases—more than 10°C for many locations (chapter 4)—will occur in high-latitude environments. Vera Alexander (chapter 17) describes changes that are likely to occur for sea ice, ocean circulation, and arctic ecosystems. This section focuses on the marine mammals that depend on annual sea ice as habitat.

The annual sea ice over the continental shelf areas of Beringia (Hopkins 1967) migrates, melts, and reforms every year over a latitudinal extent of about 1000 km from its southern maximum extent in March to its northern minimum in September (NOAA 1988). Several marine mammals migrate in response to these changes, remaining in or near ice. These include walruses (*Odobenus rosmarus*), bowhead whales (*Balaena mysticetus*), ringed seals (*Phoca fasciata*), and others. Other species inhabit Beringia mostly during the ice-free summer—for example, gray whales (*Eschrichtius robustus*)—and still others inhabit sea ice only during the winter-spring reproduction season—spotted seals (*Phoca largha*) and ribbon seals (*Histriophoca fasciata*).

A reduction in annual sea ice would be as important to species that require ice platforms for reproduction, nursing young, and molting as the clearing of tropical forests would be to howler monkeys and gorillas. It is improbable that any marine mammals would become extinct because of reduced sea ice alone, but the numbers of some would drastically change, as would community composition. Conversely, the disappearance of sea ice would enhance other species. G. C. Ray and G. L. Hufford (1989) describe six marine mammal assemblages for Beringia. Four are associated at some time of year with particular sea ice conditions, for example, the first assemblage (fig. 7.6) is associated with an ice type called broken pack. Two other assemblages occur in the absence of sea ice.

One scenario that might result from warming and sea ice depletion is that the four assemblages of marine mammals that are adapted to sea ice would be reduced and fishes might respond to a depletion in their marine mammal predators by an increase in their biomass. This shift from warm-blooded (mammalian) to cold-blooded (fish) consumers has important implications for how energy and nutrients are apportioned in the ecosystem. Another scenario concerns benthic organisms and the productivity of benthic communities. Proportional shifts among benthic-feeding marine mammals, especially walruses and gray whales that together comprise a significant portion of the total Beringian biomass but are adapted to different sea-ice conditions, would markedly alter the population dynamics of the benthic communities on which each of these marine mammals respectively feeds. In addition, walrus and gray whales disrupt massive amounts of sediment during feeding, the extent being best known for gray whales (Johnson and Nelson 1984). Sediment structure has a strong effect on the productivity and distribution of both infauna and epifauna. Should sediment structure be altered, so would dependent fauna.

In brief, a cascade of events is likely to follow global warming in polar regions. Marine mammals that are exquisitely adapted to sea ice are dominant consumers in these latitudes. Although extinctions may not occur, local to regional depletions are likely. Population shifts among marine mammals or other species such as fish would have important ecological consequences. This top-down ecological effect is likely to be of equal or even greater consequence than the bottom-up effect of altered phytoplankton production (chapter 17). Primary production is basic to all ecosystems, but we must also recognize the strong controls that species at the top of the trophic structure exert. Ocean science has traditionally been most concerned with physical controlling mechanisms, or with primary and secondary production, but perhaps more attention should be given to "the big things that run the world" (Terborgh 1988). That is, the ecological importance of marine mammals may be similar to that of large African mammals (McNaughton et al. 1988) within their analogous ecosystems.

*Figure 7.6.* Six assemblages for nine marine mammal species of the Bering, Chukchi, and Beaufort seas derived by means of principal components analysis. Four assemblages are associated with sea ice: I, broken pack and seasonal sea ice; II, summer western marginal ice (in part); III, winter marginal ice; and IV, summer eastern marginal ice (in part). Two assemblages are associated with ice-free conditions: V, nearshore environments; and VI, the Bering Strait region. For full explanation see Ray and Hufford 1989.

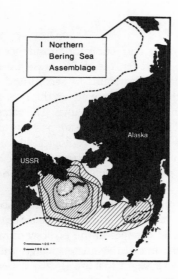

I Northern
Bering Sea
Assemblage

USSR

Alaska

0 —— 100 nm
0 —— 100 km

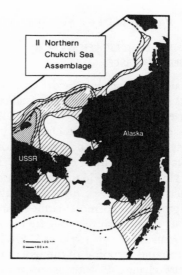

II Northern
Chukchi Sea
Assemblage

USSR

Alaska

0 —— 100 nm
0 —— 100 km

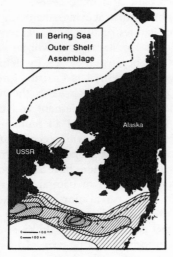

III Bering Sea
Outer Shelf
Assemblage

USSR

Alaska

0 —— 100 nm
0 —— 100 km

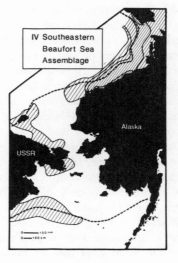

IV Southeastern
Beaufort Sea
Assemblage

USSR

Alaska

0 —— 100 nm
0 —— 100 km

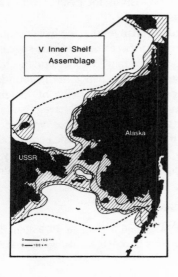

V Inner Shelf
Assemblage

USSR

Alaska

0 —— 100 nm
0 —— 100 km

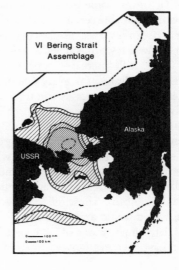

VI Bering Strait
Assemblage

USSR

Alaska

0 —— 100 nm
0 —— 100 km

## V. GENERAL IMPLICATIONS FOR MARINE DIVERSITY

Several conclusions may be drawn about the effects of global warming on coastal-marine biodiversity. First, widespread extinctions are not likely, but widespread changes in community distributions and composition are probable. Many of these changes will closely track global warming in time, because of the mobility, large ranges, high fecundity, and rapid growth rates of most marine organisms, contrary to the relatively long lag times that may occur for forests (chapter 22). With regard to rapid response of the biota to temperature change, El Niño is instructive. In 1982–83, El Niño resulted in the disappearance of an entire seabird community (Schreiber and Schreiber 1984) and also in reduction of catches of coastal pelagic and continental shelf fish species and an increase in the occurrence of tropical and temperate tuna (Smith 1985). In 1957–59, El Niño resulted in a 50% decrease in copepods, a 45% decrease in euphausiids, and a 4% decrease in thaliaceans (salps, etc.). A review of many other changes is provided in Wooster and Fluharty (1986).

Second, there will be complex impacts on coastal environments, only a few of which can be predicted in more than general terms (Thom and Roy 1988). For example, water column stratification and density currents will be altered in ways that are difficult to predict. However, some predictions may be made with reasonable certainty: for each centimeter of rise in sea level, beaches may erode a meter landward; storm surges, a major erosional force, will increase, especially for gently sloping shores; for every 10-cm rise, saltwater wedges in estuaries and tidal rivers may advance a kilometer; and any sea level rise will increase salinity intrusion into coastal freshwater aquifers (NAS 1987). Also, greater coastal erosion will increase turbidity, which is detrimental to many marine organisms, particularly filter feeders (e.g., economically important shellfish). These effects are synergistic, even catalytic, and complex.

Third, large fluctuations of sea level of 40–150 cm have occurred over the past 300

years, after correction for subsidence (NAS 1987). However, the ability of organisms to adapt and ecosystems to accommodate future change will depend on the rate, intensity, and duration of change. The rates predicted for the near future are possibly unique in this planet's recent geological history. Because it is questionable that many species can adjust rapidly enough, major disruptions are likely.

In conclusion, not only would global climate change in predicted scenarios cause profound changes in the ecology of marine systems, but also climate change will add stress to systems that are already experiencing increasing and often severe disruption from other human activities, including pollution, habitat destruction, and overharvesting of the seas. Coastal regions will be particularly at risk because they are, by definition, areas where terrestrial and marine forces interact and within which exists the highest intensity of human settlement and use on earth. More than half of all humanity has probably always lived within coastal areas of the world, and that proportion is increasing rapidly. Human activities have already caused extensive alteration of natural coastal ecosystems, which can ill afford further damage. Therefore, national and international efforts to protect coastal areas should be given the highest priority, and long-term plans should include provisions for dealing with climate change.

All of the changes that we have mentioned have consequences not only for species and ecological diversity but also for the human economic future. Change in the abundances or geographical availability of fishery resources, for example, will have great human impacts. The consequences will be even larger if warming affects basic ecological processes such as primary productivity, reef building, and the energetics of communities.

## REFERENCES

Bennett, A. F., and J. A. Rubin. 1979. Endothermy and activity in vertebrates. *Science* 206:649.

Braennaes, E. 1987. Influence of photoperiod and temperature on hatching and emergence of Baltic salmon (Salmo salar L.) Can. J. Zool. 65(6):1503.

Briand, F., and J. E. Cohen. 1987. Habitat compartmentation and environmental correlates of food chain length. Science 243:238.

Briggs, J. C. 1974. Marine Zoogeography. New York: McGraw-Hill.

Buddemeier, R. W., and S. V. Smith. 1988. Coral reef growth in an era of rising sea level: Predictions and suggestions for long-term research. Coral Reefs 7:51.

Conover, D. O., and S. W. Heins. 1987. The environment and genetic components of sex ratio in Menidia. Copeia 3:732.

Dadswell, M. J., G. D. Melvin, P. J. Williams, and D. E. Themelis. 1987. Influences of origin, life history, and chance on the Atlantic coast migration of American shad. Am. Fish. Soc. Sym. 1:313.

Di Castri, F., and T. Younes. 1989. Ecosystem function and biological diversity. Biol. International 21:1.

Dolan, R., and S. Kimball. 1988. Coastal erosion and accretion. In National Atlas of the United States of America. Denver, Colo., and Reston, Va.: U.S. Geological Survey.

Gessner, F. 1970. Temperature: Plants. In Marine Ecology: A Comprehensive, Integrated Treatise on Life in Oceans and Coastal Waters, O. Kinne, ed., vol. 1, part 1, pp. 363–406. London: Wiley-Interscience.

Grigg, R. W., and D. Epp. 1989. Critical depth for the survival of coral islands: Effects on the Hawaiian Archipelago. Science 243:638.

Hayden, B. P., G. C. Ray, and R. Dolan. 1984. Classification of coastal and marine environments. Envir. Conser. 11(3):199.

Harrison, W., R. J. Malloy, G. A. Rusnak, and J. Terasmar. 1965. Possible late Pleistocene uplift Chesapeake Bay entrance. J. Geology 73(2):201.

Hicks, S. D., and L. E. Hickman. 1988. United States sea level variation through 1986. Shore and Beach July:3.

Hopkins, D. M. 1967. The Bering Land Bridge. Stanford, Calif.: Stanford University Press.

Igumnova, L. V. 1985. Effect of temperature gradients on the embryonic development of beluga sturgeon Huso huso L. and stellate sturgeon Acipenser stellatus Pallas (Acipenseridae). Vopr. Ikhtiol. 25(3):478.

Johnson, K. R., and C. H. Nelson. 1984. Side-scan sonar assessment of gray whale feeding in the Bering Sea. Science 225:1150.

Kinne, O., ed. 1970. Marine Ecology: A Comprehensive, Integrated Treatise on Life in Oceans and Coastal Waters, vol. 1. London: Wiley-Interscience.

Knowlton, S. M. 1971. Geomorphological history of tidal marshes, eastern shore, Virginia, 1852–1966. M.S. thesis, University of Virginia.

Leggett, W. C. 1973. The migrations of the shad. Sci. Am. 228(3):92.

Levinton, J. S. 1982. Marine Ecology. Englewood Cliffs, N.J.: Prentice-Hall.

May, Robert M. 1988. How many species are there on earth? Science 241:1441.

McLeese, D. W. 1956. Effects of temperature, salinity, and oxygen on the survival of the American lobster. J. Fish. Res. Bd. Can. 18:247.

McNaughton, S. J., R. W. Ruess, and S. W. Seagle. 1988. Large mammals and process dynamics in African ecosystems. Bioscience 38(11):794.

Moyle, P. B., and J. J. Cech. 1982. Fishes: An Introduction to Ichthyology. Englewood Cliffs, N.J.: Prentice-Hall.

Mrosovsky, N., and C. L. Yntema. 1980. Temperature dependence of sexual differentiation in sea turtles: Implications for conservation practices. Biol. Conser. 18:271.

NAS (National Academy of Sciences). 1987. Responding to Changes in Sea Levels: Engineering Implications. Washington, D.C.: Marine Board, National Research Council, National Academy Press.

Nelson, J. S. 1984. Fishes of the World. New York: John Wiley and Sons.

NOAA (National Oceanic and Atmospheric Administration). 1984. East Coast Strategic Assessment Project: Living Marine Resources Data Compendium. Washington, D.C.: NOAA, Ocean Assessments Division, Strategic Assessment Branch.

NOAA. 1988. Bering, Chukchi, and Beaufort Seas Coastal and Ocean Zones, Strategic Assessment: Data Atlas. Washington, D.C.: NOAA, Ocean Assessments Division, Strategic Assessment Branch.

Ray, G. C. 1985. Man and the sea: The ecological challenge. Am. Zool. 25:451.

Ray, G. C. 1988. Ecological diversity in coastal zones and oceans. In Biodiversity, E. O. Wilson, ed. Washington, D.C.: National Academy Press.

Ray, G. C., and G. L. Hufford. 1989. Relationships among Beringian marine mammals and sea ice. Rapp. P.-v. Reun. int. Explor. Mer 188:22.

Roberts, L. 1987. Coral bleaching threatens Atlantic reefs. Science 238:1228.

Schmidt-Nielsen, K. 1983. Animal Physiology. Cambridge: Cambridge University Press.

Schreiber, T. W., and E. A. Schreiber. 1984. Central Pacific seabirds and the southern ocean oscillation: 1982 to 1983 perspectives. Science 225:713.

Seikai, T., J. B. Tanangonan, and M. Tanaka. 1986.

Temperature influence on larval growth and metamorphosis of the Japanese flounder *Paralichthys olivaceus* in the laboratory. *Bull. Jap. Soc. Sci. Fish.* 52(6):977.

Sheppard, C.R.C. 1983. *A Natural History of the Coral Reef.* Poole, Dorset: Blandford Press.

Simpson, Beryl B. 1988. Biological diversity in the context of ecosystem structure and function. *Biol. International* 17:15.

Smith, P. E. 1985. A case history of an anti-El Niño: Transition on plankton and nekton distribution and abundance. In *El Niño North*, W. S. Wooster and D. L. Fluharty, eds., pp. 121–142. Seattle: Washington Sea Grant Program, University of Washington.

Tande, K. S. 1988. The effects of temperature on metabolic rates of different life stages of *Calanus glacialis* in the Barents Sea. *Polar Biol.* 8(6):457.

Terborgh, J. 1988. The big things that run the world: A sequel to E. O. Wilson. *Conser. Biol.* 2(4):402.

Thom, B. G., and P. S. Roy. 1988. Sea-level rise and climate: Lessons from the Holocene. In *Greenhouse: Planning for Global Change*, G. I. Pearman, ed., pp. 177–188. Leiden: E. J. Brill.

Threlkeld, S. T. 1986. Differential temperature sensitivity of two cladoceran species to resource variation during a blue-green alga bloom. *Can. J. Zool.* 64(8):1739.

Udvardy, M.D.F. 1975. *A Classification of the Biogeographical Provinces of the World.* IUCN Occasional Paper 18. Morges, Switzerland: Int. Union for Conservation of Nature.

Wendland, W. M. 1977. Tropical storm frequencies related to sea surface temperature. *J. App. Met.* 16:477.

Wooster, W. S., and D. L. Fluharty, eds. 1986. *El Niño North.* Seattle: Washington Sea Grant Program, University of Washington.

Data collected by G. C. Ray on polar seas and marine mammals resulted from several contracts to the University of Virginia from the National Oceanic and Atmospheric Administration, leading to NOAA 1988 and subsequent analyses. The work of B. P. Hayden on temperate barriers, lagoons, and marshes was supported in part by a grant from the National Science Foundation (BSR8702333) for long-term ecological research on the Virginia Coast Reserve.

CHAPTER EIGHT

# A Review of the Effects of Climate on Vegetation: Ranges, Competition, and Composition

F. IAN WOODWARD

## I. INTRODUCTION

A global network of meteorological stations has been recording weather for a century or more. Analyses of annual records have provided evidence of a global warming (Jones et al. 1988) on the order of 0.5°C per century. This rise in global temperature is occurring concurrently with marked latitudinally dependent trends in precipitation (Bradley et al. 1987), such as reduced precipitation in northern Africa and increased precipitation in North America. There is thus considerable evidence that the global climate is changing, and ecologists must therefore address the question of how ecological systems will respond.

Because vegetation is a dominant factor in structuring ecosystems, a central part of understanding ecological effects of climate change is to understand how vegetation will respond. In particular, will the geographical ranges of forests and deserts increase or decrease? How will the diversity and biomass of different vegetation types be affected? This chapter will take two related approaches to those questions.

The first is to look at present-day correlations between climate and vegetation distribution to see how the different climate variables presently relate to vegetation composition and range. Such correlations cannot be certain predictors of future vegetation patterns because new climate may cause unexpected or novel responses, but correlations provide insights into climate-vegetation interactions that cannot be obtained any other way. Therefore, the search for the mechanisms that control the distribution of vegetation must start with correlations, followed by experimental and observational analyses of the mechanistic framework of the correlations (see chapter 24). Key correlations, used below, are between climate and vegetation physiognomy, which describes the temperature and moisture adaptations of leaves, and between climate and vegetation structure, which describes structural differences like those between forest and shrubland.

The second approach uses the present understanding of plant physiology to create mechanistic models of plant and vegetation behavior. Because of limitations on other methods, notably the uncertain predictive ability of climate-vegetation correlations and the technical difficulty of performing large-scale experiments on vegetation to determine response to climate variables, mechanistic models must be the key for predictions involving future and unknown environments (see chapter 4).

## II. PRESENT-DAY CORRELATIONS BETWEEN CLIMATE AND DISTRIBUTION

We have seen above that present-day correlations must be the starting point in our attempts to understand the mechanisms by which climate controls plant distribution and subsequently to predict the effects of climate change. The geographic boundaries of different vegetation types often correlate with specific climatic variables, such as temperature and precipitation, which have well-understood influences on physiology. Such correlations that have well-founded explanations at the physiological level are those between plant distribution and, respectively, growing-season length, absolute lowest annual temperature, and total precipitation. Section III presents predicted changes in global vegetation distribution based on models that contain these physiological variables. This section will now consider each variable in turn.

## A. Growing-Season Length and Temperature

Observations on the growth of a range of species from differing vegetation types in botanic gardens have indicated that plants from cold environments withstand shorter and cooler growing seasons than those from warmer environments (Lieth 1974). A conve-

nient measure of the suitability of a growing season for different plant species is the day-degree total. This is calculated by multiplying the number of days for which mean temperature exceeds an arbitrary standard of 0°C by the mean temperature over this period (chapter 21 uses a similar approach to calculating "growing degree days" but uses an arbitrary standard of 4°C instead of 0°C). For example, if there are 100 days when the temperature exceeds 0°C, and the mean temperature is 10°C, then the day-degree total is 1000. Similarly, 50 days over 0°C with a mean temperature of 20°C also gives a day-degree total of 1000. Plants growing under these two different 1000 day-degree regimes should show equal growth during the season. Because plant metabolism increases with temperature, a plant under the 20°C regimen will grow about twice as fast during each day as a plant growing at only 10°C (Woodward 1987a).

The minimum day-degree totals necessary for species of differing vegetation types to complete their vegetative and reproductive life cycles are shown in figure 8.1 (simplified from Woodward 1987a, 1988). Species in tundra vegetation have the lowest requirements, while, of the types shown, species of the broadleaved, deciduous forests have the highest. These estimates have been established by experiments and field observation. Data from A. Love and P. Sarkar (1957), for example, show that the tundra annual *Koenigia islandica* (Iceland purslane) requires a total of only 700 day-degrees to develop from a germinating seed to a mature plant producing seeds of its own. In contrast, the deciduous tree *Tilia cordata* (small-leaved lime) requires a minimum of 2000 day-degrees to complete reproductive development (Pigott 1981). Unfortunately, such detailed experimental information is limited for other vegetation types. In the absence of such experimental data, one can estimate day-degree totals by comparing present distribution limits with the day-degree totals that occur within and outside the range of the vegetation or species. For example, such analysis of tropical rain

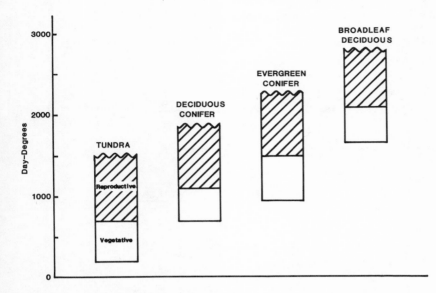

Figure 8.1. Generalized day-degree limits for vegetative development only or vegetative plus reproductive development in different vegetation types.

forest distribution suggests that a minimum total of 5000 day-degrees is required to complete development. That total, however, is probably an overestimation because the threshold temperature for the operation of tropical plant processes may be as high as 10°C (Woodward 1987a). Thus only days on which the temperature exceeds 10°C (as opposed to 0°C) should be counted when estimating the day-degree total.

## B. Minimum Temperature

Not only do plants need sufficient warm days during the year for growth and reproduction, but plant distributions are also restricted by absolute low temperatures at the poleward or upper altitudinal boundaries of the species range. In analyzing the physiological basis for such range restrictions, Woodward (1987a) and Woodward and Williams (1987) surveyed a large number of data on the ability of plant species to withstand low temperatures. They found that species growing together in any particular climatic region generally had similar abilities to withstand low temperatures. Thus tropical vegetation, composed of

what are termed chilling-sensitive species, is killed by temperatures in the range of 0°C to 10°C. Frost-resistant species, on the other hand, are divided into three types, namely broadleaved evergreen (resistant to −15°C), broadleaved deciduous (resistant to −40°C), and boreal needleleaved (resistant to all temperatures). These data were, in the main, obtained for species of trees (Woodward 1987a).

## C. Precipitation

Experiments and correlations between climate and distributions clearly show that leaf and plant biomass increase with precipitation (Grier and Running 1977, Merrill 1945, Schulze 1982). Averaging these observations on a global scale indicates that plant stature and biomass are greater where there is more precipitation, so that as precipitation increases from less than 50 mm/year, vegetation changes from desert to sparse shrub or herbaceous vegetation to parkland with scattered trees and finally to full forest (Schulze 1982). Figure 8.2 shows that vegetation with trees occurs in areas where the annual precipitation is greater than 300 mm.

As precipitation changes geographically, the characteristic leaf pattern of the vegetation changes because shape, size, and abundance of leaves are closely related to water

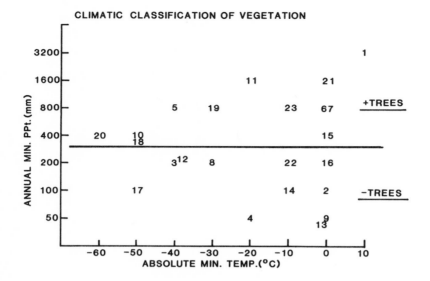

Figure 8.2. Distinction between vegetation types on the basis of tree presence. Vegetation types: 1, broadleaf evergreen forest; 2, broadleaf evergreen shrubs (warm); 3, broadleaf evergreen dwarf shrubs (cold); 4, desert; 5, broadleaf deciduous forest (cold); 6, broadleaf deciduous and evergreen forest (dry); 7, deciduous savanna (tree); 8, broadleaf deciduous shrubs; 9, broadleaf deciduous dwarf shrubs; 10, needleleaf evergreen forest (cold); 11, needleleaf evergreen forest (warm); 12, grassland; 13, grassland patches; 14, deciduous savanna (shrub); 15, evergreen savanna (tree); 16, evergreen savanna (shrub); 17, tundra; 18, mixed broadleaf deciduous and needleleaf evergreen forest (cold); 19, mixed broadleaf deciduous and needleleaf evergreen forest (warm); 20, needleleaf deciduous forest; 21, mixed broadleaf deciduous and evergreen forest; 22, mixed broadleaf deciduous and evergreen shrubs; 23, mixed broadleaf deciduous and evergreen and needleleaf evergreen forest.

balance. The leaf characteristics of the vegetation help determine how much precipitation is intercepted by leaves, how much falls through to the soil (throughfall), how much evaporates, and how much is transpired by the plants. For example, broad leaves catch more water, providing a drying surface that increases evaporation.

Loss of the plant's internal water by evaporation through leaf pores (stomata) is called transpiration, the rate of which is affected by many variables, including the size of the stomatal openings and the dryness and temperature of the leaf and air. These physical and physiological variables determine the ease with which water vapor can pass from the inside of the plant to the air. If these variables are plugged into an equation, called the Penman-Monteith equation, the result is an estimate of total transpiration by the plant (Monteith 1965; fig. 8.3). In turn, this calculated estimate of total transpiration can be used in the leaf area index (LAI) model that predicts the vegetation distribution changes in the next section. Thus, experimentally derived knowledge about the variables determining transpiration plays an essential part in modeling future changes in vegetation distribution.

Another important vegetation parameter used to model future vegetation patterns in the next section is the LAI, which is the total amount of leaf area per unit of ground. LAI units are measured in leaf layers, so an LAI of 5 means that there are five layers of leaf per unit of ground, and an LAI of 1 means that there is only one layer of leaf. Leaf area has a strong effect on both the transpiration rates just discussed and on precipitation through-

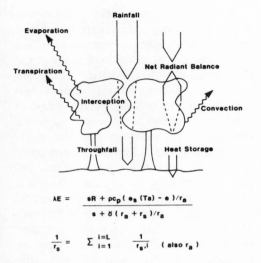

$$\lambda E = \frac{sR + \rho c_p ( e_s (Ta) - e )/r_a}{s + \delta ( r_a + r_s )/r_a}$$

$$\frac{1}{r_s} = \sum_{i=1}^{i=L} \frac{1}{r_{s,i}} \quad ( \text{also } r_a )$$

*Figure 8.3.* Schematic of the components of the hydrological and energy balance of North American forest canopy. The equation is the Penman-Monteith equation for estimating evapotranspiration, where $\lambda E$ is the energy transferred by evapotranspiration, s the slope of the curve relating saturation water vapor pressure to temperature, R net radiation, $\rho$ air density, $c_p$ specific heat of air, $e_s$(Ta) saturation vapor pressure deficit of the air at temperature Ta, e vapor pressure of the air, $\gamma$ psychrometric constant, $r_a$ canopy boundary layer resistance to water vapor, $r_s$ canopy stomatal resistance; $1/r_s$ is the inverse of the stomatal resistance for the whole canopy integrated for all the layers of leaves, i, in the canopy to a total leaf area of L (leaf area index). The canopy boundary layer resistance is estimated in the same manner. From Woodward 1987a, with permission.

fall (fig. 8.4). Transpiration is affected because the more leaf surface there is, the more water can be transpired. Therefore leafy vegetation (high LAI vegetation) has much higher transpiration rates than vegetation with a lower LAI. Not surprisingly, as the amount of leaf area increases, more rain is intercepted and the throughfall decreases. The relationship between LAI and throughfall (modeled in Woodward 1987a) makes annual precipitation one important factor in determining how leafy vegetation can be in a particular region. If precipitation in the region is low, then very leafy vegetation would catch, evaporate, and transpire too much of the rain, preventing it from reaching the soil, with the result that the leafy vegetation would either die or reduce its leafiness. In summary, high LAI adds to water stress on plants in two ways, by increasing transpiration and decreasing the amount of water reaching the soil.

Not only are transpiration and throughfall determined in part by leaf area index, but we can also turn the relationship around and use the two variables to predict the LAI. For example, in figure 8.4, the point where the two curves for throughfall and transpiration intersect represents the LAI for vegetation having these throughfall and transpiration curves. At this intersection point, transpiration equals the rate of water supply by throughfall. If we were to move to another geographic location, where transpiration remained constant but precipitation and therefore throughfall decreased, the throughfall curve would drop

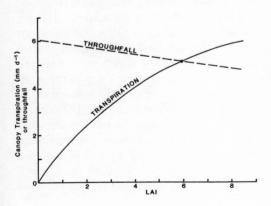

*Figure 8.4.* Predicted throughfall and transpiration, in millimeters per day, for canopies differing in leaf area index (LAI) but in otherwise identical climatic conditions. Predicted LAI for this case is the point of intersection of the two curves (LAI 6). Throughfall tends to decline with increasing LAI because the canopy intercepts more precipitation. The transpiration rate is predicted to rise with LAI because an increased number of leaves are transpiring in parallel. This increase flattens off at high LAI because of leaf self-shading and a consequent reduction in transpiration.

below its present position and there would be a net loss of water if the original LAI were maintained. The new intersection point of the lower throughfall curve and the unchanged transpiration curve would give a new expected LAI. We would expect the vegetation in the new region to have a lower LAI (less leaf cover) to compensate for the decreased amount of rain reaching the soil. This is a simplified way of thinking about how the model in section III predicts vegetation distributions.

In conclusion, it is clear that a variety of environmental variables interact with the physiology of plant species to determine vegetation ranges, structure, and physiognomy. By linking these environmental-physiological interactions with the general circulation models (GCMs), we can predict changes in broad vegetation types. In the next section, I will use this model to predict distributions in forests (LAI > 5; Woodward 1987a), continuous shrubby vegetation in which trees are infrequent (LAI > 1 and < 5), and sparse vegetation (LAI < 1).

## III. PREDICTING CHANGES IN VEGETATION DISTRIBUTION

This section presents three different projections of vegetation response that would be expected if climate changed as projected for a doubling of preindustrial $CO_2$ concentration. In each case, the results are based on climate projections from the Goddard Space Center (GISS) general circulation model (Hansen et al. 1984). This GISS model is similar to others in its prediction of changes in temperature and precipitation (see chapter 4). The GISS climate variables, notably projected changes in average seasonal temperature and precipitation, were used to drive physiologically based models of geographic distribution. These models used the observed correlations between climate variables and the key physiological variables of absolute minimum temperature, growing season day-degree total, and leaf area index (LAI) to map the resultant changes in geographical distribution.

The first model projects changes in vegetation physiognomy. Calculations were based on the global temperature changes predicted by the GISS model but did not factor in the direct metabolic effect of enhanced $CO_2$ concentration, which can decrease a plant's need for moisture. It is unlikely that this omission affected the validity of the results, since the model does not incorporate vegetation hydrology, being dependent only on minimum temperature and day-degree information. The other two projections predict changes in vegetation structure, and both use a model based on plant hydrology, using the Penman-Monteith and LAI equations described above. The two runs of the model were the same except for one important point—the first run did not factor in the direct metabolic $CO_2$ effect, while the second did. This means that plants under the second moisture regimen experienced less moisture loss due to transpiration.

## A. Predicting Changes in Vegetation Physiognomy

The term *vegetation physiognomy* is used to describe a suite of morphological and physiological characteristics that have evolved in response to climate variables, notably temperature and rainfall. The characteristics that I have considered key in defining the physiognomic type of any particular type of vegetation are leaf longevity (whether evergreen or deciduous), leaf dimension (whether broadleaf or needleleaf), and temperature sensitivity (chilling sensitive, chilling resistant, or frost resistant). Thus, physiognomic types include broadleaved evergreen species that are sensitive to chilling, broadleaved evergreen species that are frost resistant, and species that are broadleaved deciduous (table 8.1). The poleward geographic limits of these different types are substantially influenced by annual minimum temperature (Woodward 1987a). By modeling geographic distribution as a function of minimum temperature thresholds, augmented by information on day-degree thresholds for tundra and broadleaved deciduous vegetation, I was able to

*Table* 8.1. Cardinal annual minimum temperatures and day-degree totals for expected dominant physiognomies.

|  | Physiognomy |
| --- | --- |
| **Minimum temperature** | |
| > 10°C | Broadleaved, evergreen, chilling sensitive |
| 0° to 10°C | Broadleaved, evergreen, chilling resistant |
| −15° to 0°C | Broadleaved, evergreen, frost resistant |
| −40° to −15°C | Broadleaved, deciduous |
| < −40°C | Broadleaved, evergreen and deciduous, needle leaved |
| | |
| **Day-degree** | |
| < 1600 | Tundra |
| > 2000 | Broadleaved deciduous (when minimum temperature is above −40°C and below −15°C) |

project the changes in distribution that could be expected if the climate warmed as suggested by GISS (figs. 8.5 and 8.6).

In general terms the global warming should lead to a decrease in the areal extent of tundra, boreal needleleaved forest, and broadleaved evergreen and frost-resistant vegetation. The areal extent of broadleaved deciduous vegetation and of evergreen vegetation, both chilling resistant and chilling sensitive, should all increase. In some cases the expected changes are large; for example, the tundra may retreat by up to 4° of latitude, while the chilling-resistant vegetation may expand by as much as 10° of latitude. Of course these predictions are only as accurate as the GCM that drives them, and no predictions of rate have yet been done.

One feature of the GISS predictions is that winter temperatures are expected to increase more than those of summer. This may in part reflect the particular nature of the relationship between the mean and absolute minimum temperatures. In biological terms, because minimum winter temperatures presently determine the poleward limit of vegetation distributions, a relatively large increase in winter temperatures will enhance the ability of vegetation to spread in a poleward direction. Eventually, as winter minimum temperatures rise enough so that they do not present a major impediment to northward expansion, the summer temperatures, which

will not increase greatly compared with the winter warming, in turn may become the bottleneck in preventing northward expansion. The relatively low summer temperatures may provide too few day-degrees during the growing season for species to complete their life cycles, even if it becomes warm enough in the winter for survival (remember that the number of day-degrees available for growth increases with temperature).

One way to test such distribution projections is to see whether the models are able to predict actual present-day distributions of physiognomic types based on present-day climate. Woodward (1987a) and Woodward and Williams (1987) took climate data from a global network of 696 meteorological stations and used the same day-degree and minimum temperature thresholds (termed cardinal temperatures in table 8.1) described above to predict present-day distributions. Their modeled distributions compared well with published maps of global physiognomy (Polunin 1960), even though the models are based only on temperatures and do not incorporate any features of the hydrological budget.

Some limitations and assumptions of the model should be kept in mind. Although good data are available about the way in which minimum temperatures and day-degree totals set the poleward limits of the

PHYSIOGNOMY NOW

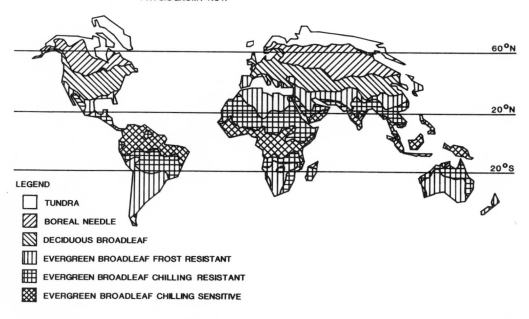

Figure 8.5. Global distribution of present-day physiognomic types of vegetation.

PHYSIOGNOMY CHANGE

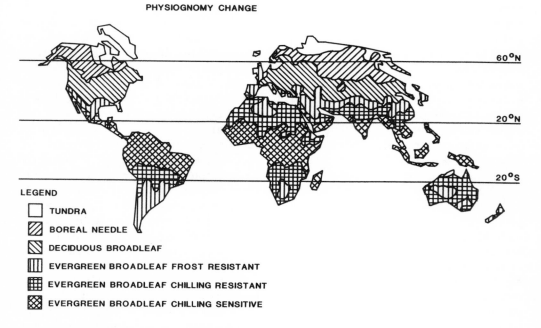

Figure 8.6. Predicted global patterns of physiognomy following $CO_2$-induced climatic change.

different physiognomic vegetation types, the mechanism that controls the equatorial limit of a species or vegetation type is poorly understood. My model assumes that the distribution of a cool-climate vegetation will be restricted in the equatorial direction by competition with a warmer-climate vegetation, as exemplified by the boreal conifer *Pinus sylvestris* in Europe, whose southern border is determined by competition with deciduous broadleaf species (Woodward 1988). In effect, once the minimum temperatures are sufficiently high to allow survival of the warm-climate vegetation, then it will occur and outcompete the cool-climate vegetation. These competitive limits to distribution are poorly understood and require much more study. In spite of that uncertainty, the model has strong predictive capability.

Another present limitation of this modeling method is the possibility that the absolute minimum temperatures used to determine the poleward limits of the vegetation types are unrealistically low, meaning that shifts in physiognomic zones may be underestimated. Since the GISS model predicts mean temperatures and not the absolute minimum temperatures, I calculated the absolute minimum temperature within a region by adding the average degree rise predicted by the GCM

to the present absolute minimum temperature within the region. (Growing-season temperatures used to calculate future day-degree totals were similarly derived by adding the GCM predictions to the present growing-season temperatures.) That approach may be conservative because changes in minimum temperatures will be determined not only by the mean temperature increase but also by the statistical variance of temperature (Wigley 1985), with the result that the absolute minimum temperature in a region may rise more than the mean temperature. Thus, a 1°C increase in mean temperature is likely to cause more than a 1°C increase in the absolute minimum temperature, so an absolute minimum calculated by adding only 1°C to the present minimum would underestimate the true value. As an example using observed data, figure 8.7 shows two different years of temperature records on Niwot Ridge, Colorado (Losleben 1983). The mean annual temperature in 1970 was 1.7°C warmer than in 1971, but the absolute minimum temperature was 7°C greater. The ratio of the mean to the standard deviation was about the same for the two years. These observations and Wigley's analysis (1985) suggest that indeed the absolute minimum temperatures within regions may increase much more than mean temperatures. Unfortunately, we cannot tell how large these increases will be until warming

Figure 8.7. Cumulative frequency diagram of daily temperature at Niwot Ridge, Colorado, in 1970 and 1971. From Losleben 1983.

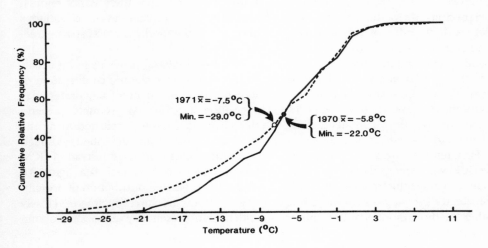

actually occurs, because estimating the magnitude of rise will depend on actual measurements of the variance of temperature.

## B. Predicting Changes in Vegetation Structure without Factoring in Doubled $CO_2$ Concentration

Vegetation structure is a term that describes the vertical construction of the vegetation, that is, whether the vegetation consists of trees or other types of plants. For the purposes of this modeling study, three structure types were defined: continuous forest vegetation, continuous vegetation in which trees are absent or infrequent, and sparse vegetation with trees absent or infrequent. Typical examples of continuous vegetation with trees absent or infrequent are African savannas, shrublands, and tall-grass prairies. Sparse vegetation is defined as having bare spots between plants, and examples include desert, tundra, and creosote chaparral.

The expected effects of climate change on vegetation structure, based on the GISS model predictions, are shown in figures 8.8 and 8.9. In America, Europe, Asia, and southern Africa the area of sparse vegetation is expected to increase, generally at the expense of the shrubby vegetation. In some cases, such as in Asia, the sparse vegetation may extend by as much as 10° of longitude. In eastern Africa, Saudi Arabia, and Australia the continuous shrubby vegetation is expected to spread into the areas of sparse vegetation, by as much as 10° in some cases. This expansion is caused by increased precipitation for these areas, which will offset the drying effect of increased temperature. Forested areas of Europe and Asia will expand by a maximum of about 4°.

Unlike the physiognomy distributions, which were derived from absolute temperature minimums, these distribution changes in structure are derived from a model that assesses changes in the hydrological budget as follows: First, temperature and precipitation from the GCM were averaged over the growing season. In geographic regions with seasonal climate, the growing season was taken to be either the three months of December,

January, and February, or the months of June, July, and August, whichever is warmer in the particular hemisphere. In tropical regions where there are no clear-cut seasons determined by temperature, the growing season was taken to be the period when precipitation is greatest. Second, the temperature data from the GCM were plugged into the Penman-Monteith equation to estimate transpiration rates for two hypothetical canopies of different leafiness. The first hypothetical canopy was defined as having an LAI of 1, meaning that there is only sparse leaf cover like that associated with shrubs. The other canopy had an LAI of 5, indicating the very leafy cover of a forest. Third, the estimated transpiration values for the two canopies and the GCM precipitation values were used in the LAI equation to predict the vegetation structure that would occur given the projected transpiration and precipitation throughfall values. Would there be forest (LAI 5), shrubland (LAI 1), or sparse vegetation (LAI less than 1)? At LAI 5, if precipitation throughfall is greater than water loss to transpiration, then adequate moisture is present to maintain a leafy canopy, and forest is predicted (Woodward 1987a). If precipitation throughfall is less than transpiration loss, there is inadequate moisture to maintain a leafy forest canopy; but at LAI 1, throughfall is greater than transpiration loss, and the maintainable LAI is somewhere between 1 and 5, so continuous shrubby vegetation is predicted. If throughfall is less than transpiration at LAI 1, there is not enough moisture to maintain even continuous shrubs, and the model predicts sparse, treeless vegetation.

Woodward (1987a) provides more details of this approach, including confirmation of its general accuracy in predicting present-day vegetation structure. For example, present-day distributions were predicted using the modeling method just described, although real climate data were used instead of GCM values. Based on these real data, figure 8.8 shows the predicted distribution of present-day vegetation structure, which agrees closely with maps of present-day vegetation (Polunin

VEGETATION NOW

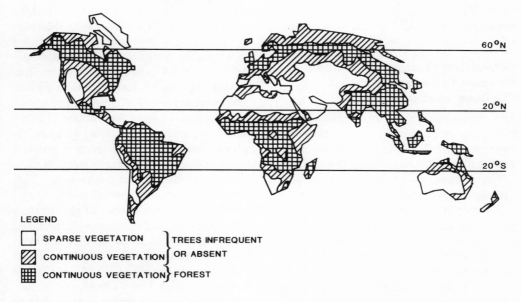

LEGEND

☐ SPARSE VEGETATION ⎫ TREES INFREQUENT
▨ CONTINUOUS VEGETATION ⎬ OR ABSENT
▦ CONTINUOUS VEGETATION ⎭ FOREST

*Figure 8.8.* Global distribution of present-day vegetation structure.

VEGETATION CHANGE ($-CO_2$)

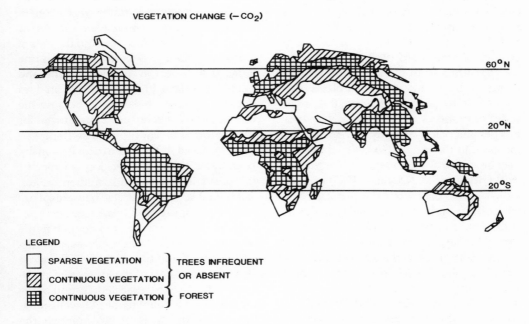

LEGEND

☐ SPARSE VEGETATION ⎫ TREES INFREQUENT
▨ CONTINUOUS VEGETATION ⎬ OR ABSENT
▦ CONTINUOUS VEGETATION ⎭ FOREST

*Figure 8.9.* Predicted global patterns of vegetation
structure following climate change (no effect of
$CO_2$ on stomatal resistance to water loss).

1960, Woodward 1987a, Woodward and Williams 1987; the figure also includes the southern limits of the tundra on the basis of table 8.1).

## C. Predicting Changes in Vegetation Structure, Factoring in Doubled $CO_2$ Concentration

The climate predictions used in this chapter to forecast vegetation change are based on a world in which $CO_2$ concentration would be double the preindustrial value. This elevated $CO_2$ will have two effects on vegetation, namely, the climate effects already described and direct effects on plant and vegetation physiology. In many small-scale experiments it has been established that increased $CO_2$ concentration decreases plant water loss by allowing the plant's stomata to take in adequate $CO_2$ for growth without opening as widely, so less water vapor escapes through the narrower stomatal openings (reviewed in Morison 1987). Additional indirect evidence that increased $CO_2$ concentrations make it easier for plants to absorb adequate $CO_2$ is that the number of stomata per unit of leaf area has actually decreased in British trees since the onset of the industrial revolution and the concomitant increase in $CO_2$ concentration (Woodward 1987b).

Generalizing from the small-scale experiments on leaves and individual plants, it is reasonable to propose that on the larger scales of canopies and regions, water-vapor loss per unit of leaf area may decrease as $CO_2$ increases in the future. However, this effect may be muted at the canopy and vegetation level (Jarvis and McNaughton 1986) because local climatic conditions, in particular the temperature and relative humidity of the air and the radiation balance, will tend to exert a stronger influence on transpiration than $CO_2$. Nonetheless, some of the antitranspirant effect of $CO_2$ should occur even at the regional scale (Jarvis and McNaughton 1986), so it is of interest to alter the climate model to take into account the lowered transpiration values obtained in experiments on individual plants grown in high $CO_2$ chambers.

Experiments have shown that the stomatal resistance of plants increases 1.67-fold with a doubling of the $CO_2$ concentration (Morison 1987), and this increased value was used to recalculate transpiration values using the Penman-Monteith equation. The recalculated transpiration values were then used in new estimates of LAI. The new distributions of vegetation structure are mapped in figure 8.10. Compare this map to that in figure 8.9, which reflects the same temperature and moisture regime but present levels of $CO_2$. The lower transpiration values under high $CO_2$ mean that a higher LAI can be maintained for the same amount of precipitation throughfall, so under warming conditions, most regions should be able to maintain more leaf cover with raised $CO_2$ than without it. This trend toward greater LAI is reflected in the geographical ranges in figure 8.10. Forests are expected to increase noticeably in America and the Soviet Union, with smaller increases in Africa and Central Asia (although it is likely that human interference may prevent some of this expansion). The increase in forest would lead to a decrease in the total area of continuous shrubby vegetation in America, Europe, Central Asia, and Africa. Under this doubled-$CO_2$ scenario the area of sparse vegetation will have about the same limits as at present in America, Europe, and Central Asia (fig. 8.8). Sparse vegetation does not spread under this scenario because the doubled $CO_2$ concentration compensates for the dryness caused by the increased temperatures modeled in both scenarios. Regionally, continuous shrubby vegetation is expected to spread in eastern and southern Africa, Saudi Arabia, and Australia because of increased precipitation in these areas.

There is an important shortcoming in the doubled-$CO_2$ model that results in an underestimation of moisture stress and therefore an underestimation of the amount of distribution change to be expected. Instead of being based on a doubling of $CO_2$ concentration alone, the GCM actually models the warming expected if other greenhouse gases, together with $CO_2$, collectively add up to as

VEGETATION CHANGE (+CO₂)

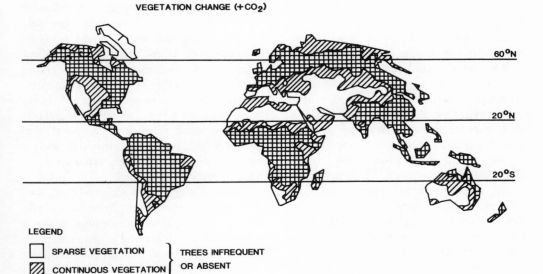

LEGEND

☐ SPARSE VEGETATION    } TREES INFREQUENT
▨ CONTINUOUS VEGETATION } OR ABSENT
▦ CONTINUOUS VEGETATION } FOREST

*Figure* 8.10. Predicted global patterns of vegetation structure following climate change, adjusted for doubled $CO_2$ concentration and stomatal resistance increased by a factor of 1.67.

much warming as if $CO_2$ alone were doubled (a doubled-$CO_2$ equivalent), with no increase in the other greenhouse gases. In reality, it is expected that at present rates of greenhouse-gas increase, $CO_2$ will make only about one-half the contribution to warming, and that other greenhouse gases collectively will contribute the other half. This means that at the time warming reaches the level projected by the GISS model for a doubled-$CO_2$ equivalent, the actual $CO_2$ level will be substantially less than double—only about 1.5 times present levels. My distribution model, which assumed an actual doubling of $CO_2$, overestimated the amount to which the direct effect of $CO_2$ would decrease transpiration and therefore water stress. The model was not corrected, however, because most physiological studies on transpiration variables were done at actual double-$CO_2$ concentrations, before the importance of the other greenhouse gases was realized. Therefore, there is a need for further laboratory experiments to determine how plant physiology responds to

more realistic $CO_2$ concentrations so the results can be used in distribution-projection models.

## IV. PREDICTING THE RATE OF MODERN-DAY VEGETATION CHANGE

The preceding predictions of climate-induced changes in the distribution of vegetation indicate the maximum extent of possible change. The generated maps represent an equilibrium state in the future when vegetation will have completed its geographic movement, dispersing to occupy any area in which the climate is suitable. In the real world, such movements take decades to centuries.

Can we predict possible rates of change in vegetation distribution? We know that the rates will be dependent on two factors: the rate of climatic change and the rate of vegetation response. Given enough computer time, it may in the future be possible to run the extremely complex models that could go beyond present equilibrium predictions to forecast the rate of warming during the transition phase. However, development of a similar model for predicting the rate at which vegetation will change is unlikely because the responses are so complex.

A wide variety of biological processes can result in either changes in species composition within a general vegetation type or in movement of the vegetation type across the landscape by expansion or retraction of its boundaries. Complex processes involved in such changes include migration by rapid seed dispersal or slow vegetative spread, the capacity of species to produce propagules, the ability to regenerate following disturbance, and the capacity to modify resource capture in the presence of other species. These life-cycle properties will be influenced by the rate at which vegetation is disturbed by a (perhaps) changing frequency of extreme events (Woodward 1987a), such as the creation of forest clearings by storms, fires, or the various activities of man. No model contains all of these features, and so it is not yet possible to predict rates on the basis of these mechanisms.

An alternative and more productive approach is to search through historical records from the last 200 years for evidence of changes in range or species composition of vegetation. Such changes integrate all the un-

*Figure 8.11.* Changes in the dominance of conifers in a mixed broadleaf deciduous and needleleaf evergreen forest (eastern United States), with changes in temperature. ●: Historical observations of stand composition. From Hamburg and Cogbill 1988.

derlying life-cycle properties mentioned above, obviating the necessity of precisely understanding each one, as would be necessary for a modeling effort. Records documenting changes during the last 100 years are particularly valuable because they can demonstrate vegetation responses not only to climatic variability but also to the increasing $CO_2$ concentrations since the industrial revolution, which may have increased rates of plant growth by stimulating photosynthesis and decreasing water need.

S. P. Hamburg and C. V. Cogbill (1988) have presented compelling evidence for the rate of change of species competition within a vegetation type. Historical evidence demonstrates, in the mixed boreal conifer and deciduous broadleaved forests of the eastern United States, that over the past 180 years canopy dominance by conifers (principally red spruce, *Picea rubens*) has been gradually declining as the growing season has been warming (fig. 8.11). Hamburg and Cogbill (1988) suggest that the replacement of spruce by deciduous species such as American beech (*Fagus grandifolia*) is controlled by climate through, presumably, an increased competitiveness of beech relative to spruce. However, as with much of the historical information collected over the last 200 years, the evidence is equivocal; the change could be ascribed to modifications in regeneration brought about entirely by the influence of

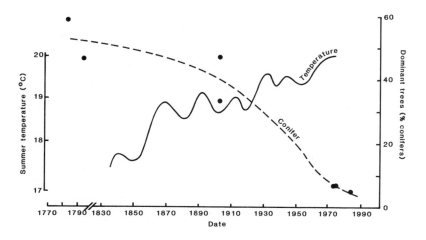

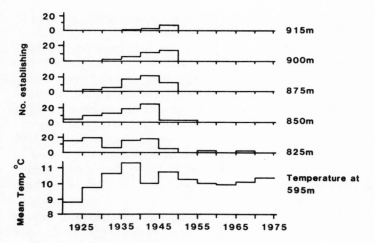

*Figure* 8.12. Changes in establishment of *Betula pubescens* at different altitudes, with changes in temperature. From Woodward 1987a, with permission.

man, although the study site does appear to be quite isolated. Taken at face value, these data indicate that an increase in the temperature of the growing season causes a noticeable change in vegetation over 100 years and an acute change (the virtual extinction of spruce) in 180 years. Because replacement of a diminishing species, like the spruce, occurs only as individual mature trees die and are replaced in the canopy by different species, changeover from one vegetation type to another will be affected by the longevity of the declining species. Given that the life span of many tree species is measured in hundreds of years, changeover from one forest type to another will always be on the order of centuries unless mature individuals are rapidly removed by extreme events such as fire or storms.

Evidence for the rate of spread of one vegetation type into another is provided by the elegant studies of L. Kullman (1979, 1983) on the spread of the birch tree (*Betula pubescens*) into tundra vegetation during a period of climatic warming between about 1920 and 1950. Kullman measured the age of existing birch saplings and trees at a range of altitudes in central Sweden and found that trees be-

came younger and younger as he sampled at increasingly higher altitudes (fig. 8.12). He concluded that upward progressive establishment of birch occurred, taking it beyond its earlier altitudinal limit (850 meters in 1920) into tundra vegetation. In the ten years from 1920 to 1930 alone, the upper limit of birch distribution rose from 850 to 900 meters, as the local temperature rose approximately 1°C. Unlike the case for spruce, this suggests that encroachment on tundra vegetation and the resulting changes in tundra distribution might be expected to begin almost immediately upon the onset of the warming predicted by GCMs, and that changes should be measurable in as little as five to ten years. These data for birch indicate that competitive exclusion of birch seedlings by the native tundra was probably rather limited, the taller tree species apparently has an advantage during periods of climatic amelioration. It appears likely, then, that the predicted reduction in the areal extent of tundra due to the extended range of the boreal forest will not be impeded by competition and will occur rapidly.

## V. VEGETATION COMPOSITION AND DIVERSITY

Perhaps the greatest human concern with the potential ecological effects of climatic change is that there may be reductions in species diversity, resulting in the extinction of species presently or potentially beneficial to humans. That climate change may cause such reductions is indicated by the global predictions of vegetation change, mapped in figures 8.6, 8.9, and 8.10, which suggest that in wide areas the present complements of species will die out, resulting in local and regional changes in diversity composition and abundance. To this may be added some historical evidence for changes in species diversity that may be explained by climatic warming (although other factors, notably atmospheric pollutants, may also be involved; Hamburg and Cogbill 1988). It is important that ecologists begin to estimate how vegetation changes may translate into losses of diversity. Unfortunately, the task is difficult because our understanding of the controls of diversity are limited and virtually restricted to temperate vegetation (Diamond and Case 1986).

We can begin attacking this problem by looking for correlations of diversity and climate, which turn out to be strong. Figure 8.13 shows the correlation between absolute minimum temperature in North and Central America and vegetation diversity, measured as the total number of families of flowering plants (estimated from the maps in Heywood 1979 and normalized to a maximum of 100). The average family diversity decreases by about 8% for every 10°C decrease in the minimum temperature. It may be shown that a similar response is also seen in Africa, although the minimum temperature range is much narrower (Muller 1982). This decrease in diversity with falling temperature has been interpreted as a measure of the increasingly limited capacities of families to endure lower and lower temperatures (Woodward 1987a). The positive aspect of the relationship is that a global warming should lead to at least temporarily increased diversity in some local areas where new species move into the prior community, as shown by the spread of birch into tundra. However, diversity will remain high, in this case, only as long as the original tundra species remain, and if they are outcompeted by the invading birch, species diversity would again drop. It appears that in most such cases, local extinctions will occur among the invaded community.

Figure 8.13. Correlation between flowering plant (angiosperm and gymnosperm) family diversity (normalized to a maximum of 100) and absolute minimum temperature (for the whole period of climatic recording at each meteorological station) for North and Central America. Each open square represents a meteorological station and an estimate of diversity at that station. The dashed line (---) shows the same relationship estimated for Africa but, for clarity, excluding the individual points.

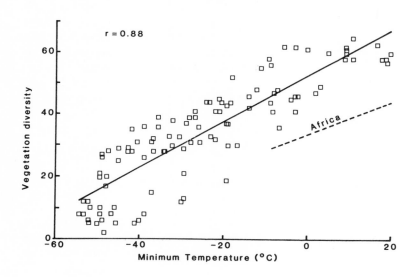

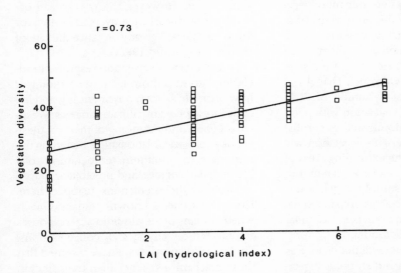

*Figure* 8.14. Correlation between family diversity and leaf area index (LAI) for Africa.

There is also a positive correlation between leaf area index and vegetation diversity (families of flowering plants) for Africa (fig. 8.14). As LAI increases along with precipitation, so does diversity. A simple explanation would be that the correlation measures the increasing inability of families to survive in areas of increasing drought. This correlation and the correlation above between decreasing critical minimum temperature and falling diversity lead to the conclusion that in general climatic adversity (assessed by comparison with an arbitrary equable temperature and precipitation) acts like a sieve, excluding the intolerant.

**VI. SUMMARY**

Several types of knowledge, including experiments on plant physiology and observed correlations between climate and present distribution patterns of vegetation, allow us to build models that predict how global warming would change vegetation patterns on a global scale. We have confidence in the accuracy of these vegetation models because, when fed with data on present-day climate, they have successfully predicted present-day vegetation patterns (Woodward 1987a). Nonetheless, even if the vegetation models are highly accurate, their predictions are ultimately based on the input of climate data from the GCMs, and these models are relatively simple and their accuracy uncertain (see chapter 4). Further refinement of climate models will increase our confidence in projections of vegetation distribution.

General conclusions of the vegetation models presented in this chapter include a northward retreat of the southern limits of tundra, by as much as 4°, from the encroaching boreal forest. Long-term observations of the invasion of birch into tundra suggest that the beginning of this change could be observable within 10 years of the onset of warming, depending on the magnitude and rate of temperature change.

Predictions by the GCM of relatively large increases in absolute minimum winter temperatures, which are the primary determinant of how vegetation physiognomy is geographically distributed, suggest that physiognomic classes will show large changes in extent. Evergreen chilling-sensitive vegetation, for example, is expected to increase its range. The area of evergreen frost-resistant vegetation will decrease, presumably through competition with cold-sensitive species moving toward the poles, and deciduous vegetation will stay relatively the same.

The relatively small changes in summer temperature, a primary determinant of vegetation structure, mean the present distribution of vegetation structural types is not expected to change as much as the distribution of physiognomic types. However, trends are still apparent, primarily through changes in precipitation. North of a latitude of 20°N the area of continuous vegetation with occasional trees (probably dominated by shrubs) is expected to decrease. This vegetation will be encroached on by the expanding areas of sparse vegetation and, to a lesser extent, forest. South of 20°N the shrubby vegetation is expected to expand into areas of sparse vegetation. Such a change could occur at a similar rate to the change in the tundra.

Areas suitable for forest will not change as markedly as the sparse and shrubby vegetation. Some areas, such as Africa, Australia, Europe, and the Soviet Union, should see an increase in the area of forest. The extent of this change could be enhanced if forests transpire less water (i.e., became more water efficient) because of partial stomatal closure in response to increased $CO_2$ levels (Morison 1987, Woodward 1987b). To date insufficient experimental data are available to know how important this direct $CO_2$ effect will be in decreasing water loss on a large scale, but if most species do increase water conservation as $CO_2$ levels rise, the effect on vegetation changes could be profound, as shown by comparing figures 8.9 and 8.10. An increase in water efficiency could, for example, greatly enhance the spread of vegetation into drought areas.

The response time of vegetation to change can vary dramatically, depending on a number of ecological conditions. Changes in local species composition will be most rapid at the boundaries between vegetation types, where noticeable changes in response to warming of 1° to 2°C may occur within a decade. In the center of distribution of a particular vegetation type, as within the mixed boreal conifer and deciduous broadleaved forest represented in figure 8.11, the response time will be on the order of at least a century if the natural processes of mortality and gap creation prevail. However, the widespread occurrence of disturbance events like fires or forest cutting will greatly enhance the rate of change (Woodward 1987a).

The accuracy of vegetation response models depends on a thorough understanding of how plant physiology, physiognomy, and life cycles are affected by climate variables. We do have a good understanding of many of these ecophysiological relationships, including the responses to minimum temperature, total growing day-degrees, and precipitation used to develop the predictions mapped here. However, some unknowns remain. One is whether many of the physiological responses are affected directly by $CO_2$ concentrations. For modeling purposes, I have assumed that these mechanisms (other than transpiration response) are insensitive to the expected increases in $CO_2$, but it is not known if this is so. Therefore, studying the responses of vegetation processes to varying $CO_2$ levels should be a focus of future work. Another unknown, not included in the present models, is the effect of competition. Competition determines limits on vegetation types in the equatorial direction, but its mechanism is poorly understood. Even though for modeling purposes we obtain relatively accurate results by ignoring competition and focusing on the effects of minimum temperatures (Woodward 1987a), it would be good to understand all the factors determining range, including competition.

## REFERENCES

Bradley, R. S., H. F. Diaz, J. K. Eischeid, P. D. Jones, P. M. Kelly, and C. M. Goodess. 1987. Precipitation fluctuations over Northern Hemisphere land areas since the mid-nineteenth century. *Science* 237:171.

Diamond, J., and T. J. Case, eds. 1986. *Community Ecology.* New York: Harper and Row.

Grier, C. C., and S. W. Running. 1977. Leaf area of mature north-western coniferous forests: Relation to site water balance. *Ecology* 58:893.

Hamburg, S. P., and C. V. Cogbill. 1988. Historical decline of red spruce populations and climatic warming. *Nature* 331:428.

Hansen, J., A. Lacis, D. Rind, G. Russell, P. Stone, I. Fung, R. Ruedy, and J. Lerner. 1984. Climate sensitivity: Analysis of feedback mechanisms. In *Climate Processes and Climate Sensitivity*, J. E. Hansen and T. Takahashi, eds., pp. 130–163. Washington, D.C.: American Geophysical Union.

Heywood, V. H., ed. 1979. *Flowering Plants of the World*. Oxford: Oxford University Press.

Jarvis, P. G., and K. G. McNaughton. 1986. Stomatal control of transpiration: Scaling up from leaf to region. *Advances in Ecol. Res.* 15:1.

Jones, P. D., T.M.L. Wigley, C. K. Folland, D. E. Parker, J. K. Angell, S. Lebedeff, and J. E. Hansen. 1988. Evidence for global warming in the past decade. *Nature* 332:790.

Kullman, L. 1979. Change and stability in the altitude of the birch tree limit in the southern Swedish Scandes, 1915–1975. *Acta phytogeographica suecica 65*.

Kullman, L. 1983. Past and present tree lines of different species in the Handolan Valley, Central Sweden. In *Tree Line Ecology*, P. Morisset and S. Payette, eds., pp. 25–42. Quebec: Centre d'études nordiques de l'Université Laval.

Lieth, H., ed. 1974. *Phenology and Seasonality Modeling*. Ecological Studies 8. Berlin: Springer-Verlag.

Losleben, M. V. 1983. Climatological data from Niwot Ridge, east slope, Front Range, Colorado, 1970–1982. Long-Term Ecological Research Data Report 83/10. Boulder, Colo.: Institute of Arctic and Alpine Research, University of Colorado.

Love, A., and P. Sarkar. 1957. Heat tolerances of *Koenigia islandica*. *Botaniska notiser* 110:478.

Merrill, E. D. 1945. *Plant Life of the Pacific World*. New York: Macmillan.

Monteith, J. L. 1965. Evaporation and environ-ment. In *The State and Movement of Water in Living Organisms*, G. E. Fogg, ed., pp. 205–234. Symposium of the Society for Experimental Biology 19. Cambridge: Cambridge University Press.

Morison, J.I.L. 1987. Intercellular $CO_2$ concentration and stomatal response to $CO_2$. In *Stomatal Function*, E. Zeiger, G. D. Farquhar, and I. R. Cowan, eds., pp. 229–251. Stanford: Stanford University Press.

Muller, M. J. 1982. *Selected Climatic Data for a Global Set of Standard Stations for Vegetation Science*. The Hague: Junk.

Pigott, C. D. 1981. Nature of seed sterility and natural regeneration of *Tilia cordata* near its northern limit in Finland. *Ann. bot. fen.* 18:266.

Polunin, N. 1960. *Introduction to Plant Geography and Some Related Sciences*. London: Longmans.

Schulze, E. D. 1982. Plant life forms and their carbon, water, and nutrient relations. In *Encyclopedia of Plant Physiology*, O. L. Lange, P. S. Nobel, C. B. Osmond, and H. Ziegler, eds., vol. 12B, pp. 616–676. Berlin: Springer-Verlag.

Wigley, T.M.L. 1985. Impact of extreme events. *Nature* 316:106.

Woodward, F. I. 1987a. *Climate and Plant Distribution*. Cambridge: Cambridge University Press.

Woodward, F. I. 1987b. Stomatal numbers are sensitive to increases in $CO_2$ from preindustrial levels. *Nature* 327:617.

Woodward, F. I. 1988. Temperature and the distribution of plant species. In *Plants and Temperature*, S. P. Long and F. I. Woodward, eds. Symposium of the Society of Experimental Biology 42. Cambridge: Company of Biologists.

Woodward, F. I., and B. G. Williams. 1987. Climate and plant distribution at global and local scales. *Vegetatio* 69:189.

I am grateful for comments on the manuscript by J. M. Adams, A. D. Friend, A. C. Newton, R. Peters, and particularly E. J. Schwarz.

CHAPTER NINE.

# Effects of Climate Change On Soil Biotic Communities and Soil Processes

WALTER G. WHITFORD

## I. INTRODUCTION

It is difficult to evaluate how global climate change will affect soil organisms and processes because there is a shortage of comprehensive data—most of the relevant research has been done in subtropical arid environments (Whitford 1989). There are some notable exceptions, reviewed in this chapter, but as a consequence of these data limitations most of the examples I will use are from arid ecosystems.

Even so, enough is known about soils to draw some general conclusions. First, we know the soil is a thermally buffered environment that responds slowly to temperature changes in the air column above. Therefore, the biota living within the soil are less likely than surface vegetation to show large responses to changes in atmospheric temperature. Second, although climate does have direct effects on soil biota, soil biota are more strongly affected by vegetation than they are by atmospheric conditions directly. This is for two reasons: (1) the physical structure of the vegetation affects soil temperature and moisture, and (2) vegetation provides organic matter to the soil. The quality and quantity of organic matter (energy) and nutrients may affect species composition and species abundance of the soil communities more strongly than abiotic factors such as temperature and moisture (Wallwork et al. 1985, Kamill et al. 1985, Steinberger et al. 1984, Parker et al. 1984a).

Given the importance of vegetation in determining biota, it is apparent that forecasting the effects of global climate change on soils is dependent, in large part, on our ability to predict changes in vegetation and, in turn, to understand how vegetation will affect the soils. This three-way relationship among soil, vegetation, and climate must be understood in order to project soil biotic response, but understanding how these variables interact is extremely complicated because feedbacks make simple unidirectional predictions of change impossible. This chapter examines the nature of these interactions and presents some experimental evidence about how specific soil communities respond to changes in soil moisture, temperature, and nutrient availability.

Structural characteristics of the vegetative cover determine the degree to which a particular vegetation modifies the thermal and hydric characteristics of the soil that affect the soil biota. For example, the soil biotic communities under a closed-canopy forest or a solid sod grassland are subject to considerably less environmental variability than soil biotic communities in an open shrub desert or savannah with scattered trees and scattered clumps of perennial grasses. The moderating effect of an overstory canopy on the soil-litter biotic community is shown by a comparison of decomposition on clear-cut, cable-logged watersheds with forested watersheds in the southeastern United States (Whitford et al. 1981a). Decomposition rates are significantly lower on the clear-cut watershed than on the closed-canopy watershed. This reduction in decomposition rates is surprising because the actual evapotranspiration is the same on both watersheds. Decomposition rates in mesic forests have been shown to vary directly as a function of actual evapotranspiration (Meentemeyer 1978), but the data from this comparison demonstrate that the thermal and hydric differences of the litter and soil exposed to direct sunlight affect the soil biota responsible for the decomposition process.

Removal of a forest canopy resulted in a 25% reduction in densities of soil microarthropods in the leaf litter and upper 5 cm of soil. However, microarthropod densities increased by more than 100% in the deeper soil horizons when compared with uncut watersheds (Seastedt and Crossley 1981). The quantitative changes in microarthropods were matched by qualitative changes. On the clear-cut, cable-logged watershed, the micro-arthropod community was dominated by prostigmatid mites, a group of mites characteristic of desert soils (Seastedt and Crossley 1981, Whitford et al. 1981b). These studies demonstrate that surface exposure to direct insolation is a critical factor affecting soil biota. Surficial soils exposed to direct insolation exhibit marked daily fluctuations in temperature and moisture that affect the soil biotic community (Whitford et al. 1981b). Thus, the importance of vegetation structure must be kept in mind when considering the relations between climate change and soil biota.

Based on the various climate change scenarios that have been published, it seems certain that higher latitudes will experience significant warming and that large areas of continental land masses will be substantially wetter or drier than the current long-term average conditions (Hansen et al. 1988, Peters and Darling 1985). These conditions will inevitably lead to changes in species composition of the vegetation and may lead to changes in the structure of the vegetation, for example, the proportion of vegetative cover that is grass, shrubs, or trees (chapter 8). Such changes will undoubtedly affect the soil biota and soil processes mediated by that biota.

Compared with what we know of flowering plants and vertebrate animals, our knowledge of soil biota is extremely limited. Soil microorganisms can rarely be classified as to species, and that is also true of soil protozoans, nematodes, and microarthropods. Therefore, the available literature on soil organisms seldom focuses on individual species and instead deals with groups of similar species. In cases where individual species have been studied, the available data are insufficient to predict how environmental changes will affect the growth or stability of that population. As a result, predictions about the consequences of climate change on soil biota and soil processes are based only on fragments of data on a minuscule fraction of the organisms that make up the soil biota.

## II. THE SOIL BIOTA

What is the soil biota and of what importance is the soil biota in ecosystems? The soil biota is that assemblage of organisms inhabiting a volume of soil. The assemblage will generally include bacteria, fungi, protozoans, nematodes, mites, and collembolans. These organisms live in the water film surrounding soil particles and aggregates, and in the pore spaces in soil aggregates (fig. 9.1). Representatives of these groups of organisms are found in soils virtually anywhere on earth. Other organisms that are not as widely distributed may also be part of the assemblage. These include earthworms, pot worms, termites, ants, millipedes, isopods, snails, rotifers, and tardigrades (Wallwork 1976). There are no examples of which I am aware where the soil biota of a specific area of land has been completely described as to species composition and relative abundance.

One critical role of the soil biota is decomposition of dead plants and animals and subsequent mineralization to ions and molecules that can be absorbed by plant roots, including complete decomposition of complex organic molecules to simple molecules like phosphates and nitrates. Decomposition is the breakdown of recognizable plant and animal tissues to amorphous organic matter that cannot be distinguished from the soil by the naked eye. Mineralization is the subsequent breakdown of the organic matter into simple molecules like ammonia, nitrate, methane, and carbon dioxide and mineral ions such as calcium, iron, and potassium. Decomposition and mineralization are essential soil processes, mediated by the soil biota, that provide most of the soil materials necessary for plant growth.

Assemblages of soil organisms in very different habitats may be taxonomically similar to each other. For example, forest soils and desert soils share a number of genera of bacteria, fungi, protozoans, nematodes, and microarthropods. The limited data available suggest that species from different habitats, but within the same genera of soil organisms,

are ecologically similar (Wallwork 1976, Ingham et al. 1985). Moreover, many of the nominally distinct species within genera may be the same species, but this cannot be verified because of the paucity of taxonomic studies of soil organisms. Given that functional similarities between species within genera may be a common phenomenon, and until there are data that demonstrate important functional differences among the "species" of general types of soil organisms, this discussion will assume functional similarity between species within genera.

Not only may taxonomic similarity exist between different soil types but the functional structure of soil biotic communities as well appears to be similar in soil ranging from subarctic to subtropical deserts (Whitford 1989, Coleman et al. 1984). Numerous studies have documented the trophic relationships among the soil biota (Coleman et al. 1984, Ingham et al. 1985). These feeding relationships, as previously mentioned, directly affect important ecosystem processes such as decomposition (Santos et al. 1981, Parker et al. 1984b, Clarholm, 1981) and nutrient cycling (Coleman et al. 1984, Parker et al. 1984b, Ingham et al. 1985).

The physiological activity of the soil bacteria and fungi is directly affected by grazers on the microflora, which include protozoans, nematodes, and microarthropods. Grazing serves to keep the populations of microflora in a rapid growth phase by removing biomass and preventing population senescence. Further, grazing reduces nutrient immobilization by freeing nutrients sequestered by soil fungi; fragments and increases the surface area of dead plant materials, thereby enhancing decomposition; and affects mineralization rates by modifying the turnover rates of microbial biomass. Populations of grazers, in turn, are regulated by predators, which include nematodes and microarthropods. Thus, the trophic structure of the soil biotic community is complex, and many of these feeding relationships are affected directly by soil physical characteristics that reflect the general climatic regime (fig. 9.2). Modifica-

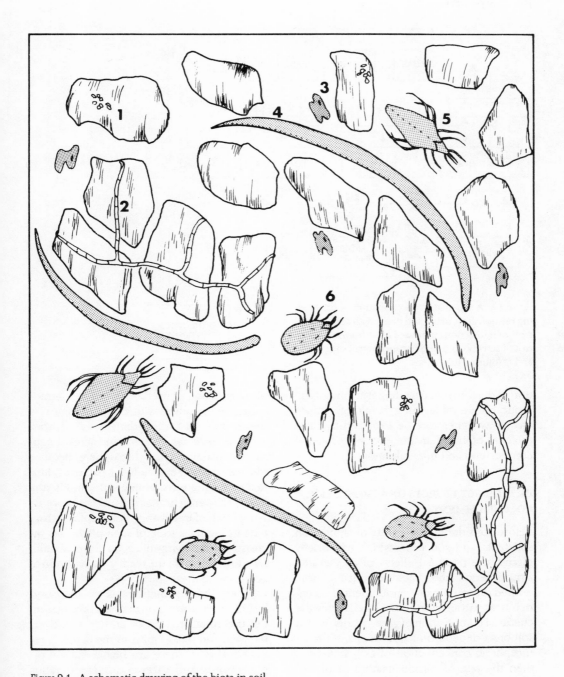

*Figure 9.1.* A schematic drawing of the biota in soil.
1. Bacteria growing on the surface of a soil particle.
2. Fungal hyphae growing on several soil particles.
3. Amoeba in a water film. 4. Nematode in a water
film. 5. Predatory mite in a pore space. 6. Fungal-
feeding mite in a pore space.

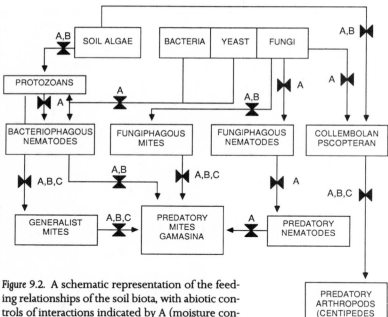

*Figure 9.2.* A schematic representation of the feeding relationships of the soil biota, with abiotic controls of interactions indicated by A (moisture control of activity), B (temperature control of activity), and C (diurnal migration).

tion of the functional feeding structure of soil biotic communities as a result of climate change will therefore be expected to affect the rates of important soil processes like decomposition and mineralization.

## III. INDIRECT EFFECTS OF ATMOSPHERIC CO₂ ENRICHMENT

Before considering the effects of climate on soil biota and soil processes, I will briefly discuss the potential effects of an elevated atmospheric carbon dioxide concentration, termed carbon dioxide enrichment. This enrichment will affect soil biota by changing the chemical composition of green plants, which soil biota depend on for nutrients. If the atmospheric concentration of $CO_2$ increases, then the rate of carbon fixation in photosynthesis also increases, which in turn affects carbon allocation patterns in plants (Potvin et al. 1984, 1985). For example, in some plants, especially arid-adapted C4 grasses, excess fixed carbon may be translocated to roots and exuded as complex carbohydrates known as mucilages and mucigels (Wullstein et al. 1979), which are substrates for free-living

nitrogen-fixing bacteria. Therefore, atmospheric enrichment with $CO_2$ that increases the quantity of carbohydrate released from roots as exudates also would increase substrate availability in the rhizosphere. Because nitrogen fixation by free-living bacteria is limited by substrate availability (Atlas and Bartha 1987), $CO_2$ enrichment should therefore favor growth of free-living, nitrogen-fixing bacteria in the root zone of such plants. An increase in free-living nitrogen fixers would, over time, increase the readily available nitrogen stores in these soils.

When photosynthetic rates are increased by $CO_2$ there is more plant tissue production, but the concentration of nutrients, such as nitrogen and phosphorus in the tissue, is reduced; that is, there are higher carbon-to-phosphorus and carbon-to-nitrogen ratios (Lincoln et al. 1986). Compared with present-day plants, $CO_2$-enriched plants will represent a rich source of complex sugars and a poor source of nutrients. As these chemically different plants or parts of such plants die and fall to the soil, the species mix of organisms that grow on that material (the decomposers and mineralizers) will change relative to present-day biota. In turn, that change will affect the rates at which essential materials are made available for plant growth. Groups of soil biota that might benefit under these future conditions are fungi and other soil organisms capable of growing on energy-rich but nutrient-poor materials. As the fungi grow, they can incorporate a large proportion of the relatively scarce, essential nutrients, like nitrogen and phosphorus, into fungal biomass, in turn reducing or eliminating the availability of these nutrients to other soil organisms and to roots, a process known as immobilization.

## IV. GENERAL RESPONSES TO TEMPERATURE AND MOISTURE

Microflora and -fauna are more sensitive to changes in the water potential of soils than they are to changes in temperature. Many soil organisms survive freezing and temperatures up to 40°C and possibly higher (W. G. Whitford and D. W. Freckman, unpublished data). Variations in temperature between 0°C and 40°C affect metabolic rates but not the survival or feeding relationships. In moist, thermally neutral environments, all of the organisms and the trophic relationships depicted in figure 9.2 are active. With these environmental conditions, trophic relationships described by models like that of Hunt et al. (1987) operate continually. However, in seasonally dry or variably dry environments, the functional relationships among the soil biota change markedly (Whitford 1989). These changes

in functional relationships are not due to changes in species composition, because the soil biota has essentially the same community structure following prolonged hot, dry periods (Cepeda and Whitford 1989). Instead, they are caused by changes in the functioning and abundance of particular species.

Many soil organisms, both microflora and microfauna, have a variety of adaptations that allow them to survive extreme heat, cold, and dryness (Crowe and Clegg 1973; table 9.1). Although a variety of terms have been applied to such adaptations, they are all essentially some form of anhydrobiosis, literally "life without water," a physiological state in which the organism contains no free water and has no measurable metabolism. Organisms in this state may be called spores, cysts, anhydrobiotes, or cryptobiotes, but all are alive and physiologically inactive. They quickly return to an active physiological state when water is again available. Since greenhouse gases affect not only atmospheric temperature but also rainfall distribution patterns, the potential long-term consequences of wetter or drier soil conditions on the functional relationships of the soil biota will be affected by the abilities of species of soil organisms to enter an anhydrobiotic state.

## V. POTENTIAL EFFECTS OF CHANGES IN SOIL TEMPERATURE AND MOISTURE

### A. Soil Microflora

In many species of plants and animals, genetically and ecologically distinct subpopula-

Table 9.1. Characteristics of soil organisms that affect responses to climate change.

| | |
|---|---|
| Bacteria | Large variability of thermal and hydric physiology within populations |
| | Ability in some species to sporulate |
| | Wide range of thermal tolerance |
| Fungi | Numerous thermal-tolerant and thermophilic species |
| | Resistant spores |
| | Possibility of high temperatures having adverse effect on mycorrhizal establishment |
| Protozoans | Ability to form cysts |
| Nematodes | Anhydrobiosis in most species |
| Collembolans | Anhydrobiosis in some species |
| Mites | Cryptobiosis in some species |

tions, known as ecotypes, can be identified. An ecotype of a species is better adapted to a particular environment than are the other ecotypes of that species. In the case of soil biota, species may be divided into ecotypes that have substantially different physiological adaptations to soil and water conditions. Species characterized by numerous ecotypes should be expected to experience relatively little range diminution or extinction due to climate change because their relatively wide ecological tolerance will preadapt them to changing conditions. Such species may be common among the soil biota, as evidenced by the known physiological races or ecotypes of some widely distributed species of soil bacteria. For example, there is a wide range in ecotypic variation in the temperature tolerances of several species of soil bacteria found in tundra, taiga, steppe, dry steppe, and subtropical soils in the Soviet Union. Temperature maxima range between 35° and 45°C in *Pseudomonas* spp. and *Mycobacterium* spp. from those environments (Mishustin and Yemtsev 1982). Several species of *Bacillus* exhibit a similar range of thermal maxima: 33°–40°C in tundra populations and 45°–47°C in populations from subtropical red soils. Species from higher latitudes are characterized by lower thermal maxima and lower temperature optima. Thus, *Pseudomonas* spp., *Mycobacterium* spp., and *Bacillus* spp. are characterized by thermal ecotypes. Also, several species of anaerobic nitrogen fixers (*Clostridium*) exhibit ecotypic variation in optimum temperature and maximum temperature. The results of this study suggest that many soil bacteria, such as *Pseudomonas* spp., *Mycobacterium* spp., *Bacillus* spp., and *Clostridium* spp., will simply shift the distribution of existing ecotypes in response to soil warming.

In another genus, *Azotobacter*, there is no thermal ecotypic variation (Mishustin and Yemtsev 1982). This suggests that soil warming could result in changes in the species composition of nitrogen fixers favoring *Clostridium* spp. over *Azotobacter* spp. While it is not possible to predict the consequences of

changing the species mix of soil bacteria because of the differences in ecotypic variation among species, there is little doubt that some soil processes will be affected.

Substantial ecotypic differentiation may exist within a relatively limited geographical area. For example, the physiological characteristics, including thermal and water tolerances, of *Rhizobium leguminosarum* isolated from soils collected from various climatic regions of Morocco varied as much within one climatic zone as among different zones (Robert et al. 1982). Therefore, even populations from a relatively localized area are likely to have high degrees of ecotypic variability, which should allow them to cope successfully with climate change.

Even within ecotypes, most species of soil microorganisms probably have wide ranges of tolerance for temperature and soil moisture. Species of the genus *Rhizobium* are nitrogen-fixing root symbionts; their presence is essential for the good growth and survival of their legume hosts. Many legumes and their rhizobia are important crop plants and are also important in agronomic practices that improve and sustain soil fertility. *Rhizobium leguminosarum* isolates have been found to be adapted to desiccation but died at temperatures above 37°C (Robert et al. 1982). Since soil temperatures in the root zone (5 cm and deeper) do not exceed 35°C, even when soil surface temperatures reach 60°C, the ability of strains of *R. leguminosarum* to survive in soils subjected to increased temperatures and dryness should not be affected. *Rhizobium japonicum* inocula exhibited poor survival in hot (ambient temperature 38°C) and dry conditions (Herridge et al. 1987). These authors, however, reported that nodulation, growth, and yield of crops were equally good the following year irrespective of the initial rhizobial status of the soil. This emphasizes the rapidity with which microbiota can build population numbers and recover from unfavorable conditions.

These limited studies suggest that nitrogen-fixing symbiotic bacteria have sufficiently wide physiological tolerances for temperature and moisture variations that changes in these variables will have little effect on their survival and metabolic activity. The biological success of root symbionts like Rhizobium spp., however, depends on the vigor of the plant species partner in the symbiosis. If the plant species does not survive, neither will the symbiont.

Do nonsymbiotic bacteria and fungi exhibit similar broad tolerances to high temperatures and variability in available soil water? No differences in nitrifying capacity have been found in a variety of soils from Egypt tested at different temperatures between 10°C and 50°C (Monib et al. 1979). Functionally, the nitrifying bacteria in the Egyptian soils are equivalent over a wide range of soil temperatures. Thus, bacteria involved in nitrogen mineralization appear to be capable of surviving and being physiologically active in thermally stressed soils.

The fluctuating thermal environment of temperate-zone soils has been found to provide sufficiently high temperatures of sufficient duration for most species of thermophilic and thermotolerant fungi to complete their life cycle (Jack and Tansey 1977). Suitable temperatures for thermophilic and thermotolerant species also occur in shaded soils. A large variety of thermophilic and thermotolerant fungi that have been studied are from soils of south-central Indiana (Jack and Tansey 1977). Since temperate soils support a wide variety of fungi that function well in soils of relatively high temperature, increased soil temperatures will probably have little effect on processes mediated by soil fungi. Another kind of symbiotic relationship that is important in many ecosystems is the mycorrhizal symbiosis, in which symbiotic fungi associate with plant roots and increase the rate at which the plants can take up various essential nutrients. Such mycorrhizal fungi appear to be more temperature sensitive than the heterotrophic microflora just discussed. Temperatures below 29.5°C have little significant effect on ectomycorrhizal and vesicular-artiscular mycorrhizal fungi, but above 29.5°C mycorrhizal formation is reduced or prevented (Parke et al. 1983). Treatment of soil at 35°C for one week did not adversely affect viability of ectomycorrhizal propagules, but young mycorrhizal hyphae exposed to 35°C appeared to be severely injured.

Although the data on thermal and hydric responses of microflora are limited, the results of the few studies cited here suggest that most species can tolerate a wide range of fluctuations in the physical environment and have sufficient ecotypic variation to continue to grow well in warmer and drier soils. Considering the short generation time of such organisms, it is highly probable that most of the soil microflora species will cope successfully with the projected rate of climate change and continue to function normally in the climate-modified ecosystems. Some groups, however, like the mycorrhizal fungi and some species of nitrogen-fixing Rhizobium bacteria, have relatively narrow physiological tolerances to thermal and hydric change and thus may be adversely affected, in turn, their decline could reduce the viability of their host plant species. These kinds of feedback may cause small climate changes to have major effects on some plant species that depend on microbial symbionts.

## B. Soil Microfauna

Studies using simulated rainfall irrigation have shown marked effects on proportions of grazers active in the soil. In a study by L. W. Parker et al. (1984a), added water had little effect on the densities of ciliate protozoans but marked effects on amoebae and flagellate protozoans, which increased by an order of magnitude 24 days after the initiation of frequent irrigation (table 9.2). The numerical responses of protozoans are not as important, however, as the percentage of the population that was in the inactive or cyst form. At soil water potentials of −0.1 megapascals (MPa), 50% of a soil protozoan population was encysted, and at −0.4 MPa, the entire protozoan population was encysted. Thus, although a soil may harbor relatively high densities of protozoans, these populations are active only when soils are moist. As soils dry, the microflora in the small pores in the soil will not be

*Table 9.2.* Effects of supplemental rainfall (12 mm every 3 days) and surface litter quantity (0 g or 150 g of litter per square meter) on population densities of soil protozoans (numbers/10 g soil) in a desert soil at 12 and 24 days after the initial irrigation. "Wet" indicates irrigated plots, "dry" the controls. From Parker et al. 1984a.

| Experimental Conditions | Flagellates | | Ciliates | | Amoebae | |
|---|---|---|---|---|---|---|
| | 12 days | 24 days | 12 days | 24 days | 12 days | 24 days |
| 0 g wet | 250 | 1144 | 250 | 250 | 5800 | 50970 |
| 0 g dry | 590 | 1004 | 350 | 360 | 2200 | 20350 |
| 150 g wet | 450 | 3050 | 380 | 2530 | 2680 | 15040 |
| 150 g dry | 710 | 5150 | 260 | 250* | 4470 | 18070 |

*Total populations 20% lower than short-grass prairie.

grazed because protozoans that are capable of moving into small pores are inactive in the cyst form. Other soil animals are too large to enter many of the small pores.

Nematodes are grazers on protozoans, bacteria, yeast, and fungi. Like protozoans, nematodes may not be active in dry soils, because they enter a state of anhydrobiosis when water is limited (Demeure and Freckman 1981, Demeure et al. 1979, Freckman 1982, Freckman et al. 1987). Basically, anhydrobiotic nematodes are desiccated worms that return to activity and can begin to breed as soon as there are free-water films to rehydrate them. In a series of simulated rainfall experiments, D. W. Freckman et al. (1987) found that only phytophagous and omnivore predator nematodes increased their population numbers in continuously moist soil. Nematodes responded like the protozoans with respect to percent of the population that was anhydrobiotic at various soil water potentials. The soil water potential at which 50% of the population was anhydrobiotic was $-0.4$ MPa, and 99% of the population was anhydrobiotic between $-3.0$ and $-5.0$ MPa. Thus, for both nematodes and protozoans, soil water potentials well above the permanent wilting point of $-1.5$ MPa will shift a major proportion of the population into an inactive state. In soils subject to frequent drying or prolonged dry periods, these most important microfloral grazers are essentially absent from the soil fauna, since they are physiologically inactive. Further, populations of nematodes may be active in surface

litter layers for only a few hours after a rain and for only 48–72 hours in the soil following a summer rain (Whitford et al. 1981b). In dry soils, the soil fauna that can graze on the microflora is reduced to a few species of soil mites (Whitford et al. 1981b). Some species of mites can remain active in surface litter even when soils reach 0% moisture for a few hours a day (Whitford et al. 1981b). Although the mites contribute to the rates of decomposition and mineralization, they are not as effective as the full complement of protozoans, nematodes, and microarthropods (Parker et al. 1984b). Thus, climate change that affects soil moisture will reduce the rates of decomposition and mineralization. Reduction in rates of these processes and disturbance of the linkage between them can produce bottlenecks in the supply of materials for plant growth.

In addition to responding to longer patterns in weather, various groups of microarthropods have also been observed to respond to daily changes in soil temperature and moisture. In the Chihuahuan Desert, microarthropods are primarily active in the surface litter during the coolest part of the day, within 1 or 2 hours after sunrise, but are largely absent from this layer at midday (Whitford et al. 1981b). Shortly before sunrise in the desert, surface litter may have up to 6% water by weight, but by midday no water is detectable (Whitford et al. 1981b). The microarthropods disappear with the water, but it is not known whether they enter an inactive stage or descend to lower soil levels. Different taxa of microarthropods display different patterns, with the tydeid mites, for example, exhibiting great diurnal variation while nanorchestids and oribatids exhibit no discernible patterns and little change throughout the day (Whitford et al. 1981b).

In experimental studies, raising soil moisture and lowering soil temperature over a period of days has had marked effects on the populations of fungiphagous and litter-feeding mites, increasing densities eight to ten times, but virtually no effect on densities of predatory mites (MacKay et al. 1985). The combination of water and shade is no more effective than shade alone on some species like the small fungivore *Tarsonemus* spp., whereas both shade and water are required to increase densities of collembolans and psocopterans. Other mites, like the algal feeder *Speleorchestes* spp., responded neither to shade nor to water. The results of such experiments show that decreasing temperature has a marked effect on fungus feeders. If climate change results in drier soils with higher mean temperatures, then populations of small fungus-feeding mites will be reduced. That will reduce the release of nutrients from fungal biomass even more than the reduction caused by a decline in the activity of protozoans and nematodes. Clearly, abiotic factors, especially temperature and soil water content, control the functional structure of soil food webs. The importance of abiotic controls is magnified at the litter-soil interface, where the thermal and hydric environment changes rapidly on a daily basis. With those relationships in mind, we can examine the following studies for additional insights into the potential effects of climate change on soil biotic processes.

## VI. ECOSYSTEM PROCESSES MEDIATED BY SOIL BIOTA

Despite the addition of 300 mm of precipitation per year by sprinkler irrigation, Whitford et al. (1986) found no significant quantitative differences in the soil biota among irrigated and nonirrigated plots in a desert ecosystem. However, even though there were no detectable changes in biomass of soil organisms, there was a difference in the rate of one of the processes mediated by these organisms—nitrogen mineralization. It is thought that although the biomass of soil biota remained constant, irrigation caused a higher turnover rate in microbial populations (i.e., both birth and death rates increased), which apparently affected the rate of nitrogen mineralization (Fisher et al. 1987). This experiment suggests that in some cases where climate change substantially affects soil processes, important characteristics like biomass may remain constant, making it difficult to detect exactly how that change has affected the soil biota.

In those same experiments, after three years of supplemental water additions, net nitrogen mineralization was significantly reduced during the summer growing season, but no effects were seen at the end of the winter dormant season (table 9.3). This seasonal difference was probably due to the rapid decomposition and mineralization of limited supplies of organic matter in the summer months. In summer, the differences in soil water availability are exaggerated by high soil temperatures and high evapotranspiration. In winter, microbial population turnover is slow and water losses by evapotranspiration are low. Thus, there are only slight differences in soil water with the different rainfall patterns. In a shorter experiment using a different approach, D. S. Schimel and W. J. Parton (1986) found higher mineralization with a single large addition of water than with frequent small additions. Both studies demonstrate that climate patterns affect the nitrogen mineralization and nitrification processes, and they suggest that long-term modification of rainfall patterns may have different effects from short-term modifications. There may also be important differences in how the soil biota respond to rainfall patterns seasonally and in relation to the vegetation.

Vegetation characteristics can markedly affect soil processes such as nitrogen immobilization and mineralization. Whitford et al. (1987) have presented a model to account for the differences in nitrogen availability in

*Table* 9.3. The effects of long-term irrigation on nitrogen mineralization. Treatments were 6 mm per week, 25 mm every fourth week, and no added water for three years. Nitrogen mineralization was measured as production of ammonium plus nitrite and nitrate in milligrams per kilogram of soil, using 28-day laboratory incubations. The summer 1984 measurements differ significantly ($P < 0.01$). Data from Fisher et al. 1987.

| Sample | Treatment | Inorganic N |
|---|---|---|
| Summer 1984 | 0 mm | 13.2 |
| | 6 mm/week | 8.9 |
| | 25 mm/month | 11.7 |
| March 1985 | 0 mm | 5.3 |
| | 6 mm/week | 6.2 |
| | 25 mm/month | 5.7 |

short-grass steppe and desert shrubland. In that model, the timing of organic matter input (which provides the energy for the microbes and ultimately for the entire soil biotic community) determines the mineralization patterns and, hence, the availability of nutrients to plants. This model reinforces the idea that the effects of climate change on plants, microbes, and animals are knit together and cannot be separated. All organisms are parts of ecosystems, affecting each other directly and indirectly, with the strength of the interactions being modified by the physical environment.

## VII. SUMMARY

What can we conclude about the consequences of the greenhouse effect on soil biota and soil processes? It is highly unlikely that the projected climate changes will result in extinctions of species of soil organisms, except possibly in the case of obligate symbionts where the plant associate fails to survive. Most components of the soil biotic community have wide ranges of tolerance to temperature and moisture fluctuations in their environment. Further, the short life cycle of most of these organisms should permit genetic adaptation to shifts in the soil microclimate. Climate change may, however, cause shifts in relative abundances of species (table 9.4). Such shifts may affect ecosystem processes both qualitatively and quantitatively; for example, dryness may cause nematode abundance to drop, thereby decreasing how efficiently the soil biota can decompose organic material. Even in the absence of major changes in species composition, changes in the abiotic environment in the soil affect trophic relationships in soil food webs (table 9.4). For example, many species may become inactive for varying periods of time when conditions become temporarily unsuitable. Since processes such as decomposition and mineralization are the result of the activities of this biota, the rates and seasonal patterns of such processes will vary with the subset of the biota that is active. Because of the many feedback loops among soil organisms, plants, and the physical-chemical environment of the soil, we can expect the functional soil communities to reflect the shifts in vegetation.

*Table* 9.4. Summary of probable effects of soil temperature and moisture changes on soil biota and ecosystem processes. Measures of change: +, ++, +++ (increase one, two, and three orders of magnitude, respectively); −−, −−− (decrease two and three orders of magnitude); 0 (no change).

| Physiological state or ecosystem process | Increased soil temperature | Increased soil moisture | Decreased soil moisture |
|---|---|---|---|
| Percent time in anhydrobiosis or crytobiosis | + | 0 | +++ |
| Number of connections in soil food webs | + | +++ | −−− |
| Root symbiosis | + | + | 0 |
| Decomposition rates | + | + | −− |
| Mineralization rates | + | ++ | −− |
| Species composition | + | 0 | ++ |

# REFERENCES

Atlas, R. M., and R. Bartha. 1987. *Microbial Ecology: Fundamentals and Applications,* 2nd ed. Menlo Park, Calif.: Benjamin/Cummings.

Cepeda, J. G., and W. G. Whitford. 1989. The relationships between abiotic factors and abundance patterns of soil microarthropods on a desert watershed. *Pedobiologia* 33:79.

Clarholm, M. 1981. Protozoan grazing of bacteria in soil: Input and importance. *Microb. Ecol.* 7:343.

Coleman, D. C., R. E. Ingham, J. F. McClellan, and J. A. Troymow. 1984. Soil nutrient transformations in the rhizosphere via animal-microbial interactions. In *Invertebrate-microbial interactions,* J. M. Anderson, A.D.M. Rayner, and D.W.H. Walton, eds., pp. 35–58. Cambridge: Cambridge University Press.

Crowe, J. H., and J. S. Clegg, eds. 1973. *Anhydrobiosis.* Stroudsburg, Pa.: Dowden, Hutchinson and Ross.

Demeure, Y., and D. S. Freckman. 1981. Recent advances in the study of anhydrobiosis in nematodes. In *Plant Parasitic Nematodes,* B. M. Zuckerman and R. A. Rohde, eds., vol. 3, pp. 204–225. New York: Academic Press.

Demeure, Y., D. W. Freckman, and S. D. Van Gundy. 1979. Anhydrobiotic coiling of nematodes in soil. *J. Nematol.* 11:189.

Fisher, F. M., L. W. Parker, J. P. Anderson, and W. G. Whitford. 1987. Nitrogen mineralization in a desert soil: Interacting effects of soil moisture and nitrogen fertilizer. *Soil Sci. Soc.* 51:1033.

Freckman, D. W. 1982. Parameters of the nematode contribution to ecosystems. In *Nematodes in Soil Ecosystems,* D. W. Freckman, ed., pp. 80–97. Austin: University of Texas Press.

Freckman, D. W., W. G. Whitford, and Y. Steinberger. 1987. Effect of irrigation on nematode population dynamics and activity in desert soils. *Biol. Fert. Soils* 3:3.

Hansen, J., I. Fung, A. Lacis, S. Lebedeff, D. Rind, R. Ruedy, G. Russell, and P. Stone. 1988. Prediction of near-term climate evolution: What can we tell decision-makers now? In *Preparing for Climate Change,* Proceedings of the first North American conference on preparing for climate change: A cooperative approach, J. C. Topping, Jr., ed., pp. 35–47. Washington, D.C.: Government Institutes.

Herridge, D. F., R. J. Roughley, and J. Brockwell. 1987. Low survival of Rhizobium-Japonicum inocula leads to reduced nodulation nitrogen fixation and yield of soybean in the current crop but not in the subsequent crop. *Aust. J. Ag. Res.* 38:75.

Hunt, H. W., D. C. Coleman, E.R. Ingham, R. E. Ingham, E. T. Elliot, J. C. Moore, S. L. Rose, C.P.P. Reid, and C. R. Morley. 1987. The detrital food web in shortgrass prairie. *Biol. Fert. Soils* 3:65.

Ingham, R. E., J. A. Trofymow, E. R. Ingham, and D. C. Coleman. 1985. Interactions of bacteria, fungi, and their nematode grazers: Effects on nutrient cycling and plant growth. *Ecol. Monogr.* 55:119.

Jack, M. A., and M. R. Tansey. 1977. Growth sporulation and germination of spores of thermophilic fungi incubated in sun heated soil. *Mycologia* 69:109.

Kamill, B. W., Y. Steinberger, and W. G. Whitford. 1985. Soil microarthropods from the Chihuahuan Desert of New Mexico. *J. Zool. Lond.* A 205:273.

Lincoln, D. E., D. Couvet, and N. Sionit. 1986. Response of an insect herbivore to host plants grown in carbon dioxide enriched atmospheres. *Oecologia* 69:556.

MacKay, W. P., S. Silva, D. C. Lightfoot, M. I. Pagani, and W. G. Whitford. 1985. Effect of increased soil moisture and reduced soil temperature on a desert soil arthropod community. *Am. Midl. Nat.* 116:45.

MacKay, W. P., S. Silva, and W. G. Whitford. 1987. Diurnal activity patterns and vertical migration in desert soil microarthropods. *Pedobiologia* 30:65.

Meentemeyer, V. 1978. Macroclimate and lignin control of litter decomposition rates. *Ecology* 59:465.

Mishustin, E. N., and V. T. Yemtsev. 1982. Ecological variability of soil microorganisms. *Zentralblatt Mikrobiologie* 137:353.

Monib, M., I. Honsy, T. T. El Hadidy, and R. El-Shahawy. 1979. Temperature adaptability of nitrifying bacteria in soils of Egypt. *Zentralblatt Bakteriol Paraditenkd Infektionskr Hyg Zweite Naturwiss Abt Mikrobiol Landwirtsch Technol Umweltschutzes* 34:528.

Parke, J. L., R. G. Linderman, and J. M. Trappe. 1983. Effect of root zone temperature on ecto mycorrhiza and vesicular arbuscular mycorrhiza formation in disturbed and undisturbed forest soils of southwest Oregon, U.S.A. *Can. J. Forest Res.* 13:657.

Parker, L. W., D. W. Freckman, Y. Steinberger, L. Driggers, and W. G. Whitford. 1984a. Effects of simulated rainfall and litter quantities on desert soil biota: Soil respiration, microflora, and protozoa. *Pedobiologia* 27:185.

Parker, L. W., P. F. Santos, J. Phillips, and W. G. Whitford. 1984b. Carbon and nitrogen dynamics during the decomposition of litter and roots of a

Chihuahuan desert annual, *Lepidium lasiocarpum*. *Ecol. Monogr.* 54:339.

Peters, R. L., and J.D.S. Darling. 1985. The greenhouse effect and nature reserves. *Bioscience* 35: 707.

Potvin, C., J. D. Goeschl, and B. R. Strain. 1984. Effects of temperature and $CO_2$ enrichment on carbon translocation of plants of the C4 grass species *Echinochloa crus-galli* (L.) Beauv. from cool and warm environments. *Plant Physl.* 75:1054.

Potvin, C., B. R. Strain, and J. D. Goeschl. 1985. Low night temperature effect on photosynthate translocation of two C4 grasses. *Oecologia* 67:305.

Robert, F. M., J.A.E. Molina, and E. L. Schmidt. 1982. Properties of *Rhizobium leguminosarum* isolated from various regions of Morocco. *Ann. Microbiol.* (Paris) 133A:461.

Santos, P. F., J. Phillips, and W. G. Whitford. 1981. The role of mites and nematodes in early stages of buried litter decomposition in a desert. *Ecology* 62:664.

Schimel, D. S., and W. J. Parton. 1986. Microclimate controls of nitrogen mineralization and nitrification in shortgrass steppe soils. *Plant and Soil* 93: 347.

Seastedt, T. R., and D. A. Crossley, Jr. 1981. Microarthropod response following cable logging and clear-cutting in the southern Appalachians. *Ecology* 62:126.

Steinberger, Y., D. W. Freckman, L. W. Parker, and W. G. Whitford. 1984. Effects of simulated rainfall and litter quantities on desert soil biota: Nematodes and microarthropods. *Pedobiologia* 26:267.

Wallwork, J. A. 1976. *The Distribution and Diversity of Soil Fauna.* London: Academic Press.

Wallwork, J. A., B. W. Kamill, and W. G. Whitford. 1985. Distribution and diversity patterns of soil mites and other microarthropods in a Chihuahuan desert site. *J. Arid Envir.* 9:215.

Whitford, W. G. 1989. Abiotic controls on the functional structure of soil food webs. *Biol. Fert. Soils* 8:1.

Whitford, W. G., V. Meentemeyer, T. R. Seastedt, K. Cromack, Jr., D. A. Crossley, Jr., P. Santos, R. L. Todd, and J. B. Waide. 1981a. Exceptions to the AET model: Deserts and clear-cut forest. *Ecology* 62:275.

Whitford, W. G., D. W. Freckman, N. Z. Elkins, L. W. Parker, R. Parmalee, J. Phillips, and S. Tucker. 1981b. Diurnal migration and responses to simulated rainfall in desert soil microarthropods and nematodes. *Soil Biol. Bioch.* 13:417.

Whitford, W. G., Y. Steinberger, W. McKay, L. W. Parker, D. W. Freckman, J. A. Wallwork, and D. Weems. 1986. Rainfall and decomposition in the Chihuahuan Desert. *Oecologia* (Berlin) 68:512.

Whitford, W. G., J. F. Reynolds, and G. L. Cunningham. 1987. How desertification affects nitrogen imitation of primary production on Chihuahuan Desert watersheds. In *Strategies for Classification and Management of Native Vegetation for Food Production in Arid Zones*, USDA Forest Service General Technical Report RM 150, pp. 143–153.

Wullstein, L. H., M. L. Bruening, and W. B. Bollen. 1979. Nitrogen fixation associated with sand grain root sheaths (rhizosheaths) of certain xeric grasses. *Physl. Plant.* 46:1.

Preparation of this chapter was supported by grant BSR 8612106 from the National Science Foundation and is a contribution of the Jornada Long-Term Ecological Research Program.

# CHAPTER TEN

# Possible Effects of Global Warming on the Biological Diversity in Tropical Forests

GARY S. HARTSHORN

## I. INTRODUCTION

Rampant deforestation, the inexorable advance of the agricultural frontier, species extinction, and the conservation of biological diversity are some of the principal problems facing those who would protect tropical forests. Global warming due to the greenhouse effect may also affect the composition and integrity of those few tropical forests that survive well into the next century. Tropical forests not only are the most species-rich ecosystems known on this planet but also are exceedingly complex ecologically. This chapter has three objectives: (1) to highlight briefly the ecological complexity and biological diversity of tropical forests; (2) to review documented effects of historically recent climatic fluctuations on tropical forest communities; and (3) to use these first two components for cautiously exploring possible effects of global warming on the biological diversity in tropical forests.

My primary theater of experience is the humid forests of tropical America; hence, I will draw heavily on such well-known sites as the Smithsonian Tropical Research Institute's Barro Colorado Island (BCI) in Panama's Lake Gatún (Leigh et al. 1982, 1990) and the Organization for Tropical Studies' La Selva Biological Station in the Caribbean lowlands of Costa Rica (Clark 1990). I am focusing on mature or old-growth forests, with minimal or no direct human disturbance; thus young secondary forests, deforestation, restoration of degraded forest lands, soil fertility, nutrient cycling, and so on are beyond the purview of this chapter.

## II. ECOLOGICAL AND BIOLOGICAL BASES

In highlighting the ecological and biological bases for considering the possible effects of global warming on the biological diversity in tropical forests, I want to review briefly three aspects of tropical forest ecology: the heterogeneity and types of tropical forests, the incredible array and diversity of interactions among species in tropical forest communities, and the extraordinary richness of species in most tropical forests.

## A. Life Zones

Most vegetation classification systems distinguish many types of tropical forests. Using simple climatic data on rainfall and temperature, L. R. Holdridge (1947, 1967) developed the life-zone classification system of world plant formations that is widely used in tropical America (cf. Tosi 1980). Holdridge's life zone system provides bioclimatic boundaries for tropical forests. Humid tropical forests occur naturally where the mean annual rainfall exceeds the potential evapotranspiration for the year. On the life zone chart, humid tropical forests occur to the right of the unity line (1.0) in the tropical and subtropical latitudinal regions (fig. 10.1).

Tropical forests also occur up to about 4000 m on high mountain ranges. At high altitudes, humid tropical forests occur with much lower mean annual rainfall, because potential evapotranspiration decreases as temperature declines with increasing altitude. Within these altitudinal and latitudinal limits, 27 life zones fall within what we generally call the humid tropical forest biome. Another way of illustrating the ecological complexity of humid tropical forests is through geographical comparisons. The tiny republic of Costa Rica, which is about the same size as West Virginia, has 12 life zones (Tosi 1969)— one more than in the entire eastern United States (Sawyer and Lindsey 1964). The republic of Peru has the highest number of life zones in the world—78 of the theoretical maximum of 120 life zones (ONERN 1976). Peru's bioclimatic richness can be attributed to its extension from tropical (and subtropical) to warm temperate latitudinal regions, and across the Andes from Pacific coastal deserts to the Amazon basin.

The many ecological life zones in the humid tropical forest biome are differentiated by bioclimatic factors, yet this does not mean that one life zone has only one type of forest.

Atmospheric, edaphic, and hydric factors interact as secondary determinants of a much greater number of forest associations or types (Holdridge 1967, Hartshorn 1988). Although detailed classification of forest types over large areas is generally lacking in the humid tropical forest biome, there must be hundreds, if not thousands, of distinctive types of tropical forest communities. The ecological complexity (cf. Janzen 1988) and heterogeneity of tropical forests may be important sources of the high local endemism of species reported by A. H. Gentry (1986).

## B. Species Interactions

Early naturalists like Bates, Darwin, Spruce, Humboldt, and Wallace, who explored tropical forests more than a century ago, reported their discoveries of incredible numbers of species, as well as the intricate and sometimes bizarre shapes, colors, and behavior of many tropical organisms. The fascination with unusual aspects of tropical species and their relationships continues, as we still unravel the coevolutionary causes and adaptations to the complex interactions that abound in tropical forests. The mimicry systems of butterflies (cf. DeVries 1987), ant-acacia mutualism (Janzen 1966), mycorrhizae (Janos 1983), plant-animal interactions (cf. Gilbert and Raven 1975), and second- and third-order interactions (e.g., Ray and Andrews 1980) are just a few examples of the array of highly evolved and intricate relations among species that live in and depend on tropical forests.

Biotic interactions in tropical forests are characterized by a high degree of specificity between species. Pollination systems (e.g., fig wasps, orchids by euglossine bees), seed dispersal (e.g., Lecythidaceae by bats, Myristicaceae by birds), and plant herbivory (e.g., *Heliconia* and hispine beetles) usually involve highly coevolved species. Many of these interactions involve multiple-species assemblages—called ecological guilds—where a suite of similar species comprises a specific component of an interaction. For example, several species of *Heliconia* flower sequentially, providing hummingbirds with year-

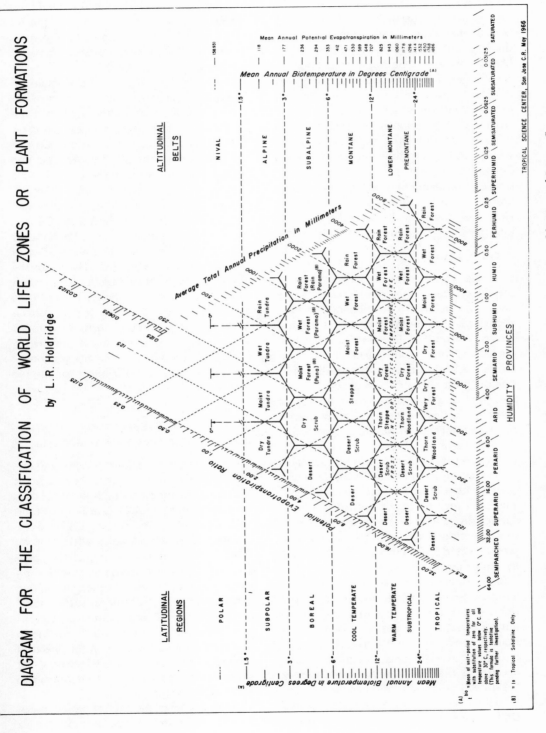

Figure 10.1. The Holdridge diagram for the classification of world life zones. Reprinted by permission of the Tropical Science Center.

round sources of nectar (Stiles 1978). From the other side of this mutualistic system, the hummingbirds raise their young during the period when there is maximum nectar production (Stiles 1980). Plant species assemblages that depend on frugivorous birds for seed dispersal (e.g., melastomes, Lauraceae) have species-specific periods of fruiting that provide a year-round source of fruits (Wheelwright 1983). In mountainous regions, there is significant altitudinal migration as frugivores track fruit availability, generally moving downslope during the mid- to late rainy season (e.g., resplendent quetzals, bare-necked umbrella birds, three-wattled bellbirds). However, migratory hummingbirds move upslope during the rainy season (Stiles 1988).

## C. Species Diversity

Until recently the estimates of the number of species on our planet were usually in the range of 3 million to 10 million, of which about half were thought to inhabit tropical forests (Wilson 1988, Raven 1988). Given that the tropical forest biome covers about 6% of the earth's land surface, those estimates indicate the great concentration of species richness in tropical forests. However, the recent work by T. L. Erwin (1988) on canopy invertebrates in the Peruvian Amazon suggests that the global estimate of species may need to be raised to 30 million. Erwin and his associates collect canopy invertebrates by fogging a tree crown with a nonpersistent insecticide. His suggested revision in the estimates of total species diversity is based on the extremely low repetition of species in the canopy samples collected from crowns of different or even the same tree species.

Biologists call the species richness encountered by Erwin alpha-diversity, that is, the richness of species in one habitat, community, or forest type. Some other outstanding examples of the high alpha-diversity in tropical forests are: 311 species of trees on one hectare in the Peruvian Amazon (Gentry 1982, 1988), which is almost half the number of native tree species in the continental United States; 236 species of vascular plants

on a 10 × 10 m plot in lowland wet forest near La Selva (Whitmore et al. 1985). Even if trees are excluded from the tally of plant species richness, tropical forests are still the most species-rich plant communities on earth (Gentry and Dodson 1987).

Not surprisingly, when we incorporate different communities or habitats in our sampling of species richness (termed beta-diversity), tropical forests are also exceptionally rich in species. La Selva has about 450 tree species, whereas the BCI flora lists 362 tree species (Hartshorn 1988). The La Selva fauna includes 63 species of bats among 102 mammalian species (Wilson 1983), 384 bird species (Stiles 1983), 45 amphibian and 74 reptilian species (Scott et al. 1983), 195 butterfly species (DeVries 1983), and 54 species of sphingid moths (Haber 1983). The seasonally dry Santa Rosa National Park in northwestern Costa Rica harbors 64 species of sphingids among the 3124 species of moths (Janzen 1988). Although comprehensive inventories are sparse, similar comparisons illustrating the high richness of species in tropical forests are to be expected for most taxonomic groups.

## III. DOCUMENTED EFFECTS OF CLIMATIC FLUCTUATIONS ON TROPICAL FORESTS

Within the tropics, seasonal patterns and distribution of rainfall are far more important than temperature in serving as proximal cues for biological activities. Only near the latitudinal and altitudinal limits of tropical forests do unusually low temperatures affect tropical forest communities. An example is the penetration of polar air masses into tropical regions with temperatures dropping to −6°C in Paraguay. The effect of Patagonian "surazos" or "friagems" on Brazilian coffee plantations is well known, but I have not been able to find published information on the effects of cold fronts or freezes on natural forests in the latitudinal transition from warm temperate to subtropical regions.

In the humid tropical forest biome, mean annual rainfall for different sites may have a range of as much as 8 m (312 inches), from a minimum of 1500 mm (59 inches) in the lowlands to about 10,000 mm (394 inches) in middle elevations due to orographic rain on front ranges. With increasing elevation, not only does the range of mean annual rainfall values decrease markedly but the minimum rainfall for humid forest life zones also decreases. For example, 250 mm is the lower limit for the tropical subalpine moist forest life zone (fig. 10.1).

One of the striking features of rainfall records from tropical forest regions is the great year-to-year variation. At La Selva, the wettest year over a 30-year period had almost double the precipitation of the driest, whereas 60 years of rainfall records on BCI show the maximum is 217% greater than for the lowest year. These extreme years are not that unusual, hence the appreciable year-to-year variation gives a large standard error to a calculated mean. Annual rainfall is more than one standard deviation from the mean for 31% of BCI records (n = 61 yr; Leigh et al. 1990) and 41% at La Selva (n = 32). Thus wetter- or drier-than-average years in humid tropical forests may not have serious repercussions on the natural communities. However, a few dry years in a row could be detrimental to many populations. This generalization cannot be extended to the dry or semiarid tropics; witness the temporary greening of the dry Galápagos Islands due to El Niño rains (Canby 1984).

Far more important than total or average rainfall to biological activities in humid tropical forests is the seasonal occurrence of dry and wet seasons. Even in areas with high total rainfall (>4 m/yr), less-rainy periods occur, and a surprisingly high percentage of these rainy areas do have a slight dry season. Generally, rainfall has to be less than 100 mm/ month to initiate an effective dry season, and even with this low level of rainfall, it may take up to one month for the soil to dry out sufficiently to cause moisture stress of plants in humid tropical forests. If the dry season continues for more than one month, forest communities begin to show effects of moisture stress, such as leaf drop in deciduous plants, flowering, and seedling mortality.

In tropical latitudes the annual rains are usually interrupted by two dry seasons of differing lengths (fig. 10.2). Toward the subtropics, one dry season tends to be longer (when the sun is in the other hemisphere), and the other dry season is shorter or occurs sporadically (i.e., not every year). Rainy seasons normally coincide with the passage of the sun, with the zenithal rains peaking approximately one month after the sun's passage. These fairly predictable patterns of rainfall seasonality in humid tropical forest regions may be interrupted or overridden by unpredictable events such as El Niño or the penetration of polar air masses into the outer tropics. Regional climatic regimes are also affected by physiography, trade winds, major volcanic eruptions, and large bodies of water that may cause striking ecological changes over a few kilometers (e.g., the Monteverde Cloud Forest, Costa Rica).

Rainfall regimes in most humid tropical forests are more complex than alternating dry and wet seasons. Brief rainless periods in the rainy season and occasional rains in the dry season are common. More important, these weather events within a season appear to influence biological activities, especially flowering. In the seasonal dry forests along the Pacific side of Central America, a shower late in the dry season will trigger massive flowering of a population of trees, such as *Tabebuia impetiginosa, T. ochracea* and *T. rosea* (Frankie et al. 1974, Hartshorn 1983). Some individual trees will flower as many as three times during the dry season. Less well documented is the potential effect of short rainless periods during the rainy season. At La Selva, *Cespedesia macrophylla* (Ochnaceae) has massive terminal inflorescences of bright yellow flowers in March during the less rainy season; I have also seen the same trees flowering again in September after 4–7 days without rain.

Even in a lowland wet tropical forest like La Selva, short rainless periods can cause moisture stress in the forest community. In April–May 1973, La Selva experienced six weeks with only one 37-mm shower, which caused many facultatively deciduous canopy

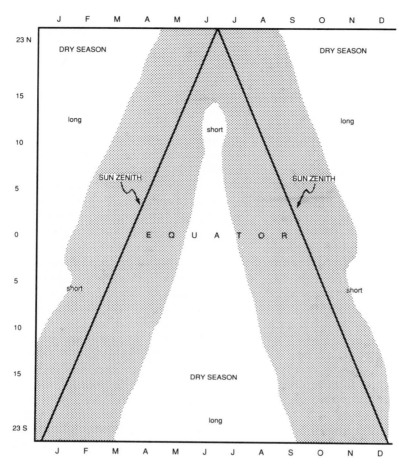

*Figure* 10.2. The occurrence of tropical wet and dry seasons with relation to latitude and position of sun. Redrawn from a handout of the Organization for Tropical Studies.

and subcanopy trees to drop their leaves. Surface-soil drying slowed litter decomposition, and the fallen leaves accumulated rapidly to as much as 10 cm deep. Moisture stress seemed to have an even more dramatic effect on the forest understory. Broadleaved monocots (palms, cyclanths, *Heliconia*) were especially affected by the greater light reaching the forest floor and the lack of moisture. The common *Asplundia uncinata* (Cyclanthaceae) literally fell over as the superficial rhizome lost turgor. With a return of the rains, the fallen *Asplundia* were able to regain turgor and their upright position. Short dry spells are a major cause of seedling mortality (Lie-

berman et al., unpublished manuscript), presumably because these small plants have not developed an extensive or deep root system.

Severe drought can also greatly increase tree mortality in tropical forests. The 1983 dry season on BCI was the most intense recorded in more than 60 years; the dry season was two months longer than usual and no rain was recorded for six months. Detailed analyses of the 1982 and 1985 inventory data from the BCI megaplot (50 ha) show 8.8% mortality of trees between inventories (Hubbell and Foster 1990). The authors attribute the very high tree mortality to the severe drought; even when converted to an annual logarithmic rate, the BCI rate is higher than for any other tropical American forest (Hartshorn 1990). An unusual feature of the BCI data is that the larger size classes of trees suffered greater mortality (Hubbell and Foster 1990).

The catastrophic drought and fires that affected 36,000 km² of eastern Borneo in 1982–83 caused high mortality of trees and lianas (Leighton and Wirawan 1986). In logged forests, ground fires killed more than 90% of the large lianas and small trees but less than 25% of the large trees. In unburned, xeric sites, large-tree mortality was much higher (37%–71%), indicating greater susceptibility to severe drought. The fire- and drought-ravaged forests "suffered other changes likely to be important to the animal community: . . . a decline in the availability of food resources drawn from plant reproductive parts, destruction of arboreal travel pathways, loss of cover and shelter from predators and climate (sun, wind, and rain), and loss in richness of habitat microsites which underlie the diversity of small organisms in tropical rain forests" (Leighton and Wirawan 1986:87).

An outstanding example of the importance of seasonality on humid tropical forest communities comes from the work of Robin Foster on Barro Colorado Island. In a multiple-year study of tree phenology and fruit production, Foster (1982a) documented that of 154 species that normally fruit during the rainy season, 55% did not produce fruit from July 1970 to February 1971. Most of these tree species usually flower at the beginning of the rainy season (April–May), but failed to do so in 1970. The dry season of 1970 was exceptionally wet, suggesting there was insufficient drought to stimulate flowering at the beginning of the rainy season. Long-term studies and observations of the mammal community on BCI have established that the late rainy season is a time of low fruit availability and increased competition for what food is available. The 1970 famine lasted much longer than in normal years.

Foster (1982b) reports the profound effects of the 1970 famine on the frugivore and omnivore guilds. During a walk of the excellent trail system on BCI, at least one dead animal was seen every 300 m. Agoutis, armadillos, coatis, howler monkeys, opossums, collared peccaries, and porcupines were the most abundant carcasses seen. Appreciable mortality also affected populations of pacas, two-toed and three-toed sloths, and white-faced monkeys. The large frugivorous birds, such as parrots and toucans, disappeared from the island; presumably they migrated in search of food. Foster also noted much damage to plants, such as gnawed bark (by peccaries), stripping of palm leaves to get at the terminal meristem, and bromeliads ripped apart. The chemical and mechanical deterrents that normally protect plants or plant parts were not effective during the severe famine.

Foster found that the weak dry season of 1970 did not affect the vegetative production of trees; because of the lack of an effective dry season, most deciduous trees stayed in leaf. Moreover, leaf litter did not accumulate, and what was on the forest floor decomposed rapidly, so that by the beginning of the typical rainy season the forest floor was essentially bare of litter. This is believed to have profoundly depressed the population of litter arthropods, the primary food source of the omnivorous coatis. During the 1958 famine, the BCI coati population declined by 50% in only 8 months, and the survivors produced no young during the next dry season (Kauffman 1962).

Using a "weakness of the dry season" index, Foster (1982b) shows good correlation with five severe famines (1931, 1956, 1958, 1960, 1970) and nine minor famine years (1932, 1935, 1942, 1951, 1952, 1955, 1963, 1972, 1978) known to have occurred over the past 50 years on BCI. Based on this analysis, Foster concludes that on BCI, tree fruiting rhythms are seriously disrupted once in ten years and mildly disrupted once in five years by a weak dry season.

## IV. POTENTIAL EFFECTS OF GLOBAL WARMING

It is unlikely that higher temperatures per se will be directly deleterious to tropical forest communities. Tropical forest regions are not nearly as hot as extratropical regions during the summer. In the humid tropical forest biome, high temperatures are modulated by cloud cover. If global warming causes increased cloud cover in tropical forest regions, it is conceivable that the global increase in average daily temperature may be modulated in the tropics. Increasing sea surface temperatures, however, could spawn more frequent tropical cyclones, with a concomitant increase in severity (Emanuel 1988). If global warming is involved in the inner tropical origin and track of tropical cyclones (such as Hurricane Joan, October 1988), appreciable areas of tropical forests could be devastated by the high winds (cf. Boucher 1990).

Global warming could contribute to a relaxation of the altitudinal and latitudinal limits of tropical forests. The paleoecological records indicate that tropical forests extended to temperate latitudes in climatically benign geological periods. Gentry (1986) argues that marginal latitudinal tropical habitats have high rates of speciation. Given the occasional penetration of polar air masses into the outer tropics, such a relaxation of latitudinal limits to tropical forests probably would not be a simple function of an increase in average temperature. The relaxation of altitudinal limits to tropical forests is even more speculative because of complex interactions of night reradiation, daily temperature regimes, cold-air drainage, and cloud cover, among others. In many parts of the high Andes, tree line has been lowered several hundred meters by human agricultural and burning practices. If global warming raises the altitudinal limits of tropical forests, the montane tropical forests could suffer appreciable decline in available habitats caused by increasing human pressure for agricultural use.

The large year-to-year variation and great range in total rainfall over the humid tropical forest biome suggest that secular changes in total rainfall should not be expected to have a major effect on the biological diversity of tropical forests. Changes in the seasonality of rainfall, however, will have noticeable effects on humid tropical forest communities. Numerous phenological studies indicate that tropical forest communities, species, and populations are sensitive to subtle fluctuations and changes in rainfall patterns. The

intricate and multifaceted food webs and the highly coevolved interactions among species mean that a seasonal or annual failure to flower or fruit will also affect second- and third-order interactions (Gilbert 1980). When flowering or fruiting failure occurs among several species in a guild or community, the consequences for frugivores are profound, as demonstrated by the BCI famines.

Increased seasonality of rainfall in humid tropical forests could have profound effects on the plant and animal communities. The effects will be due most likely to altered or more variable rainfall patterns that serve as phenological triggers for many plant species. Although most adult populations of plants flower on a more or less annual basis, a not insignificant number of species skip one or more years. At La Selva, some tree species flower in alternate years (e.g., *Hymenolobium pulcherrimum*, Fabaceae), whereas other tree species skip three to five years. For example, we have been monitoring several adult trees of *Ampelocera hottlei* (Ulmaceae?) for three years in order to collect flowers that will help confirm that this tree species belongs in a new plant family, but no flowering has yet occurred. I am not suggesting that tropical American forests will soon become as highly synchronized and sporadic as the mast-fruiting of dipterocarp forests in peninsular Malaysia and Borneo (Janzen 1974, Ng 1981). But if more species become supra-annual reproducers because of changing seasonality patterns in tropical American forests, not only the pollinators and fruit eaters of that species but also generalist flower visitors and frugivores could suffer from lack of food sources.

Despite the hundreds of tree species in a particular tropical forest, recent research findings indicate that a surprisingly low number of trees function as keystone species for frugivorous birds and mammals. Keystone species are not necessarily the most abundant trees, rather they are those species that fruit in the off-season, that is, during the seasonal period of food scarcity. Obviously, keystone species should fruit every year if the frugivore community is to be supported through the lean season. In several tropical American forests, fig trees serve as keystone species for frugivorous vertebrates. Ripening figs broke the famines on BCI. Figs are an important food of the resplendent quetzal in Monteverde, but we do not know how fruit availability and phenology influence the altitudinal movements of these large frugivores. Figs and palms comprise the majority of the 10–12 keystone species in Manu National Park in southeastern Peru (Terborgh 1986). Figs appear to be less important than palms as keystone species in the La Selva forest.

## V. CONCLUSIONS

Tropical forests are very complex ecologically, and that great complexity ranges from the ecosystem level to species interactions. Humid tropical forests are an incredible storehouse of biological diversity, with the recent suggestion that the estimated number of species on this planet should be raised to a minimum of 30 million. In the humid tropical forest biome, seasonal patterns and the distribution of rainfall are more important than temperature in cuing many biological activities. Fluctuations in the occurrence and duration of wet and dry seasons cause flowering and fruiting failures that have profound effects on animal and decomposer communities. Repetitive or more frequent failures of keystone tree species to fruit could lead to local extinction of frugivorous species. More variable seasonality in humid tropical forests could increase susceptibility to natural or human-caused disasters, such as fire.

## REFERENCES

Boucher, D. H. 1990. Growing back after hurricanes: Catastrophes may be critical to rain forest dynamics. *Bioscience* 40(3):163.

Canby, T. Y. 1984. El Niño's ill wind. *National Geographic* 165(2):144.

Clark, D. B. 1990. La Selva biological station: A blueprint for stimulating tropical research. In *Four Neotropical Rainforests*, A. H. Gentry, ed., pp. 9–27. New Haven: Yale University Press.

DeVries, P. J. 1983. Checklist of butterflies. In *Costa Rican Natural History*, D. H. Janzen, ed., pp. 654–678. Chicago: University of Chicago Press.

DeVries, P. J. 1987. *The Butterflies of Costa Rica and Their Natural History.* Princeton: Princeton University Press.

Emanuel, K. A. 1988. Toward a general theory of hurricanes. *Am. Scient.* 76(4):371.

Erwin, T. L. 1988. The tropical forest canopy: The heart of biotic diversity. In *Biodiversity*, E. O. Wilson, ed., pp. 123–129. Washington, D.C.: National Academy Press.

Foster, R. B. 1982a. The seasonal rhythm of fruitfall on Barro Colorado Island. In *The Ecology of a Tropical Forest: Seasonal Rhythms and Long-Term Changes*, E. G. Leigh, Jr., A. S. Rand, and D. M. Windsor, eds., pp. 151–172. Washington, D.C.: Smithsonian Institution Press.

Foster, R. B. 1982b. Famine on Barro Colorado Island. In *The Ecology of a Tropical Forest: Seasonal Rhythms and Long-Term Changes*, E. G. Leigh, Jr., A. S. Rand, and D. M. Windsor, eds., pp. 201–212. Washington, D.C.: Smithsonian Institution Press.

Frankie, G. W., H. G. Baker, and P. A. Opler. 1974. Comparative phenological studies of trees in tropical wet and dry forests in the lowlands of Costa Rica. *J. Ecol.* 62(3):881.

Gentry, A. H. 1982. Patterns of neotropical plant species diversity. *Evol. Biol.* 15:1.

Gentry, A. H. 1986. Endemism in tropical versus temperate plant communities. In *Conservation Biology: The Science of Scarcity and Diversity*, M. E. Soulé, ed., pp. 153–181. Sunderland, Mass.: Sinauer Associates.

Gentry, A. H. 1988. Tree species richness of upper Amazonian forests. *P. NAS* 85:156.

Gentry, A. H., and C. Dodson. 1987. Contribution of nontrees to species richness of a tropical rain forest. *Biotropica* 19(2):149.

Gilbert, L. E. 1980. Food web organization and conservation of neotropical diversity. In *Conservation Biology*, M. E. Soulé and B. A. Wilcox, eds., pp. 11–34. Sunderland, Mass.: Sinauer Associates.

Gilbert, L. E., and P. H. Raven, eds. 1975. *Coevolution of Animals and Plants.* Austin: University of Texas Press.

Haber, W. A. 1983. Checklist of Sphingidae. In *Costa Rican Natural History*, D. H. Janzen, ed., pp. 645–650. Chicago: University of Chicago Press.

Hartshorn, G. S. 1983. Plants: Introduction. In *Costa Rican Natural History*, D. H. Janzen, ed., pp. 118–157. Chicago: University of Chicago Press.

Hartshorn, G. S. 1988. Tropical and subtropical vegetation of Meso-America. In *North American Terrestrial Vegetation*, M. G. Barbour and W. D. Billings, eds., pp. 365–390. New York: Cambridge University Press.

Hartshorn, G. S. 1990. An overview of neotropical

forest dynamics. In *Four Neotropical Rainforests*, A. H. Gentry, ed., pp. 585–599. New Haven: Yale University Press.

Holdridge, L. R. 1947. Determination of world plant formations from simple climatic data. *Science* 105(2727):367.

Holdridge, L. R. 1967. *Life Zone Ecology*. San José, Costa Rica: Tropical Science Center.

Hubbell, S. P., and R. B. Foster. 1990. Structure, dynamics, and equilibrium status of old-growth forest on Barro Colorado Island. In *Four Neotropical Rainforests*, A. H. Gentry, ed., pp. 522–541. New Haven: Yale University Press.

Janos, D. P. 1983. Tropical mycorrhizas, nutrient cycles and plant growth. In *Tropical Rain Forest: Ecology and Management*, S. L. Sutton, T. C. Whitmore, and A. C. Chadwick, eds., pp. 327–345. Oxford: Blackwell Scientific Publications.

Janzen, D. H. 1966. Coevolution of mutualism between ants and acacias in Central America. *Evolution* 20:249.

Janzen, D. H. 1974. Tropical black water rivers, animals, and mast fruiting by the Dipterocarpaceae. *Biotropica* 6:69.

Janzen, D. H. 1988. Complexity is in the eye of the beholder. In *Tropical Rainforests: Diversity and Conservation*, F. Almeda and C. M. Pringle, eds., pp. 29–51. San Francisco: California Academy of Sciences and AAAS.

Kauffmann, J. H. 1962. Ecology and social behavior of the coati, *Nasua narica*, on Barro Colorado Island, Panama. *Univ. Calif. Publ. Zool.* 60:95.

Leigh, E. G., Jr., A. S. Rand, and D. M. Windsor, eds. 1982. *The Ecology of a Tropical Forest: Seasonal Rhythms and Long-Term Changes*. Washington, D.C.: Smithsonian Institution Press.

Leigh, E. G., Jr., A. S. Rand, and D. M. Windsor, eds. 1990. *Ecología de un bosque tropical: Ciclos estacionales y cambios a largo plazo*. Balboa, Panama: Smithsonian Tropical Research Institute.

Leighton, M., and N. Wirawan. 1986. Catastrophic drought and fire in Borneo tropical rain forest associated with the 1982–1983 El Niño southern oscillation event. In *Tropical Rain Forests and the World Atmosphere*, G. T. Prance, ed., pp. 75–102. Boulder, Colo.: Westview Press.

Ng, F.S.P. 1981. Vegetative and reproductive phenology of dipterocarps. *Malay. Forester* 44:197.

ONERN. 1976. *Mapa ecológico del Perú*. Lima: Oficina Nacional de Evaluación de Recursos Naturales.

Raven, P. H. 1988. Our diminishing tropical forests. In *Biodiversity*, E. O. Wilson, ed., pp. 119–122. Washington, D.C.: National Academy Press.

Ray, T. S., and C. C. Andrews. 1980. Antbutterflies: Butterflies that follow army ants to feed on antbird droppings. *Science* 210:1147.

Sawyer, J. O., and A. A. Lindsey. 1964. The Holdridge bioclimatic formations of the eastern and central United States. *Indiana Acad. Sci. Proc.* 73:105.

Scott, N. J., J. M. Savage, and D. C. Robinson. 1983. Checklist of reptiles and amphibians. In *Costa Rican Natural History*, D. H. Janzen, ed., pp. 367–374. Chicago: University of Chicago Press.

Stiles, F. G. 1978. Temporal organization of flowering among the hummingbird food plants of a tropical wet forest. *Biotropica* 9:194.

Stiles, F. G. 1980. The annual cycle in a tropical wet forest hummingbird community. *Ibis* 122:322.

Stiles, F. G. 1983. Checklist of birds. In *Costa Rican Natural History*, D. H. Janzen, ed., pp. 530–544. Chicago: University of Chicago Press.

Stiles, F. G. 1988. Altitudinal movements of birds on the Caribbean slope of Costa Rica: Implications for conservation. In *Tropical Rainforests: Diversity and Conservation*, F. Almeda and C. M. Pringle, eds., pp. 243–258. San Francisco: California Academy of Sciences and AAAS.

Terborgh, J. 1986. Keystone plant resources in the tropical forest. In *Conservation Biology: The Science of Scarcity and Diversity*, M. E. Soulé, ed., pp. 330–344. Sunderland, Mass.: Sinauer Associates.

Tosi, J. A., Jr. 1969. *Mapa ecológico de Costa Rica*. San José, Costa Rica: Tropical Science Center.

Tosi, J. A., Jr. 1980. Life zones, land use, and forest vegetation in the tropical and subtropical regions. In *The Role of Tropical Forests on the World Carbon Cycle*, S. Brown, A. Lugo, and B. Liegel, eds., pp. 44–64. Washington, D.C.: Dept. of Energy.

Wheelwright, N. T. 1983. Fruits and the ecology of resplendent quetzals. *Auk* 100:286.

Whitmore, T. C., R. Peralta, and K. Brown. 1985. Total species count in a Costa Rican tropical rain forest. *J. Trop. Ecol.* 1(4):375.

Wilson, D. E. 1983. Checklist of mammals. In *Costa Rican Natural History*, D. H. Janzen, ed., pp. 443–447. Chicago: University of Chicago Press.

Wilson, E. O. 1988. The current state of biological diversity. In *Biodiversity*, E. O. Wilson, ed., pp. 3–18. Washington, D.C.: National Academy Press.

Participation in the conference and travel support from World Wildlife Fund are gratefully acknowledged. Constructive comments by L. Hartshorn and D. Lieberman improved an earlier version of this chapter.

# CHAPTER ELEVEN

# Using Computer Models to Project Ecosystem Response, Habitat Change, and Wildlife Diversity

HERMAN H. SHUGART AND
THOMAS M. SMITH

## I. INTRODUCTION

From the point of view of an ecologist, the central question to be asked about the response of ecosystems to global climatic change is "Will the rate and magnitude of global change be ecologically significant?" Will ecosystems simply adjust to change with no apparent effects, or will changes occur too fast and be too large for such adjustments? If alterations do occur, what will be the pattern and consequences of these responses?

There is a great degree of uncertainty over the magnitude and temporal pattern of the increase in the earth's ambient $CO_2$, as well as the regional pattern of climate change that may result (Bolin et al. 1986). However, the magnitude of climatic change being predicted by models of the earth's weather machine (general circulation models, or GCMs) in response to a doubling of atmospheric carbon dioxide appears to be ecologically significant (Emanuel et al. 1985, Shugart et al. 1986) and could result in major changes in the distribution and abundance of the earth's flora and fauna.

Much emphasis has been placed on examining the potential effects of a climate change on patterns of vegetation at the regional (Smith and Tirpak 1989) and global scale (Emanuel et al. 1985, Smith et al. 1991). Much less attention has been directed toward the distribution and abundance of the earth's fauna. Although there are many facets to the question "How will patterns of animal species distribution respond to changed climate conditions?" we will address one component of this broad topic: the availability of animal habitats.

Scientific tools that can be used to assess the effects of global climate change on animal diversity include models of vegetation dynamics, habitat classification, and a theoretical framework for viewing animal habitat as a dynamic mosaic. To demonstrate these tools, we will provide concrete examples of a theory based on habitat dynamics, which we feel can be used to assess the effects of global change on biodiversity.

## II. FOREST SIMULATION MODELS BASED ON INDIVIDUAL TREES

In the mid-sixties, computer models were first developed that simulate the dynamics of a forest stand by following the fate of each individual tree, in the form of a changing map of tree positions and sizes (Newnham 1964, Lee 1967, Mitchell 1969, Lin 1970, Bella 1971a and b, Arney 1974, Hatch 1971). The individual-tree models presented here are called gap models and are based on the FORET model and its descendants (fig. 11.1; Shugart and West 1980; Shugart 1984 provides a more detailed discussion of the derivation and testing of these models). The FORET model was originally derived from the JABOWA model, an ecological version of the earlier forestry models based on individual tree dynamics (Botkin et al. 1972). One appeal of the individual approach is the considerable body of information on the performance of individual trees (e.g., growth rates, requirements for establishment, shade tolerance, allometric relationships) that can be used for estimating

Figure 11.1. Geographic distribution of gap models derived from JABOWA and FORET. 1. FORET (Shugart and West 1977, Shugart 1984). Southern Appalachian Deciduous Forest. 2. BRIND (Shugart and Noble 1981). Australian Eucalyptus Forest. 3. KIABRAM (Shugart et al. 1980). Australian Subtropical Rain Forest. 4. FORICO (Doyle 1981). Puerto Rican Montane Rain Forest. 5. SMAFS (El-Bayoumi et al. 1984). Eastern Canadian Mixed-Wood Forest. 6. CLIMACS (Dale et al. 1984). Pacific Northwest Coniferous Forest. 7. FORFLO (Pearlstine et al. 1985). Southern Floodplain Forest. 8. FORCAT (Waldrop et al. 1986). Oak-Hickory Forest. 9. FORSKA (Leemans and Prentice 1987). Scandanavian Forest. 10. FORECE (Kienast 1987). Central European Forest. 11. FOR-ANAK (Busing and Clebsch 1987). Montane Boreal Forest. 12. ZELIG (Smith and Urban 1988). Temperate Deciduous Forest. 13. LOKI (Bonan 1989). North American Boreal Forest. 14. MANGRO (in progress). Caribbean Mangrove Forest. 15. OUTENIQUA (Van Daalen and Shugart 1989). African Temperate Rain Forest. 16. JABOWA (Botkin et al. 1972). Northern Hardwood Forest. 17. FORTNITE (Aber and Melillo 1982). Wisconsin Mixed-Wood Forest. 18. SWAMP (Phipps 1979). Arkansas Floodplain. 19. SJABO (Tonu 1983). Estonian Conifer Forest. 20. SILVA (Kercher and Axelrod 1984). Mixed Conifer Forest. 21. LINKAGES (Pastor and Post 1986). Temperate-Boreal Transition.

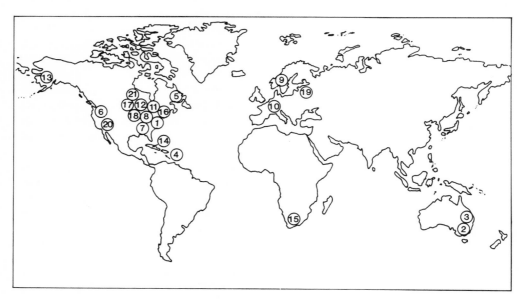

the parameters of such models. Most individual-based models of forests are parsimonious in that they are based on simple rules for birth, growth, and death of individuals.

Forest gap models simulate the establishment, diameter growth, and mortality of trees at an annual time-step on a plot of defined size (Shugart 1984). Although different models include different processes that may be important in the dynamics of a particular site (e.g., hurricane disturbance, flooding), all forest gap models share a common set of characteristics and demographic processes. Each individual plant is modeled as a unique entity with respect to the processes of establishment, growth, and mortality. This allows the model to track species- and size-specific demographic behaviors. The model structure includes two features important to a dynamic description of vegetation: the response of the individual plant to the prevailing environmental conditions, and how the individual modifies those environmental conditions.

## A. Response of Individual to Environment

The growth of an individual is calculated using a function that is species-specific and predicts, under optimal conditions, an expected growth increment given an individual's current size (fig. 11.2). This optimum increment is then modified by a set of environmental response functions, and the realized increment is added to the individual.

Environmental responses are modeled using a "constrained potential" approach. In this, an individual has a maximum potential behavior under optimal conditions (i.e., maximum diameter increment, survivorship, or establishment rate). This optimum is then reduced to reflect the environmental conditions encountered by any individual on the plot (e.g., shading, drought), to yield the realized behavior. The functions describing species response to environmental conditions tend to be generic curves that scale between 0.0 and 1.0, and species are often categorized into a small number of functional types having similar responses to the environmental gradient (fig. 11.3).

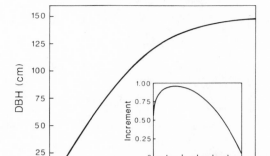

Figure 11.2. Annual diameter growth as simulated in gap models: optimal growth through time, and (inset) diameter increment as a function of current diameter. Potential increment is multiplied by environmental growth response functions (see fig. 11.3) to yield realized annual increment.

The horizontal position of each individual on the plot is not considered. The assumption is made that each individual on the plot is influenced by (and influences) the growth of all other individuals on the plot. Given this assumption of horizontal homogeneity, the size of the simulated plot is critical. The spatial scale at which these models operate is an area corresponding to the zone of influence of a single individual of maximum size. This allows for an individual growing on the plot to achieve maximum size while also allowing for the death of a large individual to significantly influence the environment on the plot (i.e., gap formation).

Competition in the model is indirect and depends on the relative performance of the individuals under the environmental conditions on the model plot. These environmental conditions may be influenced by the trees themselves (e.g., a tree's leaf area influences light available beneath it) or may be modeled as extrinsic (e.g., temperature). Competitive success thus depends on the environmental conditions on the plot, which species are present, and the relative sizes of the individuals; all vary through time in the model.

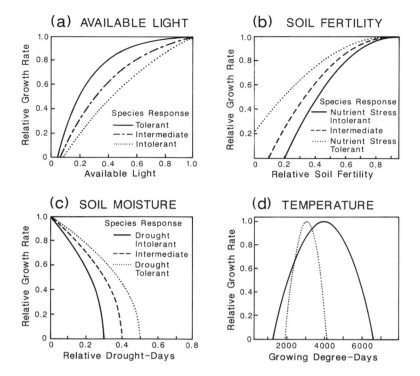

*Figure* 11.3. Response functions used to modify annual diameter growth of trees for the environmental constraints of (a) light, (b) soil fertility, (c) soil moisture, and (d) temperature. Most species can be assigned to functional groupings (e.g., shade tolerant, shade intolerant) for response to light and nutrients; temperature and moisture responses are based on the geographic range of the species.

The death of individuals is modeled as a stochastic (probability) process. Most gap models have two components of mortality: age-related and stress-induced. The age-related component applies equally to all individuals of a species and depends on the maximum expected longevity for the species. Stress is defined with respect to a minimal growth increment; individuals failing to meet this minimal condition are subjected to an increased probability of mortality.

Establishment and regeneration are largely stochastic, with maximum potential establishment rates constrained (i.e., reduced) by the same environmental factors that modify tree growth.

## B. Effect of Individual on Environment

Although the horizontal position of each individual on the plot is not considered, the vertical structure of the canopy is modeled explicitly. The height and leaf area of each individual (which are calculated based on allometric relationships relating these variables to stem diameter) are used to construct a vertical leaf-area profile. Using a light extinction equation, the vertical profile of available light is then calculated so that the light environment for each individual can be defined (fig. 11.4). All gap models simulate individual tree influence on light availability at height intervals on the plot. Plant influences on other environmental factors are incorporated to varying degrees in different versions of the model; these include soil moisture, fertility, temperature, as well as disturbances such as fires and herbivory.

In actuality, the modeling framework outlined above has two components or modules. One involves plant demographics, and the other involves the processes, both clima-

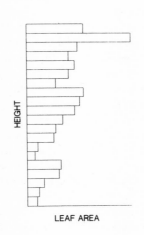

HEIGHT

LEAF AREA    % AVAILABLE LIGHT

*Figure* 11.4. The height profile of individuals on the simulated plot is used to calculate a leaf-area profile. The leaf-area profile is then used to construct an available-light profile to define available light for any given individual on the plot.

tic and biophysical, that define the environment on each plot. (An extensive discussion of the biophysical models is beyond the scope of this chapter, but examples of them and their incorporation into the gap model structure can be found in Shugart 1984, Pastor and Post 1985, and Bonan 1989.) In general, these biophysical models require the transfer of certain characteristics from the demographic module describing the vegetation of the plot and then defining the environmental conditions as a function of the geology and soils, prevailing climate, and the modifying influence of the plant structure.

## III. APPLICATIONS OF INDIVIDUAL-BASED TREE MODELS TO WILDLIFE CONSERVATION AND MANAGEMENT

Individual-based forest models can be used to predict changes in the structure and composition of forests through time. When combined with a description of the features of the vegetation correlated with the presence of (or use by) a given animal species, they can be used as habitat simulators.

Habitat simulation integrates the use of habitat classification with the predictive ability of the forest simulation model. This process requires a means of identifying the suitability (based on structure and composition) of a forest stand to provide habitat for a given animal species, and a forest model with the ability to simulate those features. With those two components, it is possible to predict habitat availability through time under changing environmental conditions. This section offers three examples of the use of forest simulation models to predict changes in habitat availability at both the population and community levels.

### A. Simulating the Dynamics of Red-Cockaded Woodpecker Nest Sites

Smith et al. (1981a) document the use of the FORAR (Mielke et al. 1977) model of ecological succession in southern pine-oak forests to simulate the dynamics of the availability of red-cockaded woodpecker (*Piciodes borealis*) habitat. The red-cockaded woodpecker prefers mature pine stands with a low, sparse understory (Cosby 1971, Jackson 1977). For nest sites, the birds select live pine trees infected with *Fomes pini*, a fungal disease that rots the heartwood of large, mature pines and is commonly called red-heart disease. The age of such nest trees varies, but rarely do the birds nest in trees younger than 60 years

(Hopkins and Lynn 1971, Jackson et al. 1979). That is considerably older than the pine forests grown in the rotation for timber production in the South, and therefore the birds are relatively rare. Since nest trees are diseased and likely to die over relatively short periods, they are temporary elements of the landscape that require a considerable time to be replaced.

The FORAR model was used to simulate the effects of several different management strategies consisting of thinning and controlled burning of forests of different ages (table 11.1). Suitable sites were defined as any stands with pine trees older than 60 years, low stocking density of trees (less than 80 square feet per acre), and a low density of understory.

The five management schemes (table 11.1) were inspected after 100 years of simulation, at which time the stand was harvested (clearcut). The management schemes represent a variety of cutting schedules with periodic controlled fires to clear the understory of invasion by hardwood species. Management scheme 3 (cuts at years 25, 40, and 60 to a basal area of 80 square feet) gave the maximum timber production as well as availability of nest trees. Management scheme 2 (thinning at year 60) resulted in larger potential nest trees but a reduction in the yield of timber. Management scheme 1 (no thinning, no burning) produced large nest trees, but because of the dense understory (a consequence of no burning) the habitat did not conform to that typically used by red-cockaded woodpeckers. Of course, the suitability of the tree as a nest site depends on the presence of the red-heart fungus (a factor strongly related to the age of the trees), and therefore these results are for *potential* habitat.

The two points to be made with respect to this example are: one can use forest simulation models to project habitat through time and over a variety of conditions; and the specialization of habitat use by some animal species, such as the red-cockaded woodpecker, tends to make them vulnerable to widespread habitat change because the habitat elements (e.g., large pine trees) may be displaced (by either management or changing environmental conditions) and may not regenerate for a long time relative to the popu-

*Table* 11.1. Simulation of potential habitat of the red-cockaded woodpecker under five different management schemes.

| Type of Management | Mean Basal Area Removed (square ft/acre) | Mean Number of Trees Cut (per acre) | Number Potential Nest Sites at Yr 100 | Basal Area of Nest Trees |
|---|---|---|---|---|
| 1. No thinning cuts. No fire. Clearcut at yr 100. | 81.11 | 17.0 | 17.0 | 4.77 |
| 2. Thinning[a] at yr 60. Periodic fire. Clearcut at yr 100. | 120.91 | 30.5 | 16.5 | 3.96 |
| 3. Thinning[a] at yrs 25, 40, and 60. | 206.35 | 187.0 | 19.5 | 1.10 |
| 4. Thinning[a] at yrs 30, 40, and 60. Periodic fire. Clearcut at yr 100. | 169.89 | 137.0 | 13.0 | 1.24 |
| 5. Thinning[b] at yrs 25, 40, and 60. Periodic fire. Clearcut at yr 100. | 165.54 | 225.0 | 15.5 | 0.74 |

a. Thinning is to level of 80 square feet per acre.
b. Thinning is to level of 60 square feet per acre.

lation dynamics of the animal species. These considerations speak against a simple monitoring of species to detect changed or changing conditions. Computer models such as those discussed here allow for the inspection of different habitat management schemes in cases where field tests of such management plans would take much too long to develop.

## B. Predicting Population Responses Based on Habitat Dynamics

The woodpecker example shows how individual-based models can simulate characteristics of individual trees that provide habitat for a given bird species. Likewise, the models can simulate a more composite description of the forest stand, which may be correlated with the suitability to provide habitat for a given animal species.

Numerous studies have correlated the presence and absence of various animal species with features of the vegetation using multivariate statistics. These procedures provide a composite picture of the vegetation in which a given species of animal is found. Combining these statistical descriptions of the habitat with a forest model that simulates changes in the features of the vegetation on which the description is based, we can model changes in the availability of habitat as a function of time and changing environmental conditions.

Smith et al. (1981a and b) used a modified version of the FORET forest model to predict the dynamics of habitat availability for a number of bird species occupying the Walker Branch Watershed, Oak Ridge, Tennessee, in the southeastern United States. As with the previous example, various management options were evaluated.

The description of the habitat for each of the bird species was based on a sample of 298 forest plots of 0.08 ha each in the watershed. Data describing the structure of the vegetation on each plot included size-class distributions of trees and various biomass estimates of branches and foliage. A census of breeding males on the plots provided presence and absence data for each of the bird species. This was used to classify each plot as habitat or unsuitable. Two-group (habitat-unsuitable) discriminant function analysis was used to develop a classification system for each species. This system could be used to classify forest stands in terms of their potential to provide habitat for any of the species found on the plots.

The FORET model was then used to predict changes in the 298 plots for unmanaged and managed scenarios. The management procedure simulated was a diameter-limit cut for all commercially valuable species (>30 cm diameter) on a 60-year rotation. Each year, the variables describing the vegetation (e.g., size-class distributions) were used to determine the suitability of each plot as habitat for each of the bird species. This was done using the classification system described above. The result is a graph showing the percentage of the original 298 forest plots that are potential habitat for each of the bird species at any given year of the simulation. Figure 11.5 shows two such graphs for the ovenbird (*Seiurus aurocapillus*) and the red-eyed vireo (*Vireo olivaceus*).

## C. Predicting Animal Community Patterns Based on Habitat Dynamics

The examples above have used features of the vegetation correlated with the presence of a given animal species to simulate the dynamics of habitat availability. Theoretically it is possible to repeat this procedure for all the species composing the animal community of a forest and, in doing so, simulate changes in properties of the community as a whole, such as species diversity. An alternative approach, however, is possible.

R. H. MacArthur and J. W. MacArthur (1961) correlated bird species diversity with an index of foliage height diversity (i.e., an index related to the range in height over which the foliage is distributed), which could be measured for different forest types. This approach implies that there is a relationship between the number of bird species present in a forest and the structural diversity of that

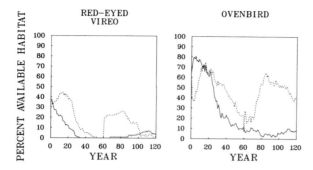

Figure 11.5. Percent available habitat for the red-eyed vireo and ovenbird as predicted for both undisturbed and timber harvest conditions. Percentage of available habitat expressed as a percentage of the 298 sample plots simulated. Simulations for undisturbed conditions are shown as a solid line and managed conditions as a dashed line. From Smith et al. 1981a.

forest or, viewed another way, the diversity of habitats provided by the forest. D. L. Urban and T. M. Smith (1989) have expanded this approach, viewing the forest as a dynamic habitat mosaic varying in time and space, with bird species diversity being a function of that mosaic. Defining a relationship between veg-

etation structure and the presence of a hypothetical set of bird species (similar to the ovenbird and red-eyed vireo examples), the authors used a spatially explicit version of the FORET forest model (ZELIG: see Smith and Urban 1988) to examine the spatial and temporal patterns of bird species diversity. By sampling the simulated forest at differing spatial scales and at differing points in time, the authors were able to match observed patterns in species diversity throughout succession and the relationship between spatial area sampled and the number of species found in an area (i.e., species-area relationship) (fig. 11.6). The development of such a dynamic habitat-mosaic approach to animal community patterns combined with forest simulation models could provide a powerful tool to examine the influences of changing environmental conditions or management practices on faunal diversity.

Figure 11.6. Patterns of (a) species diversity with succession, and (b) species-area relationship (relationship between number of species found and area sampled) simulated using an individual-based forest model with defined habitat requirements of species based on vegetation structure. From Urban and Smith 1989.

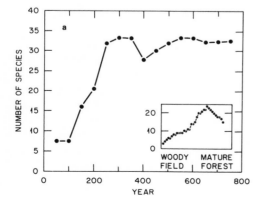

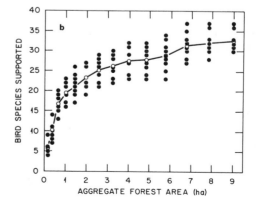

## V. SUMMARY

We have presented here a framework for predicting the dynamics of animal populations and patterns of species diversity as a function of the dynamics of the habitats required by the animals. This approach does not actually simulate changes in the populations of the animal species; rather, it predicts the availability of habitat as defined by the structure of the vegetation. Numerous factors other than the availability of habitat may limit the population dynamics of a species (e.g., predation, climate, food availability), and in many cases areas of marginal habitat may be utilized during periods of high population density. The population of a species, however, will ultimately be constrained by the existence of habitat; that is, although the availability of habitat does not mean a species will be present, if habitat is not available the species cannot exist in that location. Therefore, habitat availability provides an index for the potential presence of a species in any location. Second, this approach provides a direct link between the response of vegetation to changing climate patterns and the potential effects of climate change on patterns of animal diversity.

The forest gap models that we have discussed as habitat simulators are widely used to examine the response of forested ecosystems to scenarios of global climate change (e.g., Pastor and Post 1986, Solomon 1986, Urban and Shugart 1989). Given the possible changes in the structure of the forest as a function of climate change, the potential of the forest to provide habitat for a given species can be evaluated in the same manner as in the examples for the various management schemes.

Although providing only one index of the response of animals to changing climate conditions, the approach of linking patterns of diversity with vegetation structure can potentially provide an important tool for examining the possible effects of climate change on patterns of biodiversity.

## REFERENCES

Aber, J. D., and J. M. Melillo. 1982. FORTNITE: A computer model of organic matter and nitrogen dynamics in forest ecosystems. Univ. Wisconsin Res. Bull. R3130.

Arney, J. D. 1974. An individual tree model for stand simulation in Douglas-fir. In *Growth Models for Tree Stand Simulation*, J. Fries, ed., Research Notes 30, Dept. of Forest Yield Research. Stockholm: Royal College of Forestry.

Bella, I. E. 1971a. Simulation of growth, yield and management of aspen. Ph.D. diss., University of British Columbia, Vancouver.

Bella, I. E. 1971b. A new competition model for individual trees. *Forest Sci.* 17:364.

Bolin, B., B. R. Doos, J. Jager, and R. A. Warrick. 1986. *The Greenhouse Effect: Climate Change and Ecosystems*. (SCOPE 29) New York: John Wiley and Sons.

Bonan, G. B. 1989. A computer model of the solar radiation, soil moisture, and soil thermal regimes in boreal forests. *Ecol. Model.* 45:275.

Botkin, D. B., J. F. Janak, and J. R. Wallis. 1972. Some ecological consequences of a computer model of forest growth. *J. Ecol.* 60:849.

Busing, R. T., and E.E.C. Clebsch. 1987. Application of a spruce-fir forest canopy gap model. *Forest Ecol. Manag.* 20:151.

Cosby, G. T. 1971. Home range characteristics of the red-cockaded woodpecker in north-central Florida. In *The Ecology and Management of the Red-Cockaded Woodpecker*, R. L. Thompson, ed. Tallahassee, Fla.: Bureau of Sport Fisheries and Wildlife, U.S. Dept. of Interior and Tall Timbers Research Station.

Dale, V. M., M. Henstrom, and J. Franklin. 1984. The impact of disturbance frequency on forest succession in the Pacific Northwest. *Proc. Soc. Am. Foresters Annual Convention* 1983:300.

Doyle, T. W. 1981. The role of disturbance in the gap dynamics of a montane rain forest: An application of a tropical forest succession model. In *Forest Succession: Concepts and Applications*, D. C. West, H. H. Shugart, and D. B. Botkin, eds. New York: Springer-Verlag.

El-Bayoumi, M. A., H. H. Shugart, and R. W. Wein. 1984. Modeling succession of eastern Canadian mixed wood forest. *Ecol. Model.* 21:175.

Emanuel, W. R., H. H. Shugart, and M. P. Stevenson. 1985. Climatic change and the broad-scale distribution of terrestrial ecosystem complexes. *Clim. Change* 7:29.

Hatch, C. R. 1971. Simulation of an even-aged red

pine stand in Minnesota. Ph.D. diss., University of Minnesota, St. Paul.

Hopkins, M. L., and T. E. Lynn, Jr. 1971. Some characteristics of red-cockaded woodpecker cavity trees and management implications in South Carolina. In The Ecology and Management of the Red-Cockaded Woodpecker, R. L. Thompson, ed. Tallahassee, Fla.: Bureau of Sport Fisheries and Wildlife, U.S. Dept. of Interior and Tall Timbers Research Station.

Jackson, J. A. 1977. Red-cockaded woodpeckers and pine red-heart disease. Auk 94:160.

Jackson, J. A., M. R. Lennartz, and R. G. Hooper. 1979. Tree age and cavity initiated by red-cockaded woodpeckers. J. Forestry 77:103.

Kercher, J. R., and M. C. Axelrod. 1984. Analysis of SILVA: A model for forecasting the effects of $SO_2$ pollution and fire on western conifer forests. Ecol. Model. 23:165.

Kienast, F. 1987. FORECE: A forest succession model for southern Central Europe. ORNL/TM-10575. Oak Ridge, Tenn.: Oak Ridge National Laboratory.

Lee, Y. 1967. Stand models for lodgepole pine and limits to their application. Forestry Chron. 43:387.

Leemans, R., and I. C. Prentice. 1987. Description and simulation of tree-layer composition and size distributions in a primeval Picea-Pinus forest. Vegetatio 69:147.

Lin, J. Y. 1970. Growing space index and stand simulation of young western hemlock in Oregon. Ph.D. diss., Duke University.

MacArthur, R. H., and J. W. MacArthur. 1961. On bird species diversity. Ecology 42:594.

Mielke, D. L., H. H. Shugart, and D. C. West. 1977. User's manual for FORAR, a stand model for composition and growth of upland forests of southern Arkansas. ORNL/TM-5767. Oak Ridge, Tenn.: Oak Ridge National Laboratory.

Mitchell, K. J. 1969. Simulation of growth of even-aged stands of white spruce. Yale School of Forestry Bull. 75:1.

Newnham, R. M. 1964. The development of a stand model for douglas-fir. Ph.D. diss., University of British Columbia, Vancouver.

Pastor, J., and W. M. Post. 1985. Development of a linked forest productivity–soil carbon and nitrogen model. ORNL/TM-9519. Oak Ridge, Tenn.: Oak Ridge National Laboratory.

Pastor, J., and W. M. Post. 1986. Influence of climate, soil moisture, and succession on forest carbon and nitrogen cycles. Biogeochemistry 2:3.

Pearlstine, L., H. McKeller, and W. Kitchens. 1985. Modeling the impacts of river dimension on bot-

tomland forest communities in the Santee River floodplain, South Carolina. Ecol. Model. 29:283.

Phipps, R. L. 1979. Simulation of wetland forest dynamics. Ecol. Model. 7:257.

Shugart, H. H. 1984. A Theory of Forest Dynamics. New York: Springer-Verlag.

Shugart, H. H., and I. R. Noble. 1981. A computer model of succession and fire response of the high-altitude eucalyptus forest of the Brindabella Range, Australian Capital Territory. Aust. J. Ecol. 6:149.

Shugart, H. H., and D. C. West. 1977. Development of an Appalachian deciduous forest succession model and its application to assessment of the impact of chestnut blight. J. Envir. Manag. 5:161.

Shugart, H. H., and D. C. West. 1980. Forest succession models. Bioscience 30:308.

Shugart, H. H., M. S. Hopkins, I. P. Burgess, and A. T. Mortlock. 1980. The development of a succession model for subtropical rainforest and its application to assess the effects of timber harvest at Wiangaree State Forest, New South Wales. J. Envir. Manag. 11:243.

Shugart, H. H., D. C. West, and W. R. Emmanuel. 1981. Patterns and dynamics of forests: An application of forest succession models. In Forest Succession: Concepts and Application, D. C. West, H. H. Shugart, and D. B. Botkin, eds., pp. 74–94. New York: Springer-Verlag.

Shugart, H. H., M. Y. Antonovsky, P. G. Jarvis, and A. P. Sandford. 1986. $CO_2$, climatic change and forest ecosystems. In The Greenhouse Effect, Climatic Change, and Ecosystems, B. Bolin, B. R. Doos, J. Jager, and R. A. Worrick, eds., pp. 475–521. Chichester: John Wiley and Sons.

Smith, J., and D. Tirpak, eds. 1989. The Potential Effects of Global Climate Change on the United States. Washington, D.C.: Environmental Protection Agency.

Smith, T. M., and D. L. Urban. 1988. Scale and resolution of forest structural pattern. Vegetatio 74:143.

Smith, T. M., H. H. Shugart, and D. C. West. 1981a. Use of forest simulation models to integrate timber harvest and nongame bird management. Trans. 46th North American Wildl. and Natural Res. Conf., Washington, D.C., Wildlife Management Institute.

Smith, T. M., H. H. Shugart, and D. C. West. 1981b. FORHAB: A forest simulation model to predict habitat structure for nongame bird species. In The Use of Multivariate Statistics in Studies of Wildlife Habitat, D. E. Capen, ed., USDA Forest Service General Tech. Report RM-87.

Smith, T. M., R. Leemans, and H. H. Shugart. 1991. Sensitivity of terrestrial carbon storage to $CO_2$-

induced climate change: Comparison of four scenarios based on general circulation models. *Clim. Change.* (in press).

Solomon, A. M. 1986. Transient responses of forests to $CO_2$-induced climate change: Simulation modeling experiments in eastern North America. *Oecologia* 68:567.

Tonu, Oja. 1983. Metsa suktsessiooni ja tasandilise struktuuri imiteerimisest. *Yearbook of the Estonian Naturalist Soc.* 69:110.

Urban, D. L., and H. H. Shugart. 1989. Forest response to climate change: A simulation study for Southeastern forests. In *The Potential Effects of Global Climate Change on the United States*, J. Smith and D. Tirpak, eds., EPA-230-05-89-054, pp. 3-1–3-45. Washington, D.C.: Environmental Protection Agency.

Urban, D. L., and T. M. Smith. 1989. Microhabitat pattern and the structure of forest bird communities. *Am. Naturalist* 113:811.

Van Daalen, J. C., amd H. H. Shugart. 1989. OUTENIQUA: A computer model to simulate succession in the mixed evergreen forests of the southern Cape, South Africa. *Landscape Ecol.* 24: 255.

Waldrop, T. A., E. R. Buckner, H. H. Shugart, and C. E. McGee. 1986. FORCAT: A single-tree model of stand development on the Cumberland Plateau. *Forest Sci.* 32:297.

This research was supported in part by the National Aeronautics and Space Administration (grant NAG-5-1018) and the National Science Foundation (grants BSR-8702333 and BSR-8807882).

# CHAPTER TWELVE

# Physiological Responses of Animals to Higher Temperatures

WILLIAM R. DAWSON

## I. INTRODUCTION

The prospect of global warming prompts an examination of how the physiology of different types of animals responds to high temperatures and related factors. This examination is a formidable task because species vary substantially in characteristics that affect physiological response, including ecology, site tenacity, body size, general levels of energy expenditure, thermal sensitivity, locomotor ability, migratory tendencies, capacities for regulation of internal state, activity patterns, breeding biology, and general behavior. Such a wide review faces an obstacle in that most of the relevant physiological literature tells us more about the short-term effects of high temperatures on individual physiological processes than about the survival of particular species of animals during a projected warming trend in nature. This chapter will look beyond the short-term effects to assess not only the overall modes of response of adult animals to temperature but also how global warming might affect the breeding biology of various species and development of their young. To reduce the scope of this review to manageable proportions, I shall focus on terrestrial vertebrates (reptiles, birds, and mammals), with some mention of aquatic animals.

## II. MODES OF RESPONSE TO TEMPERATURE

Temperature exerts important effects on biological systems (e.g., Precht 1973). Its influence on rates of biochemical reactions, which reflects its role in determining the fraction of molecules in a population that are reactive, is pervasive. These rates may more than double with a rise in body temperature of 10°C. Therefore, the basic metabolic and other bodily functions of organisms whose body temperatures are primarily determined by external thermal conditions (ectotherms) may change dramatically in response to fairly small changes in local temperature. Consequences include changed requirements for energy and for materials such as protein, oxygen, and water. Some animals (endotherms, primarily mammals, and birds) protect their internal environment by physiologically regulating their body temperatures within narrow limits (Whittow 1973, Dizon and Brill 1979, Heinrich 1981), although this thermoregulation increases metabolic costs in a manner dependent on external thermal conditions.

Temperature is also an important influence on the equilibrium constants of biochemical reactions, particularly those involved in the formation of weak (noncovalent) chemical bonds (Hochachka and Somero 1984). These bonds affect a variety of features (e.g., protein structure, viscosity of lipids, nucleic acid structure, and binding of hormones to receptors), and they are highly sensitive to thermal perturbation. Resultant changes in such things as enzyme function, structure of cell membranes, and endocrine coordination can have profound effects on the physiological capacities of organisms undergoing temperature changes.

The same thermal conditions that can influence the requirements of ectotherms and endotherms for physiologically important resources such as energy, oxygen, and water can also affect the availability of those resources. For example, in aquatic species such as fish, rising water temperatures increase metabolic requirements for oxygen while decreasing its solubility (Fry and Hart 1948). High temperatures also can impose potentially conflicting requirements on terrestrial animals. For example, evaporation of water is the primary physiological means available to land animals for preventing overheating in hot environments, but extensive loss of water through evaporation can make it difficult to maintain adequate hydration in their tissues.

A battery of responses encompassing behavioral as well as physiological activities assists animals in dealing with the potentially adverse effects of high temperatures and intense solar radiation.

## A. Behavioral Responses

Behavior allows many animals to evade potentially adverse effects of extreme physical conditions. On a geographic scale, many northern hemisphere birds and some mammals evade extreme cold by migrating south in the fall. (The fossil record indicates that more gradual latitudinal or altitudinal movements have been important in allowing various plants and animals to deal with changes in the biotic and physical environment produced by long-term climatic trends, such as those involving glaciation in the Quaternary [Brown and Gibson 1983; see also chapters 5 and 22].) Range expansions over the past century by such species as the cattle egret, *Bubulcus ibis*, and the opossum, *Didelphis virginiana* (Brown and Gibson 1983), remind us that animal distributions are dynamic and that the potential for movement in response to altered environmental conditions still exists.

On a much more local scale, species can move between microclimates to find those that help maintain optimum body temperatures. Endotherms like birds and mammals can reduce demands on their physiological thermoregulation by moving into less thermally stressful areas (e.g., Buttemer 1985). Among reptiles, which have limited capacities for physiological thermoregulation (Bartholomew 1982), many species maintain their body temperature at suitable levels during activity by basking and, where risk of overheating exists, by use of shade (Huey 1982). Fish may move to areas of their habitat where temperatures allow various physiological processes to proceed optimally (Fry 1947). That option, however, may be severely limited in shallow ponds and streams, which tend to be thermally homogeneous. Instances of heat stroke in salmon and trout inhabiting Canadian streams have been recorded during hot weather (Huntsman 1942, 1946).

Most terrestrial vertebrates reduce activity

during the middle of hot days, and many desert species are nocturnal (Schmidt-Nielsen 1964). Various desert rodents, many of which have surprisingly limited heat resistance, survive under summer conditions by remaining in underground burrows until cooler nighttime conditions allow them to be active above ground (Dawson 1955, Schmidt-Nielsen 1964). Where water is available, bathing can assist certain animals in avoiding overheating (Dawson and Bartholomew 1968). Certain large African birds are known to soar to higher, cooler elevations during the middle of the day, thereby avoiding ground-level heat (Madsen 1930).

The recent thermal experience of animals can affect their behavioral responses to temperature, probably as a result of temperature compensation. Goldfish (Carassius auratus), for example, preferentially move to water with a temperature matching that at which they have previously been maintained (Fry 1947). Temperature selected by tadpoles of bullfrogs (Rana catesbiana) varies at some stages with season and maintenance temperature, being higher following adjustment of the animals to warmer conditions (Wollmuth and Crawshaw 1988).

## B. Physiological Regulation

Physiological regulation is another means of protecting an animal's internal environment from stressful thermal conditions. Birds and mammals, in particular, possess excellent capacities for monitoring their internal temperature and initiating compensatory physiological activity that maintains it within a narrow range (Simon et al. 1986). At effective ambient temperatures that are warm but lower than body temperature, adjustment of the insulation provided by fur or feathers and peripheral vasodilation permit loss of excess heat produced by metabolism. On the other hand, at temperatures exceeding body temperature, where heat gain from the environment adds to the metabolic heat load, some reptiles, birds, and many mammals prevent body temperature from rising to injurious levels by evaporative cooling through such

mechanisms as panting, sweating, or, as mentioned above, bathing (Dawson and Bartholomew, 1968; see Whittow 1973).

Active heat defense can involve the evaporation of substantial amounts of water. For example, in the galah (Cacatua roseicapilla), a cockatoo inhabiting interior parts of Australia, the rate of evaporative water loss at an ambient temperature of 48°C is about 40 times that at 21°C (Dawson and Fisher 1982). Thus, maintenance of water balance is a crucial challenge for terrestrial vertebrates that contend with heat by resorting to evaporative cooling. Many desert birds (e.g., parrots, pigeons, and finches) depend on drinking during the hotter parts of the year, as documented in the interior of Australia by C. D. Fisher et al. (1972). Some of them, such as hawks, replenish water expended in evaporation by ingesting succulent food (Dawson 1984).

Global warming is likely to be accompanied by reduced precipitation in some regions (Wigley et al. 1980, Hansen et al. 1981). Under low rainfall regimes, watering points can be widely separated, and the activities of animals that must drink tend to center on them. This can lead to depletion of food in the vicinity. Visiting such watering points to drink is not without risk, for they are also convenient places for predators to seek their prey. Therefore, potential prey species often are circumspect in approaching water, as T. J. Cade (1965) notes for various columbiform species (doves, pigeons, sandgrouse) in Africa. K. Schmidt-Nielsen (1964) observes that animals such as dromedary camels (Camelus dromedarius) minimize this risk by being able to ingest large amounts of water quickly.

## C. Relaxation of Regulatory Limits under Extreme Conditions

Physiological thermoregulation can be costly in terms of energy and water. In hot, arid environments, many animals sometimes reduce these costs by relaxing the precision of such regulation under extreme conditions (e.g., Bartholomew 1964). For example, some birds and mammals engaging in vigorous activity in hot environments can allow their body temperature to rise several degrees (Schmidt-Nielsen et al. 1957, Dawson 1984), a response termed hyperthermia. The rise is controlled so that heat injury is avoided, and special vascular arrangements maintain brain temperature at safe levels (Baker and Hayward 1968, Taylor and Lyman 1971, Bernstein et al. 1979). Allowing hyperthermia to develop in this way decreases somewhat the need for water for evaporative cooling, and this device appears to become increasingly important as water stress increases. For example, dromedary camels denied water show progressively greater hyperthermia on succeeding days as dehydration proceeds (Schmidt-Nielsen et al. 1957). This is also the case for rock doves (Columba livia), which evaporate water at a high rate when fully hydrated but reduce the rate (primarily by lowering cutaneous evaporation) and rely more on hyperthermia during dehydration (Arad et al. 1987).

Another type of thermoregulatory relaxation occurs during dormancy. Dormancy is normally thought of as a response benefiting animals in cold winter conditions, but some mammals, notably rodents, become dormant, or estivate, during warmer parts of the year when food and water may be in short supply (Hudson and Bartholomew 1964). The rodents remain underground, where overheating is not a problem, and lower their body temperature, thereby reducing their metabolic rate. This reduction decreases the need for food or water, thus enhancing survival during summer scarcity of resources.

## D. Temperature Compensation

Behavior, physiological regulation, and relaxation of regulatory limits represent immediate responses of animals to thermal challenges. Long-term physiological adjustments are also possible within individual animals, as they gradually adjust themselves to prevailing climatic conditions. Examples of such adjustments include shifts from summer to winter pelage in mammals, hormonal changes, winter fattening, vascular changes, and alterations in concentrations of various enzymes. These represent a form of temperature compensation referred to as thermal acclimatization.

In endotherms thermal acclimatization tends to enhance the ability to maintain a constant body temperature in very warm or cold environments. For example, American goldfinches (Carduelis tristis) adjusted to winter conditions in the north-central United States can maintain a normal body temperature for at least three hours during exposure to ambient temperatures below −60°C. None of these birds can do this during spring or summer (Dawson and Carey 1976). On the other hand, in ectotherms, whose body temperatures tend to track the external thermal conditions, adjustments are more likely to modify physiological performance at warmer or colder body temperatures and alter lethal temperatures (see Fry 1958). For example, heart rates of some species of marine littoral mollusks tend to be higher at a given temperature during winter than in summer (Segal 1956). Such a circumstance could indicate enhanced physiological activity at the cooler water temperatures prevailing during winter and adjustments leading to energy conservation at the warmer temperatures occurring in summer. The seasonal adjustment of heart rate appears to be evoked by continued exposure to particular water temperatures (Segal et al. 1953, Segal 1956). The alteration of lethal temperatures through thermal acclimatization has been well documented in fish (Fry 1947). In the goldfish (Carassius auratus), in which the effect is particularly impressive, the upper lethal temperature in individuals experimentally acclimatized to 0° and 40°C, respectively, is 27° and 41°C (Fry et al. 1942). There is a limit to acclimatization capacity and exposure of goldfish to temperatures above 40°C does not produce additional heat tolerance.

Thermal acclimatization can arise as a direct response to changing temperature conditions, and numerous studies indicate that conditioning of organisms under particular thermal regimes can modify the responses to heat or cold. For example, maintenance of deer mice (*Peromyscus maniculatus*) at warm temperatures improves their resistance to heat stress (Sealander 1951). However, more complex situations exist in which thermal acclimatization is actually stimulated by some environmental variable other than temperature. For example, some species use day length as a cue. Day length is especially useful because it changes over the year in a predictable manner, whereas environmental temperature, often fluctuating rapidly or in an unseasonable manner, is a far less reliable cue. The lethal temperatures for certain fish in Canadian lakes begin to change in the spring before any change in water temperature occurs, and Fry (1958) suggests that the animals are responding to increasing day length.

The underlying events controlling thermal acclimatization are not well understood, but they probably involve both shifts in cellular environment through modification of acid-base balance and activation of particular genes as a result of the response of the organism to some thermal, photoperiodic, or other environmental cue. Such activation would ultimately lead to the shifts in structural and functional capacities described above. Some shifts appear to be linked with synthesis of alternate forms of key enzymes and other proteins that are more suited to altered thermal conditions, or with synthesis of additional amounts of existing enzymes (Hochachka and Somero 1984).

One intriguing type of adjustment shown by ectotherms in particular is a change in cellular function for operation at warm or cold body temperatures. Prominent among these are so-called homeoviscous adjustments (Hazel 1988), which involve alterations in the chemical composition of phospholipids, important components of cell membranes. The alterations contribute to maintenance of proper fluidity of the membranes under altered temperature conditions. Such maintenance is crucial for proper function of various membrane-bound enzymes (Hochachka and Somero 1984), and homeoviscous adjustments are probably critical both for modifications of performance at noninjurious temperatures and for alterations in lethal temperatures.

## III. POTENTIAL EFFECTS OF GLOBAL WARMING ON ANIMALS

### A. Interspecific Variation in Thermal Responses of Adult Animals

At least initially, animals may be able to cope with global warming through a combination of existing behavioral and physiological responses by individuals and, if the fossil record is any guide, gradual shifts in distribution of the species to newly suitable habitat. One cannot have great confidence in this assessment, however, for temperature has been a powerful environmental factor in the evolution of these organisms, and even if warming is not immediately injurious to individual adults, it may exert subtle effects on habitat requirements, energetics, competitive interactions, temporal relationships, and other key facets of their biology ultimately important for the persistence of the species. An example of such subtle effects is afforded by C. R. Tracy and K. A. Christian's (1986) observations on the interrelations between thermal requirements of lizards and the spatial and temporal aspects of their habitat utilization.

Estimates of the direct physiological and ecological impact of any rapid warming trend are complicated not only by the capacity of individual animals for various modes of compensatory activity but also by variation among species in temperature responses. Some show biochemical and physiological specialization for operation under particular sets of thermal conditions, and their distributions might be expected to track any shifts in environmental temperature quite closely. For example, barracuda (*Sphyraena* spp.) in the eastern Pacific are represented by one species in temperate waters south of the equator, a second in temperate waters north of the equator, another in slightly warmer waters in the Gulf of California, and a fourth in warm tropical waters. There is a 6°–8°C difference in temperature between the typical temperate and tropical habitat. The four species have forms of the enzyme $M_4$-lactate dehydrogenase with different kinetic and electrophoretic properties, and each form seems adapted to the local water temperature of its habitat (Graves and Somero 1982).

Compared with the barracudas, some other animals appear much less specialized in their thermal requirements (i.e., they are eurythermal) and perhaps will be less immediately affected by any shift in thermal conditions. For example, one subspecies of the raven, Corvus corax sinuatus, is resident from the mountains of south-central British Columbia through the deserts of the southwestern United States to northwestern Nicaragua (AOU 1957). Some ectotherms like the lizard Stellio stellio are also eurythermal. This animal is active at body temperatures extending from 24° to approximately 40°C and shows similar running performance over the entire range (Huey and Hertz 1984).

Estimation of the possible biological impact of global warming from information on temperatures required to produce heat injury in various animals appears difficult. Documented instances of such injury under natural conditions, such as those described by A. G. Huntsman (1942, 1946) for Atlantic salmon (Salmo salar) and other fishes in streams in the maritime provinces of Canada, are rare. Some interspecific variation in thermal tolerances does appear to correlate with habitat temperature, and P. F. Scholander et al. (1953) showed that the highest temperatures that certain arctic fish and crustaceans could survive approximated the lowest temperatures tolerated by their tropical counterparts. Thermal tolerances also may reflect evolutionary relationships, so that species occupying the same general environment but representing different taxonomic groups may have quite different lethal temperatures. For example, Australian skinks and geckos generally are less heat tolerant than agamid lizards (Licht et al. 1966). Similar variation was noted by R. M. Bailey (1955) among fishes subjected

to heat stress in a shallow pond in southeastern Michigan. Construction of a causeway had created the pond, isolating a large number of fish within it from the lake in which they had been living. Water temperature reached 38°C on a hot July day, and a mass die-off occurred. However, not all species were affected. Umbrid (mudminnows), catostomid (suckers), cyprinid (minnows), ictalurid (catfishes), and percid (perch and darters) species were mostly killed, whereas cyprinodontid (top minnows and killifish) and most centrarchid (sunfish) fish survived, as did some recently introduced gambusia (Gambusia affinis). These differences in thermal tolerance at a single locality raise the possibility of differential effects of warming on species assemblages, complicating risk assessment. Variation in thermal tolerance with stage of life (or with some correlated variable, such as size) further complicates matters. Adults of some animals show greater sensitivity to high temperatures than do immature individuals, as Bailey (1955) noted in the heat-sensitive fish species he observed. Huntsman (1942) and J. S. Hart (1952) reached similar conclusions for Atlantic salmon and two species of minnows, respectively.

If it is difficult to extrapolate from information on thermal tolerance to estimates of the effects of global warming, can distributional information assist us? Coincidence of range boundaries of certain species with particular isotherms suggests that temperature can play a role in setting distributional limits in some instances. For example, the northern boundary of the area of high density for winter populations of the phoebe (Sayornis phoebe) in eastern North America lies along the 4.4°C January isotherm (Root 1988). A northward movement of this range boundary would be a reasonable prospect with any global warming.

## B. Implications of Global Warming for Reproductive Biology

In addition to the possibility of direct effects of warming on individual adults, there is the matter of the sensitivity to high temperatures of the reproductive biology of many species. A glimpse is provided by observations on reproduction of seabirds during the 1982–83 El Niño event. Higher than normal sea temperatures were recorded off the coast of Oregon, and the concurrent reproductive success of three species breeding there was adversely affected (Hodder and Graybill 1985). It is difficult to generalize concerning the extent of the anticipated adverse effects of warming on reproduction, if for no other reason than that reproductive patterns are so diverse. Animals differ in timing and duration of breeding season, in mode of reproduction, in duration of the developmental period, in the maturity of the young at hatching or birth, and in the extent of nurture provided the young. It is beyond the scope of this discussion to examine all the ways in which warming conditions might affect reproductive biology. Nevertheless, it is possible to specify several areas in vertebrates in which the effects of high temperature might be especially important.

1. *Parental Behavior*  The effects of high temperatures on parental behavior appear especially critical for birds nesting in exposed situations or depending on food and water sources distant from the nesting site. In hot climates, their primary problem is preventing their eggs and young from overheating, rather than keeping them warm.

G. A. Bartholomew and W. R. Dawson (1979) note for the Heermann's gull (*Larus heermanni*) and other birds that the functional thermoregulatory unit is an attentive adult and its eggs or young. Parent birds of this species show an elaborate set of capacities, including adjustments of plumage, shading of the eggs, and evaporative cooling, that act to produce a suitable nest environment for their eggs under the intense solar radiation and warm ambient temperatures that prevail in their breeding colonies on desert islands in the Gulf of California, Mexico.

Even more impressive capacities have been documented in columbiform birds. For example, white-winged doves (*Zenaida asiatica*) and mourning doves (*Zenaidura macroura*) nest-

ing in deserts actually keep their eggs cooler than the surrounding environment even on very hot days, thereby protecting them from injurious temperatures (Russell 1969, Walsberg and Voss-Roberts 1983). Laboratory tests have established that rock doves (*Columba livia*) share this capacity. These birds maintained their eggs between 39.8° and 41.7°C during severe heat challenges involving exposure to ambient temperatures of 50°–60°C. The incubating parent accomplishes this by absorbing and then dissipating much of the heat that the eggs gain from the environment (Marder and Gavrieli-Levin 1986). Incubating birds maintain their body temperatures 1°–2.5°C below those of nonincubating individuals (Arieli et al. 1988). The heat from the eggs plus that gained directly by the adult from its metabolism and the environment is dissipated primarily by cutaneous evaporation (Marder and Gavrieli-Levin 1986).

In hot situations where continuous behavioral regulation of egg or chick temperature by parent birds is essential, the task is complicated immensely when the nest site is remote from drinking water or feeding areas. Male sandgrouse (*Pterocles* spp.) in arid regions of the Old World are notable for transporting water to their young, who require it for proper fluid balance (Cade and Maclean 1967). The males may travel up to 50 km to obtain water (Heim de Balsac 1936). Although this helps maintain fluid balance in the chicks, it requires extended periods of attentiveness on the part of the female parent while the male is enroute between the nesting area and water. Ethiopian plovers (*Pluvianus aegyptius*) have also been shown to transport water from nearby to cool their eggs and chicks (Howell 1979).

The need to protect eggs and chicks from intense solar radiation and high temperatures can impose special burdens on the parent birds, which must sometimes give attentiveness precedence over behavior serving to reduce their own heat stress. This is well illustrated by G. S. Grant's (1982) observations on several species (terns, black skimmer, shorebirds) during their nesting on the strand of the Salton Sea in the Colorado Desert of southern California. Were these birds to nest in late winter and early spring, they would avoid severe heat stress. Their breeding schedules, however, did not differ from those of members of the same species breeding in milder climates. This suggests that modification of the time of breeding may not be a readily available option for birds dealing with any long-term warming trend. The Salton Sea birds had to shade their eggs nearly continuously on the hot days that prevailed during the nesting season. Their habit of soaking their belly feathers in the nearby water assisted in cooling both parents and eggs. The adults shared incubation duties and further cooled themselves during relief periods by immersing their feet and legs in the nearby Salton Sea. Grant's observations suggest that loss of a member of a breeding pair of these birds would make it difficult for the survivor to hatch a clutch of eggs by itself. The lengthened attentive periods that protection of the eggs would require in the absence of a partner would simply expose the remaining parent to excessive heat stress.

Attentiveness of parent birds to their eggs and chicks may also be jeopardized when feeding areas are remote from nest sites. Displacement of fishing areas from rookeries has been linked with breeding failures of sea birds during El Niño conditions. The protracted absence of the foraging parent from the nest ultimately leads to abandonment of the nest by the attentive partner. Such problems might be intensified in any long-term warming trend if patterns of planetary circulation and ocean currents are affected.

2. *Embryos and Chicks*  If adult birds are unsuccessful in maintaining proper temperatures for their eggs or chicks, the young can perish. Only a few degrees separate incubation temperatures from upper lethal ones (Webb 1987). For example, incubating Heermann's gulls maintain their eggs near 37°C (Rahn and Dawson 1979). Experimental cooling of week-old embryos of this species to 23°–30°C below this level led to cessation of the heart beat. However, it was reestablished after rewarming, even following one hour at 6°C. On the other hand, a 6°C rise in temperature to 43°C, which could easily occur if the eggs were left unshaded during the middle of the day, was fatal to the embryos after one hour (Bennett and Dawson 1979). The early stages of development appear to be the most heat sensitive in avian embryos (Webb 1987). Adult birds of many species tolerate 43°C indefinitely (Dawson and Hudson 1970). Moreover, older avian embryos also surpass young ones in tolerance of elevated temperatures (Webb 1987).

Hot weather can put chicks as well as embryos at risk, and this is well illustrated by A. G. Salzman's (1982) observations of a nesting colony of western gulls (*Larus occidentalis*) on Santa Barbara Island off the coast of southern California. The colony was subjected to a brief but intense heat wave in early June 1979, resulting in 90% mortality of chicks in some parts of the nesting area. Mortality was affected by temperature and wind conditions at the nest site. Higher mortality occurred where ground and air temperatures were hotter and wind velocities were lower. Climatic records indicate that heat waves during the breeding season have been rare on Santa Barbara Island and nearby islands. One wonders whether such episodes will become more common as the anticipated global warming proceeds.

3. *Reproductive Processes*  High temperatures are known to have adverse effects on reproductive processes generally in domestic birds and mammals (e.g., Brody 1945). Processes affected include fertility and fecundity (Cowles 1965, Robinson and Van Niekirk 1978, Joshi et al. 1980, Hermes et al. 1983, Francos and Mayer 1983, Akpokodje et al. 1985, Cavestany et al. 1985), intensity and duration of estrus (Reddy et al. 1987, Kanai et al. 1987), and egg production (Madkour and Mahmoud 1974).

In nature, the situation is obviously more complex, for changes in reproductive functions may reflect not only thermal conditions but also such problems as shortages of food or water. D. M. McLean (1981) raises the possibility that late Pleistocene extinctions of large mammals may have resulted from the effects of rapid (by geological standards) global warming. He notes that evidence from domesticated mammals indicates that high environmental temperatures can depress synthesis and release of hormones governing reproductive cycles, reduce fertility and litter size, and increase rates of spontaneous abortion. Pregnant female mammals respond to heat stress by decreasing uterine blood flow. A recent example is found in a study of pregnant Hereford cows by L. P. Reynolds et al. (1985). Exposure of the animals to heat stress in midgestation reduced uterine and umbilical blood flow, and their fetuses weighed less than those of control animals. The livers of the fetuses of the heat-stressed mothers also were lighter, both absolutely and in relation to fetal weight, and total liver protein and RNA were lower than in the controls. Heat stress also increases the incidence of retained placentas in dairy cows (DuBois and Williams 1980).

Because large bodies have proportionally less body surface for heat dissipation, McLean (1981) hypothesizes that large female mammals would have experienced more heat stress than small ones in a rapid warming trend such as occurred at the end of the Pleistocene. Large females would thus have been especially prone to hyperthermia during the warmer parts of the year, increasing the risk of damage to embryos and of reduced fecundity, developments that could have led to extinction of their respective lines. Moreover, he suggests that those species would have had insufficient time to achieve evolutionary changes in body size. McLean's argument appears deficient, in that it does not consider any aspects of thermoregulation besides surface-volume ratios, that large animals will be less affected by convective heat exchange than smaller ones, and that other factors such as timing of breeding may be involved. There can be no doubt, though, that high temperatures can adversely affect reproductive processes, embryonic development, and, ultimately, Darwinian fitness.

4. *Sex Determination in Reptiles*    Recent findings on environmental determination of sex in various animals afford a further example of how changing thermal conditions can affect the welfare of individual species. Unlike the situation in most animals, in which sex is determined primarily by the combination of sex chromosomes established at conception, it has become apparent over the past couple of decades that temperature is a key factor for sex determination in reptiles (earlier work reviewed by Bull 1980), though moisture relations may also be involved in some species (Bull 1983). The crocodilians, turtles, and lizards in which temperature is important all place their eggs in the ground or in constructed mounds (some crocodilians). As J. J. Bull (1983) notes, the hatchlings of these animals do not mature for months or years, so mating would never occur in the nest. In the majority of species studied, most temperatures at which development can proceed result in only one sex. Production of both sexes occurs only over a narrow thermal range. For example, in map turtles (*Graptemys pseudogeographica*) incubation at 23°–28°C, 28°–30°C, and above 30°C produces only males, males and females, and essentially females, respectively (Bull and Vogt 1979, Bull et al. 1982). The opposite situation is evident in lizards and certain crocodiles, in which temperatures below and above the narrow range where both sexes are produced result in females and males, respectively (see Bull 1983). A third pattern is evident in snapping turtles (*Chelydra serpentina*) and at least one species of crocodile (Webb and Smith 1984), where higher and lower temperatures produce females, and males develop at intermediate temperatures. Finally, environmental sex determination appears not to have been found in snakes (Bull 1983).

Thermal effects on sex determination do not appear to be due to differential mortality, and they seem to persist through the lifetime of the organism. Bull (1983) suggests that environmental sex determination would have a definite advantage over genetic sex determination for life histories in which environmental factors such as temperature have different effects on males and females. Iron-ically, the thermal sensitivity of the sex-determining mechanism in green sea turtles (*Chelonia mydas*) has complicated conservation efforts on behalf of this species, for the temperatures occurring during artificial incubation of its eggs tend to favor males (Mrosovsky 1982).

## IV. PHYSIOLOGICAL CONSEQUENCES AND SOME EVOLUTIONARY CONSIDERATIONS

### A. Physiological Responses of Animals to Higher Temperatures

The diversity of contemporary faunas makes it difficult to provide a general assessment of the physiological effects of global warming. It is clear from consideration of vertebrates, however, that adults, whether endotherms or ectotherms, have a number of potential means for enhancing survival at higher temperatures. Among these, thermoregulation by terrestrial vertebrates in the heat requires expenditure of substantial amounts of water in evaporative cooling. For species that engage in this activity, water balance may rival overheating as the most serious problem confronting them. The threat of inadequate water supplies could have serious implications for the welfare of animals in parts of the world where anticipated warming trends are expected to be accompanied by reduced precipitation (see Wigley et al. 1980, Hansen et al. 1981). Even where precipitation remains constant, higher temperatures will affect water availability by increasing evaporation and transpiration at the same time that animals' need for water is increasing. Any effective conservation efforts on their behalf will surely require assurance of dependable sources of water, an issue of vital concern to agricultural interests as well.

In at least the initial phase of the anticipated global warming, it appears that available behavioral and physiological options for dealing with high temperatures may afford adult animals reasonable prospects for survival, particularly in widely distributed species that already encounter hot weather in some part of their present distributions. Still, we cannot rule out subtle effects of chronic exposure to heat on behavior, physiology, and biotic relationships that may work to the disadvantage of particular species. Just as adults could be at greater risk in a prolonged warming trend, reproductive biology may be affected, too. High temperatures can adversely affect a number of aspects of reproduction ranging from fertility to rearing of young.

### B. Evolutionary Changes in Behavioral and Physiological Capacities

With the probability of major global warming, one cannot help but wonder whether various animals will evolve capacities better suited to the altered environmental conditions, thus enhancing their prospects for long-term survival. It is certainly true that breeds or species of animals that do especially well at warmer temperatures already exist (e.g., Scholander et al. 1953, Macfarlane 1964, Schmidt-Nielsen 1964, Arad and Marder 1982), either as the result of natural evolutionary processes or artificial selection by man. New evolutionary adjustments to any warming scenario, however, appear to be a remote prospect for most species for several reasons. The further buildup of $CO_2$ and other greenhouse gases and the resultant warming are likely to occur over the next decades, whereas evolutionary change usually takes much longer. Second, it is by no means certain that any given species will have a genetic potential allowing production of variants that would be favored by natural selection under warming conditions. Third, any contemporary organism already represents a highly evolved entity in which any possible evolutionary developments are likely to be constrained by ontogeny and tradeoffs involving other functional requirements (see Sibly and Callow 1986). This probably accounts for the conservatism in physiological adaptations commented on by Peters and Darling (1985). Extinction is a far more likely outcome than survival through evolutionary change, should present-day capacities of various species prove inadequate for dealing with the challenge of rapid global warming.

## V. SUMMARY

Adult animals may use several modes of response to higher temperatures: behavior serving to minimize thermal stress, physiological thermoregulation, hyperthermia or dormancy involving a relaxation of thermoregulatory control, and temperature compensation. Each enhances chances of survival, but they cannot eliminate subtle effects of warming on animal biology. In birds and mammals, the substantial amounts of water used in evaporative cooling can create problems for the animal concerning maintenance of proper hydration. Future conservation efforts will need to take this into account in areas where warming increases animals' need for water and decreases water availability, particularly where warming is accompanied by reductions in precipitation. High temperatures exert substantial direct effects on breeding biology of various animals, which in a period of global warming may place certain species at greater risk than any direct effect of heat on adults. The prospect of evolutionary solutions to physiological problems imposed by such warming appears remote for several reasons.

## REFERENCES

Akpokodje, J. V., T. I. Dede, and P. I. Odili. 1985. Seasonal variation in seminal characteristics of West African dwarf sheep in the humid tropics. *Trop. Vet.* 3:61.

AOU (American Ornithologists' Union). 1957. *Check-list of North American Birds*, 5th ed. Baltimore: American Ornithologists' Union.

Arad, Z., and J. Marder. 1982. Strain differences in heat resistance to acute heat stress, between the Bedouin desert fowl, the White Leghorn and their crossbreeds. *Comp. Biochem. Physl.* 72A:191.

Arad, Z., I. Gavrieli-Levin, U. Eylath, and J. Marder. 1987. Effect of dehydration on cutaneous water evaporation in heat-exposed pigeons (*Columba livia*). *Physl. Zool.* 60:623.

Arieli, Y., L. Peltonen, and J. Marder. 1988. Reproduction of rock pigeon exposed to extreme ambient temperatures. *Comp. Biochem. Physl.* 90A:497.

Bailey, R. M. 1955. Differential mortality from high temperature in a mixed population of fishes in southern Michigan. *Ecology* 36:526.

Baker, M. A., and J. N. Hayward. 1968. The influence of the nasal mucosa and the carotid rete upon hypothalamic temperature in sheep. *J. Physl. (Lond.)* 198:561.

Bartholomew, G. A. 1964. The roles of physiology and behaviour in the maintenance of homeostasis in the desert environment. *Symp. Soc. Exp. Biol.* 18:7.

Bartholomew, G. A. 1982. Physiological control of body temperature. In *Biology of the Reptilia*, C. Gans and F. H. Pough, eds., vol. 12, Physiology C, pp. 167–211. London: Academic Press.

Bartholomew, G. A., and W. R. Dawson. 1979. Thermoregulatory behavior during incubation in Heermann's gulls. *Physl. Zool.* 52:422.

Bennett, A. F., and W. R. Dawson. 1979. Physiological responses of embryonic Heermann's gulls to temperature. *Physl. Zool.* 52:413.

Bernstein, M. H., M. B. Curtis, and D. M. Hudson. 1979. Independence of brain and body temperature in flying American kestrels, *Falco sparverius*. *Am. J. Physl.* 237:R58.

Brody, S. 1945. *Bioenergetics and Growth*. New York: Reinhold.

Brown, J. H., and A. C. Gibson. 1983. *Biogeography*. St. Louis: Mosby.

Bull, J. J. 1980. Sex determination in reptiles. *Quart. Rev. Biol.* 55:3.

Bull, J. J. 1983. *Evolution of Sex Determining Mechanisms*. Menlo Park, Calif.: Benjamin/Cummings.

Bull, J. J., and R. C. Vogt. 1979. Temperature-dependent sex determination in turtles. *Science* 206:1186.

Bull, J. J., R. C. Vogt, and C. J. McCoy. 1982. Sex-determining temperatures in emydid turtles: A geographic comparison. *Evolution* 36:326.

Buttemer, W. L. 1985. Energy relations of winter roost-site utilization by American goldfinches (*Carduelis tristis*). *Oecologia* (Berlin) 68:126.

Cade, T. J. 1965. Relations between raptors and columbiform birds at a desert water hole. *Wilson Bull.* 77:340.

Cade, T. J., and G. L. Maclean. 1967. Transport of water by adult sandgrouse to their young. *Condor* 69:323.

Cavestany, D., A. B. El Wishy, and R. H. Foote. 1985. Effect of season and high environmental temperature on fertility of Holstein cattle. *J. Dairy Sci.* 68:1471.

Cowles, R. B. 1965. Hyperthermia, aspermia, mutation rates and evolution. *Quart. Rev. Biol.* 49:341.

Dawson, W. R. 1955. The relation of oxygen consumption to temperature in desert rodents. *J. Mamm.* 36:543.

Dawson, W. R. 1984. Physiological studies of desert birds: Present and future considerations. J. Arid Envir. 7:133.

Dawson, W. R., and G. A. Bartholomew. 1968. Temperature regulation in desert mammals. In Desert Biology, G. W. Brown, Jr., ed., vol. 1, pp. 395–421. New York: Academic Press.

Dawson, W. R., and C. Carey. 1976. Seasonal acclimatization to temperature in cardueline finches: I. Insulative and metabolic adjustments. J. Comp. Physl. 112:317.

Dawson, W. R., and C. D. Fisher. 1982. Observations on the temperature regulation and water economy of the galah (Cacatua roseicapilla). Comp. Biochem. Physl. 72A:1.

Dawson, W. R., and J. W. Hudson. 1970. Birds. In Comparative Physiology of Thermoregulation, G. C. Whittow, ed., vol. 1, pp. 223–310. New York: Academic Press.

Dizon, A. E., and R. W. Brill. 1979. Thermoregulation in tunas. Am. Zool. 19:249.

DuBois, P. R., and D. J. Williams. 1980. Increased incidence of retained placenta associated with heat stress in dairy cows. Theriogenology 13:115.

Fisher, C. D., E. Lindgren, and W. R. Dawson. 1972. Drinking patterns and behavior of Australian desert birds in relation to their ecology and abundance. Condor 74:111.

Francos, G., and E. Mayer. 1983. Observations on some environmental factors connected with fertility in heat stressed cows. Theriogenology 19:625.

Fry, F.E.J. 1947. Effects of environment on animal activity. Univ. of Toronto Studies, Biological Series 55. Publ. Ontario Fish. Res. Lab. 68:1.

Fry, F.E.J. 1958. Temperature compensation. Ann. Rev. Physl. 20:207.

Fry, F.E.J., and J. S. Hart. 1948. The relation of temperature to oxygen consumption in the goldfish. Biol. Bull. 94:66.

Fry, F.E.J., J. R. Brett, and G. H. Clausen. 1942. Lethal limits of temperature for young goldfish. Rev. Can. Biol. 1:50.

Grant, G. S. 1982. Avian incubation: Egg temperature, nest humidity, and behavioral thermoregulation in a hot environment. Ornith. Monogr. 30.

Graves, J. E., and G. N. Somero. 1982. Electrophoretic and functional enzymatic evolution in four species of eastern Pacific barracudas from different thermal environments. Evolution 36:97.

Hansen, J., D. Johnson, A. Lacis, S. Lebedoff, P. Lee, D. Rind, and G. Russell. 1981. Climate impact of increasing atmospheric carbon dioxide. Science 213:957.

Hart, J. S. 1952. Geographic variations of some physiological and morphological characters in certain freshwater fish. Univ. Toronto Studies, Biological Series 60. Publ. Ontario Fish. Res. Lab. 72:1.

Hazel, J. R. 1988. Homeoviscous adaptation in animal cell membranes. In Advances in Membrane Fluidity, R. C. Aloia, C. C. Curtain, and L. M. Gordon, eds., pp. 149–188. New York: Liss.

Heim de Balsac, H. 1936. Biogéographie des mammifères et de oiseaux de l'Afrique de Nord. Bull. Biol. France Belg. Suppl. 21:1.

Heinrich, B. 1981. Insect Thermoregulation. New York: John Wiley and Sons.

Hermes, J. C., A. E. Woodward, P. Vohra, and R. L. Snyder. 1983. The effect of ambient temperature and energy level on reproduction in red-legged partridges Alectoris graeca. Poult. Sci. 62:1160.

Hochachka, P. W., and G. N. Somero. 1984. Biochemical Adaptation. Princeton, N.J.: Princeton University Press.

Hodder, J., and M. R. Graybill. 1985. Reproduction and survival of seabirds in Oregon, U.S.A., during the 1982–1983 El Niño. Condor 87:535.

Howell, T. R. 1979. Breeding biology of the Egyptian plover, Pluvianus aegyptius (Aves: Glareolidae). Univ. Calif. Publ. Zool. 113:1.

Hudson, J. W., and G. A. Bartholomew. 1964. Terrestrial animals in dry heat: Estivators. In Handbook of Physiology, D. B. Dill, ed., sect. 4, Adaptation to the Environment, pp. 541–550. Washington, D.C.: American Physiological Society.

Huey, R. B. 1982. Temperature, physiology, and ecology of reptiles. In Biology of the Reptilia, C. Gans and F. H. Pough, eds., vol. 12, Physiology C, pp. 25–91. London: Academic Press.

Huey, R. B., and P. E. Hertz. 1984. Is a Jack-of-all-trades a master of none? Evolution 38:441.

Huntsman, A. G. 1942. Death of salmon and trout with high temperature. J. Fish. Res. Bd. Can. 5:485.

Huntsman, A. G. 1946. Heat stroke in Canadian Maritime stream fishes. J. Fish. Res. Bd. Can. 6:476.

Joshi, P. C., B. Panda, and B. C. Joshi. 1980. Effect of ambient temperature on semen characteristics of White Leghorn male chickens. Indian Vet. J. 57:52.

Kanai, Y., T. Abdul Latief, and H. Shimizu. 1987. Estrus and some related phenomena in Shiba goats under hot environmental conditions. Jap. J. Zootech. Sci. 58:781.

Licht, P., W. R. Dawson, and V. H. Shoemaker. 1966. Heat resistance of some Western Australian lizards. Copeia 1966:162.

Macfarlane, W. V. 1964. Terrestrial animals in dry heat: Ungulates. In Handbook of Physiology, D. B. Dill, ed., sect. 4, Adaptation to the Environment, pp. 509–539. Washington, D.C.: American Physiological Society.

Madkour, Y. H., and T. H. Mahmoud. 1974. The influence of environmental temperature on egg production. *Ag. Res. Rev.* (Cairo) 52:107.

Madsen, H. 1930. Quelques remarques sur la cause pourquoi le grands oiseaux au Soudan planent si haut au milieu de la journée. *Videnskab. Medd. Dansk Naturhist. Forh. Kobenhaven* 8:301.

Marder, J., and I. Gavrieli-Levin. 1986. Body and egg temperature regulation in incubating pigeons exposed to heat stress: The role of skin evaporation. *Physl. Zool.* 59:532.

McLean, D. M. 1981. Size factor in late Pleistocene mammalian extinctions. *Am. J. Sci.* 281:1144.

Mrosovsky, N. 1982. Sex ratio bias in hatchling sea turtles from artificially incubated eggs. *Biol. Conser.* 23:309.

Peters, R. H., and J. D. S. Darling. 1985. The greenhouse effect and nature reserves. *Bioscience* 35:707.

Precht, H., ed. 1973. *Temperature and Life.* New York: Springer-Verlag.

Rahn, H., and W. R. Dawson. 1979. Incubation water loss in eggs of Heermann's gull. *Physl. Zool.* 52:451.

Reddy, A. O., V. N. Tripathy, and V. S. Raina. 1987. Effect of climate on the incidence of estrus and conception rate in Murrah buffaloes. *Indian J. Anim. Sci.* 57:204.

Reynolds, L. P., C. L. Ferrell, J. A. Nienaber, and S. P. Ford. 1985. Effects of chronic environmental heat stress on blood flow and nutrient uptake of the gravid bovine uterus and fetus. *J. Ag. Sci.* 104:289.

Robinson, R.D.V., and B.D.H. Van Niekirk. 1978. Effect of ambient temperature on farrowing rate in pigs. *S-Afr. Tydskr. Veekd.* 8:105.

Root, T. 1988. Energy constraints on avian distributions and abundances. *Ecology* 69:330.

Russell, S. M. 1969. Regulation of egg temperatures in incubating white-winged doves. In *Physiological Systems in Semi-arid Environments,* C. C. Hoff and M. L. Riedesel, eds., pp. 107–112. Albuquerque: University of New Mexico Press.

Salzman, A. G. 1982. The selective importance of heat stress in gull nest location. *Ecology* 63:742.

Schmidt-Nielsen, K. 1964. *Desert Animals: Physiological Problems of Heat and Water.* London: Oxford University Press.

Schmidt-Nielsen, K., B. Schmidt-Nielsen, S. A. Jar-

num, and T. R. Houpt. 1957. Body temperature of the camel and its relation to water supply. *Am. J. Physl.* 188:103.

Scholander, P. F., W. Flagg, V. Walters, and L. Irving. 1953. Climatic adaptation in arctic and tropical poikilotherms. *Physl. Zool.* 26:67.

Sealander, J. A., Jr. 1951. Survival of *Peromyscus* in relation to environmental temperature and acclimation at high and low temperatures. *Am. Midl. Nat.* 46:257.

Segal, E. 1956. Microgeographic variation as thermal acclimation in an intertidal mollusc. *Biol. Bull.* 111:129.

Segal, E., K. P. Rao, and T. W. James. 1953. Rate of activity as a function of intertidal height within populations of some littoral molluscs. *Nature* 172:1108.

Sibly, R. M., and P. Calow. 1986. *Physiological Ecology of Animals: An Evolutionary Approach.* Oxford: Blackwell Scientific Publications.

Simon, E., F.-K. Pierau, and D.C.M. Taylor. 1986. Central and peripheral thermal control of effectors in homeothermic temperature regulation. *Physl. Rev.* 66:235.

Taylor, C. R., and C. P. Lyman. 1971. Heat storage in running antelopes: Independence of brain and body temperatures. *Am. J. Physl.* 222:114.

Tracy, C. R., and K. A. Christian. 1986. Ecological relations among space, time, and thermal niche axes. *Ecology* 67:609.

Walsberg, G. E., and K. A. Voss-Roberts. 1983. Incubation in desert-inhabiting doves: Mechanisms for egg cooling. *Physl. Zool.* 56:88.

Webb, D. R. 1987. Thermal tolerance of avian embryos: A review. *Condor* 89:874.

Webb, G.J.W., and A.M.A. Smith. 1984. Sex ratio and survivorship in the Australian freshwater crocodile *Crocodylus johnstoni. Symp. Zool. Soc. Lond.* 52:319.

Whittow, G. C. 1973. *Comparative Physiology of Thermoregulation,* vols. 1–3. New York: Academic Press.

Wigley, T.M.L., P. D. Jones, and P. M. Kelly. 1980. Scenario for a warm, high $CO_2$ world. *Nature* 283:17.

Wollmuth, L. P., and L. I. Crawshaw. 1988. The effect of development and season on temperature selection in bullfrog tadpoles. *Physl. Zool.* 61:461.

The work of W. R. Dawson and associates cited in this chapter was supported in part by grants from the National Science Foundation, most recently BSR 84-07952.

# CHAPTER THIRTEEN

# Ecological Responses of Animals to Climate

C. RICHARD TRACY

## I. INTRODUCTION

What are the consequences to the earth's biota of climate change of the kinds described in chapter 4? (See also chapter 3 and Hansen et al. 1981, Hansen et al. 1988; for a cautionary note see Maduro 1989.) In this chapter, I review some of the evidence bearing on the ways in which animal populations and communities respond to changes in climate and associated changes in physical environments.

Increases in greenhouse gases have the potential to change many attributes of environments of animals. For example, an increased greenhouse effect theoretically will increase the radiant energy received by the earth from its atmosphere, and this, of course, could elevate temperatures of the air, ground, and water. Alternatively, changes to the energy budget of the earth might change cloudiness and precipitation patterns, possibly offsetting global warming. Nevertheless, if warming does occur, the resulting increases in temperatures of the earth would have numerous effects on the physical environments of animals. For example, increases in the temperatures of water necessarily reduce the amount of oxygen that can be dissolved in water (fig. 13.1a) and thus reduce the availability of oxygen to aquatic organisms. Also, increases in air temperature necessarily increase the amount of water that air can hold (fig. 13.1b), which, in turn, could potentially change the rates of water loss from organisms and change the amount of water in food items for animals.

In addition to climate effects, carbon dioxide, the primary greenhouse gas, can directly affect plant growth and, thus, the animals that depend on plants. Increases in carbon dioxide can "fertilize" some plants with needed carbon, resulting in increased photosynthetic rates and increased water-use efficiency (Oechel and Strain 1985; chapter 8). Therefore, it is not surprising that plants in

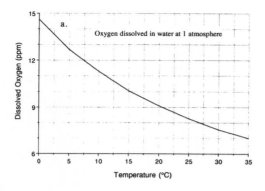

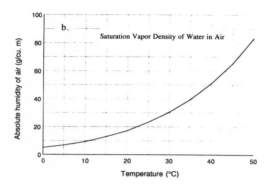

*Figure* 13.1. The maximum amount of oxygen that can be dissolved in water as a function of water temperature (a), and water that air can hold at different air temperatures (b).

atmospheres enriched with carbon dioxide grow to a larger size than plants grown in atmospheres with today's level of carbon dioxide. When herbivorous insects eat these high-$CO_2$ plants, however, they consume a greater volume of plant material than they do when eating normal-$CO_2$ plants (Strain 1987). The high-$CO_2$ plants contain less nitrogen per unit of volume than plants grown under present-day $CO_2$ levels, and the insects apparently need to eat more of the high-$CO_2$ plants to satisfy their nitrogen requirements. Although such simple experiments relating increased carbon dioxide to plant growth and to plant-animal interactions give fairly straightforward results, it is harder to predict the overall consequences of increased carbon dioxide to complex ecosystems (Strain 1987).

All in all, an increased greenhouse effect

that includes changes in atmospheric gases and increases in temperatures can affect numerous attributes of the physical environments of animals. These environmental changes clearly could affect the fitness of individual organisms, and possibly the distributions and abundances of species.

## II. EFFECTS OF TEMPERATURE ON COMMUNITIES

### A. Historical Thinking on the Importance of Temperature to Living Organisms

Historically, ecologists have disagreed about the role of the physical environment in determining the composition of communities. For example, R. M. MacArthur (1972) believed physical environment had little to do with determining the presence or absence of species in a community. He offered our ability to grow exotic plants in gardens and raise exotic animals in zoos as proof. Because most organisms can be successfully raised in physical environments different from those in which they naturally occur, MacArthur concluded that plants and animals can tolerate a wider range of physical conditions than they experience in their natural areas of distribution. He further concluded that something else, namely interactions with other organisms, must ultimately determine the presence or absence of species in communities. Conversely, J. Grinnell (1904, 1917, 1924, 1928) stressed that the physical environment, including climate, can play an important part in defining the niche of a species (James et al. 1984).

Present thinking is that, indeed, both interactions with other species and with physical environments can limit distributions and abundances of species (note F. I. Woodward's observation in chapter 8 that physiognomic types of vegetation are limited toward the poles by temperature and toward the equator by competition). In the following sections, I present examples demonstrating the ways in which temperature can influence community composition and regulate species abundances and ranges.

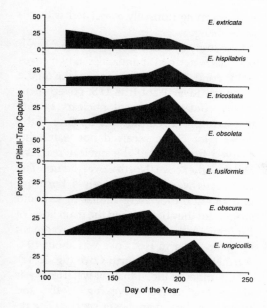

*Figure 13.2.* Abundance of seven species of beetles (genus *Eleodes*) captured during the summer in pitfall traps in the short-grass prairie in northeastern Colorado on the Pawnee National Grasslands. Abundance is expressed as a percentage of the number of beetles caught throughout the year. Time is in Julian days (days from January 1).

*Figure 13.3.* Means and standard deviations of the times during which beetles were seen to be active, and body temperatures of tenebrionid beetles from the short-grass prairie in northeastern Colorado on the Pawnee National Grasslands.

## B. Patterns of Temperature Change Appear to Create Patterns in the Abundances and Activities of Species within a Community

The responses of different species to different temperature conditions may determine the relative abundances and activities of animal species within a community on both a daily and an annual basis. One community that appears to have properties molded by temperature is a darkling beetle community in the short-grass prairie in northeastern Colorado (Whicker and Tracy 1987). This community contains seven beetle species of the genus *Eleodes*, each of which seems to do best under different thermal conditions. Seasonally, abundances of the different species rise and fall at different times throughout the year as average temperatures change (fig. 13.2). Although the patterns of abundance correlate with temperature, it is not known whether temperature directly affects species abundances or whether it affects other species that in turn affect the beetles. Nevertheless, these beetle species also show patterns in their daily activity (fig. 13.3). Specifically, individuals of some species are active both early and late in the day at times when ambient temperatures are relatively low, and these species are found to have relatively lower body temperatures when active. Other species are active nearer to the middle of the day, when

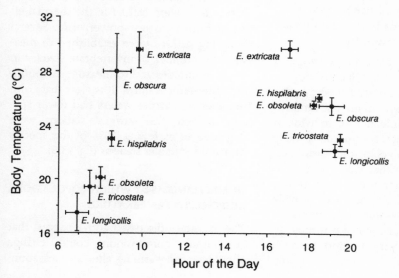

ambient temperatures are high and the beetles have higher body temperatures. Again, the possibility exists that the beetles are responding to changes in the behavior of other organisms, which in turn affect the beetles. In the laboratory, however, these beetle species move about within a thermal gradient so as to select body temperatures essentially identical to the body temperatures measured in the field. (Not surprisingly, the species that are active at the hottest times of the day are also most abundant at the warmest times of the year.)

We still do not understand the mechanisms producing these patterns of daily and seasonal abundance in darkling beetles, but it appears that temperature has a direct role in molding the distribution and abundance of species within this community. Therefore, large changes in temperature caused by global warming could conceivably change the abundance of the species in ways suggested by present daily and annual response patterns. In particular, warming might favor those species, like E. extricata, that prefer high temperatures and could cause population declines, or range reductions, in those like E. longicollis that appear to prefer low temperatures. (See also Tracy and Christian [1983] for a discussion of mechanisms by which climate works to produce patterns in ecological systems.)

## C. Short-term Temperature Increases Can Cause Reductions in the Number of Species within a Community

Insofar as each species performs best within a particular temperature range, it is not surprising that numerous case studies show that short-term changes in temperature can produce changes in abundance of individuals and species in ecological communities. Many examples come from ecological research on aquatic communities that have been artificially heated with hot water from power plants. Such artificial warming can provide large-scale experiments that might mimic the local effects of potential global warming, although it should be born in mind that such artificial warming is short-term compared

with warming normally associated with climate change.

One study by I. L. Brisbin (1974) measured seasonal waterfowl abundances in two coves of Par Pond. Par Pond is a cooling lake on the Savannah River Plant (the U.S. government facility containing several nuclear reactors that have produced plutonium). One of the coves (Hot Cove) received hot water discharged from one of the nuclear reactors, and the other cove (North Cove) did not. The two coves were approximately 4 km from one another, they were similar in size and shape, and they both joined the main body of Par Pond on the northwest side. Throughout the year, water in Hot Cove was about 10°C warmer than that in North Cove (fig. 13.4a). The number of individuals of waterfowl was always less in Hot Cove (fig. 13.4b), and moreover, there were fewer species. The reduced species density primarily reflected the absence of some species of dabbling and diving ducks in the hot cove, even though these species were seen in North Cove and other parts of Par Pond. Thus, increases in temperatures resulted in a community with reduced species richness and reduced abundance of individual waterfowl.

A similar pattern of species degradation was found in a study of the invertebrate communities found on mangrove roots (Kolehmainen and Castro 1974). These mangrove-root communities on the south coast of Puerto Rico were bathed in the thermal effluent from a nuclear power plant. Several sampling stations were established to measure the number of invertebrate species in the communities and to measure surrounding water temperatures. It is clear that communities in warmer waters had fewer species. Indeed, the warmest stations were degraded to as few as 10% of the species measured at cooler stations (fig. 13.5).

## III. MECHANISMS BY WHICH POPULATIONS RESPOND TO TEMPERATURE

The examples discussed above suggest that community composition could change should global warming alter species abun-

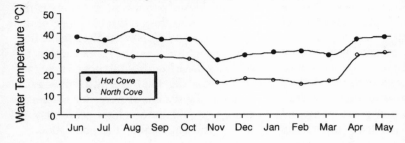

a. Water temperatures from two coves in Par Pond

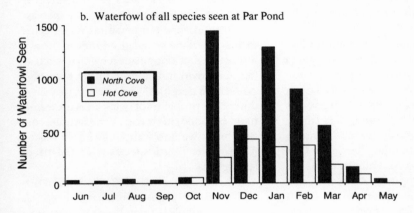

b. Waterfowl of all species seen at Par Pond

Figure 13.4 (*above*). Water temperatures (a) and numbers of individual waterfowl (b) seen throughout the year on North Cove and Hot Cove of Par Pond. Par Pond is a reservoir on the Savannah River Plant, South Carolina. Water from the reservoir is used for cooling a nuclear reactor, and warm water from the reactor is returned to Hot Cove.

Figure 13.5 (*below*). Number of species of invertebrates found on the roots of mangroves at seven different sites on the south coast of Puerto Rico. The sites differed in their proximity to a thermal discharge from a nuclear power plant and thus differed in water temperature. Data were collected for two years.

Mangrove Root Communities, South Coast of Puerto Rico

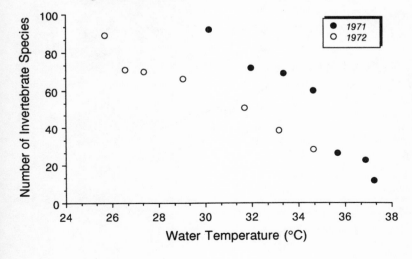

dances and geographic ranges. What are the mechanisms causing such changes at the species level? The physiological responses of organisms to temperature are well known (chapter 12), but the ecological responses of populations (particularly terrestrial animal populations) are not well understood. Nevertheless, there are many examples illustrating how populations are affected by temporal and spatial patterns in ambient temperature. Such examples serve as a basis for predicting responses of populations to changes in climate.

One interesting example of how a healthy population could be vulnerable to changes in climate comes from a study of a population of chuckwallas at Amboy Crater in California (Zimmerman and Tracy 1989). Chuckwallas are herbivorous lizards in the iguana family, confined to the Mojave Desert extending from the region surrounding the Sea of Cortez north to southeastern California and the southwestern tip of Utah. Amboy Crater, which has an extensive lava flow in a desert valley northeast of Los Angeles, is an extremely hot and dry habitat with particularly low plant diversity. In spite of the dramat-

ically hostile environment (in human terms), the chuckwalla population at Amboy is the densest within the geographic range of the species (Case 1976). Nevertheless, in the hottest time of the year (July), the coolest retreats for these lizards are so hot that they are nearly lethal (fig. 13.6). Clearly, a temperature increase of just a few degrees, due to climate change, would have a high probability of causing the extinction of this population.

Why would chuckwallas live in an environment that is so hostile that a modest increase in environmental temperatures could cause extinction of the population? Indeed, how is it possible that a population living in such an inimical environment could have such high population densities? The answers may lie in the special problems of being an ectothermic (cold-blooded) herbivore. For example, consider what we know about a closely related herbivorous lizard species, the Galápagos land iguana.

K. A. Christian et al. (1983) discovered that individual Galápagos land iguanas from Isla Santa Fe establish their home ranges near cliffs along the edges of plateaus (fig. 13.7). At the end of each day, the lizards move to the edge of these cliffs where they can, and do, keep their body temperatures elevated later into the evening than they could if they stayed on the plateau (fig. 13.8). This behavior allows the lizards to be warm for about 20%

*Figure* 13.6. Body temperatures of a male chuckwalla throughout three different days in March, May, and July. Midday body temperatures (especially in July) were near temperatures that are lethal to chuckwallas.

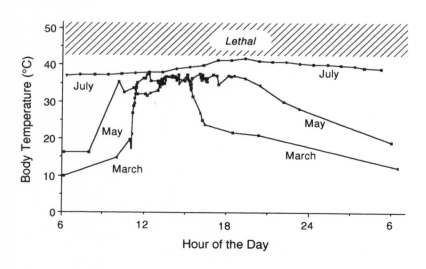

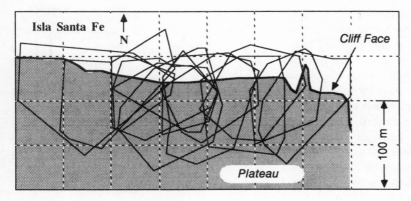

Figure 13.7. Home ranges of lizards in a population on Isla Santa Fe, Galápagos, in the rainy season. All lizards had home ranges that included a cliff face.

longer during the cooler season of the year (Christian et al. 1983). Given that lizards never established home ranges in areas without cliffs, it appears that the bulk of Isla Santa Fe is simply not suitable iguana habitat unless it is near a cliff. Why would it be important for

Figure 13.8. Measured body temperatures (connected dots) of a 4.2-kg female Galápagos land iguana on Isla Santa Fe during the cool season in relation to the maximum and minimum body temperatures for this animal predicted from a biophysical model. By moving to the cliff face, this land iguana was able to remain at approximately 34°C approximately 20% longer than it would have been able to had it remained on the plateau.

land iguanas to maintain elevated body temperatures for long periods of the day? The answer apparently lies in the mechanisms by which land iguanas acquire energy from their environment. Furthermore, the answer bears on the questions we had about chuckwallas.

Land iguanas are herbivores, and herbivorous reptiles have a particular problem assimilating energy from their food (Zimmerman and Tracy 1989). Plant material is less energy dense than the food of carnivorous lizards. Furthermore, digestion of plant material is a slow, temperature-dependent process. Chuckwallas simply do not digest food when their body temperature falls below 28°C (Zimmerman and Tracy 1989), and it is likely that land iguanas similarly need to be

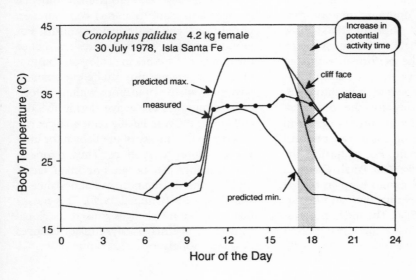

warm to digest their herbivorous diet. Thus, a longer period of warmth may allow the lizards to acquire more energy each day. In the case of land iguanas, it may be that the cliffs allow them to acquire 20% more energy than would be possible in areas without cliffs.

The land iguana example is interesting because climate change toward warmer temperatures in the Galápagos presumably would increase the time that lizards could be warm. Remaining warmer for longer periods of time, seemingly, could allow land iguanas to establish home ranges on parts of Isla Santa Fe that are currently too cool. Air temperatures, however, would have to increase by more than 6°C before land iguanas could maintain elevated body temperatures on plateaus without the aid of cliffs (Tracy, unpublished biophysical analysis)—a substantially larger climate change than climate modelers have predicted for the tropics (chapter 4). Thus, global warming due to the greenhouse effect is not likely to enhance the environment for Galápagos land iguanas. Indeed, such a large increase in temperatures would likely adversely affect other ecological interactions important to land iguanas (e.g., it might change the abundance and composition of plants growing on Isla Santa Fe).

The land iguana example also points to an important interaction between physiological function (e.g., digestion) and the ecological opportunities to achieve particular physiological states (e.g., optimum physiological function). Tracy and Christian (1983) show that an organism's habitat can be characterized according to the performance that an organism can achieve in that habitat. This characterization can be integrated into an index of habitat quality for the organism, roughly akin to summing up the time and space in which the organism can perform well in its environment. For whiptailed lizards, *Cnemidophorus tigris*, in Putah Creek (a tributary to the Sacramento River in central California), this index quantifies an interesting pattern (Tracy 1982). The index progressively declined during a three-year period (between 1963 and 1965) in response to cli-

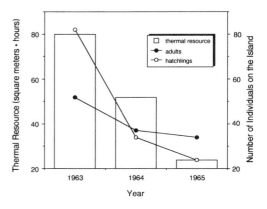

Figure 13.9. Thermal resource index of habitat quality and number of whiptailed lizards (adult and hatchling) seen on Putah Creek, California, in 1963 through 1965.

mate change (fig. 13.9). Correlated with the yearly change in the index were a progressive decline in the number of adult lizards seen on the island, a decline in the number of hatchlings produced by the adults, and a decline in the number of hatchlings produced per adult lizard seen. Clearly, habitats in which organisms can interact with physical factors facilitating good physiological performance can be important to animals.

## IV. SUMMARY

Increases in greenhouse gases in the atmosphere can cause numerous environmental changes important to animal communities, populations, and individual organisms. For example, increased temperatures can reduce the number of species in ecological communities. Temperature also can be important in governing when populations within ecological communities are active during the day and during the year. Finally, temperatures can determine the quality of the habitat for individuals within populations. Thus, temperature certainly can be an important factor molding the distribution and abundance of animal species (Grinnell 1928). This means that temperature changes caused by global climate change may catalyze major changes in animal populations and communities.

## REFERENCES

Brisbin, I. L. 1974. Abundance and diversity of waterfowl inhabiting heated and unheated portions of a reactor cooling reservoir. In *Thermal Ecology*, J. W. Gibbons and R. R. Sharitz, eds., CONF-730505, pp. 579–593. Washington, D.C.: U.S. Atomic Energy Commission.

Case, T. J. 1976. Body size differences between populations of the chuckwalla, *Sauromalus obesus*. *Ecology* 57:313.

Christian, K. A., C. R. Tracy, and W. P. Porter. 1983. Seasonal shifts in body temperature and use of microhabitats by the Galápagos land iguana. *Ecology* 64(3):463.

Grinnell, J. 1904. The origin and distribution of the chestnut-backed chickadee. *Auk* 21:375.

Grinnell, J. 1917. The niche relationships of the California thrasher. *Auk* 34:427.

Grinnell, J. 1924. Geography and evolution. *Ecology* 5:225.

Grinnell, J. 1928. Presence and absence of animals. *Univ. California Chronicle* 30:429.

Hansen, J., D. Johnson, A. Lacis, S. Lebedeff, P. Lee, D. Rind, and G. Russell. 1981. Climate impact of increasing atmospheric carbon dioxide. *Science* 213:957.

Hansen, J., I. Fung, A. Lacis, S. Lebedeff, D. Rind, R. Ruedy, G. Russell, and P. Stone. 1988. Prediction of near-term climate evolution: What can we tell decision-makers now? In *Preparing for Climate Change*, Proceedings of the first North American conference on preparing for climate change, J. C. Topping, Jr., ed., pp. 35–47. Washington, D.C.: Government Institutes.

James, F. C., R. F. Wamer, N. O. Niemi, and G. J. Boecklen. 1984. The Grinnellian niche of the wood thrush. *American Naturalist* 124:17.

Kolehmainen, T. M., and R. Castro. 1974. Mangrove-root communities in a thermally altered area in Guayanilla Bay, Puerto Rico. In *Thermal Ecology*, J. W. Gibbons and R. R. Sharitz, eds., CONF-730505, pp. 371–390. Washington, D.C.: U.S. Atomic Energy Commission.

MacArthur, R. M. 1972. *Geographical Ecology*. New York: Harper and Row.

Maduro, R. A. 1989. The greenhouse effect is a fraud. *21st Century Science and Technology* 2:14.

Oechel, W. C., and B. R. Strain. 1985. Native species responses to increased atmospheric carbon dioxide concentration. In *Direct Effects of Increasing Carbon Dioxide on Vegetation*, B. R. Strain and J. D. Cure, eds., US DOE/ER-2038, pp. 117–154. Washington, D.C.: National Technology Information Service.

Strain, B. R. 1987. Direct effects of increasing atmospheric $CO_2$ on plants and ecosystems. *Trends in Ecology and Evolution* 2:18.

Tracy, C. R. 1982. Biophysical modelling in reptilian physiology and ecology. In *Biology of the Reptilia*, C. Gans and F. H. Pough, eds., vol. 12, pp. 275–321. London: Academic Press.

Tracy, C. R., and K. A. Christian. 1983. Ecological relations among space, time and thermal niche vectors. *Ecology* 67(3):609.

Whicker, A. D., and C. R. Tracy. 1987. Tenebrionid beetles in the shortgrass prairie: Daily and seasonal patterns of activity and temperature. *Ecological Entomology* 12:97.

Zimmerman, L. C., and C. R. Tracy. 1989. Interactions between the environment and ectothermy and herbivory in reptiles. *Physiological Zoology* 62:374.

I am particularly grateful to R. Peters and J. A. Sugar for offering suggestions on several versions of the manuscript. This chapter was written while I was on sabbatical from Colorado State University as an Associated Western Universities Fellow at the University of California at Los Angeles.

CHAPTER FOURTEEN

# The Greenhouse Effect and Changes in Animal Behavior: Effects on Social Structure and Life-History Strategies

DANIEL I. RUBENSTEIN

## I. INTRODUCTION

The primary purpose of this chapter is to review selected examples of animal behaviors, such as feeding or reproduction, that vary with climate and therefore are likely to change substantially if global warming occurs. Because these behaviors are closely tied to survival and reproduction, many changes in behavior may have substantial ecological and evolutionary repercussions. Changes in mating behavior, for example, may alter the genetic structure and demography of populations. Behavior likely to have ecological and evolutionary consequences will be the focus of this chapter.

Since adjusting behavior is often an animal's first means of coping with changing environmental conditions, climatic changes wrought by the greenhouse effect should change substantially the quality or frequency of many behaviors. Indications of possible future changes can be observed when seasonal or other short-term changes in temperature or rainfall cause animals to alter their ranging, feeding, and mating behavior. Thus, a drop in temperature often induces small mammals to change their activity patterns, spending more time huddling, while a short-term decline in food abundance often causes foragers to spend less time searching in particular patches (Krebs and Davies 1987).

The problem of projecting future behavioral changes is complicated because only some of the effects of climate change will directly affect behavioral response. Many other effects will be indirect and therefore harder to predict. For example, large mammals, such as elephants, horses, and topi, respond to changes in landscape pattern induced by climatic changes. Similarly, ecological stresses that change the balance between adult and juvenile mortality may change reproductive strategies and thus reproductive behavior (Horn and Rubenstein 1986).

## II. DIRECT EFFECTS

Even a brief perusal of H. G. Andrewartha and L. C. Birch's classic text *The Distribution and Abundance of Animals* (1954) indicates that activity and reproductive patterns of most species are directly affected by climatic conditions. Vertebrates, for example, may huddle for warmth or show shifts in seasonal activity. Direct responses are perhaps most easily seen in ectothermic species, whose reliance on external sources of heat for metabolism makes them sensitive to changes in ambient temperature, while endothermic animals may buffer themselves from climate effects by maintaining constant body temperatures regardless of ambient temperatures. This section examines two types of direct effects: first, changes in insect metabolism, behavior, and resultant geographical ranges, and second, possible changes in reproduction and life histories for a wider range of organisms.

### A. Climate as a Determinant of Geographic Range in Insects

Because the metabolism of insects is directly affected by ambient temperature, climate change can directly affect the population densities and geographic distributions of insects by changing the ambient temperatures within their ranges. An increase in local temperature, for example, can speed up insect metabolism, with resultant effects on local population density. Predicting effects on insects is difficult because insects can buffer the effects of temperature behaviorally—by seeking shelter, for example—and herbivorous insects are also limited by the presence of host plants they need. Nonetheless, it is possible to draw some general conclusions about how climate might change insect distributions by looking at their basic physiological and related behavioral responses to climate variables (see chapter 8 for physiologically derived models of plant distribution).

Insects' metabolic activity and resultant behavior patterns fall into distinct types, depending on how their performance, as measured by traits such as basic metabolism and fecundity, responds to climate variables, no-tably temperature and humidity. The first type of response is linear, where continued increase in a climate variable, such as temperature, results in continued increase in the biological response. An example is shown in part A of figure 14.1, where locust (*Locusta migratoria*) activity increases linearly as long as temperature increases (until lethality). In general, as activity and feeding rates increase linearly in response to temperature, they will in turn enhance growth and fecundity.

A second type of response occurs when an increase in a climate variable increases the biological response up to a point, but thereafter further increase in the variable decreases the response. In such cases, where optimum performance occurs at moderate temperatures (or humidities), the graph of response is

*Figure* 14.1. Relationships between climate variables and fitness-related responses by locusts. A. The effects of temperature (°C) on movement rates. B. The effects of relative humidity on the average number of eggs individual females lay.

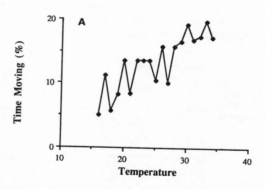

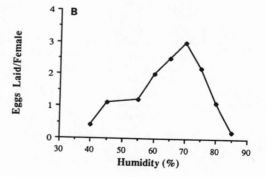

not linear but dome-shaped. In locusts, for example, the rate of egg production is highest only at moderate humidities (fig. 14.1B), dropping off as humidity increases or decreases past the optimum.

*Figure 14.2.* Range change by a hypothetical insect with a linear response to temperature. Ranges are mapped for two climate states, the present and the future under conditions of global warming with regional temperatures 3°C higher than present values. This model assumes that the insect's response to temperature is dominated by a single trait (for example, basic metabolism or fecundity), whose activity increases with temperature. As the trait increases in value, so does insect population density. On the figure, more cross-hatching indicates greater density. Insects with such linear responses will continue to occupy their present range, at the same time expanding into cooler areas, in this case toward the north. They will have denser populations at the southern end of their ranges than at present. This figure assumes that the southern boundary of the species' range under both present and future conditions is determined not by temperature but by some other constraint, such as a boundary with a superior competitor or the presence of unsuitable soil or vegetation. The northern limit is determined by temperature.

These physiological responses and the behaviors they determine are complex: it is difficult to predict ahead of time how a particular trait will respond to a climatic variable, whether it will show a linear function or one that is domelike. Further, some species may have a linear response pattern for one trait and a nonlinear pattern for another, as typified by the locust.

Understanding these fundamental physiological relationships is important for predicting how insect distributions and ecological relationships would be altered by climate change. Although future distributions would depend on a variety of effects, including insect interactions with vegetation and other animals, we can begin to understand how ranges might change by making the simplifying assumption that ranges of ectothermic species are determined by a single dominant response to a climate variable, a response that is either linear or dome-shaped. Given this assumption, the shape of the response curve will determine the pattern of climate-induced range change. If the dominant response by the hypothetical species to a climate variable is linear, then population den-

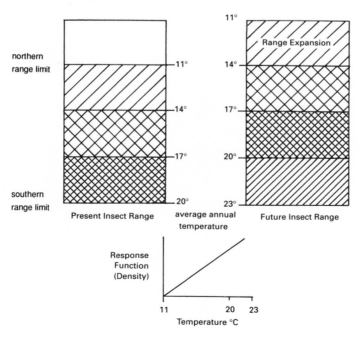

sities of the species could rise within its present range. This is because, other factors like competition or predation being constant, a speedup in the life cycle translates into greater production of individuals.

In addition, as shown in figure 14.2, a linear-response species should expand its range into cooler habitat—northward in the northern hemisphere—as habitat there warms enough to be suitable (provided that nonclimatic factors do not restrict expansion), and the species would also maintain its present southern distribution. Insect species with domelike response curves would behave differently. Figure 14.3 shows such a species expanding into warming areas to the north, but

*Figure 14.3.* Range change for a hypothetical insect with a dome-shaped response to temperature. Climate states are the same as in figure 14.2. Highest density is found at the center of the range, and global warming causes the range to shift toward the north while maintaining its present size. Insect density would remain the same throughout the range, unless affected by factors other than temperature. The model assumes that for this species the northern and southern range limits are both determined by temperature.

its southern boundary also moves northward because the newly raised temperatures in the south exceed physiological limits of the dominant response feature.

Although these scenarios are useful in thinking about a major component of insect response to climate, they make several simplifying assumptions. First, they assume that present-day southern boundaries for linear-response species are determined by nonclimatic factors. This is probably the case for many species, given that at present lethal temperatures are unlikely to be reached at the southern boundary. These scenarios also assume that for species with dome-shaped curves, southern boundaries are now temperature limited, which again is likely but not certain in any particular case.

Second, the outcomes assume that the ranges and densities of the host plants of these insects will respond to climatic changes in exactly the same way as the insects, so that distribution of host plants moves northward and does not limit the insects' temperature-induced range changes. In the real world, there may be changes in plant distribution or

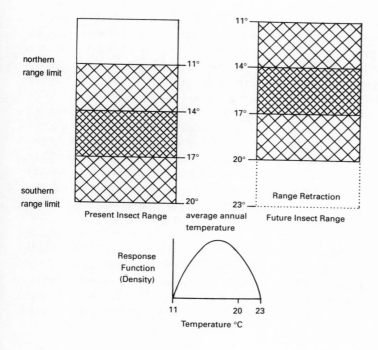

in the timing of critical events, such as flowering and seed set, so that the insect's population growth or range expansion may be limited or possibly even enhanced.

Third, the scenarios assume that individuals of species with dome-shaped response curves will not avoid southern range retractions caused by chronic stressful conditions through the evolution of new behaviors or physiological changes. Obviously, much will depend on the magnitude of the population's genetic variation.

Genetic adaptation could play a role in northward expansion as well, for species with either type of response function. For example, with longer days and warmer climates species that now have a single life cycle per year (univoltine) could evolve the capability for more than one (multivoltine). S. Masaki's (1978) study on crickets suggests how easily proximate control of diapause and nymphal development could be altered to change the number of generations. And with two or more generations of insect pests devouring important crops, the economic impact of such changes could be considerable.

That also has implications for community stability. Because species differ from each other in how key traits respond to climate, and because the functional responses of these traits can determine geographic range, climate change will cause species to move about the landscape independently of each other. This means direct climate effects will be one factor (others include competition and differential dispersal capability) that would cause present insect communities to break up in response to global warming, in a manner analogous to that described by Thompson Webb (chapter 5) and Russell W. Graham for mammals (chapter 6).

Even though we can predict that communities would be disrupted by climate change, it is difficult, given the complexity of dealing with a single species, to project the aggregate response of a community of species. Such a community was investigated in a series of experiments by W. P. McKay and his coworkers

(1986), who varied soil moisture and temperature to assess how these factors influenced the foraging behavior of soil arthropods and their effectiveness in breaking down litter. The arthropods showed increased behavioral activity when, from their point of view, environmental conditions improved, but different groups were favored by different conditions (see chapter 13 for further examples of thermal niche partitioning). Whereas an increase in soil moisture increased termite activity, it decreased the activity of ants. And whereas temperature had no effect on either of these macroarthropods, it did have profound effects on the microarthropods. For both fungivores and predators activity was enhanced by lower soil temperatures. Clearly, what is an environmental constraint for one species may not be for another. And this diversity of responses makes it difficult to assess the long-term ecological consequences, including those that are economically important, of environmentally induced changes in behavior.

## B. Reproduction and Life Histories

In addition to changes in activity and ranging patterns, animal reproductive patterns and life histories are also likely to be altered by large-scale changes in climate. For most reptiles an individual's sex is determined by environmental conditions (Bull 1983). In most turtle species only females are produced from eggs incubated at high soil temperatures, whereas in crocodiles and alligators and many lizards the converse occurs (fig. 14.4). Such variations arise because high temperatures during early development accelerate growth and increase final adult size. There is therefore selection pressure for the warmest nests to produce whichever sex achieves the greatest reproductive success by growing rapidly to large size. As long as threshold temperatures for determining male and female development are narrowly and rigidly defined, then large and consistent changes in temperature could cause local extinctions by eliminating one or the other sex from a region. Even if one sex was not eradicated but

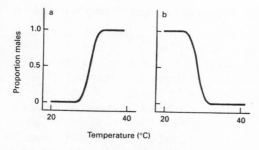

*Figure 14.4.* Generalized relationships depicting the response of sex ratio to incubation temperature for different types of reptiles. Females develop at low temperatures, as in many lizards and alligators (a). Males develop at low temperatures, as in most turtles (b).

severely limited, the ensuing skewed sex ratios would produce small effective population sizes in the rarer sex, thus limiting a population's genetic diversity. Future potential for adapting to subsequent environmental changes might be so limited that local extinction would occur anyway.

Complete switches in strategies of reproduction might also occur in response to global climatic change. Life-history theory predicts that animals should vary their investment in reproduction depending on their survival prospects relative to those of their young (Horn and Rubenstein 1986). In general, when mortality of juveniles is high but adult mortality is low, it makes little sense for adults to expend much energy producing many small young who are likely to die. Instead, they should produce only a few young at a time, young that often are large or otherwise well endowed to maximize their chance of survival. On the other hand, when juvenile mortality is low, there is an incentive to bear many small young, with a smaller investment in each. Such a life-history variation is exhibited by shad, a fish that grows and matures at sea but returns to rivers to breed and spawn (Gilebe and Leggett 1981). In the southern edge of its range it reproduces copiously, but only once in its lifetime. In the more north-

ern streams of New England and Canada it breeds repeatedly but produces only a few young each time. As theory predicts, massive and exhaustive reproduction occurs in the consistently mild southern streams where juvenile mortality is low, but not in the northern streams where yearly variations in icy conditions make juvenile survival prospects questionable (Shaffer and Elson 1975). Any global environmental change that warmed the northern rivers and made the aquatic environment more consistently benign would switch the northern population from a repeated, or iteroparous, life history to a semelparous one, with a one-shot reproductive effort. The effects that such a shift could have on the structure of entire aquatic communities and the relative abundances of species or the economics of the shad or even the salmon fishery (Gross 1985) could be profound.

The direct effects of global changes in weather need not be limited to temperature and rainfall. Gross changes in patterns of circulation will alter patterns of animal dispersal. Locusts, for example, move from spring to summer breeding grounds and back again with oscillations in the trade winds that are produced by seasonal shifts in the position of the intercontinental convergence (Mikkola 1986). This enables eggs laid during dry periods to hatch after the rains and developing larvae to feast on natural and agricultural greenery. Swarms form and the landscape is devastated. Even though the general patterns of movement are known, it is difficult to predict precisely where locusts will alight and swarms will develop. Changes in global circulation patterns will not only alter where the winds will take such migrants but they could also complicate control measures. In the past, effective, long-lasting, and hence ecologically damaging chemicals could be used to saturate likely feeding areas. Thus adults could be killed whenever they landed in these traps and swarms could be eliminated (Mackenzie 1988). With the shift to safer pesticides that break down more quickly, control can be effected only by spraying the young and destroying enough of them to prevent the formation of swarms. The imprecise, wide-ranging pest control strategies of the past might have been able to cope with changes in global circulation patterns that either broaden breeding areas or move them to agriculturally richer regions. But any major changes in circulation patterns in conjunction with the new target-sensitive strategies will only make the problem of control more difficult.

## III. INDIRECT EFFECTS

Global changes in temperature, rainfall, and atmospheric and oceanic circulation patterns will clearly alter the behavior and life histories of animals. Many changes in behavior, however, occur as indirect responses to previous changes in other components of an ecological community. And although such indirect responses are less easy to detect and make it more difficult to identify cause-and-effect relationships than the direct ones examined above, indirect changes in behavior are common and can have important ecological or evolutionary consequences. For example, the time and activity budgets of animals can be disrupted if the nutrient quality of their forage changes. Ordinarily, if the impala can obtain vegetation composed of 30% water, it does not need to drink and its movements are not restricted to the vicinity of watering sources. In periods of drought, however, the water content of the vegetation drops precipitously, yet the impala's watering requirements remain unchanged. As a result, it must alter its behavior to minimize its chances of incurring water stress. It tends to forage during the coolest times of the day and rest more often in the shade, and if it must forage into the night to maintain its daily ration, it may be forced to feed in open habitats where visibility is greatest (Jarman and Jarman 1973). Clearly, any global change that resulted in the warming and drying of landscapes could lead to massive reorganizations of the time, activity, and movement patterns of animals.

## A. Reproduction and Mating Systems

Major adjustments of time and activity budgets often lead to changes in social and reproductive behavior as well. Populations of Alaskan red foxes, after experiencing altered environmental conditions during 1982–84 induced by El Niño–Southern Oscillation in the Bering Sea, exhibited a variety of behavioral shifts (Zabel and Taggart 1989). Before El Niño most of the foxes on an island in the Bering Sea bred polygynously. One male had many mates, as well as a few nonbreeding adult females at the den who helped raise young. As a result, litter sizes and pup survival rates for polygynous parents were equivalent to or higher than those of monogamous ones. After El Niño, however, a major shift occurred in the mating system. Triggered by a reproductive failure of large seabirds, the red foxes shifted their diet to smaller, less abundant, and harder to catch prey. Because this dietary shift led to food shortages, fewer females were able to breed. And those that did, did so monogamously and produced fewer and smaller litters. Changes in behavior in response to environmental changes clearly enabled some animals to make the best of a bad situation. But its effect on the population's ability to maintain itself was large and, in this case, detrimental.

El Niño had dramatic effects on Andean condors as well. In a study that encompassed three populations, M. P. Wallace and S. A. Temple (1988) showed that after the heavy El Niño–induced rains the number of dead and decaying animals increased dramatically along the coast. The population inhabiting this area, unlike the two inhabiting the higher grasslands, normally fails to fledge many young. But after such an influx of resources, it is not surprising that a pulse of reproduction followed. Enough young were produced that the enlarged population provided immigrants to neighboring populations, supplying new genes and linking otherwise isolated populations. As with the red foxes, the ecological and demographic consequences of these global climatic changes may be dramatic.

## B. Foraging Behavior

Indirect behavioral and life-history changes are not limited to large, slowly reproducing species, since drought is known to affect many plant processes, which in turn affect insect behavior and reproductive performance (Mattson and Haack 1987). Leaves often turn yellow as chlorophyll production is reduced (Kramer 1983); cavitation, or breaking of water columns, often increases as plants dehydrate (Pena and Grace 1986); and leaf temperatures often rise as stomata remain closed to limit water loss (Mattson and Haack 1987). Each of these changes could make plants significantly more attractive to phytophagus insects, since many are attracted to yellow hues (Prokopy and Owens 1983), have heat or infrared receptors (Altner and Loftus 1985), or can readily detect vibrations (Barr 1969, Carlson and Knight 1969) even in the region of ultrasound (Prosser 1973), which is the type of sound emitted by breaking water columns (Sandford and Grace 1985). Thus drought may increase the likelihood that insects can find suitable food.

And once insects have alighted, other drought-induced changes may even enhance their foraging behavior. Increases in water and heat stress alter the biochemical composition of plants. Typically concentrations of amino acids, nitrates, sugars, and alcohols tend to increase, and many insects have contact chemoreceptors that are sensitive to these compounds. Moreover, they exhibit feeding responses that are proportional to the concentrations of these substances (Städler 1984, Mattson and Haack 1987, Visser 1986). Both the spruce budworm (Albert et al. 1982) and the locust (Mattson and Haack 1987) have peak feeding responses at sucrose concentrations higher than those found in their normal unstressed host plants. In addition, the higher concentrations or a better balance of sugars and amino acids, when coupled to higher leaf or ambient temperatures, may make drought-stressed plants more suitable for the growth and reproductive success of such insects as butterflies, moths, and grasshoppers (Mattson and Haack 1987). For other sapsucking arthropods, however, increased sap viscosity often lowers feeding ability (Auclair 1963). There is even some evidence that the altered chemical composition of plants enhances the ability of insects to detoxify some compounds (Wargo 1981). High temperatures and low humidities will also limit the ability of pathogens to control populations of phytophagous insects (Wilson 1974, and chapter 16).

Thus drought may indirectly alter the feeding behavior and reproductive performance of insects by increasing water and temperature stress on plants. Compromised defenses and increased attractiveness leave plants susceptible to many types of insects, but usually only for a short time. Normally, as climate improves so does the state of the affected plants, and populations of insect pests ordinarily decline. Of course the return to normal insect levels may be delayed because even normal plants may remain overwhelmed for some time by the outbreak. But if global climate change is not episodic, then the natural plant responses to stress may lead to massive insect outbreaks and chronically large populations that persist until plant abundances decline.

Climate change will not only change the appearance and nutrient balance of individual plants, but it will also change the quality, density, and composition of entire plant communities. For organisms higher up in the food chain, the consequences are likely to be profound. S. Riechert (1986) has shown that the aggressive tendencies of spiders vary dramatically with habitat. In riparian areas where shade keeps the habitat cool, 90% of the ground is available for web sites. Insect prey is abundant, and competition for resources or costs associated with defense of territories is minimal. Moreover, there is no correlation between web-site quality and reproductive success. It appears that most spiders can make a living fairly peacefully and are performing about equally well. In open grassland areas, however, temperatures are extremely high and winds exert a significant drying effect. Only about 12% of the substrate is suitable for web sites and insect abundances are low. Aggressive competition is severe, and a strong correlation exists between web-site quality and reproductive output. The simplest game-theory models predict that the probability of severe aggression should increase as the value attached to securing a resource increases or as the costs of fighting, either in terms of energy expended or damage incurred, decrease. Since in the grassland habitat the opportunity costs of not having a web site are greater than those in the riparian habitat, it is not surprising that aggression is much more frequent and intense in the grasslands (table 14.1). Whereas contests in riparian habitats are likely to escalate to display, those in grassland areas often proceed beyond display to threat. Thus, if the effect of global climate change is to make critical resources scarce, not only should levels of specific aggressive behaviors increase, but also should the nature of individual social interactions become qualitatively different.

## C. Social Behavior

On a larger scale, climate changes that affect vegetation quality often lead to changes in the patchiness of habitats and large-scale alterations of the landscape. Altering the number, size, and juxtaposition of habitats has profound effects on the social systems of a variety of large-bodied and wide-ranging species such as ungulates, and these in turn influence the genetic structure and recruitment abilities of populations. In elephants, for example, group sizes and composition change seasonally in response to seasonal differences in vegetation abundance (Western and Lindsay 1984). After the rains, when grasses begin growing rapidly, elephants graze on the plains. Because the grasslands stretch more or less continuously for hundreds of kilometers, little competition occurs when feeding. As a result, elephant females aggregate in large herds. With the flush of quality vegetation many come into estrus and become reproductively active. Typically, males remain sexually active for a short period of the year and advertise their condition, termed musth, with a variety of secretions (Poole and Moss 1981, Poole 1989a). Only the most dominant males come into musth immediately after the rains. Because the reproductive value of gaining access to the female groups is so high, the formation of massive female aggregations effectively incites male-male competition, with only the most capable being able to attach themselves to these large groups and being able to aggressively drive away interlopers. In the dry season, however, after the grasses on the plains cease growing and become cropped so that there is too little biomass to support such large-bodied ungulates as elephants, the elephants retreat to the swamps, where other grasses and browsable leaves remain. The patchy and closed nature of these

*Table* 14.1. Comparison of agelenopsid spider contests between opponents from sites of varying quality. Opponents were equal in weight, within 10%. Adapted from Riechert 1986.

| Site quality | Cost<br>Estimated energy expended | Length<br>Number of acts | Length<br>Number of bouts | Outcome<br>Probability that owner will win |
|---|---|---|---|---|
| Grassland | | | | |
| Poor: Surface | 123.3 | 11.9 | 1.9 | 0.56 |
| Average: holes | 344.0 | 31.4 | 3.0 | 0.76 |
| Excellent: holes | 556.7 | 51.6 | 3.7 | 0.92 |
| Riparian | | | | |
| Excellent: rocks | 126.2 | 13.5 | 1.8 | 0.92 |
| Excellent: grasses | 146.2 | 14.5 | 2.1 | 0.60 |
| Excellent: leaf litter | 185.5 | 16.1 | 2.1 | 0.89 |

habitats forces the large elephant herds to fragment into smaller family units. In addition, declining vegetation abundance and quality reduce the number of females that remain reproductively active. With fewer reproductive opportunities, dominant males go out of musth at this time of year while subordinate males enter it (Poole 1989a).

The consequences of these seasonal shifts are many. First, more females are sexually receptive after the rains, and they are also more localized than in the dry season. And second, given these seasonal differences in the distribution and abundance of sexually active females, dominant males have greater access to, and control over, reproductive females than do subordinate males. As a result, the reproductive success and Darwinian fitness of dominants are likely to exceed those of subordinates (Poole 1989b). Thus any global change in climate that would either affect the timing or duration of the rainy season could have profound effect on the genetic structure of elephant populations, as long as dominance is not strictly an age-related phenomenon. If, for instance, dry-season conditions came to predominate, elephants would be spending more time in smaller and fragmented groups. The ability of one or a few dominant males to control matings and sire a disproportionate share of the offspring would then be limited. More subordinates would be able to mate and a more equitable distribution of genotypes would be disseminated among the young. The effect of such a dramatic shift in the genetic structure of the population is hard to ascertain, but the change itself will be very real.

Similar social changes occur in many other ungulates. Topi, for example, exhibit lek mating systems, in which males defend small territories where females come to mate in a landscape characterized by sharp gradations in relief (Gosling 1986). In such habitats males can control elevated open sites where predation is unlikely because of excellent visibility, and females can feed free of male interference in the more productive intervening valleys. In areas of rolling grasslands where relief is more shallow and habitat differentiation is less distinct, small groups of females are instead defended by single males. Thus if global climate change alters the patterning of the landscape, it is likely that mating and social systems of animals living off that landscape will change.

Social organization can directly affect the reproductive success of a population. Horses live in a variety of societies (Rubenstein 1981, Berger 1986), and this variability illustrates how the social relations of different populations affect population dynamics. Despite significant ecological and social differences, all horse societies can be characterized by one general rule: unrelated females initiate associations with males and live in so-called harem groups whenever topography permits (Rubenstein 1986). The males they prefer are generally the most dominant, because only such males limit harassment from other males seeking sexual access. And protection increases the likelihood that females will rear offspring to the age of independence (Rubenstein 1986).

The following example shows not only the effect of social structure on reproductive success but also that social structure itself can be determined by the patterning of resources. On a barrier island off the North Carolina coast before 1980 one population of horses lived in traditional harem groups, whereas another did not. The population with harems ranged over an area of ten square kilometers where grazing swards were continuously distributed. The other lived in an area where tall dunes and dense maritime forest fragmented grassy areas into patches of variable size. In the largest patches large groups could form, but in the small ones intense competition for food caused groups to fragment. Thus a fission-fusion society developed and made it impossible for males to associate permanently with any group of females. Males wandered solitarily, and aggressive contests occurred whenever two or more males encountered reproductive females. Since the contests occurred near the females, harassment of females was high and their ability to forage was reduced. When the per capita reproductive successes of females from the different populations were compared, those in the harem society were higher than those in the fission-fusion society despite the increased vegetation available to females living in the latter. Thus before 1980 it was not surprising that the number of horses in the fission-fusion population remained unchanged, while the number in the harem population kept increasing.

After 1980 the social organization of the population in the patchy grasses changed. Dredging of the channel separating the island from the mainland increased the area of the island in the vicinity of the fission-fusion population. Because the new land was flat, the grasses that soon covered it were distributed continuously. Within two years harems developed there, harassment of females declined, per capita reproductive success improved, and the population began to increase in size. Again, a change in landscape altered a population's social organization. But even more important, the social change led to a significant change in the population's dynamics and its ability to expand.

This type of indirect effect can not only alter the shape of an animal's society, but it can also influence the composition of animal communities or even humankind's conservation efforts and ability to manage endangered species. If rare species find themselves in or are reintroduced into altered habitats, the societies they develop may make it impossible for them to maintain themselves. Such outcomes would complicate conservation efforts.

## IV. CONCLUSIONS

These examples illustrate that global climate change is likely to alter the behavior and life histories of many types of animals. Whether these effects are the direct or indirect result of changes in temperature, moisture, or circulation patterns, the consequences for the genetic structure and demographic properties of populations are likely to be dramatic. The examples and the changes they illustrate re-

flect natural responses to existing levels of environmental variability or change. There is no reason to suspect that these variations are any more drastic than those envisioned for future global changes in climate. In fact, they might underestimate the changes animals are likely to experience. They certainly are more analogous to anticipated changes in average tendencies than to changes in the extremes.

Actual studies that specifically examine the expected effects of global climate changes due to the greenhouse effect are rare. It is hoped that the speculative accounts and hypothetical predictions can be tested by direct experimentation. By perturbing ecological microcosms, behavioral changes and their ecological impact can be monitored. But more important, biologists need more finely tuned predictions from the climatologists as to what conditions animals at a local level are likely to experience. This is especially important for forecasting indirect effects, since small differences in vegetation or landscape can produce markedly different alternative behavioral outcomes.

At the same time, however, ecologists must begin identifying a few key ecological habitats worthy of detailed ecological study. Important species, those that carry out key functions in different locales, must be studied so that the strength and direction of indirect effects on behavior and life history can be evaluated. By identifying these patterns as well as the major variables, global modelers will gain insights into what variables will make their climate predictions more useful.

## V. SUMMARY

Changes in climate will profoundly affect the behavior and life histories of animals. Increases in temperature and changes in humidity will often be direct. For many insects, as temperature and humidity increase, developmental rates, speed and distance of movement, and fecundity will generally be enhanced, and these changes should have pronounced effects on many ecological processes. For the many reptiles whose sex is environmentally determined, slight changes in temperature will alter the relative proportion of sexes in a particular locale and thus could potentially disrupt population dynamics. Moreover, global changes in wind circulation will affect the migratory routes of many insects, such as locusts, and any change that alters the balance between adult and juvenile mortality will also change a species' life history, sometimes as dramatically as changing it from a repeated to a one-time breeder.

Other effects will be indirect, since changes in climate will change features of the landscape and they in turn will alter the behavior of the animals residing in those habitats. At the moment, however, the exact ecological consequences of these changes are difficult to predict; the magnitude of the temporal and spatial variation associated with anticipated climatic change is unknown, and even small variations could eventually cause major swings in behavioral outcomes. Nevertheless, some trends can be perceived. For example, changes in global circulation systems often disrupt the population dynamics of predator species because critical features of the biology of their prey are disrupted. Moreover, wholesale stressing of plant communities often enhances the foraging ability of herbivorous insects, which in some instances will reinforce the stresses already being experienced by the plants. Major changes in the landscape, including shifting of biome boundaries, will also occur and lead to changes in the social behavior of a variety of animals. Whereas increases in aridity tend to heighten the intensity of aggression displayed by such species as spiders, the same types of landscape change will lead to major adjustments in the daily cycle, time budgets, and mating systems of such large-bodied species as grazing ungulates. The long-term consequences of these behavioral changes on the genetic structure or demographic dynamics of populations are likely to be great.

## REFERENCES

Albert, P. J., C. Cearley, F. Hanson, and S. Parisella. 1982. Feeding responses of eastern spruce budworm larvae to sucrose and other carbohydrates. *J. Chem. Ecol.* 8:233.

Altner, H., and R. Loftus. 1985. Ultrastructure and function of insect thermo and hygroreceptors. *Ann. Rev. Entomol.* 30:273.

Andrewartha, H. G., and L. C. Birch. 1954. *The Distribution and Abundance of Animals.* Chicago: University of Chicago Press.

Auclair, J. L. 1963. Aphid feeding and nutrition. *Ann. Rev. Entomol.* 8:439.

Barr, B. A. 1969. Sound production in Scolytidae (Coleoptera) with emphasis on the genus *Ips*. *Can. Entomol.* 101:636.

Berger, J. 1986. *Wild Horses of the Great Basin.* Chicago: University of Chicago Press.

Bull, J. J. 1983. *Evolution of Sex-Determining Mechanisms.* Menlo Park: Benjamin/Cummings.

Carlson, R. W., and F. B. Knight. 1969. Biology, taxonomy, and evolution of four sympatric *Agrilus* beetles. *Contrib. Am. Entomol. Inst.* 4(3):1.

Gilebe, B. D., and W. C. Leggett. 1981. Latitudinal differences in energy allocation and use during the freshwater migrations of American shad (*Alosa sapidissima*) and their life history consequences. *Can. J. Fish. Aquat. Sci.* 38:806.

Gosling, M. 1986. The evolution of mating systems in male antelopes. In *Ecological Aspects of Social Evolution,* D. I. Rubenstein and R. W. Wrangham, eds., pp. 244–281. Princeton: Princeton University Press.

Gross, M. R. 1985. Disruptive selection for alternative life histories in salmon. *Nature* 313:47.

Horn, H. S., and D. I. Rubenstein. 1986. Behavioral adaptations and life history. In *Behavioral Ecology,* J. R. Krebs and N. B. Davies, eds., 2nd ed., pp. 279–300. Oxford: Blackwell Scientific.

Jarman, P. H., and M. V. Jarman. 1973. Daily activity of impala. *E. Afr. Wildl. J.* 11:75.

Kramer, P. J. 1983. *Water Relations of Plants.* Orlando, Fla.: Academic Press.

Krebs, J. R., and N. B. Davies. 1987. *An Introduction to Behavioral Ecology.* Oxford: Blackwell Scientific.

Mackay, W. P., S. Siva, D. C. Lightfoot, M. I. Pagai, and W. G. Whitford. 1986. Effect of increased soil moisture and reduced soil temperature on a desert soil arthropod community. *Am. Midl. Nat.* 116:45.

Mackenzie, D. 1988. Call to unleash dieldrin on locust plague. *New Scientist* 15:26.

Masaki, S. 1978. Seasonal and latitudinal adaptations in the life cycles of crickets. In *Evolution of Insect Migration and Diapause,* H. Dingle, ed., pp. 72–100. New York: Springer-Verlag.

Mattson, W. J., and R. A. Haack. 1987. The role of drought in outbreaks of plant-eating insects. *Bioscience* 37:110.

Mikkola, K. 1986. Direction of insect migrations in relation to wind. In *Insect Flight*, W. Danthanaragana, ed., pp. 152–171. New York: Springer-Verlag.

Pena, J., and J. Grace. 1986. Water relations and ultrasound emissions of *Pinus sylvestris* L. before, during, and after a period of water stress. *New Phytologist* 103:515.

Poole, J. H. 1989a. Announcing intent: The aggressive state of musth in African elephants. *Anim. Behav.* 37:140.

Poole, J. H. 1989b. Mate guarding, reproductive success and female choice in African elephants. *Anim. Behav.* 37:842.

Poole, J. H., and C. J. Moss. 1981. Musth in the African elephant, *Loxodonta africana. Nature* 292:830.

Prokopy, J. R., and E. D. Owens. 1983. Visual detection of plants by herbivorous insects. *Ann. Rev. Entomol.* 28:337.

Prosser, C. L., ed. 1973. *Comparative Animal Physiology*. New York: W. B. Saunders.

Riechert, S. 1986. Spider fights as a test of evolutionary game theory. *Am. Scient.* 74:604.

Rubenstein, D. I. 1981. Behavioral ecology of island feral horses. *Equin. Vet. J.* 13:27.

Rubenstein, D. I. 1986. Ecology and sociality in horses and zebras. In *Ecological Aspects of Social Evolution*, D. I. Rubenstein and R. W. Wrangham, eds., pp. 282–302. Princeton: Princeton University Press.

Sandford, A. P., and J. Grace. 1985. The measurement and interpretation of ultrasound from woody stems. *J. Exp. Bot.* 36:283.

Shaffer, W. M., and P. F. Elson. 1975. The adaptive significance of variations in life history among local populations of Atlantic salmon in N. America. *Ecology* 56:577.

Städler, E. 1984. Contact chemoreception. In *Chemical Ecology of Insects*, W. J. Bell and R. T. Carde, eds., pp. 3–35. Sunderland, Mass.: Sinauer Associates.

Visser, J. H. 1986. Host odor perception in phytophagous insects. *Ann. Rev. Entomol.* 31:121.

Wallace, M. P., and S. A. Temple. 1988. Impacts of the 1982–1983 El Niño on population dynamics of Andean condors in Peru. *Biotropica* 20:144.

Wargo, P. M. 1981. Defoliation and secondary-action organism attack: With emphasis on *Armillaria mellea. J. Arboric.* 7:64.

Western, D., and W. K. Lindsay. 1984. Seasonal herd dynamics of a savanna elephant population. *Afr. J. Ecol.* 22:229.

Wilson, G. F. 1974. The effects of temperature and UV radiation on the infection of *Choristoneura fumiferana* and *Malacosoma pluviale* by a microsporidian parasite, *Nosema* (Perzia) *fumiferanae* (Thom.). *Can. J. Zool.* 52:59.

Zabel, C. J., and S. J. Taggart. 1989. Shift in red fox, *Vulpes vulpes*, mating system associated with El Niño in the Bering Sea. *Anim. Behav.* 38:830.

I thank R. Peters for his encouragement throughout this project and for drafting figures 14.2 and 14.3. This research was supported by National Science Foundation grant BSR8352137 and National Institutes of Health grant PSNMH34890.

CHAPTER FIFTEEN

# Double Jeopardy for Migrating Animals: Multiple Hits and Resource Asynchrony

J. P. MYERS AND ROBERT T. LESTER

## I. INTRODUCTION

Our mandate here is to extend climatic predictions into the biotic world. It is no minor challenge, even if we minimize the uncertainties in the climatic predictions about magnitude and pace. To begin with, our data are mismatched. The climate models work best at large spatial scales: the narrower the geographic focus, the more uncertain the climate prediction. Our ecological data, by contrast, are most relevant to specific sites and particular processes.

The challenge is further compounded by the nature of our ecological knowledge. Particularly for most animal species, links between climate and ecology are poorly understood. And in those realms of ecology where the links are relatively well known, the evidence is that ecology is often affected less by mean climatological conditions—something the climatological models can handle—than by the frequency and severity of extreme events—parameters the models are only now being asked to yield.

The net result of these discordances between ecology and climatology is that any biological predictions are at best circumscribed by wide margins of error, and any conclusions are unavoidably vulnerable to legitimate scientific skepticism, as well as to abuse by those with a political agenda. That does not mean, however, that preliminary efforts to extrapolate potential effects are without value. Given the first-cut assessments of what climatic disruption may entail for biotic systems (e.g., Peters and Darling 1985; Emanuel et al. 1985a, 1985b; EPA 1988; Lester and Myers 1989), we must begin to lay the scientific groundwork now for rigorous predictions, even if that entails climbing out together on the limb of data-free speculation.

One starting point—that taken in this chapter—is to search for organisms and processes that are particularly vulnerable to climate change. Using this approach, even without committing ourselves to specific projections, we can identify systems that might be expected to suffer early and severe effects of climate change. This should help set priorities for monitoring and for conservation action.

This paper focuses on shorebirds (Aves: Charadrii), which may be especially sensitive to climatic instability because of their migratory way of life and relatively narrow habitat preference—the coastal zone. We identify key points in their life histories and ecology that may be affected by climatic change. Among other vulnerabilities, this assessment reveals a heretofore unidentified complication of global change: migratory species may respond not only to how much climate changes at a particular site but also to relative differences in the amount of warming among sites. As biological musings on climate change proliferate, we should expect to encounter increasing numbers of these and other second-order effects.

## II. BIOLOGICAL BACKGROUND ON SHOREBIRD MIGRATIONS

Most shorebirds migrate away from breeding sites during the winter. All but one of 49 species breeding in North America are either completely migratory or have at least some migratory populations, and of those, 31 species fly to South America, with the annual migrations of some exceeding 25,000 km (Morrison 1984, Myers et al. 1987). The majority of all shorebirds breed in subarctic wetlands and arctic tundra. During migration and the nonbreeding season, most occur principally in interior wetlands and coastal habitats, particularly intertidal mudflats. Of the remainder, two species are largely pelagic, and five use upland grasslands.

Migration occurs in steps. Individuals fly from one staging site to the next, pausing for varying lengths of time to replace fat burned during flight (fig. 15.1). The flights last two to three days and cover 2000 kilometers or more. The timing of these flights is influ-

Figure 15.1. Northbound (toward breeding ground) migration routes of shorebirds in western hemisphere. Adapted from Morrison 1984 and Myers et al. 1987.

enced by the availability of resources. Northbound flights, in particular, are timed to match the opening of breeding habitat in the arctic and the emergence of food resources along the migration pathway (e.g., Myers 1986).

For most species, migration southward is spread over much of the summer, with birds traveling from late June through October. This lengthy migration period derives from a variety of factors, including latitudinal differences in breeding phenology, as well as a staggered departure schedule in most species. Depending on mating system, males or females depart first, followed by the other sex (Myers 1981). Unsuccessful breeders may depart even earlier in the season.

The major migration corridors for New World shorebirds are located along the Atlantic and Pacific coasts of North and South

America and through the western Gulf of Mexico and the American Great Plains.

## III. SHOREBIRD POPULATIONS AND CLIMATE DISRUPTION

This life-history pattern renders arctic-breeding shorebirds vulnerable to climate change in at least three different phases of their life cycle.

### A. Effects on the Breeding Ground

Breeding species will be affected by the magnitude of warming in the arctic, both through direct physiological impacts on individuals and through effects on the habitats and food items used by these species in the arctic. Some direct effects may be beneficial. Although shorebird chicks are precocial, on relatively cold days soon after hatching they still must be brooded by adults to maintain body temperature (Norton 1973, Chappell 1980, Pienkowski 1984). Warmer temperatures will probably enhance chick survivorship, on average.

Several indirect effects would be deleterious. First, the quantity, character, and availability of food will change with global warming. Adult shorebirds during the breeding season feed principally on subsurface insect larvae as well as surface-crawling insects and arachnids. Chicks depend on surface forms and especially flightless craneflies and midges (Tipulidae and Chironomidae). With warming, over time the invertebrate fauna will change. One trend may be toward fewer flightless species, if indeed the insect response to warming follows the current geographic trend for flightless forms to decrease in warmer climates. The tremendous abundance of flightless insects for foraging chicks, in fact, is one hypothesis advanced for the numerical success of arctic-breeding shorebirds. Hence, a warmer climate may mean less efficient foraging for chicks, all other things being equal.

Second, as warming shifts climatic zones northward, we should anticipate large declines in populations as a result of substantial reductions projected for the amount of tundra suitable for these species' breeding efforts. Two species breed almost exclusively above 75°N (sanderlings, *Calidris alba*, and knot, *C. canutus*). At least fourteen others are restricted largely to latitudes above 70°N and, in the subarctic, to higher elevation patches of tundra vegetation (e.g., Kessel 1989). Based on existing climate-vegetation relationships at these latitudes, simple extrapolations built on climate warming predictions and concomitant changes in precipitation indicate that the current habitats used by these species would be substantially reduced with moderate global warming (Emanuel et al. 1985a, Edlund 1986). Although the precise links between habitat size and population size are unknown for any shorebird species, basic ecological work with several species suggests density-dependent population regulation (Holmes 1970, Goss-Custard 1979). Major decreases in habitat extent should therefore have an effect on population size.

One key question is how vegetation stature responds to warming climates. Shorebirds show striking habitat selectivity in their choice of breeding sites (Pitelka and Myers 1980). Vegetation stature is a key variable because the mobility of young chicks is hampered by luxuriant vegetation. Open wetland habitats dominated by graminoids or lichens with vegetation heights below 10 cm are strongly preferred. Warming temperatures should lead to changes in the productivity, structure, and composition of arctic vegetation (Edlund 1986). The length of time needed for the changes to occur depends on the responses of existing vegetation to altered climatic regimes as well as the rate at which new flora invade northward.

### B. Effects on Migration and Wintering Sites

Shorebirds' dependence on wetlands renders them vulnerable to two indirect effects of global warming during the nonbreeding season: drought and sea-level rise.

During migration large fractions of several North American populations pass northward and southward through the Great Plains states, pausing at wetland staging sites to replenish fuel reserves. This region of the United States and Canada is expected to experience drier conditions as the climate warms, increasing the frequency of droughts. Observation of historical drought periods can illuminate likely effects on shorebirds. For example, many sites used by shorebirds in the Great Plains were rendered useless for feeding purposes during the summer drought of 1988. Birds abandoned them without sufficient fuel reserves to reach the next major stopovers (Castro et al. 1990). Data are as yet unavailable on the long-term population impact of this disruption in the migration system, but based on current understanding of the role of stopovers in shorebird migration (Myers et al. 1987), it appears likely that recurring events of this magnitude would reduce population sizes. Disruptions in habitat conditions along migratory routes in fact are speculated to be one cause of irregular but widespread breeding failures in arctic shorebirds.

Sea-level rise is also expected to have deleterious effects, principally through loss of habitat. Recent work in England demonstrates unequivocally that loss of wetland habitat reduces shorebird population size (e.g., Goss-Custard and Moser 1988), and EPA projections anticipate a net loss of 25%–80% of U.S. wetlands by the year 2100 with moderate sea-level rise (EPA 1988). It may be even worse in West Coast intertidal mudflats, where onshore human developments will prevent any landward expansion of crucial foraging habitats. These sites, particularly in the San Francisco Bay, are among the most important wintering and migration areas for shorebirds in the United States.

## C. Second-Order Effects

Climatologists have identified a number of physical, second-order effects to be expected from global warming. These derive from changes in physical gradients across the earth's surface as the amount of warming differs latitudinally and regionally. Among potential second-order effects that have been identified are changes in ocean currents resulting from a reduced equator-polar temperature gradient (Broecker 1987). Biologists have yet to focus on analogous changes to ecological systems, perhaps because the first-order effects are difficult enough as it is. Yet in contemplating the vulnerability of migratory birds to climate change, two possible second-order effects are immediately evident: disruptions in relative timing and differences in the physiology of timing mechanisms.

*1. Disruptions in Relative Timing* The previous problems of breeding and migration would result from climatic changes at specific sites, regardless of the relative magnitude of change elsewhere. Second-order changes, analogous to the physical examples above, however, may emerge as a result of the differences expected to develop latitudinally in degree of warming. Differences in climatic change among stopover sites at different latitudes may substantially alter the present relative timing of resource availabilities to which shorebirds have fine-tuned their migrations. If warming is substantially greater at higher latitudes, as projected by all models, and if the timing of food availability is determined by the phenology of temperatures, the present relationships between food availability at different sites and shorebird migrations may become asynchronous.

Presently, birds migrate to the arctic by gathering in large numbers at successive stopping points along the migration pathway. Overall, the northbound migration can be described as a wave of birds moving from one high point of resource availability to another. Particularly in the northern parts of this passage, a narrow window in time is defined by three factors: the receding of winter conditions, the emergence of food resources at specific stopover sites, and the need, given the short time available to complete the breeding cycle in the arctic summer, to reach arctic breeding sites as soon as they become occupiable during the spring thaw.

For example, the largest passage site on the

U.S. east coast in spring is Delaware Bay (Myers 1986). More than a million shorebirds move through this site in May. Before that month, however, the food resources of the bay are insufficient to support large numbers of birds. Only when bay temperatures warm and large numbers of horseshoe crabs (*Limulus polyphemus*) migrate onto local beaches to lay eggs, usually beginning in the first week of May and peaking in the third week, does the site function as a refueling area for shorebirds. Arctic breeding sites used by these same birds begin to open (with snow melt) in early June. To reproduce successfully, newly arrived migrants must occupy the breeding sites by early June to ensure that the young hatch immediately before the flightless insects emerge. With a month or more (depending on species) required to establish a territory, lay eggs, and incubate eggs to hatching, little leeway in timing is available.

It is this relative timing that may be disrupted by the latitudinal gradient in global warming, making it likely that ecological events in the arctic would occur sooner in the annual cycle relative to those at midlatitude (i.e., sooner compared with their current temporal relationships). Midlatitudes, in turn, would be affected more than those at the equator (fig. 15.2). In the case of migration from Delaware Bay to the arctic, moderate warming might cause relatively little change in when horseshoe crabs spawn but could cause the arctic insects to emerge substantially earlier than they do now. Therefore, shorebirds might not have enough time both to fatten in Delaware Bay and to reach the arctic before the insect flush has peaked and declined. If the time available for migration is shortened too much, the temporal pattern of resource availability may no longer support the migration system. With present data it is difficult to know at what point such a temporal compression would cause significant breeding failure.

2. *Differences in the Physiology of Timing Mechanisms*
The annual rhythm of migratory birds is regulated by a combination of internal and exter-

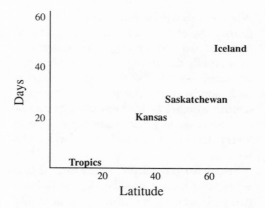

*Figure 15.2.* Approximate number of days by which global warming would advance the onset of the growing season at different latitudes. Calculations based on GISS model for $CO_2$ doubling, with no direct effects of $CO_2$. Compiled from data in Perry et al. 1988 and Rosenzweig 1989.

nal factors. For long-distance migrants, particularly, the onset of migration is sensitive to day length but insensitive to local temperature. This is interpreted as a evolutionary response to the fact that weather conditions in Tierra del Fuego during February, for example, where red knots spend the nonbreeding season, are not good predictors of when in early June breeding sites will become snow-free in the Canadian arctic. Red knots migrate at the same time of year regardless (within limits) of weather conditions.

In contrast, the onset of spring activity by insect larvae in tundra soils and emerging adults is acutely sensitive to local temperature. Emergence is earlier when early summer temperatures are warmer and later if they have been colder.

Because the prey, insects, respond to local temperatures while the predator, shorebirds, respond to day length, the annual cycles of prey and predator may diverge to the point that the prey's life cycle is out of phase with the predator's pattern of resource use. In other words, if spring thaw and summer insect emergence advance too much, the emergence may already have occurred by the time that shorebird chicks have begun to hatch.

Whether this will become a problem for migratory shorebirds is a matter of speculation. It would depend, among other things, on how rapidly the timing of insect emergence advanced—would it shift 2 weeks in 10 years or 2 months in 30? The pace of shift would control whether an evolutionary adjustment of shorebirds' migratory timing (i.e., leaving Tierra del Fuego at a different day length) were possible.

The more general point, however, is that ecosystems are composed of organisms from many phyletic origins with a host of timing mechanisms adjusted to climate and various patterns of resource availability. As global warming shifts the timing of temperature changes relative to the annual cycle of day lengths and to other cues used by plants and animals to control their annual cycles, we might expect to see some unraveling of current ecological processes that are time-dependent.

## IV. CONCLUSIONS

No quantitative predictions can be made currently for the effects of climate change on shorebird populations, nor are such predictions more feasible in other migration systems, for example, passerine birds, waterfowl, or, for that matter, whales or wildebeests. The links between vertebrate population ecology and climatology are too poorly understood for those predictions to withstand careful scrutiny. It should be clear from the qualitative assessment offered here, however, that there are many ways that migratory species will take hits from global warming. Because of their use of coastal habitats and the enormous latitudinal breadth of their migrations, shorebirds are likely to suffer severely.

### A. Research Needs

Several research programs would reduce the uncertainties now preventing quantitative predictions. None are trivial, and they include both continued investigations into shorebird population biology as well as

efforts to predict what habitats and resources will be available, where, when, and how much. It will be important, moreover, to distinguish the physiological effects of warming (e.g., potential benefits from higher chick survivorship) from long-term effects such as habitat modifications through migrations of plant and insect communities. The net impact of the physiological effects may be zero if the birds' breeding distribution shifts northward in association with plant communities.

The most basic questions are: what will be the areal extent of habitat that is available and capable of supporting shorebird reproduction or winter survivorship, and will the amount of area in breeding or winter (or neither) be the limiting factor for shorebird population size? These are old questions in migratory bird population ecology (cf. Morse 1980) that have resisted definitive answers even during a period of stable climate. It is entirely plausible that the relative contributions of survivorship and reproduction to population limitation may change dramatically if there are major changes in the relative amounts of breeding and wintering habitats.

In the arctic breeding grounds it would be particularly useful to know whether chick survivorship to fledging varies as a function of vegetation density. If survivorship decreases with increased density—and if vegetation density is enhanced by $CO_2$ enrichment, rising temperatures, and a long-term northward shift of southern plant species—then a northward shift of shorebird breeding ranges seems likely. On the wintering ground the main variable will be, beyond simply the areal extent and configuration of available habitat, the effect of warming on the invertebrate food base.

### B. Conservation Implications

As some warming appears inevitable (based on the GCM results using greenhouse gas concentration changes that have already occurred), the question then becomes: what steps might be taken to minimize the effects of future warming on migratory shorebird populations? Although Paul Ehrlich's "24-

hour, mandatory, world-wide hold-your-breath day" may be the only simple policy that would work with certainty, it would be prudent to consider more practical mitigations focused on biological conservation policies.

Peters and Darling (1985) offer a series of general considerations, several of which are pertinent to shorebird conservation. Of them, it would appear that the most practical at this stage is to work to maximize the extent of habitat available in the future: coastal reserves should be designed to accommodate sea-level rise, including provisions for the formation of new intertidal mudflats at sea levels above current; buffer zones should be protected that ensure above-tide roosting sites are still available once current ones are submerged; interior wetland protection systems should be diversified in a network of sites rather than concentrated around single wetlands.

## C. Monitoring Needs

Shorebirds also offer opportunities as an early-warning system for the biological effects of climate change. An immense amount of qualitative information is available about the limits of species' ranges, especially during midwinter. This information should be organized to detect systematic multispecies range shifts consistent with warming trends. It would also be useful to return periodically (perhaps once per decade) to sites in the arctic where quantitative studies on shorebird breeding communities have been conducted over the last 40 years (e.g., J. R. Jehl's work near Churchill, Manitoba [Jehl and Smith 1970], and F. A. Pitelka's studies at Barrow, Alaska [Pitelka and Myers 1980]).

Large amounts of quantitative data are also available on timing of migrations and numbers of migrants from volunteer monitoring programs (such as Manomet Bird Observatory's International Shorebird Survey or comparable work by the British Trust for Ornithology and the international Wader Study Group). These programs have established long-term data bases that track fluctuations in shorebird population levels. The surveys are particularly interesting because they obtain data on arctic species, but in the temperate zone, and the populations sampled come from across broad areas within the arctic and hence integrate population trends over large expanses of the breeding ground. It may be possible to design a monitoring program that could reveal climate change signals based on a carefully selected subset of species. Among the variables important to selection would be ease of accurate monitoring, wide geographic range, and heterogeneous habitat choice.

## REFERENCES

Broecker, W. S. 1987. Unpleasant surprises in the greenhouse? *Nature* 328:123.

Castro, G., F. L. Knopf, and B. A. Wunder. 1990. The drying of a wetland. *Am. Birds* 44:204.

Chappell, M. A. 1980. Thermal energetics of chicks of Arctic-breeding shorebirds. *Comp. Biochem. Physl.* 65A:311.

Edlund, S. A. 1986. Modern arctic vegetation distribution and its congruence with summer climate patterns. In *Proceedings, Impact of Climate Change on the Canadian Arctic, March 3–5, 1986, Orillia, Ontario,* H. M. French, ed., pp. 84–99. Ottawa: Canadian Climate Program.

Emanuel, W. R., H. H. Shugart, and M. P. Stevenson. 1985a. Climatic change and the broad-scale distribution of terrestrial ecosystem complexes. *Clim. Change* 7:29.

Emanuel, W. R., H. H. Shugart, and M. P. Stevenson. 1985b. Response to comment: Climatic change and the broad-scale distribution of terrestrial ecosystem complexes. *Clim. Change* 7:457.

EPA (Environmental Protection Agency). 1988. *The Potential Effects of Global Climate Change on the United States,* vol. 2: National studies. Washington, D.C.: EPA.

Goss-Custard, J. D. 1979. Role of winter food supplies in the population ecology of some common British wading birds. *Verh. Orn. Ges. Bayern* 23:124.

Goss-Custard, J. D., and M. E. Moser. 1988. Rates of change in the numbers of dunlin, *Calidris alpina,* wintering in British estuaries in relation to the spread of *Spartina anglica. J. Appl. Ecol.* 25:95.

Holmes, R. T. 1970. Differences in population density, territoriality and food supply of Dunlin on arctic and subarctic tundra. *Symp. British Ecol. Soc.* 10:303.

Jehl, J. R., Jr., and B. A. Smith. 1970. *Birds of the Churchill Region, Manitoba.* Manitoba Museum of Man and Nature Special Publication 1, Winnipeg: Manitoba Museum of Man and Nature.

Kessel, B. 1989. *Birds of the Seward Peninsula, Alaska: Their Biogeography, Seasonality, and Natural History.* Fairbanks: University of Alaska Press.

Lester, R. T., and J. P. Myers. 1989. Global warming, climate disruption, and biological diversity. *Audubon Wildlife Report:* 177.

Morrison, R.I.G. 1984. Migration systems of some New World shorebirds. *Behav. Marine Anim.* 6:125.

Morse, D. H. 1980. Population limitation: Breeding or wintering grounds. In *Migrant Birds in the Neotropics*, A. Keast and E. S. Morton, eds., pp. 505–516. Washington, D.C.: Smithsonian Institution Press.

Myers, J. P. 1981. On cross-seasonal interactions in the evolution of sandpiper social systems. *Behav. Ecol. Sociobiol.* 8:195.

Myers, J. P. 1986. Sex and gluttony on Delaware Bay. *Nat. Hist.* 95:68.

Myers, J. P., R.I.G. Morrison, P. D. McLain, P. Z. Antas, B. A. Harrington, T. E. Lovejoy, M. Sallaberry, S. E. Senner, and A. Tarak. 1987. Conservation strategy for migratory species: An example with shorebirds (Charadrii). *Am. Scient.* 75:18.

Norton, D. W. 1973. Ecological energetics of calidridine sandpipers breeding at Barrow, Alaska. Ph.D. diss., University of Alaska, Fairbanks.

Pienkowski, M. W. 1984. Behaviour of young ringed plovers, *Charadrius hiaticula*, and its relationship to growth and survival to reproductive age. *Ibis* 126:133.

Perry, M. L., R. R. Carter, and N. T. Konijn. 1988. *The Impact of Climatic Variation on Agriculture.* Dordrecht: Kluwer Academic.

Peters, R. L., and J.D.S. Darling. 1985. The greenhouse effect and nature reserves. *Bioscience* 35: 707.

Pitelka, F. A., and J. P. Myers. 1980. The effect of habitat conditions on spatial parameters of shorebird populations. Tech. Rep. to U.S. Dept. of Energy, contract EY-76-S-03-0034. Washington, D.C.: Dept. of Energy.

Rosenzweig, C. 1989. Potential effects of climate change on agricultural production in the Great Plains: A simulation study. In *The Potential Effects of Global Climate Change on the United States*, J. Smith and D. A. Terpak, eds., appendix C—Agriculture. Washington, D.C.: Environmental Protection Agency, Office of Policy, Planning, and Evaluation.

# CHAPTER SIXTEEN

# Global Warming and Potential Changes in Host-Parasite and Disease-Vector Relationships

ANDREW DOBSON AND
ROBIN CARPER

## I. INTRODUCTION

Parasitology has always been a discipline in which purely academic studies of the evolution of parasites and their life cycles have progressed as a necessary complement to the study of the pathology and control of the major tropical diseases of humans and their livestock. Indeed, the most striking feature of parasitology is the diversity of parasites in the warm tropical regions of the world and the frightening levels of debilitation and misery they cause. Determining how long-term climatic changes will affect the distributions of different parasites and pathogens at first seems a daunting task that almost defies quantification. Nevertheless, as parasitologists have always been concerned with the influence of climatological effects on different parasite species, it is possible to begin to speculate on the ways that global warming might affect the distributions of some specific tropical diseases. Similarly, the study of parasite population dynamics has developed within a solid theoretical framework (Anderson and May 1979, May and Anderson 1979). This permits the development of quantitative speculation in more general studies concerned with how parasite-host interactions may respond to perturbation.

This chapter addresses both general questions about the response of parasite-host systems to long-term climatic changes and the specific response of one particular pathogen, *Trypanosoma*, to the changes in climate predicted for the next hundred years.

### A. Macroparasites and Microparasites

Current estimates suggest that parasitism of one form or another may be the most common life-history strategy in at least three of the five major phylogenetic kingdoms (May 1988, Toft 1986). The enormous array of pathogens that infect humans and other animals may be conveniently divided on epidemiological grounds into microparasites and macroparasites (Anderson and May 1979, May and Anderson 1979). The former include the viruses, bacteria, and fungi and are characterized by their ability to reproduce directly within individual hosts, their small size and relatively short duration of infection, and the production of an immune response in infected and recovered individuals. Mathematical models examining the dynamics of microparasites divide the host population into susceptible, infected, and recovered classes. In contrast, the macroparasites (the parasitic helminths and arthropods) do not multiply directly within an infected individual but instead produce infective stages that usually pass out of the host before transmission to another host. Macroparasites tend to produce a limited immune response in infected hosts; they are relatively long-lived and usually visible to the naked eye. Mathematical models of the population dynamics of macroparasites have to consider the statistical distribution of parasites within the host population.

## B. Direct and Indirect Life Cycles

A second division of parasite life histories distinguishes between those species with monoxenic life cycles and those with heteroxenic life cycles. The former produce infective stages that can directly infect another susceptible definitive host individual. Heteroxenic species utilize a number of intermediate hosts or vectors in their transmission between definitive hosts. The evolution of complex heteroxenic life cycles permits parasite species to colonize hosts from a wide range of ephemeral and permanent environments, while also permitting them to exploit host populations at lower population densities than would be possible with simple direct transmission (Anderson 1988, Dobson 1988, Mackiewicz 1988, Shoop 1988). However, heteroxenic life cycles essentially confine the parasite to areas where the distributions of all the hosts in the life cycle overlap.

Shifts in the distribution of these host species due to climatic changes, will therefore be important in determining the areas where parasites may persist and areas where parasites may be able to colonize new hosts.

## C. Aquatic and Terrestrial Hosts

Climatic changes are likely to have different effects on aquatic and terrestrial environments (chapter 24). The heteroxenic life cycles of some parasite species often allow them to utilize hosts sequentially from either type of habitat. It is thus important to determine the different responses of the terrestrial and aquatic stages of a parasite's life cycle to climatic change. That, along with an examination of other parasite responses to climatic change, demands a quantitative framework within which to discuss parasite life-history strategies.

## II. PARASITE LIFE-HISTORY STRATEGIES

The complexities of parasite host population dynamics may be reduced by the derivation of expressions that describe the most important epidemiological features of a parasite's life cycle (Anderson and May 1979, May and Anderson 1979, Dobson 1988). Three parameters are important in describing the dynamics of a pathogen: the rate it will spread in a population, the threshold number of hosts required for the parasite to establish, and the mean levels of infection for the parasite in the host population.

*Basic reproductive rate of a parasite, $R_0$:* The basic reproductive rate, $R_0$, of a microparasite may be formally defined as the number of new infections that a solitary infected individual is able to produce in a population of susceptible hosts (Anderson and May 1979). In contrast, $R_0$ for a macroparasite is defined as the number of daughters that are established in a host population following the introduction of a solitary fertilized female worm. In both cases the resultant expression for $R_0$ usually consists of a term for the rates of parasite transmission divided by an expression for the rate of mortality of the parasite in each stage in the life cycle (Dobson 1989). Increases in host population size or rates of transmission tend to increase Ro, and increases in parasite virulence or other sources of parasite mortality tend to reduce the spread of the pathogen through the population.

*Threshold for establishment,* $H_T$: The threshold for establishment of a parasite, $H_T$, is the minimum number of hosts required to sustain an infection of the pathogen. An expression for $H_T$ may be obtained by rearranging the expression for Ro to find the population density at which Ro equals unity. This may be done for both micro- and macroparasites with either simple or complex life cycles. The resultant expressions suggest that changes in the parameters that tend to increase Ro tend to reduce $H_T$, and vice versa. Although many virulent species require large populations to sustain themselvles, reductions in the mortality rate of transmission stages may allow parasites to compensate for increased virulence and maintain infections in populations previously too small to sustain them.

*Mean prevalence and burden at equilibrium:* It is also possible to derive expressions for the levels of prevalence (proportion of the hosts infected) and incidence (mean parasite burden) of parasites in the host populations. In general, parameters that tend to increase Ro also tend to give increases in the proportion of hosts infected by a microparasite and increases in the mean levels of abundance of any particular macroparasite (Anderson and May 1979, May and Anderson 1979, Dobson 1988). Most important, increases in the size of the host population usually lead to increases in the prevalence and incidence of the parasite population (fig. 16.1).

These expressions, which characterize the most important features of a parasite's interaction with its host at the population level, can be used to ascertain how parasites with different life cycles will respond to long-term climatic changes. This may best be undertaken by determining which stages of the life cycles are most susceptible to climatic variation and by quantifying the response of those stages to climatic change.

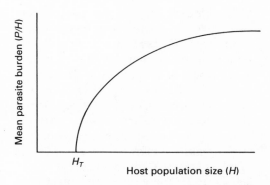

Figure 16.1. The theoretical relationship between mean parasite burden and host population size for a direct life-cycle macroparasite. Most epidemiological models produce this type of relationship between parasite abundance and host density. $H_T$ indicates the threshold number of hosts, below which the parasite is unable to establish. Note the asymptotic leveling off of parasite burdens in large host populations. After Dobson 1990.

## III. EFFECT OF TEMPERATURE ON PARASITE TRANSMISSION RATES

The physiology of adult parasites is intimately linked with the physiology of their hosts. Providing the hosts can withstand environmental changes, it seems unlikely that the within-host component of the parasite life cycle will be significantly affected. However, any form of increased stress on the host may lead to increases in rates of parasite-induced host mortality (Esch et al. 1975). In the absence of data from the specific experimental studies that could throw considerable light on these relationships, this study will concentrate on the effect of changes in meteorological factors on the free-living infective stages of different groups of parasites.

### A. Parasites with Aquatic Transmission Stages

Several detailed laboratory studies have examined the effect of temperature on the transmission success of parasites with aquatic infective stages. The parasitic trematodes are probably the most important class of parasites to utilize an aquatic stage for at least part of their life cycle. The data presented in figure 16.2 are for an echinostome species that is a parasite of ducks. Increased temperature leads to increased mortality of the larval infective stages of the parasite. It also leads to increased infectivity of the larval stage. The interaction between larval infectivity and survival means that net transmission efficiency peaks at some intermediate temperature but remains relatively efficient over a broad range of values (16°–36°C for *Echinostoma liei* cercariae; fig. 16.2). These synergistic interactions between the different physiological processes determining survival and infectivity allow the aquatic parasites to infect hosts at a relatively constant rate over the entire spectrum of water temperatures that they are likely to experience in their natural habitats (Evans 1985).

## B. Poikilothermic Hosts

The effect of temperature on the developmental rate of parasites in both aquatic and terrestrial hosts has been examined for several of the major parasites of humans in the tropics. In contrast to the effect on transmission efficiency, increases in temperature usually lead to reduced development times for parasites that utilize poikilothermic hosts (fig. 16.3). As with many physiological processes, a 10° increase in temperature seems to lead to a halving of the developmental time. This may allow parasite populations to build up rapidly following increases in temperature.

## C. Parasite Populations in Thermal Cooling Streams

The expressions for $Ro$ and $H_T$ derived in the first part of this chapter, suggest that increases in transmission efficiency and reductions in development time induced by temperature changes allow parasites to establish in smaller populations and grow at more rapid rates. This is observed to some extent in a pair of long-term studies that compare the parasite burdens of mosquito fish (*Gambusia affinis*) populations in artificially heated and control sections of the Savannah River in South Caro-

lina. The data for the trematode *Ornithodiplostomum ptychocheilus* show significant differences between heated and ambient sites during the earlier period of the study when temperature differences were most pronounced. Infection by the parasites starts several months earlier each year in the thermally altered sites (fig. 16.4). However, infection rates decline in the summer in the artificially heated sites when populations of hosts decline in response to high water temperatures (Camp et al. 1982). This effect may be compounded by the movement of the waterfowl that act as definitive hosts for the parasite. These birds tend to prefer the warmer water in winter and cooler water in the summer. Similar but less clearly defined patterns are observed in the data for *Diplostomum scheuringi* from the same site (Aho et al. 1982).

These studies illustrate the important role of host population density in the response of a parasite's transmission rate to thermal stress, while also demonstrating the ability of parasites to capitalize on improved opportunities for transmission and to establish whenever opportunities arise. Obviously the data are open to several interpretations, but they do emphasize the importance of long-term experiments in determining the possible effects of global warming on the distribution of parasites.

## D. Terrestrial Hosts

The survival rates of the infective stages of the parasites of most terrestrial species tend to decrease with increasing temperature (fig. 16.5a). Although little evidence is available to determine how the infectivity of these larvae is affected by temperature, rates of larval development tend to increase with increasing temperature (fig. 16.5b). These two processes again interact synergistically—as an increase in temperature depresses survival, develop-

*Figure 16.2.* (a) The effect of water temperature on the survival rate of the cercariae of *Echinostoma liei* (after Evans 1985). (b) The influence of water temperature on the infectivity of *E. liei* cercariae. (c) The net effect of temperature on the transmission efficiency of *E. liei* cercariae (Evans 1985).

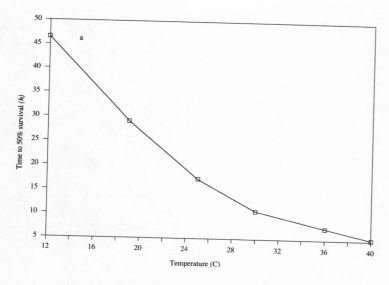

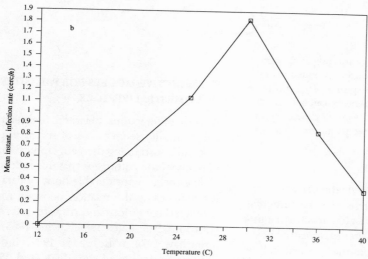

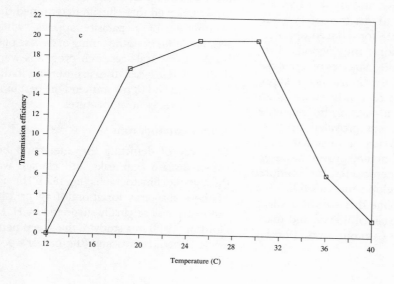

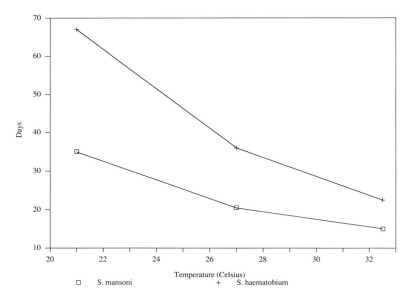

Figure 16.3. The duration of the prepatent period (the time between infection and production of infective cercariae) for two species of schistosome (*Schistosoma mansoni* and *S. haematobium*) in their snail hosts (*Planorbis pfeifferi* and *Physopsis globosa*, respectively) at three different temperatures. After Gordon et al. 1934.

ment speeds up—allowing the parasite to establish at a broad range of environmental temperatures. In contrast to parasites that utilize aquatic hosts, parasites of terrestrial hosts have transmission stages that are susceptible to reduced humidity, and these stages are highly susceptible to desiccation (Wallace 1961). To compensate for reduced opportunities for transmission during periods of severely adverse climate, parasites of terrestrial hosts have evolved adaptations such as hypobiosis, the ability to remain in a state of arrested development within the relatively protected environment provided by their hosts until such time as transmission through the external environment proves more effective. Terrestrial nematodes, for example, can arrest their development. This ability is a heritable trait and one that seems to adapt rapidly to different climatological and management regimes (Armour and Duncan 1987).

## IV. PREDICTIVE MODELS FOR PARASITES OF DOMESTIC LIVESTOCK

Because interactions between temperature and humidity seem to be of major importance in constraining the geographical range of many of the pathogens that infect domestic livestock, a considerable body of data exists concerning the relation between meteorological conditions and parasite outbreaks (Gordon 1948, Kates 1965, Levine 1963, Ollerenshaw 1974, Wilson et al. 1982). Indeed the parasitologists of the 1950s and 1960s firmly believed that climate determined the distribution of a parasite species, while weather influenced the timing of disease outbreaks. Large-scale research programs were designed to forecast disease outbreaks in different areas and to recommend the best time to administer control measures.

### A. Bioclimatographs

One way of depicting the interaction between disease outbreaks and climate was through bioclimatographs (fig. 16.6). The use of these diagrams for monitoring parasite outbreaks was originally suggested by H. M. Gordon (1948) in a study of the sheep nematode *Haemonchus contortus* (the barber's pole worm). Bioclimatographs are constructed by plotting the climatological conditions under which a parasite is able to exist and under which outbreaks occur onto a graph of mean

monthly temperature and rainfall. When that plot is compared with the observed mean weather data for a specific geographical location, it is possible to determine the time of year when outbreaks of the parasite are likely. Although the initial production of a diagram requires a long-term study of the parasite in any region, once the conditions for establishment and optimal development have been

Figure 16.4. The mean parasite burdens of (a) Ornithodiplostomum ptychocheilus and (b) Diplostomum scheuringi in Gambusia affinis from an artificially warmed and a control stream. After Camp et al. 1982 and Aho et al. 1982.

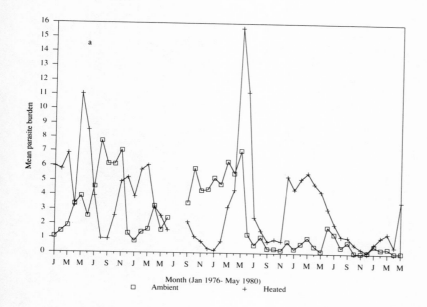

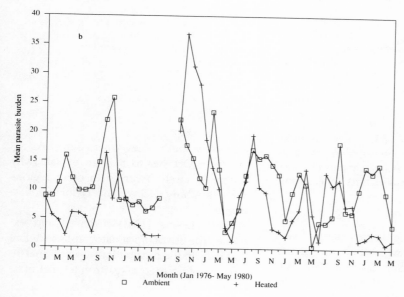

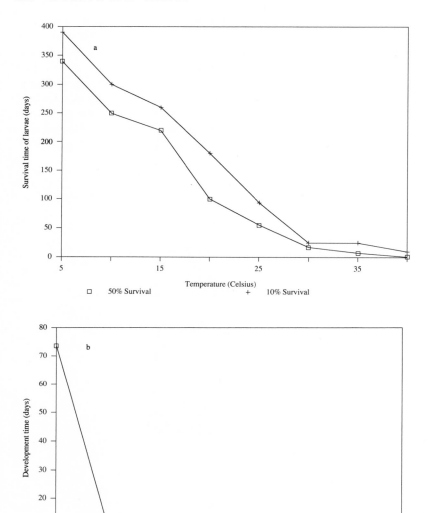

Figure 16.5. (a) Survival of the larvae of *Tricho-strongylus retortaeformis*, a parasite of rabbits, at a range of temperatures. (b) Development time of *T. retortaeformis* at the same temperatures. After Levine 1963.

described, then extrapolations may be made to other regions for which only the climate data are available. Thus Gordon (1948) was able to use his data for *H. contortus* in Armi-dale, New South Wales, where outbreaks occur from October to May, to explain why outbreaks rarely occurred in other regions such as Albury, N.S.W., and Deloraine, Tasmania.

N. D. Levine (1963) reviewed and extended the use of bioclimatographs to define and explain the distribution and seasonal incidence of a variety of gastrointestinal para-

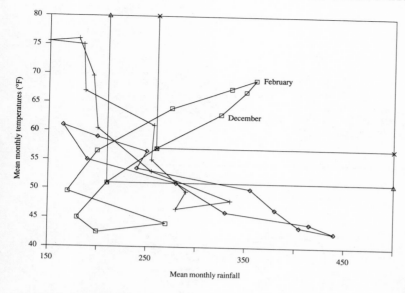

Figure 16.6. Bioclimatograph for *Haemonchus contortus* in three regions of Australia: Armidale, N.S.W. (□), Albury, N.S.W. (+), and Deloraine, Tasmania (◇). The isoclines at 51°F and 210 pts rainfall delineate the meteorological conditions at which *H. contortus* can just establish; the isoclines at 57°F and 260 pts rainfall outline the conditions under which epidemic outbreaks occur. Haemochosis season in Armidale lasts from September till April. In contrast, although *H. contortus* is recorded from Albury and Deloraine, the climate conditions for an epidemic are rarely attained. After Gordon 1948.

sites of sheep and cattle. Being based on mean temperature and rainfall data, bioclimatographs are usually only partially successful in predicting parasite outbreaks in any specific year. Similarly, bioclimatographs are seldom derived from laboratory determinations of a parasite's development constraints, because the climate conditions experienced by the parasite larvae in the soil are often different from those measured by the local weather station. However, bioclimatographs remain useful tools for determining whether a parasite will establish in a region. They may prove invaluable in determining whether long-term climatic changes will permit specific parasites of domestic livestock to establish in regions where they are not at present a problem.

## B. Effect of Temperature on Transmission Stages of Microparasites

Our focus so far on parasitic helminths reflects the available literature. Data on the effects of temperature, humidity, and ultraviolet light on the survival and infectivity of viral and bacterial transmission stages have been hard to locate, possibly because work with this material is beset with technical difficulties. There are, however, data suggesting that the development time of microparasite infections depends on ambient temperature, and there is evidence that the infectivity of some vector-transmitted pathogens is determined by the temperature at which their insect hosts are raised (Ford 1971:104). Temperature may also indirectly affect transmission rates by altering the behavior of insect vectors.

## V. EFFECT OF CLIMATE CHANGE ON THE DISTRIBUTION OF TRYPANOSOMIASIS IN AFRICA

Trypanosomiasis, or sleeping sickness, is one of the major diseases of humans and their domestic animals in Africa (Ford 1971). The disease is of particular importance to conservation in Africa as its presence may exclude

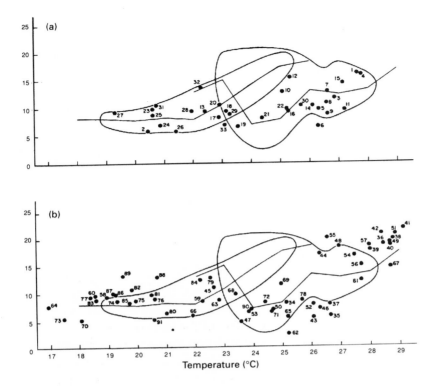

Figure 16.7. The predicted bioclimatic limits and annual means of monthly meteorological conditions for 33 G. morsitans areas (a) and 58 nontsetse areas (b).

humans and their domestic livestock from areas where wild animals act as a reservoir of the disease (Molyneux 1982, Rogers and Randolph 1988). The pathogen may be classified as a microparasite; it is transmitted by an insect vector, the tsetse fly (Glossina spp.). D. J. Rogers and S. E. Randolph have made an extensive study of the meteorological conditions that determine the distribution of three species of tsetse flies, Glossina morsitans, G. palpalis, and G. tacinoides (Rogers 1979, Rogers and Randolph 1986). Their study is complemented by two models of the dynamics of the different Trypanosoma species, one by Rogers (1988) and one by P.J.M. Milligan and R. D. Baker (1988). The former derives expressions for Ro and $H_T$ that provide some useful general insights into the processes that

are most important in determining the conditions that allow the pathogen to establish; the latter develops a more specific analytical model for trypanosomiasis based on detailed parameter estimates from a study of Trypanosoma vivax in Tanzania.

Rogers's (1979) analysis of the bioclimatic tolerances of tsetse flies may be used to determine how predicted patterns of climate change in tropical Africa might affect the distribution of tsetse flies and trypanosomiasis. Using data from several long-term studies of two subspecies of tsetse flies in Nigeria (Glossina morsitans submorsitans) and Zambia (G. m. morsitans), Rogers shows that the mean monthly density-independent mortality rates for these flies are most closely related to mean monthly saturation deficit (an index of humidity) and, to a lesser extent, mean monthly temperature. Those analyses allow Rogers to identify an environmental optimum for each subspecies of *G. morsitans*. When the data for 91 sites throughout tropical Africa are examined in terms of these climatological conditions, 94% of the sites within the present known distribution of *G. morsitans* fall within the predicted bioclimatic limits, while only 50% of nontsetse areas do so (fig. 16.7). These data can be used to compare the present distribution of *G. morsitans* with the possible distribution given a mean 2° increase in temperature for sub-Saharan Af-

rica (fig. 16.8). Because the bioclimatic data correlate better with the presence of *G. morsitans* than they do with its absence, greater confidence may be placed in the prediction

Figure 16.8. (a) Weather stations listed by Rogers (1979): +, *G. morsitans* present; ◇, *G. morsitans* absent. (b) The potential change in the distribution of *G. morsitans* following a mean 2° increase in temperature. Copyright © 1979 by Blackwell Scientific Publications.

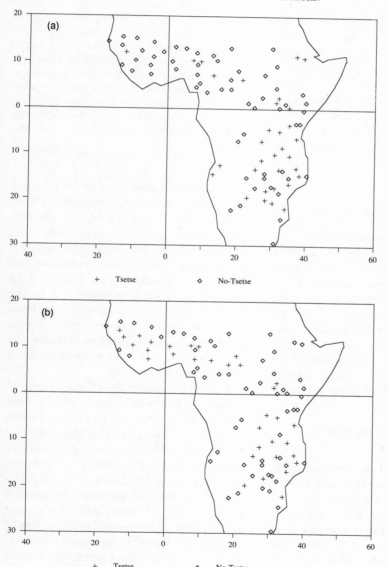

for where *G. morsitans* may decrease in abundance than for regions where it might establish. Keeping that in mind, the analysis suggests that *G. morsitans* may become less common in West Africa and across the main sub-Saharan zone of central Africa. This pattern may be matched by a spread farther south of the vector and its parasite in East Africa.

Although the approach we have adopted in this analysis is rather coarse, the data are available to make more sophisticated and detailed analyses for this and other pathogens. From a conservation perspective it remains important to determine to what extent trypanosomiasis is at present maintaining areas as refuges for wild animals by excluding humans and their livestock (Molyneux 1982, Rogers and Randolph 1988). If a change of climate reduces tsetse levels, then pressure for the exploitation of the areas would increase with their subsequent loss as a wildlife refuge.

## VI. THE STRUCTURE OF PARASITE-HOST COMMUNITIES

So far we have concentrated on simple one-host, one-parasite relationships, a sensible approach to systems that are dominated by one particularly prevalent pathogen. Many host populations, however, maintain a community of several parasite species. The diversity of such a community and the abundance of its constituent parasite species are intimately linked not only to the density of the host population but also to the presence of other host species that act as reservoirs for other parasite species.

## A. Communities with One Host and Many Parasite Species

It is possible to extend the basic one-host, one-parasite models to examine the dynamics of more complex communities (fig. 16.9). Preliminary analysis of models for such communities suggests that parasite species diversity is a direct function of host density and that the relative abundance of each parasite

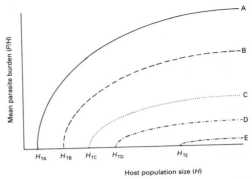

Figure 16.9. The theoretical relationship between relative abundance of different parasite species (A, B, C, D, and E) and host population density in a simple one-host, five-parasite community. Each of the parasite species requires a different threshold number of hosts ($H_T$). The number of parasite species in the community thus depends on the size of the host population. The relative abundance of each parasite varies inversely with the relative magnitude of its threshold for establishment. After Dobson 1990.

species is determined more by the parasite's life-history attributes that determine its transmission success than by interactions with other parasite species (Dobson 1986, 1989). This suggests that changes in host density due to changes in meteorological conditions will be crucial in determining the diversity of the community of parasites supported by the hosts. Increases in the density of some hosts will allow them to support a more diverse parasite fauna, while decreases in the density of other hosts will reduce the diversity of their parasite community.

A study comparing the effects of artificial heating on the parasite fauna of an aquatic snail presents some corroborative evidence in support of this model. C. S. Sankurathi and J. C. Holmes (1976a,b) studied a population of *Physa gyrina* and its parasites and commensals in Lake Wabamun in Alberta, Canada. A section of the lake was used for cooling by a power station and consequently was warmer than the rest of the lake and relatively free of ice in winter. The effects on the population of snails were pronounced when both density and population structure are compared for heated and control sites, with population density often several orders of magnitude higher in the heated areas (fig. 16.10). That, and the continual presence of the vertebrate definitive hosts in the parasites' life cycle, allowed a considerable increase in both the prevalence and diversity of the parasite community living in the snail population (fig. 16.11). The increased water temperature also had a detrimental effect on the two species of commensal chaetogasters that live in the mantle of the snails. Laboratory experiments showed that these commensals live as predators, attacking and ingesting the infective stages of parasites that try to infect their snail host (Sankurathi and Holmes 1976b). When

the temperature rises, the chaetogasters abandon the snail and die, leading to further increases in the rates of parasitism of the snail hosts.

## B. Communities with Two Hosts and Many Parasite Species

A more complex pattern emerges if we consider the community structure of parasites in two host species that share parasites. When parasites are able to use more than one species as a definitive host, their ability to establish in any one host species depends on the density of all the potential host species present in an area. Because different host species may have different susceptibilities to the parasite and different parasite species may reproduce at different rates in different host species, the density of different host species will be crucial to the composition of the parasite assemblage (Dobson 1989). Variations in the population density of different host species may thus lead to variations in the parasite burdens of other host species; in some cases this may allow pathogenic parasites to estab-

Figure 16.10. The effect of artificial heating on the density of a mollusk population and its community of parasites in a Canadian lake. Top: Surface water temperatures in the control (A) and heated (B) areas of the lake. Bottom: Population density of the snail *Physa gyrina* in the control and heated areas. After Sankurathi and Holmes 1976a,b; courtesy of National Research Council of Canada.

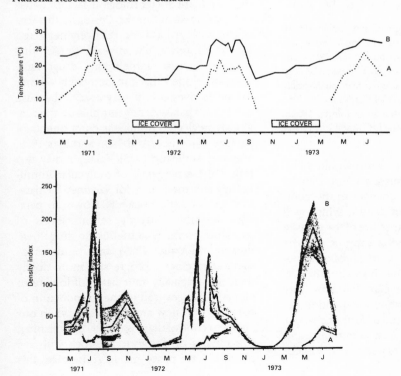

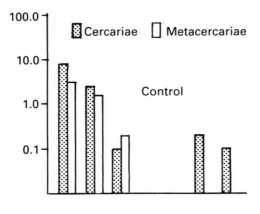

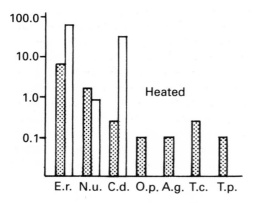

*Figure* 16.11. The percentage of snails infected with cercariae and metacercariae in the control and heated areas of Lake Wabamum. E.r., *Echinopary-phium recurvatum*; N.u., *Notocotylus urbanensis*; C.d., *Cercaria douglasi*; O.p., *Ornithodiplostomum ptychocheilus*; A.g., *Apatemon gracilis*; T.c., *Trichobilharzia cameroni*; T.p., *T. physellae*. After Sankurathi and Holmes 1976b; courtesy of National Research Council of Canada.

lish in populations of hosts that would otherwise be too small to sustain them. Climate changes could lead to changes in the composition of host communities, which will lead to changes in the structure of the parasite community that the hosts support and the possible introduction of parasites not previously present in the host population. Where members of the parasite community are important in mediating competition between hosts, this may lead to further changes in the structure of the host community and the possible extinction of particularly susceptible hosts.

## VII. CONCLUSIONS

The geographical distributions of most parasite species are limited by the distributions of potential host species or by environmental constraints on the parasite's rates of development. Although developmental rates in vertebrate hosts may be comparatively unaffected by changes in environmental temperature, the available evidence suggests that the free-living stages of parasites and those that live in invertebrate poikilothermic hosts are susceptible to prevailing meteorological conditions. J. D. Gillett (1974) suggests that many vector-transmitted diseases are limited in their range because the development time of the parasite exceeds the average life expectancy of the insect vector. But increases in environmental temperature are likely to speed up development for those stages in the parasite life cycle, so long-term increases in temperature are likely to lead to increases in the ranges of many diseases transmitted by insects, such as malaria and filariasis.

Up until the mid-1970s parasitologists believed that temperature and moisture were the dominant meteorological factors determining disease outbreaks. Curiously, this area of parasitology has been relatively neglected for the last ten to fifteen years. In part, that may be because anthelmintic drugs have been developed that can be readily administered to livestock. It may also be because models for parasites now emphasize the previously neglected nonlinear components of parasite dynamics (Anderson and May 1979, May and Anderson 1979). Finally, it may also reflect the emergence of molecular immunology and the search for vaccines for parasites of domestic livestock. However, parasites are now showing serious levels of resistance to many anthelmintic drugs (Anderson and Waller 1985), and the development of vaccines is progressing more slowly than was originally anticipated. If long-term climatic changes lead to the introduction of parasites into new areas at a time when our ability to control them is rapidly diminishing, many types of domestic livestock will face major disease problems. In some cases this will lead to the abandonment of present pasture lands, which may then be set aside for nature reserves. In other regions an increasingly hungry human population will exert pressure to utilize present reserves as grazing areas. It seems unlikely that the net result of this exchange will favor wildlife.

A considerable body of literature is already available that deals with the climatic responses of a variety of parasites (Kates 1965, Levine 1963, Wilson et al. 1982). We now also have much better models for examining the dynamics of parasites at all stages of their life cycles (Anderson and May 1979, 1986; May and Anderson 1979). Although there are problems of scale associated with extrapolating between the physiological processes of parasites measured under controlled laboratory conditions and the coarser predictions available for longer-term climate change, it should be possible to merge these various sources of information to produce a quantitative synthesis of the way global climate change may affect the distribution of many parasites. It thus seems likely that global warming will give new prominence to an area of parasitology that had fallen into relative neglect.

The examples given above are mainly from well-studied species in little danger of extinction. Assessment of the potential effects of global warming on the parasites of endangered species can really only be undertaken by extrapolation from these examples and the models used to explain the more general features of parasite-host population dynamics.

A number of possible scenarios are likely to arise as host populations respond to long-term climate changes.

Consider first an endangered species whose population density has declined to such low levels that it is present only in a single nature reserve. Under these conditions it seems likely that a further decline in population size due to global warming will reduce the effects of the parasites already present in that population. However, the immigration of new host species into the area, as a response to climate change, may lead to the introduction of novel pathogens. If the endangered host has had no previous contact with these parasites, they may fail to establish, if the host is sufficiently novel, or they may establish and produce significant levels of mortality. Under these conditions, increases in the density of the immigrant hosts will lead to increases in the rates of parasite transmission, and constraints may have to be placed on interactions between the endangered species and the newly immigrating species.

Where endangered species are tolerant to increases in temperature and humidity, they are still likely to face increased assault by parasites whose transmission efficiency improves with increases in temperature and humidity (e.g., tropical diseases such as hookworm may become more important in temperate zones). Furthermore, those host populations that increase as temperatures rise are likely to suffer an increase in parasite prevalence and diversity.

If the population sizes of host species decline because of climatic changes, their rarer species of parasites and mutualists may become extinct. These species have their own intrinsic value, and they often perform a valuable function, such as the commensal chaetogasters living in the snail mantles discussed above. The absence of a parasite may be as important as its presence; some species of hosts may grow to become pests in the absence of pathogens that are now regulating their numbers.

Parasites and disease will do well on a warming earth. They are, by definition, organisms that colonize and exploit. Those species of parasite that are already common will be able to spread and perhaps colonize new susceptible hosts that may have no prior genetic resistance to them. Parasite species that are rare and have more specialized requirements may be driven to extinction. In general, these effects are likely to be worse in the temperate zone, where parasites from the tropics can colonize new hosts, than in the the tropics, where parasites will have to adapt or evolve. Rare parasites that are adapted to extreme temperature, however, may become common; changes in the ranges and sizes of some host populations may allow some hitherto unimportant pathogens to become more widespread.

## REFERENCES

Aho, J. M., J. W. Camp, and G. W. Esch. 1982. Long-term studies on the population biology of *Diplostomum scheuringi* in a thermally altered reservoir. *J. Parasitol.* 68:695.

Anderson, R. C. 1988. Nematode transmission patterns. *J. Parasitol.* 74:30.

Anderson, R. M., and R. M. May. 1979. Population biology of infectious diseases: Part I. *Nature* 280: 361.

Anderson, R. M., and R. M. May. 1985. Helminth infections of humans: Mathematical models, population dynamics and control. *Advances in Parasitol.* 24:1.

Anderson, N., and P. J. Waller. 1985. *Resistance in Nematodes to Anthelmintic Drugs.* Glebe, Australia: CSIRO Division of Animal Health Publications.

Armour, J., and M. Duncan. 1987. Arrested larval development in cattle nematodes. *Parasitol. Today* 3:171.

Camp, J. W., J. M. Aho, and G. W. Esch. 1982. A long-term study on various aspects of the population biology of *Ornithodiplostomum ptychocheilus* in a South Carolina cooling reservoir. *J. Parasitol.* 68:709.

Dobson, A. P. 1986. The population dynamics of competition between parasites. *Parasitology* 91: 317.

Dobson, A. P. 1988. The population biology of parasite-induced changes in host behavior. *Quart. Rev. Biol.* 63:139.

Dobson, A. P. 1990. Models for multi-species parasite-host communities. In *The Structure of Parasite Communities*, G. Esch, C. R. Kennedy, and J. Aho, eds., pp. 261–288. New York and London: Chapman and Hall.

Esch, G. W., J. W. Gibbons, and J. E. Bourque. 1975. An analysis of the relationship between stress and parasitism. *Am. Midl. Nat.* 93:339.

Evans, N. A. 1985. The influence of environmental temperature upon transmission of the cercariae of *Echinostoma liei* (Digenea: Echinostimatidae). *Parasitology* 90:269.

Ford, J. 1971. *The Role of the Trypanosomiases in African Ecology: A Study of the Tsetse Fly Problem.* Oxford: Clarendon Press.

Gillett, J. D. 1974. Direct and indirect influences of temperature on the transmission of parasites from insects to man. *Symp. British Soc. Parasitol.* 12:79.

Gordon, H. M. 1948. The epidemiology of parasitic diseases, with special reference to studies with nematode parasites of sheep. *Aust. Vet. J.* 24:17.

Gordon, R. M., T. H. Davey, and H. Peaston. 1934. The transmission of human bilharziasis in Sierra Leone, with an account of the life cycle of the schistosomes concerned, *S. mansoni* and *S. haematobium. Ann. Trop. Med. Parasitol.* 28:323.

Kates, K. C. 1965. Ecological aspects of helminth transmission in domesticated animals. *Am. Zool.* 5:95.

Levine, N. D. 1963. Weather, climate and the bionomics of ruminant nematode larvae. *Advances in Vet. Sci.* 8:215.

Mackiewicz, J. S. 1988. Cestode transmission patterns. *J. Parasitol.* 74:60.

May, R. M. 1988. How many species are there on earth? *Science* 241:1441.

May, R. M., and R. M. Anderson. 1979. The population biology of infectious diseases: II. *Nature* 280: 455.

Milligan, P.J.M., and R. D. Baker. 1988. A model for tsetse-transmitted animal trypanosomiasis. *Parasitology* 96:211.

Molyneux, D. H. 1982. Trypanosomes, trypanosomiasis and tsetse control: Impact on wildlife and its conservation. *Symp. Zool. Soc. Lond.* 50:29.

Ollerenshaw, C. B. 1974. Forecasting liver fluke disease. *Symp. British Soc. Parasitol.* 12:33.

Rogers, D. J. 1979. Tsetse population dynamics and distribution: A new analytical approach. *J. Anim. Ecol.* 48:825.

Rogers, D. J. 1988. A general model for the African trypanosomiasis. *Parasitology* 97:193.

Rogers, D. J., and S. E. Randolph. 1986. Distribution and abundance of tsetse flies (*Glossina* spp.). *J. Anim. Ecol.* 55:1007.

Rogers, D. J., and S. E. Randolph. 1988. Tsetse flies in Africa: Bane or boon? *Conserv. Biol.* 2:57.

Sankurathi, C. S., and J. C. Holmes. 1976a. Effects of thermal effluents on the population dynamics of *Physa gyrina* Say (Mollusca: Gastropoda) at Lake Wabamun, Alberta. *Can. J. Zool.* 54:582.

Sankurathi, C. S., and J. C. Holmes. 1976b. Effects of thermal effluents on parasites and commensals of *Physa gyrina* Say (Mollusca: Gastropoda) and their interactions at Lake Wabamun, Alberta. *Can. J. Zool.* 54:1742.

Shoop, W. L. 1988. Trematode transmission patterns. *J. Parasitol.* 74:46.

Toft, C. A. 1986. Parasites etc. In *Community Ecology*, J. M. Diamond and T. J. Case, eds., pp. 445–463. New York: Harper and Row.

Wallace, H. R. 1961. The bionomics of the free-living stages of zoo-parasitic nematodes: A critical survey. *Helminth. Abstr.* 30:1.

Wilson, R. A., G. Smith, and M. R. Thomas. 1982. Fascioliasis. In *The Population Dynamics of Infectious Diseases: Theory and Applications*, R. M. Anderson, ed., pp. 262–319. London: Chapman and Hall.

We would like to thank R. Peters for encouraging us to attend the symposium. The manuscript benefited considerably from discussions with T. Anderson, N. Georgiadis, A. Lyles, R. May, C. Toft, and D. Rogers.

# PART IV

# Specific Regions and Sites

# CHAPTER SEVENTEEN

# Arctic Marine Ecosystems

VERA ALEXANDER

## I. INTRODUCTION

The possibility that the earth's climate is undergoing substantial warming is generating widespread concern; the inevitable consequences could eventually even include threats to the earth's habitability. Models currently predict a rate of temperature change much greater than has previously been experienced and that may exceed the ability of plants to respond by migration. The former point was made by Erich Bloch, director of the National Science Foundation, in a talk to the National Research Council Committee on Global Change in June 1987. He stated that the biological productivity and continued habitability of our planet are being threatened by a number of factors, that we are in the middle of conducting a global experiment—an uncontrolled one—and our planet has become a global laboratory. In this chapter I address the potential effects on arctic marine ecosystems, but I wish to emphasize strongly that any scenario at this time is speculative and in no sense predictive. In the context of an uncontrolled and unintended global experiment, such speculative predictions can be viewed as the first step in formulating hypotheses against which we can observe and evaluate manifestations if and as they occur.

It is not at all clear that a significant long-term increase in temperature would result in an increase in arctic marine biological productivity. On the contrary, a decrease in production is likely, accompanied by the loss or at least reduction in numbers of many of the currently dominant and characteristically arctic species. That expectation is supported by the past climatic record and some of the major characteristics of arctic marine ecosystems and their faunal and floral constituents. Based on the potential environmental changes, we can evaluate probable responses by the system and its components.

Global climate models suggest that climate change induced by increasing atmospheric concentrations of carbon dioxide and other radiatively important gases will be largest at high latitudes (MacCracken and Luther 1985, Semtner 1987). Modeling by J. Hansen et al. (1988) confirms that ice-covered areas of the polar sea as well as polar land areas will experience the greatest warming, and it further predicts that at high latitudes the warming will be greater in winter than in summer. The estimated rate of this temperature increase is about 0.5°C per decade. As is the case for past regimes, this increase might occur in major steps rather than as a continuous trend (Broecker 1987). W. Broecker believes that we have been lulled into complacency by model simulations that suggest a gradual warming over a period of about 100 years. The polar ice, ocean sediment, and bog muck record suggest that historically the climate has responded in sharp jumps that involve large-scale reorganization of the earth's system. The strongest evidence for this comes from the holes drilled in the Greenland ice cap, which provide not only a long-term record of atmospheric carbon dioxide and other greenhouse gas concentrations but also a record of the temperature as traced in the ratio of isotopically heavy water to isotopically light water in the ice, and also a record of the dustiness of the air over the ice cap. The evidence from these cores shows that during glacial time the climate changed frequently and in

great jumps. The typical leap involved a 6°C change in air temperature, a fivefold change in atmospheric dust content, and a 20% change in $CO_2$ content of the air (Broecker 1987). The large-scale marine implications of such warming extend far beyond the biological considerations. For example, the present role of northern high-latitude seas in deep saline water formation would certainly be altered in the absence of zones of sea ice formation, and with major effects on large-scale ocean circulation. In this way, the relatively extreme arctic manifestation of global warming would have worldwide consequences. In the marine biological area, consequences could range from loss of the world's major commercial fish populations to severe hardships on typically arctic marine mammals such as the walrus (*Odobenus rosmarus*; fig. 17.1) and polar bear (*Ursus maritimus*).

The response of the arctic marine ecosystem to climate warming depends on the fate of the sea ice, since the present-day biological system is adapted to ice cover, whether seasonal or permanent. Figure 17.2 shows the current average maximum and minimum sea ice for the Arctic Ocean and adjacent seas. Although some experts believe that the Arctic Ocean has been ice-free in the geological past, there are many unknowns in predicting the future response. The thermal structure of the contemporary central Arctic Ocean is determined by the inflow of water from the Atlantic Ocean through the West Spitzbergen

*Figure 17.1.* Walrus on sea ice, by G. B. Threlkeld.

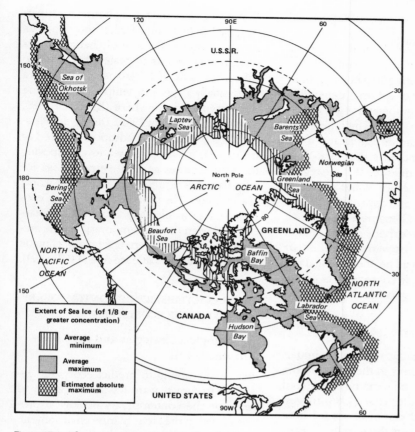

*Figure* 17.2. The arctic seas, showing the absolute and normal maximum and minimum annual ice cover.

Current and from the North Pacific Ocean through the Bering Strait, as well as by the large riverine inflow. Thermohaline processes—interactions between waters of different temperatures and salinities—can create water layers of different densities. The thermohaline processes within the Arctic Ocean create a stable water column with a layer of cold water of relatively low salt content overlying warmer and more saline waters. A question under dispute is whether this structure could be perturbed sufficiently to mix the deeper warm waters into the surface. If this happened, there could be large changes in ice cover because, at present, stratification allows the cold surface layer to shield the ice and atmosphere from the relatively warm lower ocean. If this layering broke down, however, the warm lower waters could mix with surface waters, raising the temperature of the ocean surface and thus keeping the surface ice from reforming in winter. Efforts to model this layering have not agreed on the likelihood that ice cover would permanently disappear. For example, the mechanisms for maintenance of this layering have been considered in models by P. D. Kilworth and J. M. Smith (1984) and A. J. Semtner (1984), and there is general agreement that if the ice cover disappeared, it would be difficult to refreeze the surface. On the other hand, C. L. Parkinson and W. W. Kellogg (1979) have found, through five-year simulations using a 5°C surface atmospheric temperature increase, that the ice pack disappeared in August and September, but reformed in the central ocean in mid-fall, and that even with a 6°–9°C temperature increase, sea ice still reappeared in winter. Cloud cover had only a minor influence. (For a further discussion of these issues, see Aagard and Coachman 1975.)

Although the data are limited, there is a clear trend for the southernmost limit of winter sea ice over the Bering Sea shelf to extend and retract from year to year, depending on whether winter temperatures are cooler or warmer (Niebauer 1988). The differences among years in maximum extent are of the order of several hundred kilometers. With warming, there is no question that the region covered by seasonal sea ice would at least be reduced significantly, and apparently the Arctic Ocean itself could become a seasonal ice zone. Thus, although experts are not certain how large and how fast warming will be, or what the responses will be, the potential implications of large-scale consequences are serious and cannot be ignored.

## II. THE PAST RECORD

The climate of the arctic regions has undergone major fluctuations in the past, and the Arctic as we know it now can be considered ephemeral. L. D. Carter et al. (1986) suggest that permafrost and perennial sea ice have been present for about three million years, although the mollusk species present in the deposits indicate that marine conditions have at times during that period been substantially warmer than they are today. Carter et al. equate the climate of northern Alaska at that time to that of Anchorage in south central Alaska today, with the treeline extending all the way to the northern coast. There appears to be controversy about the exact timing of the formation of ice cover on the Arctic Ocean, a dispute that cannot be addressed here. Carter et al. (1986) think permafrost was probably initiated at some time during the climatic deterioration that began about 2.5 million years ago and led into the first major glaciation of the northern hemisphere, and that extensive perennial sea ice may have been related in time.

Since those warmer times, there have been periods of extreme cold. Simulation under the CLIMAP program shows that at about 18,000 years B.P., the sea ice boundary apparently had shifted much farther south than its present position. Because of changes in sea level, the eastern arctic waters became isolated from those of the Atlantic, resulting in a substantially larger sea temperature decline than was found elsewhere (Kutzbach 1985, Kutzbach and Wright 1985). During the Pleistocene ice ages, glacial advances resulted in major (100-m) drops in sea level, exposing the arctic shelves as land. Perpetual sea ice may have been present in current seasonal sea ice zones. The modern arctic marine ecosystem is probably only about 10,000 years old, and even the contemporary marine arctic ecosystem has experienced major climatic events such as the Little Ice Age (1450–1859).

## III. CHARACTERISTICS OF CONTEMPORARY ARCTIC MARINE ECOSYSTEMS

### A. Ecological Strategies and the Lower Trophic Levels

Before attempting to predict ecosystem response to climate warming, it is necessary to examine the contemporary ecological processes within the arctic marine environment.

At the lowest trophic level, the marine food web is universally dependent on downwelling light and its absorption by phytoplankton. In this regard, arctic systems face extreme conditions, with a long period of winter darkness, ice cover that affects light transmission into the seawater, and low solar angles even in the summer. Based on that, we might assume that lack of light is a major constraint on arctic biological productivity, and indeed this is the case throughout a substantial part of the year. Some subarctic marine systems, however, are among the most productive in the world and are certainly much more productive than tropical seas, unless upwelling is present. Other factors besides the limited duration and intensity of incoming solar radiation must compensate.

Arctic and subarctic phytoplankton production takes place over a relatively short period during the year, not only because of the long, dark winter with insufficient light for photosynthesis but also because it occurs in the form of marked blooms. A particularly striking phenomenon is the spring bloom, which typically occurs immediately upon the onset of density stratification in the water column. Even before then, in ice-covered seas there is ice-algae growth on the underside of the ice as soon as there is sufficient light in spring. Described as an inverted benthos, analogous to algae growing on the surface of the sediment on the sea bottom, this layer of plants may derive inorganic nutrients from the seawater below the ice or from the interstitial water within the ice. In any event, it is maintained at the relatively well-lighted interface of ice and water. Most of these arctic waters are not particularly nutrient-rich, but they still have more nutrients than the permanently stratified tropical seas.

Little information exists on the actual light availability at the bottom of the sea ice layer. Columnar ice appears to act as a weakly channeling medium, with light strongly peaked in a vertical direction, so that even sunlight falling on ice at a low angle will still propagate with a strong peak in the vertical direction (Beckley and Trodahl 1987). The point is that in a region of extended darkness with low solar angles, ice might be a factor enhancing, rather than inhibiting, primary production. As in all areas, more definitive work is needed. Snow cover, on the other hand, does limit light transmission, resulting in a patchy distribution of ice algae (Clasby et al. 1976).

The overall quantitative contribution of the ice algae to the total primary production in arctic seas is not known but has been identified as substantial (Alexander 1980; Demers et al. 1986, Horner 1985, Horner and Schrader 1982). Even more important, it takes place earlier in spring than any growth in the water, provides a concentrated food source for many grazing organisms, and is particularly important to some fishes such as arctic cod, Boreogadus saida (Gulliksen et al. 1986). In many arctic coastal regions, the ice algae appear in the bottom layer of the ice as it forms during the fall and are maintained during the winter, but they begin to grow rapidly in spring (summarized by Horner 1985). Pennate diatoms form the bulk of the biomass, with some species highly characteristic of the ice community (e.g., Nitzchia frigida).

A community of organisms develops in conjunction with the ice algae layer, especially in some regions of multiyear ice (Gulliksen and Lønne 1989). The under-ice fauna, sometimes termed sympagic fauna, consists of organisms closely associated with the undersurface of the ice. In the Barents Sea, this community includes three species of amphipods believed to form a link in energy transport from primary production through arctic cod to seabirds and mammals (Gulliksen et al. 1986). Current evidence confirms that sea ice is beneficial to arctic marine biological productivity, given the light regime at far northern latitudes. Loss of the ice cover due to warming would not markedly improve the light climate but would significantly reduce the length of the growing season by delaying biological spring.

Ice enhances arctic and subarctic biological production in yet another way. In many regions, especially at the southern limits of sea ice in the seas adjacent to the Arctic Ocean proper, a dramatic spring bloom takes place within the area of melting ice in spring, triggered by the onset of vertical stability in the water column. There is a clear relationship between water column density structure and phytoplankton blooms in northern temperate and high-latitude areas (Sverdrup 1953). Theoretical models confirm that the important mechanism is the tendency for phytoplankton to remain circulating within a well-lit surface layer, the so-called mixed layer. This layer can, and at most latitudes does, result from solar heating. At high latitudes, with low seawater temperatures, thermal gradients are not very effective in establishing density differentials, and salinity differences become more important. The same amount of solar heat can produce density gradients through melting ice approximately ten times more efficiently than through heating the surface seawater, and therefore, the salinity gradient produced by melting sea ice at the sea ice edge produces strong stratification. As a result, the spring bloom occurs earlier at a given latitude in the presence of seasonal sea ice than in its absence. H. J. Niebauer (1983) has discussed the interannual variation in sea ice extent on the Bering Sea and its biological implications. Seeding by algae from the ice also plays a role in the rapid growth of phytoplankton at the ice edge, so that this bloom develops more rapidly than most spring blooms.

The bloom at the ice edge supplies a substantial portion of the annual energy input (i.e., organic carbon created by photosynthesis) in some regions, and we have concluded that much of this material falls to the bottom sediments, at least in the Bering Sea, where it supports a benthic-based food chain. Blooms at an ice edge cannot last indefinitely, since the stability produced as a result of the density gradient from melting ice is strong, and plant growth is restrained as nutrient elements become exhausted in the surface waters. In the Bering Sea, the frequent storms reinject nutrients through enhanced mixing at the ice edge, but regeneration of nutrient elements in situ is also important (Müller-Karger and Alexander 1987).

On a totally different scale, vertical density stratification of the seawater based on salinity exerts a permanent constraint on primary production in the Arctic Ocean basin. The extensive inflow of fresh water into the Arctic Ocean through rivers ($0.15 \times 10^6 m^3$/second) and the melting of ice result in a thick low-salinity surface layer (Semtner 1984), which limits nutrient injection from deeper waters. A cold halocline is found between 50 m and 200 m, sealing off the deep water. A layer of warm water derived from the Atlantic Ocean is found between 300 m and 500 m (Aagaard et al. 1981). In the Arctic Ocean, most nutrients are supplied to surface waters via upwelling along the continental shelves (in the Beaufort Sea) and by rivers and coastal erosion. Only such coastal areas are biologically productive, but since one third of the Arctic Ocean consists of shelf areas, the total biological production is not insignificant.

Within the arctic seas (defined as seas subject to seasonal or multiyear sea ice) the dominant invertebrate communities tend to be benthic, and consequently, shellfish are relatively more important than finfish. This may be partly because many of the waters covered by sea ice outside the Arctic Ocean basin lie over continental shelves, in some cases quite shallow shelves. G. H. Petersen and M. A. Curtis have suggested that northern high-latitude shelves tend to allocate a relatively large proportion of primary production to the bottom. This allocation probably results in a relatively efficient system, in that the organic matter is accumulated in sessile long-living animals. The importance of zooplankton as a major link in the food web is reduced in the Arctic because of the absence of phytoplankton during the long winter. Furthermore, their reproductive cycles are extended in the Arctic, and their population growth at low temperatures may be too slow for effective grazing during a spring bloom. There is evidence, however, that arctic pelagic animals can grow rapidly and accumulate large lipid reserves during periods with an excess of food, an adaptive mechanism that would allow survival during periods of inadequate food (Falk-Petersen et al. 1986a,b). Such periods of accumulation appear to be tied to the annual phytoplankton production cycle. Furthermore, in spite of the importance of the benthos, pelagic euphausiids and amphipods are a significant link in the arctic marine food web.

The ice cover of the Arctic Ocean is not continuous even in winter, but it is intersected by numerous leads and is in constant motion. Coastal leads are used for migration by bowhead whales (Balaena mysticetus). Biologists are particularly interested in structures known as polynyas, which are recognized as important feeding grounds for marine mammals and birds, although little is known about their biological or physical regimes. Polynyas are defined as areas of open water in regions covered by sea ice. Whereas ice-edge zones are known to be productive biologically and to be critical to many arctic birds and mammals, polynyas serve as outposts of enhanced activity within pack ice removed from the effects of marginal ice zones. They appear to have both biological and oceanographic significance far in excess of their size and extent, and without doubt, the life cycles of many arctic animals have evolved around polynyas. Historically, the coastal arctic Inuit people have used polynyas as hunting areas for at least 4000 years. Polynyas are sensitive to climatic change, since their size is determined by temperature and wind relationships, and their spatial and temporal extent varies from year to year. Polynyas tend to occur in specific areas as quasipermanent structures during the period of sea ice coverage, although some open up only in the spring each year. Massive mortality of birds and mammals has been documented when a polynya fails to open.

The compression of biological production into a single peak early in the growing season or, at the most, an ice algae peak followed by a spring bloom with only minor secondary pulses later in the summer would seem to shorten an already abbreviated growing season. Yet there seems to be an advantage to the system in the large amount of organic carbon injected in a short time. This short benthic food chain appears to support large populations of fishes, birds, and mammals, many of which have evolved to take advantage of the food associated with sea ice. R.G.B. Brown (1989) has stated that essentially all arctic marine birds feed on organisms associated with ice. Many arctic mammals are capable of diving for extended periods and are able to reduce their metabolic rates to reduce oxygen consumption while under water. Many seals use ice for breeding; among them, the spotted seal (*Phoca largha*) in the Bering Sea breeds exclusively at the ice edge in spring (Burns pers. comm.). The harp seal (*Pagophilus groenlandicus*) lives at the edge of the drift ice belt, from Newfoundland via Greenland to Svalbard and the White Sea (Vibe 1967). Probably the highest biomass of seals associated with sea ice is that of the ringed seal (*Phoca hispida*). Other ice-related arctic species include the ribbon seal (*Phoca fasciata*) and the bearded seal (*Erignathus barbatus*). The narwhal (*Monodon*

*monoceros*) and beluga whale (*Delphinapterus leucas*) are strictly arctic species associated with ice.

Many arctic animals and birds are migratory. Bowhead whales feed along the Bering Sea ice edge in spring as they migrate northward to their summer feeding grounds on the Beaufort Sea continental shelf. They spend the winter months in the Bering Sea. Gray whales (*Eschrichtius robustus*) migrate even farther from their winter habitat in southern California. Even nonmigratory animals range over large areas. Polar bears use the ice as a means of transportation and cover large distances. Clearly, arctic marine mammal populations are abundant and diverse, and they are well adapted for life in ice-covered seas.

## B. Fish Populations

The southern boundaries of the arctic seas are the sites of some of the world's major fisheries, which occur at the confluence of polar and north temperate waters. These fisheries do not take place in truly arctic waters, although the Barents and Bering Sea fisheries extend into waters covered seasonally by sea ice.

A number of fish species occur in the coastal regions of the Beaufort Sea, including salmon (*Oncorhynchus* spp.), arctic char, arctic cisco, least cisco, broad and humpback whitefishes, fourhorn sculpin, and arctic flounder. Essentially all of them are anadromous and potentially can venture from fresh water to marine water. An anadromous fish such as the arctic char faces extreme conditions but has evolved strategies to cope with the constraints of the environment. Cold water temperatures and reduced areas of stream habitat in winter and low prey densities in the rivers are major problems (Craig 1989).

P. C. Craig (1989) has analyzed the adaptation of arctic char to the severe Beaufort Sea climate, and he points out that migration between marine and freshwater environments provides the advantage of exploiting feeding grounds in the near-shore coastal waters and spring waters where the densities of prey are five times higher than those found in tundra and coastal-plain streams. Consequently, carbon of marine source is the major source of energy (Schell 1983, Ziemann 1986). Thus, the char concentrate their feeding and growth in the near-shore marine waters during the summer, whereas there is little feeding or growth during the winter. Winter stream habitat is limited, and a single spring-fed system may harbor practically all members of a particular char population, from eggs to adult fish, during 8 to 9 months in the winter period (Craig 1978). Such overwintering areas for arctic fishes represent critical habitats, which are affected by changes in quality and distribution of source water. In a mild winter, the rivers might freeze less and perhaps inundate portions of the spawning bed with 0°C water, which could slow egg development, and consequently the fry might not emerge before spring breakup, when flooding scours the spawning beds (Craig 1989).

## C. Summary

We can summarize the distinctive properties of arctic marine ecosystems as follows:

- The principal constraints to biological productivity are low water temperatures and the brief summer period of biological activity at the primary production trophic level. The overall limitation is light availability on an annual basis. The tendency for a strong salinity-based vertical structure further restricts primary production during the summer period through nutrient limitation, although its initial establishment triggers the spring phytoplankton bloom at ice edges.
- Mesoscale and even small-scale oceanographic structures and processes play a major role in the primary production regime. These include ice edges, polynyas, oceanographic fronts, and ice-seawater interfaces. In the Arctic Ocean itself, the near-shore environments are the primary feeding areas for secondary production and, therefore, provide much of the sustenance for anadromous fishes and for birds and marine mammals, although complex communities may develop away from shore associated with multi-year ice, based on ice algae.
- There is a tendency for a strong benthic involvement in the food chain on the arctic continental shelves.
- The top of the food chain is characterized by large animals. Bowhead and gray whales are examples of large consumers. In many cases these animals make use of the ice as a platform for locomotion and reproduction (e.g., walrus, polar bears, and seals). The food chains supporting these animals are quite short and often based on benthic systems and a benthic shunt for primary production. Another way of looking at these populations is that there is a tendency toward k-selected species, with long-lived adults, slow growth, delayed maturity, and multiple reproductive occasions. This is true also for the coastal anadromous fishes. Such animals are able to store biomass over long periods of time.
- The arctic marine system has evolved over a long time, extending 200,000 years (Craig 1989), although conditions similar to the present have existed for only 10,000 years. Nevertheless, the organisms have had time to adapt to and even take advantage of the unique arctic conditions. Arctic conditions are indeed unique, since in the Antarctic (Southern Ocean) there is no truly polar ocean, even though there is an extensive subantarctic seasonally ice-covered region. With the exception of light conditions, the environment there is very different, notably because in the north there is a vast ocean over the pole, whereas in the south there is a huge glaciated land mass over the pole.

## IV. THE POTENTIAL CONSEQUENCES OF GLOBAL WARMING

The first clear biological response to warming in the arctic will be an extension northward of the range for many subarctic and northern temperate species and a movement northward of arctic species. Such responses can be quite rapid. For example, C. Vibe (1967), in his introduction to his work on arctic animals in relation to climate fluctuations, pointed out that the history of Greenland is characterized by periods of prosperity and periods of decline, each lasting at the most a few hundred years. Sea mammals and birds of Greenland were forced to migrate, leaving humans to starve. In our case, the north coast of Alaska would become subarctic, first with seasonal sea ice and finally with limited or no sea ice. The Canadian High Arctic, Greenland, and Svalbard would persist longer as arctic lands, providing a refuge for arctic species.

The scenarios that have been proposed for future warming predict a dramatic temperature increase for the Arctic. Although a few degrees' increase in the seawater temperature may not seem critical, the consequences would, in fact, be devastating to the arctic marine ecosystem as we know it today. Among the major environmental changes, the following consequences seem likely:

• There would be a decrease in sea ice cover. Seasonally ice-covered areas will become ice-free throughout the year, and the ice over much of the Arctic Ocean will become seasonal. Eventually, an ice-free Arctic Ocean might result. The consequences of this will include a self-reinforcing increase in surface temperature due to the reduction in albedo, an increase in evaporation, and an increase in wind-mixing of the surface waters, thereby bringing warmer water to the surface from below. An increase in cloudiness is also likely, as a result of the increased evaporation.

• The immediate effects of an ice-free Arctic Ocean will include the elimination of thermohaline convection, presently believed to be an important mixing process. This will reduce nutrient input to coastal regions and result in lower primary production. Even though wind-mixing will increase, it will probably not be sufficient to break down the low salinity surface layer, since river input of fresh water may even increase. The increase in evaporation will increase precipitation, although not necessarily in the immediate area or nearby coastal areas.

• A change in sea level is already occurring

and may accelerate if major ice caps melt. This is a global process and not restricted to the Arctic. In the arctic, however, the consequences include effects on the freshwater environments presently used by anadromous fishes.

• Increased coastal erosion in arctic tundra may result from thawing of the permafrost and other inland processes (transport by rivers). This may provide an alternate source of carbon for near-shore areas.

The consequences of some of these changes are not well understood; for example, the consequences of the rapid disappearance of arctic sea ice. Would the increased evaporation and precipitation resulting from this disappearance increase glaciation and counterbalance the increase in sea level? Melting the sea ice may be a self-reinforcing process, since once sea ice has melted, the reduction in surface albedo would make it difficult for the water to refreeze.

With respect to primary production, there would be a reduction and perhaps ultimately a loss of ice algae and elimination of the entire ice-associated community, including the dependent species, such as polar cod. The animals that depend on ice as a platform, such as the seals, walrus, and polar bears, would be left vulnerable by the loss of their habitat. Blooms at the edge of the seasonal sea ice would be restricted to those areas still subject to seasonal sea ice cover, further restricting the length of the growing season by the delay in the timing of the spring bloom. The spring bloom would then be dependent on water-column stability produced by surface heating, which is ineffective at the low water temperatures that would still be extant in spite of the warming. This would affect not only the early spring primary production but possibly also the efficiency of the energy transfer through the benthos.

There is no reason to believe that the salinity-based density stratification of the Arctic Ocean would change, unless there is a major change in river inflow or a major change in overall Arctic Ocean circulation (Kilworth and Smith 1984). In either case, primary production would remain low, because of nutrient limitation in the first case and deep mixing and low light angles in the second. With an open Arctic Ocean, precipitation and river flow are likely to increase, and the salinity stratification would be maintained. Thermohaline circulation would be eliminated, and no mechanisms would exist to inject nutrient-rich deeper water into the surface, even in the coastal regions. There would probably be an increased input of inorganic nutrients as well as peat through the rivers, and there might therefore remain some pockets of high primary production dependent on this allochthonous source.

Thus it seems likely that the overall biological productivity of the arctic seas would be severely reduced by a temperature increase. Essentially all the distinctive arctic animals would disappear. Probably the first to suffer would be the walrus populations, dependent over much of their range on seasonal sea ice, on polynyas, and on a strong benthic community. Floating ice provides a means of transportation for walruses, and this mobility allows them to feed over a large area. Sea otters (Enhydra lutris) could fill the walrus niche and are, in fact, already extending their range northward. Polar bears, clearly adapted to sea ice, evolved from grizzly bears (Ursus sp.) in response to the arctic ice cover. On the other hand, the gray whale, which feeds on benthic amphipods in the northern Bering Sea, is found in an area of high pelagic production caused by transport of nutrient-rich water from the Bering Sea basin northward via the Gulf of Anadyr. Assuming that the circulation pattern remains unaltered, primary production over this particular shallow shelf region could remain sufficiently high to allow survival of these animals. However, they are also known to feed along the ice edge earlier in spring, and this environment would become increasingly rare. Furthermore, we do not know to what extent the current global circulation would remain active. It presently depends on a difference in sea level between the Atlantic and Pacific oceans.

The initial response of benthic communities on the Beaufort Sea shelf and other shelves around the Arctic Ocean rim might be an increase, since ice scouring would be eliminated, making more habitat available for colonization. Increasing water depths would extend the shelves inland also.

Changes in river flow would affect the anadromous fishes by increasing, or possibly eliminating, their habitat. Salmon may be able to colonize streams and beaches currently unavailable to them, and the small salmon run in the Bering Sea could become more significant. The consequences to fisheries farther south in the Bering Sea, the Icelandic area, or the Norwegian Sea are uncertain but may be related more to the changes in large-scale oceanic circulation than to local in situ warming, per se. They do, however, seem to depend on the confluence of arctic and north temperate waters.

Although it seems likely that the system would not adapt as well to the high seasonality without an ice cover, in terms of fisheries, this may not be true. The North Sea is a major fishery ground. It, too, is subject to high seasonality, suffers from nasty weather (which seems to be characteristic of all major fishing areas), and is not very deep. It is hard, with our present state of knowledge, to second-guess what will happen with respect to subarctic commercial fish stocks. It is clear, though, that the distinctive large arctic marine mammals, the abundant bird populations, and the stark ice-dominated scenery that we have come to know as arctic probably will suffer. The response of many species to a reduction in sea ice and a warming cannot be simple migration northward, because most arctic animals are either adapted to feeding over shallow continental shelves (such as the walrus) or to feeding on ice-related organisms (ringed seal, polar bear, arctic birds, polar cod). As ice recedes to areas underlain by deep water, these would suffer a loss of habitat. The primary productivity of those shallow shelves that have adequate nutrient supply, such as the northern Bering Sea–Chukchi Sea area, would increase, however, and here we could expect a new and possibly rich biological community to develop in response to the warmer temperatures.

## REFERENCES

Aagaard, K., L. K. Coachman, and E. Carmack. 1981. On the halocline of the Arctic Ocean. *Deep-sea Res.* 28A:529.

Aagard, K., and L. K. Coachman. 1975. Toward an ice-free Arctic Ocean. *Trans. Am. Geophys. Union* 56:484.

Alexander, V. 1980. Interrelationships between the seasonal sea ice and biological regimes. *Cold Regions Sci. Technol.* 2:157.

Beckley, R. G., and H. J. Trodahl. 1987. Scattering and absorption of visible light by sea ice. *Nature* 326:867.

Broecker, W. 1987. Unpleasant surprises in the greenhouse? *Nature* 328:123.

Brown, R.G.B. 1989. Seabirds and the arctic marine environment. Proc. Sixth Conf. Com. Arct. Int., L. Rey and V. Alexander, eds., pp. 179–200. Leiden: E. J. Brill.

Carter, L. D., J. Brigham-Grette, J. Marincovich, Jr., V. L. Pease, and J. W. Hillhouse. 1986. Late Cenozoic Arctic Ocean sea ice and terrestrial paleoclimate. *Geology* 14:675.

Clasby, R. C., V. Alexander, and R. Horner. 1976. Primary productivity of sea-ice algae. In *Assessment of the Arctic Marine Environment: Selected Topics*, D. W. Hood and D. C. Burrell, eds., pp. 289–304. Fairbanks: Institute of Marine Science, University of Alaska.

Craig, P. C. 1978. Movement of stream-resident and anadromous arctic char (*Salvelinus alpinus*) in a perennial spring on the Canning River, Alaska. *J. Fish. Res. Bd. Can.* 35:48.

Craig, P. C. 1989. An introduction to anadromous fishes in the Alaskan Arctic. *Univ. Alaska Biol. Papers* 24:27.

Demers, S., L. Legendre, J. C. Therriault, and R. G. Ingram. 1986. Biological production at the ice-water ergocline. In *Marine Interfaces Ecohydrodynamics*, J.C.J. Nihoul, ed., pp. 31–54. Amsterdam: Elsevier.

Falk-Petersen, S., I. B. Falk-Petersen, and J. R. Sargent. 1986a. Structure and function of an unusual lipid storage organ in the arctic fish *Lumpenus maculatus* Fries. *Sarsia* 71:1.

Falk-Petersen, I. B., S. Falk-Petersen, and J. R. Sargent. 1986b. Nature, origin, and possible roles of lipid deposits in *Maurolicus muelleri* (Gmelin), and *Benthosema glaciale* (Reinhart) from Ullsfjorden, Northern Norway. *Polar Biol.* 5:235.

Gulliksen, B., and O. J. Lønne. 1989. Distribution, abundance, and ecological importance of marine sympagic fauna in the Arctic. Oceanography and Biology of Arctic Seas, G. Hempel, ed. *Rapp. P.-v. Reun. Cons. Int. Explor. Mer.* 188:133.

Gulliksen, B., W. Vader, and O. J. Lønne. 1986. Under-ice fauna in the Arctic. *Pro Mare Ann. Rep.*: 53.

Hansen, J., I. Fung, A. Lacis, D. Rind, S. Lebedeff, R. Ruedy, and G. Russell. 1988. Global climate changes as forecast by Goddard Institute for Space Studies three-dimensional model. *J. Geo. Res.* 93(D8):9341.

Horner, R. A. 1985. Ecology of sea-ice microalgae. In *Sea ice biota*, R. A. Horner, ed., pp. 83–103. Boca Raton, Fla.: CRC Press.

Horner, R. A., and G. C. Schrader. 1982. Relative contribution of ice algae, phytoplankton, and benthic microalgae to primary production in nearshore regions of the Beaufort Sea. *Arctic* 35: 485.

Kilworth, P. D., and J. M. Smith. 1984. A one-and-a-half dimensional model for the arctic halocline. *Deep-sea Res.* 31:271.

Kutzbach, J. E. 1985. Modeling of paleoclimates. *Advances in Geophysics* 28A:159.

Kutzbach, J. E., and H. E. Wright, Jr. 1985. Simulation of the climate of 18,000 years B.P.: Results for the North American/North Atlantic/European sector and comparison with the geologic record of North America. *Quat. Sci. Rev.* 4:147.

MacCracken, M. C., and F. M. Luther, eds. 1985. Projecting the climatic effects of increasing carbon dioxide. Rept. DOE/ER-0237. Washington, D.C.: Dept. of Energy.

Müller-Karger, F., and V. Alexander. 1987. Nitrogen dynamics in a marginal sea-ice zone. *Cont. Shelf Res.* 7:805.

Niebauer, H. J. 1983. Multiyear sea ice variability in the eastern Bering Sea: An update. *J. Geo. Res.* 88:2733.

Niebauer, H. J. 1988. Effects of El Niño-Southern Oscillation and North Pacific weather patterns on interannual variability in the subarctic Bering Sea. *J. Geo. Res.* 93:5051.

Parkinson, C. L., and W. W. Kellogg. 1979. Arctic sea ice decay simulated for a $CO_2$-induced temperature rise. *Clim. Change* 2:149.

Petersen, G. H. 1988. Benthos, an important compartment in northern aquatic ecosystems. In *Proc. Sixth Conf. Marine Living Systems of the Far North*, L. Rey and V. Alexander, eds. Fairbanks, Alaska: Comité Arctique International; Leiden: E. J. Brill.

Petersen, G. H., and M. A. Curtis. 1980. Differences

in energy flow through major components of subarctic, temperate, and tropical marine shelf ecosystems. *Dana* 1:53.

Schell, D. 1983. Carbon-13 and carbon-14 abundance in Alaskan aquatic organisms: Delayed production from peat in arctic aquatic foodwebs. *Science* 219:1068.

Semtner, A. J., Jr. 1984. The climatic response of the Arctic Ocean to Soviet river diversions. *Clim. Change* 6:109.

Semtner, A. J., Jr. 1987. A numerical study of sea ice and ocean circulation in the Arctic. *J. Phys. Oceanog.* 17:1077.

Sverdrup, H. U. 1953. On conditions for the vernal blooming of phytoplankton. *J. Cons., Cons. Perma. Int. Explor. Mer* 18:287.

Vibe, C. 1967. Arctic animals in relation to climatic fluctuations. *Meddelelser om Gronland* 170:1967.

Ziemann, P. 1986. Study of aquatic foodwebs using natural carbon isotopes. M.S. thesis, University of Alaska, Fairbanks.

# CHAPTER EIGHTEEN

# Some Possible Effects of Climatic Warming on Arctic Tundra Ecosystems of the Alaskan North Slope

W. DWIGHT BILLINGS AND
KIM MOREAU PETERSON

## I. INTRODUCTION

Arctic tundra ecosystems are dominated by plants that are unique in being able to grow, metabolize, and reproduce at temperatures only slightly above freezing (Billings 1987a). On the North Slope of Alaska, these taxa are almost all perennial herbaceous or dwarf shrubby vascular plants, lichens, and bryophytes. The percentage of nonvascular plants is high compared with warmer regions of the earth. For example, the flora at Point Barrow (lat. 71°23′ N) consists overwhelmingly of nonvascular taxa, 73.3% nonvascular to 26.7% vascular. All dominant tundra species, both vascular and nonvascular, photosynthesize by the $C_3$ pathway, with peak rates between 15° and 29°C or lower; they have the ability to acclimate to temperature; and their photosynthetic rates respond rapidly to changes in the amount of available sunlight (Billings et al. 1971, Tieszen 1978b).

Although the tundra biome has been relatively stable for thousands of years, it is surprisingly vulnerable to warming conditions. Relatively slight amounts of warming may cause melting of permafrost, invasion of tundra communities by woody species, and thermokarst erosion. Thermokarst is a dramatic type of erosion, caused by soil warming, that results in loss of soil and vegetation in those areas of wet coastal tundra characterized at present by permafrost, soil polygons, and thaw lakes. Another effect of such temperature increase is that it can change the global carbon balance by melting frozen peat in which carbon compounds are then respired by microorganisms to atmospheric carbon dioxide ($CO_2$) or methane ($CH_4$). So, the possibility exists that a substantial increase in the release of carbon from warming arctic peat deposits could raise atmospheric carbon concentrations enough to add significantly to the greenhouse effect (Billings et al. 1982, 1983, 1984; Billings 1987b).

This chapter begins with an overview of the climate, soils, and vegetation types presently found on the North Slope of Alaska. It then focuses on two of the most important effects of warming on arctic ecosystems: possible future contributions of arctic peat deposits to the atmospheric concentration of carbon dioxide, and thermokarst erosion.

## II. CHARACTERISTICS OF ALASKA'S NORTH SLOPE

### A. Vegetation

North Slope vegetation can be divided into two major types: wet coastal tundra and tussock tundra. The wet coastal tundra forms a strip 50–100 km wide along the coastal plain, extending from Icy Cape (long. 162° W) eastward to beyond Kaktovic (long. 143°30′ W). Inland, the wet coastal tundra grades into the tussock tundra, which covers most of the remainder of the North Slope, extending up into the foothills of the Brooks Range (see Billings and Peterson 1980, Peterson and Billings 1980, Komarkova and Webber 1980 for more precise descriptions of these tundras).

The species composition of the wet coastal tundra varies from place to place but is characteristically dominated by rhizomatous sod graminoids (sedges and grasses), mosses, and lichens. The principal sedges and grasses are *Carex aquatilis, Eriophorum angustifolium, E. scheuchzeri, Dupontia fisheri,* and *Arctophila fulva.* These graminoids reproduce largely by rhizomes that spread to form clonal colonies. Reproduction by seeds, although possible, is usually less successful, except in small tufted grasses such as arctic bluegrass (*Poa arctica*) and the herbaceous and dwarf shrubby dicots. The areas of graminoid vegetation are interspersed with thousands of thaw lakes (small, shallow lakes) and wet low-center polygons. Such polygons are actually early stages in the thaw lake cycle.

The tussock tundra has a different appearance from the wet coastal tundra. Instead of low sod graminoids, there are tussocks of sedges (*Eriophorum vaginatum*) that occasionally

can be high enough to impede walking. There are some thaw lakes, as in wet coastal tundra. Compared with wet coastal tundra, plant species diversity is greater and there are more types of life forms represented. The graminoid part of the inland tundra vegetation is dominated by the tussock sedge, which occupies the slightly drier sites, while the rhizomatous graminoids are near the wetter end of the moisture gradient in low-center polygons and tundra ponds. Larger shrubs such as the willows *Salix alaxensis* and *S. glauca,* not often present in the coastal wet tundra, become prominent along the rivers and streams. This "tussock tundra" prevails at Toolik Lake, about 150 km south of Prudhoe Bay, and at Atkasook, about 100 km south of Barrow, and over all the wild country north of the Brooks Range, for about 900 km from Cape Thompson (long. 166° W) to the Arctic National Wildlife Refuge near longitude 145° W.

### B. Weather and Climate

The weather and climate of the North Slope are very cold compared with the rest of North America. Mean daily temperatures at Barrow are below −20°C from December through March, with extreme means to at least −49°C. Only June, July, and August have mean temperatures above 0°, with the warmest month, July, reaching a mean of only 3.7°C. Prudhoe Bay is slightly colder in winter but warmer in summer than Barrow: means of −25° to −30°C from December through March vs. about 6°C mean temperature for July. Temperatures between Barrow and the Brooks Range in the tussock tundra are colder in winter than the coastal stations but much warmer in summer. Precipitation along the coast is scanty. The 30-year mean annual figure for Barrow is about 125 mm, most of which falls as snow. The sea ice that lies immediately onshore throughout the year at Barrow contributes to the cold, foggy conditions there in summer.

### C. Soils, Hydrology, and Permafrost

The soils of the coastal tundra, derived from uplifted marine sediments, remain frozen for all but July and August, and even then, the maximum depth of thaw to the top of the permafrost ranges down to only 25–45 cm. Permafrost itself underlies all the coastal tundra, extending as deep as 600 m (Brown et al. 1980). The soils are mostly nutrient-poor, acidic pergelic cryohemists or pergelic cryoaquepts, with the upper peat horizon ranging from 10 cm to 20 cm deep. Most of the plant biomass in the wet coastal tundra is below ground in the form of roots and rhizomes in the active layer, or summer thawed zone (Billings et al. 1978). Maximum live plant biomass is between 0 cm and −10 cm deep (Dennis and Johnson 1970). The annual roots of cottongrass (Eriophorum angustifolium), however, follow the thaw down and finally grow in contact with and along the permanently frozen soil and ice of the top of the permafrost (Billings et al. 1976) at temperatures near 0°C or slightly below.

The surface of the coastal tundra is mostly water, frozen for nine to ten months of the year. Most of this water is in thousands of shallow thaw lakes formed in the upper meter or two of the permafrost by thermokarst, the result of wind and wave action in small tundra ponds where the water is slightly warmer than air temperature. Such thermokarst erosion is an integral part of the thaw-lake cycle hypothesized by M. E. Britton (1957) and described in detail by Billings and Peterson (1980). In early stages of the thaw-lake cycle, wedges of pure ice form in contraction cracks in the uplifted frozen marine sediments or windblown sands during the coldest times of the winter (Leffingwell 1919, Lachenbruch 1962). These ice wedges continue to grow year by year and become progenitors of the polygon patterns so characteristic of the North Slope tundra. Ice-wedge polygons play an important role in the development of tundra ponds and thaw lakes through gradual thermokarst of the upper permafrost in the short arctic summers. Thermokarst caused by small streams and rivers also has an important role in the erosional development of drowned creeks and small estuaries along the coastlines. The North

Slope tundra cannot be understood as an ecosystem without a knowledge of the thaw-lake cycle as diagrammed in figure 18.1 (Billings and Peterson 1980) and illustrated by the aerial photograph in figure 18.2.

## III. PHOTOSYNTHETIC CARBON CAPTURE AND ITS STORAGE IN COLD SOILS

Even though the tundra summer is short and cold, the sun is continuously above the horizon at Barrow from 11 May to 2 August. Though the sun is never high above the horizon (48° noon solar angle on 22 June and 19° midnight solar angle on the same date), the continuous light adds up to considerable energy in photosynthetic quanta in clear weather, as much as 32 Megajoules per square meter per day (Dingman et al. 1980). Since snowmelt and the onset of rapid plant growth do not occur until the later part of June, a time when solar radiation is dropping, the rate of incoming shortwave solar radiation from clear skies has dropped to about 25 MJ per $m^2$ per day by the time of peak plant biomass and leaf area in late July, when, on a leaf-area basis, 200–400 mg $CO_2$ per square decimeter per hour are fixed in photosynthesis (Tieszen et al. 1980). Foggy weather can cut down on photosynthetic rates, but the net amount of carbon fixed by the end of the growing season in August is still surprisingly high at about 174 g C per $m^2$ per season (Chapin et al. 1980).

Much of this fixed carbon, the product of photosynthesis, moves below ground in the plants, where it is stored as carbohydrates in roots, rhizomes, and stem bases. This is particularly true of the graminoids. Dwarf evergreen shrubs store some of these products as lipids in the leaves. As daylengths shorten in late summer, such translocation to belowground organs is hastened. In the Barrow tundra, to a depth of 20 cm in the summer-thawed soil layer, there was nearly six times as much carbon stored in the live roots and rhizomes as in the aboveground live leaves, stems, and reproductive structures (Chapin et al. 1980).

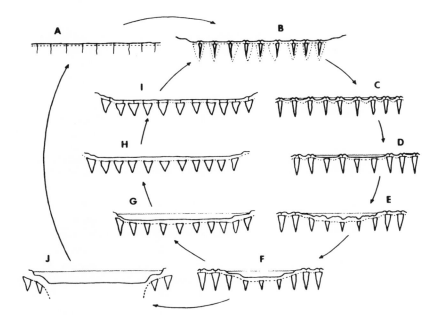

*Figure* 18.1. The thaw-lake cycle diagrammed in cross-section. A. Newly exposed marine sediments with contraction cracks into the permafrost. Dotted line indicates upper limit of permafrost. B. Growing ice wedges and low-center polygons in these sediments; dotted-line ice wedges are older and have gone through at least one cycle. C. An older stage of B. D. Formation of thaw ponds by erosion and coalescence of low-center polygons. E. A young thaw lake. F. A small, shallow thaw lake with bench. G. A mature but shallow thaw lake with bench; polygons remain visible on bench because of persistence of old ice wedges. H. A drained, shallow thaw lake. I. A drained, shallow thaw-lake basin with old ice wedges being rejuvenated. J. A deep, large thaw lake with permafrost and ice wedges melted below it. Reprinted with permission from Billings and Peterson 1980; copyright © 1980, Pergamon Press.

The tundra is essentially a below-ground ecosystem as far as carbon is concerned. Carbon is transferred from live below-ground organs to dead roots and rhizomes at a rate of 45 g C per m² per year. That rate through the years has resulted in a carbon pool of about 347 g C per m² in dead structures to a depth of 20 cm. The dead plant material in the soil decomposes very slowly into humus and peat, which remains a cold pool of carbon averaging 18,992 g C per m² in the Barrow tundra. As this organic matter accumulates, the permafrost table rises because of the insulating effect of the peat itself, and the peat becomes frozen and can remain so for many centuries under present environmental conditions.

Because of incorporation of the carbon-rich peat into the permafrost, the turnover rate of carbon in the tundra is slow. The result is the accumulation of great frozen reserves of carbon compounds in tundra soils. This has been, over the centuries, a major sink for atmospheric $CO_2$. W. M. Post et al. (1982) conclude that the mean carbon density of the circumpolar tundra is on the order of 21.8 kilograms per square meter, which totals $191.8 \times 10^{15}$ grams of carbon for this entire biome, or 13.75% of the earth's soil carbon. Another 13% has accumulated in the often permafrosted peaty soils of the taiga, or boreal forest. This means that roughly 27% of the earth's soil carbon is tied up in cold northern soils. Most of that carbon has been incorporated into the frozen soils in the last 10,000 or even fewer years. Such soils make up about 17% in area of total global soils

Figure 18.2. Aerial photograph of low-center polygons and thaw lakes near Barrow, Alaska. Photograph by W. D. Billings.

(Post et al. 1982, Emanuel et al. 1985b, Billings 1987b).

With a warming arctic climate, the frozen peats would be prone to thaw, and increasing decomposition rates would release large amounts of carbon dioxide and methane to the already rapidly escalating carbon content of the global atmosphere. The result could be a positive feedback from soil to atmosphere, exacerbating the greenhouse effect (Billings et al. 1982).

Calculation of present carbon balances in the tundra is fraught with problems, the main one being the lack of long-term measurement of carbon dioxide and methane fluxes in a variety of sites. F. S. Chapin et al. (1980) and Billings (1987b) have taken data from the Barrow tundra and produced a rough carbon budget for the 1970s, but their results are short-term approximations only. That budget (presented here as table 18.1) estimates that about half the carbon captured in the Barrow tundra remains there, at present, in a positive carbon balance of 109 g C per m$^2$ per year. There are some uncertainties in these measurements and calculations. More quantitative measurements are needed.

The prediction of future carbon balances in the Arctic is even more difficult, not only because data are lacking for the necessary models but also because we have no real assurance of how the climate is going to change globally, regionally, or locally. Such future balances in these carbon-rich systems could be quite different from those in the past. As suggested (Billings et al. 1982), tundra and taiga peats could become carbon sources in the near future as decomposition accelerates, rather than remaining as carbon sinks. Much depends on what will happen to the upper permafrost under a warming climate. That in turn depends on the health and environmental tolerances of the graminoid plants dominating the vegetation that now insulates the permafrost from accelerated melt during the short summers. Disturbance or loss of the native sedges and grasses could result in severe thermokarst erosion in the coastal tundra.

## IV. EXPERIMENTAL SIMULATION OF FUTURE CARBON BALANCES USING MICROCOSMS

In an attempt to predict future carbon situations, we have studied intact natural cores (microcosms) of tundra soils and plants in

*Table* 18.1. Annual carbon budget (g per m² per year) of the wet coastal tundra at Barrow, Alaska. From Billings 1987b.

| Process | Carbon capture | Carbon loss | Carbon balance |
|---|---|---|---|
| Gross photosynthesis | | | |
|    vascular plants | 202 | | |
|    mosses | 12 | | |
| Respiration | | | |
|    vascular plant leaves and stems | | 38.0 | |
|    mosses | | 2.0 | |
|    vascular plant roots and rhizomes | | 40.0 | |
|    invertebrates | | 1.0 | |
|    herbivores | | 1.0 | |
|    carnivores | | 0.1 | |
|    decomposers | | 21.0 | |
| Export by water and wind | | 2.0 (?) | |
| Total | +214 | −105.0+ | +109 |

the Duke Phytotron under simulated twenty-first-century environments of the Alaskan tundra (Billings et al. 1982, 1983, 1984; Peterson et al. 1984). The microcosms were obtained by coring into the frozen soil through the snow at Barrow in the winter, and the cores were kept frozen until use. In the phytotron, one set of cores was subjected to typical arctic summer environments, including continuous photoperiod (controls). The experimental microcosms were subjected to various combinations of temperature, water table levels, nutrients, and atmospheric $CO_2$ concentrations in accordance with predictions for these parameters from models for arctic climates and ecosystem processes in the next century (Manabe and Stouffer 1979, Strain et al. 1983, and others). Our technique was to measure $CO_2$ exchange between the tops of the cores, with their growing tundra plants, and the controlled atmosphere by using an infrared gas analysis system. We also measured net $CO_2$ fixation and relative growth rates under the controlled environmental conditions.

Before starting the experiments, we hypothesized that the postulated increase in summer temperature at Barrow that has been predicted for doubled atmospheric $CO_2$ would change that tundra from a sink for atmospheric $CO_2$ to a source of $CO_2$. Also, we suggested that lowering the water table (now at the surface) would expose additional soil to oxygen and thus increase the rate of $CO_2$ efflux from the tundra to the atmosphere. If true, both would cause breakdown of the peaty surface soil, which is at present such a large carbon pool. Both hypotheses are supported by our resulting data. The data indicate that sooner or later the integrity of the wet tundra is likely to break down in a warmer $CO_2$-rich climate. The signal for this may be apparent in the incipient warming of the permafrost (Lachenbruch and Marshall 1986). Emanuel et al. (1985a,b) have predicted in their model that at the doubling of atmospheric $CO_2$, the earth's boreal forest will have decreased in area by 37% and the tundra by 32% because of the predicted warming of their environments and changes in species composition.

Our experiments with the wet coastal tundra microcosms have shown the following results:

• Net ecosystem $CO_2$ uptake was almost twice as great at present summer temperatures (4°C) in Barrow microcosms as at 8°C, demonstrating that net $CO_2$ release to the atmosphere is likely if warming occurs in the arctic.

- Lowering the water table in Barrow microcosms from the surface to $-5$ cm and $-10$ cm resulted in a net loss of $CO_2$ from the system of about 300–500 g $CO_2$ per m² per 85 days.

- Doubling atmospheric carbon dioxide concentrations over Barrow microcosms from 400 ppm to 800 ppm at 8°C resulted in only a 48% net ecosystem carbon gain with the water table at the surface. With the water table at $-10$ cm and $CO_2$ concentration at 800 ppm, there was a net loss of 311 g $CO_2$ per m² over an 85-day season. That loss, however, was 35% less than at 400 ppm atmospheric $CO_2$, indicating that the doubling of atmospheric carbon dioxide concentrations at higher temperatures has a slight compensating effect. But 400 ppm at 8°C is an unlikely scenario for Barrow because of the cold seas on both sides of the region.

- In Barrow microcosm cores, added nitrogen significantly increased leaf area, phytomass, and net ecosystem carbon dioxide uptake. But increased atmospheric $CO_2$ concentration, in addition, had little or no effect on mean ecosystem carbon uptake by these vegetated microcosms.

- We conclude that atmospheric $CO_2$ is not now limiting net ecosystem primary production in the wet coastal tundra and that its direct effects on the ecosystem will be slight even at double the present concentration. The most probable effects of $CO_2$ on the coastal tundra will be through its indirect effects on temperature, water table, peat decomposition, thermokarst, and availability of soil nutrients. It is these nutrients that are most limiting at present.

- D. T. Tissue and W. C. Oechel (1987) also found essentially the same thing in tussock tundra at Toolik Lake. In field plots, long-term exposure of the tussock sedge (*Eriophorum vaginatum*) to elevated $CO_2$ did not increase photosynthesis or growth. They conclude that in tussock tundra, as in wet coastal tundra, plant growth is not limited by $CO_2$, temperature, water, light, or carbohydrate availability but by the current low nutrient status of the vegetation.

From field observations and the results of our controlled microcosm experiments, we can come to the following tentative conclusions for the tundras on a warmer North Slope:

- The plant growing season would probably be snow-free for a longer time. The plants could take advantage of those days in May and the first half of June when the photoperiod is already continuous but during which, at present, the land remains snow-covered and frozen.

- Maximum summer air temperatures would be higher, thus resulting in higher evaporation and transpiration rates. This could result in higher net photosynthesis rates but perhaps lower net carbon storage because of higher microbial respiration rates.

- Some of the plant components of this coastal ecosystem may be replaced eventually by populations of taxa not present there now. Such species have not yet appeared at Barrow; there are no alien taxa now, and diversity remains low, primarily because the Barrow climate still has not warmed in the last several decades.

- There will be higher rates of peat decomposition, resulting in the release of plant nutrients, particularly nitrogen and phosphorus, that could increase primary production. The outward flux of carbon dioxide would increase and probably also that of methane.

- Water temperatures would rise in the tundra ponds and lakes.

- An increase in water temperature would likely result in accelerated thermokarst erosion, including loss of much of the ice-wedge network.

- Higher air and water temperatures, including resultant decomposition of peat, would inevitably lead to greater depth of thaw, retreat of the permafrost table, and loss of the upper 1 or 2 meters of the permafrost.

- Melting of permafrost could result in a loss of structural integrity (by thermokarst) of the coastal tundra ecosystem, which now is held together by roots, rhizomes, and the upper permafrost.

## V. A PROTOTYPE OF WARM WATER EROSION: FOOTPRINT CREEK

The history of the Voth Drainage Tundra, below Upper and Lower Footprint lakes and Upper and Lower Dry lakes, not far east of the village of Barrow, provides an example of what could happen in the wet coastal tundra under a warmer climate. In this drainage, a considerable area of wet coastal tundra, some 70 hectares, was flooded by relatively warm water from the artificial drainage of four thaw lakes in 1950. This water flow still continues and provides an experimental view of what might happen to similar tundra subjected to global warming. Before the ditch was dug between the lakes and the Voth Drainage Tundra (hereafter Voth Tundra), water draining from Upper and Lower Footprint lakes spread out widely over the Voth Tundra, which was covered by very wet and unspoiled tundra vegetation, dominated by water sedge (Carex aquatilis), tundra grass (Dupontia fisheri), pendant grass (Arctophila fulva), and mosses. This wet area of the Voth Drainage forms the headwaters of Footprint Creek, which ultimately carries the waters into Middle Salt Lagoon, a brackish thaw lake, and thence into the Chukchi Sea by another man-made ditch through the littoral gravel bar. That latter lowering of the original deltalike flow across the bar has accelerated the headward thermokarst erosion of Footprint Creek below the Voth Tundra.

To understand how drainage changes can cause thermokarst erosion, it is important to recognize that the Footprint lakes contain relatively warm water during two summer months but are frozen the rest of the year. In general, mean water temperatures in tundra ponds and lakes during the summer are higher than soil temperatures in the wet tundra. Higher water temperatures are particularly characteristic of small, shallow tundra ponds and lakes (Miller et al. 1980), such as the Footprint lakes. These two lakes have a water depth of about 0.5 to 1.0 meter, while water in the Voth Drainage is even shallower. The Footprint and Dry lakes and the Voth Tundra become snow- and ice-free by late June. Although continuously recorded water temperatures are not available for these locations, temperatures are probably quite similar to those in Miller's ponds, which are nearby and comparable in depth. The ponds, on sunny days, commonly showed diurnal changes of 8°C and reached maximum temperatures in late June and July of 15°–16°C (Miller et al. 1980).

The drainage ditches of 1950 changed the patterns of water flow substantially, bringing larger amounts of warmer lake water into contact with the cold tundra of the Voth Drainage. The changes were initiated when gas wells were drilled into the tundra in the 1940s near the juncture of Upper Footprint Lake and Lower Dry Lake in the Naval Petroleum Reserve. At thaw each year, the lakes overflowed and caused flooding problems in and around the wells. In 1950 an attempt was made to improve the drainage by digging ditches to lead water from Lower Dry Lake into Upper Footprint Lake and from Lower Footprint Lake into the Voth Drainage Tundra and thence into Footprint Creek. That improved the situation at the gas wells but resulted in increased flow and warmer water from the lakes on and through the Voth Tundra. The warmer water has caused extensive thermokarst erosion by melting underground ice wedges, which has caused the overlying soil and vegetation to collapse (see fig. 18.3).

By the late 1950s and early 1960s, accelerated erosion was becoming obvious, expanding upstream in the Voth Tundra from the area where the bed of Footprint Creek began—the knickpoint. Scientists from the U.S. Army Cold Regions Laboratory marked the position of the knickpoint with stakes in certain years. Also, we (Billings and Peterson) started making measurements in 1973 and mapped the areas above and below the eroding knickpoint in 1975 and at intervals of several years through 1988.

In our opinion, it is the relatively warmer water from the shallowly drained Footprint and Dry lakes, through the drainage ditches and across the shallow-water wetlands of the Voth Tundra, that is causing the accelerated

Figure 18.3 (above). Early stage of thermokarst erosion at the knickpoint of Footprint Creek in the wet coastal tundra as it appeared in midsummer 1974. The drapelike vegetation that is relictual over the blocks of soil is dominated by *Dupontia fisheri* and *Carex aquatilis*. Fifteen years later (1989), this soil and vegetation have long since washed out to sea, and the site is a barren gully. Photograph by W. D. Billings.

Figure 18.4 (below). Destruction of tundra vegetation by thermokarst erosion at the head of Footprint Creek mapped in the summers of 1977 and 1988. The numbers refer to the principal dominants of the plant communities of the lower Voth Tundra as they existed in 1977. The communities mapped in 1977 have been destroyed by continued warmwater thermokarst.

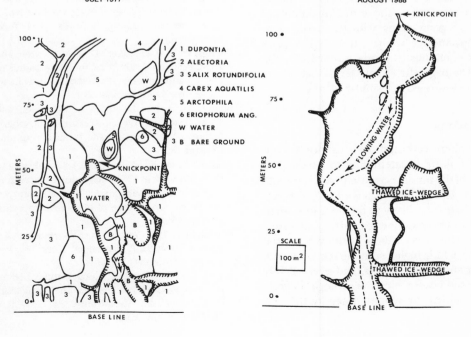

THERMOKARST EROSION, FOOTPRINT CREEK, BARROW, AK.
JULY 1977

THERMOKARST EROSION, FOOTPRINT CREEK, BARROW, AK.
AUGUST 1988

1 DUPONTIA
2 ALECTORIA
3 SALIX ROTUNDIFOLIA
4 CAREX AQUATILIS
5 ARCTOPHILA
6 ERIOPHORUM ANG.
W WATER
B BARE GROUND

thermokarst erosion at the head of Footprint Creek, shown in figure 18.3. To illustrate the speed and extent of this erosion, figure 18.4 shows a pair of our maps of the knickpoint area on the creek as it appeared in 1977 and eleven years later in 1988. The water channel crosses the baseline from the Voth Drainage flowing from south to north. On the 1971 aerial photographs, the head of erosion (knickpoint) was about 60 m to the north of the baseline; in the six years from 1971 to 1977, thermokarst erosion had proceeded about 115 m upstream; and by 1988 it had gone another 55 meters. The erosion appears to have slowed, partially because its head has entered the broader and shallower waters of the Voth Drainage, with the attendant destruction of that pristine tundra and its ice wedges.

As the ice wedges are exposed by erosion, they melt rapidly and open up new erosional branches in a dendritic pattern, allowing further loss of peat and sediments by thermokarst. The loss of peat and sediments to the sea from this deep (2–3 m) and wide erosional gully has been considerable (about 1175 square meters, 2940 cubic meters, 1050 metric tons of soil and ground ice) within the eleven years from 1977 to 1988. All the tundra plant communities above the knickpoint of 1977 are gone, along with their soils, permafrost, and ice wedges, too. As Footprint Creek continues to erode upward into the Voth Drainage, the remainder of that tundra will be destroyed in a similar manner.

## VI. SUMMARY

Atmospheric carbon dioxide cannot be considered alone in attempting to predict the future of the tundra ecosystem, or of any ecosystem. The indirect effects of $CO_2$ and carbon balance are likely to be more important in regard to ecosystem responses than direct effects. If indirect effects in the wet coastal tundra include higher summer temperatures, longer snow-free seasons, and even slightly depressed water tables, the ecological effects could destroy the system's integrity, through decomposition of peat and the loss of the upper permafrost. The upper permafrost has bound this ecosystem together for at least the last 10,000 years. If the upper 2–3 m of permafrost are lost, the wet coastal tundra is lost. To some extent, this could be true of the tussock tundra, too.

## REFERENCES

Billings, W. D. 1987a. Constraints to plant growth, reproduction, and establishment in arctic environments. *Arctic and Alpine Res.* 19:357.

Billings, W. D. 1987b. Carbon balance of Alaskan tundra and taiga ecosystems: Past, present, and future. *Quat. Sci. Rev.* 6:165.

Billings, W. D., and K. M. Peterson. 1980. Vegetational change and ice-wedge polygons through the thaw lake cycle in arctic Alaska. *Arctic and Alpine Res.* 12:413.

Billings, W. D., P. J. Godfrey, B. F. Chabot, and D. P. Bourque. 1971. Metabolic acclimation to temperature in arctic and alpine ecotypes of *Oxyria digyna. Arctic and Alpine Res.* 3:277.

Billings, W. D., G. R. Shaver, and A. W. Trent. 1976. Measurement of root growth in simulated, and natural temperature gradients over permafrost. *Arctic and Alpine Res.* 8:247.

Billings, W. D., K. M. Peterson, and G. R. Shaver. 1978. Growth turnover and respiration rates of roots and tillers in tundra graminoids. In *Vegetation and Production Ecology of an Alaskan Arctic Tundra,* L. L. Tieszen, ed., pp. 415–434. New York: Springer-Verlag.

Billings, W. D., J. O. Luken, D. A. Mortensen, and K. M. Peterson. 1982. Arctic tundra: A source or sink for atmospheric carbon dioxide in a changing environment? *Oecologia* 53:7.

Billings, W. D., J. O. Luken, D. A. Mortensen, and K. M. Peterson. 1983. Increasing atmospheric carbon dioxide: Possible effects on arctic tundra. *Oecologia* 58:286.

Billings, W. D., K. M. Peterson, J. O. Luken, and D. A. Mortensen. 1984. Interaction of increasing atmospheric carbon dioxide and soil nitrogen on the carbon balance of tundra microcosms. *Oecologia* 65:26.

Britton, M. E. 1957. Vegetation of the arctic tundra. In *Arctic Biology,* H. P. Hansen, ed. Corvallis: Oregon State University Press.

Brown, J., P. C. Miller, L. L. Tieszen, and F. L. Bunnell, eds. 1980. *An Arctic Ecosystem: The Coastal Tundra at Barrow, Alaska.* Stroudsburg, Pa.: Dowden, Hutchinson, and Ross.

Chapin, F. S. III, P. C. Miller, W. D. Billings, and P. I. Coyne. 1980. Carbon and nutrient budgets and their control in coastal tundra. In *An Arctic Ecosystem: The Coastal Tundra at Barrow, Alaska,* J. Brown et al., eds., pp. 458–482. Stroudsburg, Pa: Dowden, Hutchinson, and Ross.

Dennis, J. G., and P. L. Johnson. 1970. Shoot and rhizome-root standing crops of tundra vegetation at Barrow, Alaska. *Arctic and Alpine Res.* 2:253.

Dingman, S. L., et al. 1980. Climate, snow cover, microclimate, and hydrology. In *An Arctic Ecosystem: The Coastal Tundra at Barrow, Alaska,* J. Brown et al., eds., pp. 30–65. Stroudsburg, Pa: Dowden, Hutchinson, and Ross.

Emanuel, W. R., H. H. Shugart, and M. P. Stevenson. 1985a. Climatic change and the broad-scale distribution of terrestrial ecosystem complexes. *Clim. Change* 7:29.

Emanuel, W. R., H. H. Shugart, and M. P. Stevenson. 1985b. Response to comment: Climatic change and the broad-scale distribution of terrestrial ecosystem complexes. *Clim. Change* 7:457.

Komarkova, V., and P. J. Webber. 1980. Two low arctic vegetation maps near Atkasook, Alaska. *Arctic and Alpine Res.* 12:447.

Lachenbruch, A. H. 1962. Mechanics of thermal contraction cracks and ice-wedge polygons in permafrost. *Geol. Soc. Am. Special Paper 70.*

Lachenbruch, A. H., and B. V. Marshall. 1986. Changing climate: Geothermal evidence from permafrost in the Alaskan Arctic. *Science* 234:689.

Leffingwell, E. D. 1919. The Canning River region, northern Alaska. *U.S. Geological Survey Professional Paper 109.*

Manabe, S., and R. J. Stouffer. 1979. A $CO_2$–climatic sensitivity study with a mathematical model of the global climate. *Nature* 282:491.

Miller, M. C., R. T. Prentki, and R. J. Barsdate. 1980. Physics. In *Limnology of Tundra Ponds,* J. E. Hobbie, ed., pp. 51–75. Stroudsburg, Pa: Dowden, Hutchinson, and Ross.

Peterson, K. M., and W. D. Billings. 1980. Tundra vegetational patterns and succession in relation to microtopography near Atkasook, Alaska. *Arctic and Alpine Res.* 12:473.

Peterson, K. M., W. D. Billings, and D. N. Reynolds. 1984. Influence of water table and atmospheric $CO_2$ concentration on the carbon balance of arctic tundra. *Arctic and Alpine Res.* 16:331.

Post, W. M., W. R. Emanuel, P. J. Zinke, and A. G. Stangenberger. 1982. Soil carbon pools and world life zones. *Nature* 298:156.

Strain, B. R., F. A. Bazzaz, et al. 1983. Terrestrial plant communities. In *$CO_2$ and Plants: The Response of Plants to Rising Levels of Atmospheric Carbon Dioxide,* E. R. Lemon, ed., AAAS Selected Symposium 84. Boulder, Colo.: Westview Press.

Tieszen, L. L., ed. 1978a. *Vegetation and Production Ecology of an Alaskan Arctic Tundra.* New York: Springer-Verlag.

Tieszen, L. L. 1978b. Photosynthesis in the principal Barrow, Alaska, species: A summary of field and laboratory responses. In *Vegetation and Production Ecology of an Alaskan Arctic Tundra,* L. L. Tieszen, ed., pp. 241–268. New York: Springer-Verlag.

Tieszen, L. L., P. C. Miller, and W. C. Oechel. 1980. Photosynthesis. In *An Arctic Ecosystem: The Coastal Tundra at Barrow, Alaska,* J. Brown et al., eds., pp. 102–139. Stroudsburg, Pa: Dowden, Hutchinson, and Ross.

Tissue, D. T., and W. C. Oechel. 1987. Response of *Eriophorum vaginatum* to elevated $CO_2$ and temperature in the Alaskan tussock tundra. *Ecology* 68:401.

This research was funded by National Science Foundation grants DEB-8103490 to Duke University and DPP-8520730 to Clemson University, for which we express our appreciation.

CHAPTER NINETEEN

# Effects of Global Climatic Change on Forests in Northwestern North America

JERRY F. FRANKLIN,
FREDERICK J. SWANSON,
MARK E. HARMON, DAVID A. PERRY,
THOMAS A. SPIES,
VIRGINIA H. DALE, ARTHUR MCKEE,
WILLIAM K. FERRELL,
JOSEPH E. MEANS,
STANLEY V. GREGORY,
JOHN D. LATTIN,
TIMOTHY D. SCHOWALTER,
AND DAVID LARSEN

*"This is how it happens," the Dinosaur said.
"Drought, fire, hurricanes and floods. Throw in a
little radon, and the next thing you know, you're
extinct."*
—Steve Palay in The Oregonian, Sept. 25, 1988

## I. INTRODUCTION

Dense coniferous forests characterize the Pacific Coast of northwestern North America. Such species as Douglas fir (*Pseudotsuga menziesii*), western hemlock (*Tsuga heterophylla*), western red cedar (*Thuja plicata*), and Pacific silver fir (*Abies amabilis*) are dominant trees. Both the species and the forest stands are noted for their potential longevity and productivity (Franklin and Dyrness 1973, Franklin 1988). The natural forests in this region have, on average, the largest organic matter accumulations per unit area of any major plant formation in the world (Franklin and Waring 1981).

It has been suggested that the humid temperate climatic regime of the Pacific Northwest buffers the forests against global climatic change (e.g., Woodman 1987). In fact, northwestern forests are probably quite vulnerable to major climatic shifts. This is because the most important environmental variable affecting the composition and function of these forests is the effective moisture regime (a product of temperature and precipitation) during the relatively dry summers (e.g., Waring and Franklin 1979, Zobel et al. 1976). Differences in moisture regime produce dramatic local and regional gradients in the species composition and ecosystem functioning (e.g., rates of productivity and nutrient cycling) of these forests. Locally this is evidenced in sharp differences between slopes having different aspects. Regional gradients, from coastal Sitka spruce (*Picea sitchensis*) to interior ponderosa pine (*Pinus ponderosa*) forests, are also related primarily to moisture regime (e.g., Grier and Running 1977, Gholz 1982). These climatic considerations suggest great potential sensitivity to global climate change in northwestern forests. The most important catastrophic forest disturbances on the Pacific Coast—wildfire and windstorm—are also climatically driven, indicating significant additional mechanisms for indirect influences of global climate change (e.g., Hemstrom and Franklin 1982, Agee and Flewelling 1983, Henderson and Peter 1981).

This chapter explores the potential effects of some proposed changes in global climate on the coniferous forests of the Pacific Northwest. We have accepted the scenarios for global climatic change provided by several existing global circulation models (Hansen et al. 1988) as the basis for our exercise; the uncertainties associated with these models are discussed in chapter 4. We consider vegetational shifts along local and regional environmental gradients, output from forest successional simulators, paleobotanical data on change during the Holocene, and effects of altered disturbance regimes. Benefits of increased atmospheric $CO_2$ receive limited consideration in our analysis but are discussed in chapter 8. Our objective is a collective judgment about probable effects of the global climatic change scenarios on forest ecosystems in northwestern North America and their implications for management of both commodity and preserved lands.

We propose that altered disturbance regimes, including intensities and frequencies of wildfires, storms, and outbreaks of pests and pathogens, will produce much more rapid changes in forest conditions than the direct effects of increased temperature and moisture stress (fig. 19.1). Hence, the potential indirect effects of climate change on biotic change should receive more attention from physical and biological scientists. Forest management can either exacerbate or reduce the effects of climate change on the productivity and biological diversity of northwestern forestscapes.

## II. PERSPECTIVES ON THE PROJECTED CLIMATIC CHANGES

Forest environments in the Pacific Northwest will become significantly warmer and drier under the three global climate change scenarios provided (Hansen et al. 1988). These

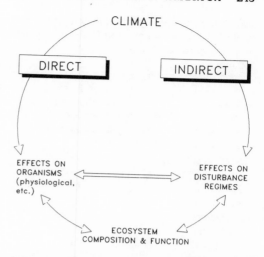

Figure 19.1. Climatic changes have both direct and indirect effects on ecosystem composition and structure.

models are consistent in projecting an increase in northwestern North America of 2°–5°C in mean temperature (both winter and summer) and little (−1 mm/day summer) or no change in precipitation. There are no seasonal shifts, so the pattern of relatively dry summers and mild, wet winters will persist.

Available moisture for plant growth will, of course, decline as the temperature increases, with no corresponding change in precipitation (table 19.1). Currently, northwestern forests grow under a wide range of temperature and moisture regimes; the range in mean annual temperatures in this forest region is about 5°C, and precipitation varies by an order of magnitude (approximately 3000 mm). The overall effect of warming without increased precipitation in this region will be increased potential evapotranspiration. A 5°C increase at Hoh River, Washington, for example, increases potential evapotranspiration from 537 mm/year to 881 mm/year. That temperature increase would also increase the number of months that potential evapotranspiration exceeds precipitation from 1 to 3 months at Hoh River, Washington, and from 6 to 8 months at Medford. Changes in actual evapotranspiration would depend on soil moisture storage; however, assuming that

*Table* 19.1. Hypothetical changes in evapotranspiration for three weather stations within forest zones of the Pacific Northwest, assuming an increase in temperature and no change in precipitation. PET, potential evapotranspiration (Mather 1974). AET, actual evapotranspiration (Mather 1974) with soil storage of 150 mm.

| Station | Hoh River, Washington | H. J. Andrews, Oregon | Medford, Oregon |
|---|---|---|---|
| Vegetation Type | Spruce-hemlock | Hemlock | Mixed Conifer–Oak Savanna |
| Temperature (°C) | 8.9 | 8.5 | 12.1 |
| Precipitation (mm) | 3492 | 2302 | 481 |
| Current | | | |
| PET (mm) | 537 | 547 | 822 |
| AET (mm) | 524 | 530 | 354 |
| 2.5°C increase | | | |
| PET (mm) | 708 | 720 | 997 |
| AET (mm) | 675 | 687 | 408 |
| 5.0°C increase | | | |
| PET (mm) | 881 | 895 | 1174 |
| AET (mm) | 818 | 840 | 448 |

such storage is constant between sites, we can conclude that changes in actual evapotranspiration are most likely to be greatest at wetter sites, such as Hoh River and H. J. Andrews (table 19.1).

Obviously, the climate change scenarios represent a massive climatic shift. For example, they can be viewed as equivalent to moving current climatic conditions at the base of a mountain upward between 500 m and over 1000 m in elevation; this is based on calculated lapse rates of about 4.8°C per 1000 m in southwestern Oregon and 4.0°C at Mount Rainier, Washington (Greene and Klopsch 1985). The climatic change is also equivalent to shifting current climates 200–500 km north of their current locations, that is, moving the climate of northern California into northern Oregon.

A simple assessment of an environmental shift of this magnitude can be made by relating it to forest community gradients or classifications for the Pacific Northwest (Franklin 1988), even though forest communities will probably not shift as intact multispecies units. Major plant series or vegetation zones differ by 1.5°–2.0°C in mean annual temperature at Mount Rainier, Washington (Franklin et al. 1988), and by about 2.5°C in southwestern Oregon (Atzet and Wheeler 1984); hence, a 4°C temperature increase could shift a given forest site by as much as two vegetation zones. Sites currently occupied by communities typical of the mountain hemlock (*Tsuga mertensiana*) zone could be replaced by communities characteristic of the western hemlock zone (fig. 19.2). In southwestern Oregon some sites currently occupied by vegetation assignable to the white fir (*Abies concolor*) series could be replaced with vegetation representative of the ponderosa pine (*Pinus ponderosa*) series while other forested sites would shift to nonforested (e.g., chaparral) ecosystems.

Elevational shifts of this magnitude will, of course, produce massive changes in the proportion of regional landscape occupied by different vegetation types (fig. 19.2, table 19.2). For example, on the western slopes of the central Oregon Cascade Range the proportion of the landscape characterized by dry coniferous forest (area of Douglas fir series) would increase from 8% to 39% or 27% depending on the temperature increase; the area occupied by the productive and commercially valuable western hemlock series would, on the other hand, decline from 56% to 38% or 24% of the landscape. Loss of forested area to juniper savanna and sagebrush steppe would predictably be massive on the eastern slopes of the central Oregon Cascade Range—a decline from 58% of the current landscape to 12% under the 5°C temperature increase.

Paleobotanical evidence suggests that vegetative shifts will be more complex and individualistic than simple shifts of intact vegetation communities and mosaics along environmental gradients. The simple projections provided here have assumed that intact forest communities (i.e., all species) will shift location as a unit as the environment changes. Recent studies strongly suggest that species shift independently, producing new combinations of plant species (Davis 1986, Delcourt and Delcourt 1987). For example, in the Pacific Northwest, dominance by Douglas fir is a phenomenon of the present interglacial, but hemlocks and true firs (*Abies*) have been around for much longer periods (Tsukada 1982). Analyses based on physiological considerations also suggest varying responses by tree species to climatic change (e.g., Leverenz and Lev 1987).

Direct paleobotanical evidence of vegetative responses during the Holocene in the Pacific Northwest cannot be used to make precise projections of future vegetative change (e.g., Brubaker 1986, 1988). That is because every warming event causes complex and unique changes and because, in the case of the Pacific Northwest, paleobotanical studies are relatively few and confined mainly to lowland areas, especially the Puget Lowland and valleys of the Okanogan Highlands (Barnosky et al. 1987); almost no studies exist for major mountain regions. It is clear, however, that the projected changes will produce an environment hotter than any previously experienced during this interglacial. Mean regional temperature during the Holocene has varied by only about 5°C, and current temperatures were exceeded by a maximum of only 2°C during the early Holo-

*Figure* 19.2. Percent of area in major vegetation zones on the western and eastern slopes of the central Oregon Cascade Range (latitude 44°30′ north) under current climate and with temperature increases of 2.5° and 5°C. Major shifts are predicted in elevational boundaries and the total area occupied by vegetation zones under global climatic change.

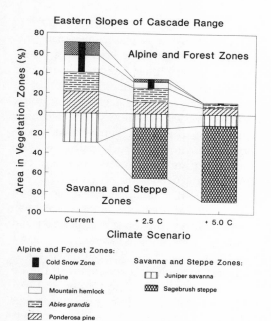

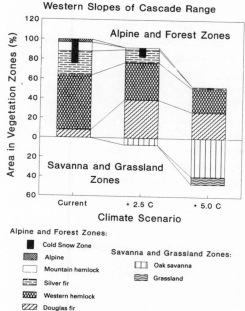

Table 19.2. Percent of area in various vegetation zones in the central Oregon Cascade Range (latitude 44°30′ north) under current climate and with increases of 2.5° and 5°C.

| Zone | Climate | | |
|---|---|---|---|
| | Current | +2.5°C | +5.0°C |
| **Western slopes** | | | |
| Nonforested | 0 | 0 | 8 |
| Oak savanna | 0 | 8 | 39 |
| Douglas fir | 8 | 39 | 27 |
| Western hemlock | 56 | 38 | 24 |
| Silver fir | 24 | 13 | 2 |
| Mountain hemlock | 9 | 2 | 0 |
| Alpine | 3 | 0 | 0 |
| Total forested | 97 | 92 | 53 |
| Cold snow zone[1] | 24 | 9 | 1 |
| | | | |
| **Eastern slopes** | | | |
| Sagebrush steppe | 0 | 51 | 77 |
| Juniper | 29 | 14 | 11 |
| Ponderosa pine | 22 | 12 | 7 |
| Grand fir | 19 | 14 | 4 |
| Mountain hemlock | 17 | 6 | 1 |
| Alpine | 13 | 3 | 0 |
| Total forested | 58 | 32 | 12 |
| Cold snow zone[2] | 30 | 9 | 1 |

[1]Includes half of silver fir zone and all of the mountain hemlock and alpine zones.

[2]Includes mountain hemlock and alpine zones.

cene, the period of maximum temperature during this interglacial (Brubaker 1988).

This section has focused primarily on community composition and structure. Shifts in the functional properties of the ecosystems, such as net primary productivity and decomposition and nutrient cycling, will be associated with the compositional changes. Functional shifts are important to consider, especially since they provide a feedback that can reduce or increase the effects of climate and changes in greenhouse-gas concentrations.

### III. EFFECTS ON ESTABLISHED FORESTS

We began our analysis by considering the effects of global change on established forests. An existing forest is expected to be more tolerant of environmental changes than a forest that is reestablishing itself after a disturbance. The established forest is able to buffer the effects of altered conditions, such as increased temperature or decreased moisture, so they are felt only slowly. Tree seedlings attempting to establish themselves in a nonforested environment, such as after a wildfire or logging, are much more sensitive to current climatic conditions.

A model that simulates forest succession (CLIMACS) was used to examine effects of altered climates on forest composition and structure. This model is based on the JABOWA and FORET paradigm (Shugart 1984) but is adapted to ecological conditions in the Pacific Northwest (Dale and Hemstrom 1984). Fundamentally, it is a tree population model that tracks birth, growth, and death of individual trees based on environmental conditions. Birth and death are both stochastic functions. CLIMACS has been used successfully to examine effects of disturbances on forest conditions on the Olympic Peninsula (Dale et al. 1986). Simulations were run on several different forest environments using current and projected climatic scenarios.

Several lines of evidence, including results from these model exercises, lead us to propose that responses of established forests to climate change should be muted and have substantial time lags, assuming that the forests are not destroyed or significantly altered by human activities or natural catastrophe. The slow, muted response is due to the ability of established trees to accommodate significant change and to their long life spans. As noted by Brubaker (1986), "[Mature] trees can survive long periods of marginal climate." Furthermore, established forests ameliorate onsite climatic conditions, which also shields them against the effects of the external environment. Hence, barring destruction of the existing stands, changes should be gradual. Model projections confirm this scenario; changes in forest composition diverge only gradually from those projected under current climate for typical sites. Earlier modeling efforts showed similar results (Dale and Franklin 1989).

The most environmentally sensitive stage for western tree species is at the time of seedling establishment, reflected in the muted responses of established forests. As noted by Brubaker (1986), "The long lifespans of trees can slow the retreat of range boundaries, if adults remain in the vegetation several centuries after climatic deterioration makes local conditions unsuitable for seedling establishment." High rates of mortality are typical of germinants and small seedlings, mostly as a result of adverse environmental conditions, such as heat, drought, and frost. Indeed, mortality at the seedling stage may effectively preclude any tree establishment where environmental conditions are severe.

Despite the inertia, gradual compositional shifts can be expected even in established

forests. These will tend to occur along environmental gradients—upward in elevation, northward in latitude, and from southerly to northerly aspects within a locale. Shifts in elevation and aspect are likely to be more rapid and to occur with greater predictability than latitudinal shifts; this is because of the proximity of vegetation zones along topographic transects, whereas valley and mountain barriers may block or slow latitudinal shifts. As others have suggested, we expect that species will typically shift independently, since this is what appears to have happened in the past and it reflects their differing physiological tolerances and susceptibilities to pests and pathogens. As suggested by our model results, typical shifts might be the replacement of Pacific silver fir by western hemlock as the major climax species at middle elevations on the western slopes of the Cascade Range and Olympic Mountains and the loss of western hemlock at lower elevations from decreased moisture availability.

Shifts will not always be simple displacements of one or a few species, however. Surprises (exceptions to our logical predictions) will probably be the rule, as species and environments interact unexpectedly. To a large extent this apparent unpredictability is due to our inability to understand or model all important environmental variables. For example, Woodman's (1987) prediction that Douglas fir will grow to higher elevations under altered climate does not take account of the Douglas fir's aversion to wind-exposed environments, even when moisture and temperature conditions are well within its tolerances. Hence, Douglas fir may not always provide appropriate replacements for the true firs (*Abies* spp.) lost from exposed upper-slope and ridgetop environments.

Competitive interactions between species will also alter predicted responses of individual species based on physiological attributes. Freshwater fish communities provide an example related to species' differing thermal tolerances. In the Pacific Northwest, redside shiners (*Richardsonius balteatus*) compete more effectively with juvenile steelhead trout

(*Salmo gairdneri*) as water temperatures increase (Reeves 1987); trout dominate in waters less than 15°C, but shiners dominate at temperatures greater than 19°C, even though temperatures are well within the physiological limits of both species. Competitive interactions of this type could lead to species extinctions long before physiological tolerances are approached.

There will doubtless be community-level as well as individual species responses. One striking change could be the elimination of much of the cold snow zone, the mountain hemlock zone, in the Cascade and northern Sierra Nevada ranges and Olympic Mountains. A deep winter snowpack and short snow-free season characterize this zone, which is important hydrologically. Unfortunately, because mountain tops are smaller than bases, extensive land surfaces are not available to replace the acreage lost by the mountain hemlock zone as temperatures force it up in elevation (e.g., table 19.2). Even where new land is available, substantial lags in forest establishment can be expected at upper timberline; hence, the vegetative mosaic of the subalpine parkland, the upper subzone of the mountain hemlock zone, will not quickly replace itself as the warmer climate moves upslope.

Another community-level change will involve the loss of forest to nonforested communities at lower timberline (i.e., at the hot, dry lower elevation sites). Many sites in northern California and southwestern Oregon (and a scattering of locales farther north) are already near the environmental limits of forest growth. Hotter and drier conditions, coupled with disruption of the forest through fire or other human activities, could shift vegetation on these sites from forest to non-forested types, such as chaparral, or to woodlands dominated by hardwoods, such as oaks (*Quercus* spp.). Sites currently characterized as belonging to the ponderosa pine and Douglas fir series would be candidates for conversion to nonforested conditions.

Some aspects of ecosystem function may respond more rapidly to global change than composition and structure. Functional changes, such as in productivity, will occur both as the climate warms and dries and as communities and species migrate along environmental gradients. Moreover, functional responses will be linked and multidirectional. For example, aboveground forest production at the dry end of the environmental gradient is likely to decrease with increased temperature. On the other hand, evidence from controlled environment studies indicates that higher levels of $CO_2$ may result in improved water-use efficiency; hence, tree species that have good stomatal control may maintain or even increase forest productivity by decreasing stomatal aperture while still receiving sufficient influx of $CO_2$. Increased evapotranspiration and summer drought probably will increase allocation of photosynthate to below-ground production. In contrast, decomposition in soils is likely to increase with warming because respiration rates will increase and more carbon will be available to decomposers from greater below-ground production. Decomposition under global climate change should lead to reductions in total carbon storage in soils, assuming that the effects of increased temperature are greater than those of decreased soil moisture.

Generally, we would expect the responses of established (intact) forests to be greater in the following cases: in dry rather than moist habitats, at more southerly than northerly latitudes, and in interior than in coastal environments. Changes may be greatest in southwestern Oregon and northern California, where many species are at their physiological limits already, and at high elevations in the area currently characterized by a permanent winter snowpack. We can expect that because of disturbance regimes, the lower elevation (hot and dry) timberline will retreat much more rapidly than upper timberline will advance by colonization of presently unforested slopes. It is certain that there will be many surprises as we are blindsided by unknown interactions between shifting environments and biota. Still, changes in intact, relatively undisturbed forests will be dwarfed by those occurring in forests stressed by altered disturbance regimes.

## IV. NEW CATASTROPHIC SCENARIOS: EFFECTS OF ALTERED DISTURBANCE REGIMES

We believe that the most rapid and extensive biotic changes in forests from climate change will be caused by altered disturbance regimes. Disturbances create the conditions for change in ecosystems, effectively doing the work of eliminating the established forest with its inertia, or tolerance of altered climatic conditions. As noted by Brubaker (1986), "[Disturbances] should also mitigate the lagging effects of long tree lifespans by accelerating rates of population decline when climatic change makes conditions unfavorable for seedling establishment."

Altered frequencies, intensities, and locales of catastrophic disturbances are probable under the proposed scenarios of global change. These would include wildfire, storms of all types (including extreme wind conditions and rain-on-snow events), and outbreaks of pests. In fact, the combination of disturbance and climate change will provide a double whammy for the forests of the Pacific Northwest. First, a disturbance destroys the existing forest, which has an ability to resist change. Second, the environment under global change provides conditions for forest reestablishment that may be much more severe than those that existed previously (i.e., drier and hotter environments). We again note that tree regeneration is the stage of forest succession most sensitive to moisture and temperature conditions.

### A. Wildfire

Increased frequency of fire is certain under the climate change scenario, and greater intensities are probable, at least during a transitional period. A latitudinal gradient in fire frequency exists in the Pacific Northwest because of associated climatic changes. For example, fire return intervals of 114 to 166 years

have been described for the central Oregon Cascade Range (Morrison and Swanson 1990) while intervals of 425 years are reported in the central Washington Cascade Range (Hemstrom and Franklin 1982). Hence, fire frequencies more characteristic of northern California might be expected throughout western Oregon, and frequencies characteristic of west central Oregon could migrate to western Washington. Such shifts could create severe fire threats to resources and problems in fire control as forests with large fuel loadings (high biomass) are subjected to more frequent and, for an interim period, more intense burns.

Frequency and subsequent recovery from fires will probably be strongly influenced by synergisms between human activities and the changed environments. Human uses of the forestscapes, especially timber cutting, increase ignition probabilities, as well as the need for fire control. The effect of intensive management of forest lands is at least as important; young managed forests are apparently more susceptible to catastrophic fire than old-growth forests. For example, during the 1987 wildfires in southwestern Oregon, plantations were affected much more drastically than older forests (Perry 1988). In the Galice and Longwood fires most plantations between the ages of 5 and 25 years (generally the oldest plantings present) were destroyed, while many old-growth stands survived (Perry 1988). A mixed young stand of conifers and hardwoods, which is characteristic of natural stands, may be less vulnerable to fire than pure young conifer stands.

Intensive utilization also reduces or eliminates the biological legacies (such as green trees and coarse woody debris) that contribute significantly to the speed and completeness of forest recovery on burned-over sites. For example, some shrubby angiosperms form mycorrhizae with the same fungal species as many conifers; eliminating these shrubs by forestry practices may reduce the recovery potential of forested sites following catastrophic fire. This occurs in southwestern Oregon where ericaceous shrubs and conifers host some common mycorrhizal fungi; conifer seedlings are disproportionately associated with the shrubs, suggesting that the rich concentration of mycorrhizal inocula enhances their survival (Amaranthus and Perry 1987, Perry et al. 1987).

Hence, higher fire frequencies coupled with the simplifying effects of intensive management could contribute significantly to shifts from forest to nonforest conditions under a hotter, drier climatic regime. Increased fire frequency could also shift some sites from conifer to hardwood dominance, since many hardwood species reproduce readily following fire by sprouting (e.g., tanoak, *Lithocarpus densiflorus*, and Pacific madrone, *Arbutus menziesii*) or seed (e.g., red alder, *Alnus rubra*).

## B. Storm Events

Storm events are important disturbances that have catastrophic effects on both terrestrial and aquatic ecosystems. Shifts in the frequency, intensity, and location of storm events seem likely under global climate change. For example, increased contrasts between oceanic and continental temperatures should produce more intense storms.

Storms with high winds may cause extensive tree mortality by uprooting and breaking trees. Entire stands can be eliminated over hundreds or even thousands of acres, as exemplified by the 1921 blowdown on the western Olympic Peninsula (Buchanan and Englerth 1940), the 1952 blowdowns in the Oregon Coast Ranges (Ruth and Yoder 1953), and the Columbus Day windstorm in 1962 (Lynott and Cramer 1966). Henderson and Peter (1981) identify windstorm, not wildfire, as the primary forest catastrophe on the western Olympic Peninsula, which is also similar to coastal Alaska (Ruth and Harris 1979).

High flows and debris avalanches associated with winter storms are major forces that disrupt and reset stream and river ecosystems and their associated riparian zones (Gregory et al. 1991) to early successional states. Major flood events in the Pacific Northwest are invariably associated with rain-on-snow storms (Harr 1981). Extensive

mountain regions from central California to coastal British Columbia fall into an elevational band known as the "warm snow zone." Significant snow accumulates in these areas during cold periods, but the relatively warm snowpacks melt rapidly when warm wet air masses move in from the Pacific Ocean, producing large snowmelts concurrently with heavy rain. High stream runoffs result in extensive flooding, as in December 1964.

Human activities can accentuate the damage of such storm events. For example, the network of clearcuts and roads created by logging activities contributed significantly to the blowdown suffered in a 1983 windstorm in the Bull Run River drainage of the northern Oregon Cascade Range; approximately 80% of half a billion board feet of blowdown timber was associated with boundaries of clearcuts and road rights-of-way (Franklin and Forman 1987). Similarly, large areas of clearcutting can magnify the effects of rain-on-snow flood events (Harr 1981); typically snow accumulations are larger and more susceptible to warming and melting in recently cutover areas than on forested sites.

Hence, alterations in the frequency, intensity, and location of major storm events as a result of global warming could have profound consequences for forests and associated rivers.

## C. Pests and Pathogens

New and intensified problems with insects and diseases are probable under global change. In some cases these will result from more favorable environments for the establishment and spread of a particular pest. In other cases problems will result from the indirect effects of warming as reflected in increased tree stress.

The possible expansion of the balsam woolly aphid (*Adelges piceae*) into stands of subalpine fir (*Abies lasiocarpa*) is a good example of how altered climates could produce a major insect epidemic. The aphid is an introduced pest that has been a serious problem on Pacific silver fir and low-elevation occurrences of subalpine fir in the Pacific Northwest (Mitchell 1966, Franklin and Mitchell 1967). It has been effectively restricted to low and middle elevations by its temperature requirements; the second generation of the aphid must reach the first instar stage to survive the winter. The subalpine zone of the coastal mountains rarely provides the necessary heat, so too few of the aphid attain the critical stage during most years to produce dense populations (Mitchell 1966).

A 2.5°C increase in mean temperature (the low end of the climate change scenario) would allow the aphid to reproduce and spread at the higher elevations where subalpine fir is a major stand component. Mature subalpine fir have low resistance to the aphid (Mitchell 1966). Consequently, high levels of subalpine fir mortality are probable, perhaps comparable to those experienced in the Fraser fir (*Abies fraseri*) populations in the southern Appalachian Mountains—nearly 100% of adult individuals (Dale et al. 1991).

Increased numbers and intensities of pest outbreaks can be expected as established forest stands are subjected to increasing physiological stresses associated with global warming (Mattson and Haack 1987). Even under conditions of stable climate the majority of pest outbreaks are associated with increased host stress. Furthermore, pests and pathogens are often highly vagile; hence, they can shift rapidly in response to altered environmental conditions (Schowalter et al. 1986).

As with fire and storm, undesirable synergisms between altered disturbance regimes and human activities are probable with insects and disease. Forestry practices have predisposed many forests to outbreaks. One example is the recent outbreak of southern pine beetle (*Dendroctonus frontalis*) in the extensive, pure, even-aged stands of southern pines that have been created in the southeastern United States (Schowalter et al. 1981). The simplified forest stands created by intensive management tend to be more vulnerable to outbreaks of pests than natural stands (Franklin et al. 1989, Schowalter 1988). In

comparison, J. D. Lattin and P. Oman (1983) have noted how established forests, especially old-growth forests, are buffered from rapid floral and faunal change.

## D. Summary on Disturbances

We conclude that altered disturbance regimes will interact with global warming to produce major change in the forestscapes of the Pacific Northwest long before climate change alone would produce significant change in established forests. Disturbances will create opportunities for change by reducing the inertia of established forests. Hence, disturbances can be viewed positively as events that speed adjustments of vegetation to current environmental conditions. On the other hand, these changes could be highly disruptive in the short- and mid-term. Global warming creates more severe conditions for forest reestablishment, which may cause drastic shifts in the composition and function of the postdisturbance forests. Some probable overall effects of these changes include a net shift in area from forest to nonforest vegetation, net loss of biotic diversity as some species fail to track suitable environments, and minor additions to atmospheric $CO_2$ as organic matter accumulations decrease on forest sites. Transitions will be a problem because forest destruction will almost certainly occur more rapidly than forest reestablishment, especially at the lower forestlines. In general, natural forest ecosystems, with their greater compositional and functional redundancy, are expected to show greater resistance to change and to recover more rapidly following disturbance than intensively managed forests.

Research is critically needed on some currently obscure aspects of climate change and ecological responses. Information on changes in factors affecting the frequency, intensity, and locale of major disturbances is essential. For example, what are predicted magnitudes and intensities of summer drought periods? What are the probabilities for repeated drought years? Are changes in the frequency and tracks of storms with high winds likely? Information on seasonal changes in climate are also important to assess both direct and indirect effects of climate change.

## V. IMPLICATIONS FOR FOREST MANAGERS AND APPLIED ECOLOGISTS

Environmental changes such as those proposed here under global warming have profound implications for managers and applied ecologists. Significant changes will be necessary in the way we perceive and manage those lands devoted to commodity production, as well as those devoted to preservation, if we are to have some reasonable hope of success in preserving biological diversity.

First, lands that are devoted to production of commodities—the managed timberlands, rangelands, and agricultural lands—dominate our landscape; yet, in planning for biological diversity we have often overlooked these lands and instead focused on reserves or primeval habitats. Typically, conservationists and land managers both view biological diversity as a question of setting aside resources. It is critical that we begin to take, as scientists and a society, a more critical and holistic view of how the commodity lands, our seminatural landscape matrix, are managed. These are the bulk of our lands, and as such, they form a vast ocean in which ecological preserves are located and also contain much, perhaps most, of the remaining biological diversity, at least in the temperate zones. Consequently, the commodity lands are the battleground where the war for biodiversity will ultimately be won or lost, especially with the drastic environmental shifts that global warming will bring. We need to alter management practices that produce ecologically depauperate landscapes and thereby reduce indigenous diversity and maximize the isolation of preserves.

Traditional forest management, for example, tends to reduce genetic, structural, temporal, and spatial diversity by emphasizing efficient timber production through simplification (Franklin et al. 1989). Traditional sil-

vicultural systems have focused almost exclusively on the trees—removal of wood products while providing for reestablishment of the next tree crop. Foresters have been innovative in developing practices to accommodate other ecological values, such as provision of dead trees for wildlife, but this has been a largely piecemeal adaptation to the emerging needs of ecological forestry. Foresters need to develop and implement a new forestry that has as its philosophical underpinning the maintenance of complex forest ecosystems and not simply the reestablishment of trees. Examples of specific silvicultural practices might include maintenance of structural diversity by providing for coarse woody debris (standing dead trees and downed boles) and development of stands of mixed composition and structure (Franklin et al. 1989). Such practices are already being applied on some federal, state, and private lands in the Pacific Northwest. At the landscape level foresters can adopt spatial patterns for cutting that reduce forest fragmentation, select patch sizes relevant to the needs of interior species, and incorporate natural patches and corridors (Franklin and Forman 1987). Approaches of this sort are being tried in several forested regions of the United States; for example, an approach called minimum fragmentation is being tested in western Oregon (Hemstrom 1989).

Our perspectives on the ecological reserves—national parks, wilderness, research natural areas, and nature reserves—also may need some readjustment. These areas remain critical to maintenance of biological diversity, but they will not provide adequately for biological diversity in the face of the climate scenarios that have been presented. In fact, even without climate change they probably could not do the job alone, since their representation of diversity is incomplete, their total acreage is small, and, most important, they are increasingly isolated within hostile landscapes. These reserves will provide society with rich reservoirs of ecological diversity to conserve and utilize in coping with global warming.

If environmental changes reach the magnitude proposed in the climate scenarios, the objectives and management approaches on ecological reserves must be drastically revised. Under global warming, using naturalness or natural processes as a guiding principle will be of limited value in maintaining biological diversity. To speak of mitigating such effects is ridiculous. As habitats and environments migrate from reserves, as catastrophes destroy relict ecosystems, as species disappear, we must become first-class ecological engineers or risk losing large components of the diversity that now resides in the reserved lands.

We conclude that any strategy for preserving biological diversity, particularly in the face of global warming, must involve a comprehensive approach to both commodity and reserved lands. Commodity lands, the seminatural matrix, must be managed with more consideration for ecological values, including those on the reserved lands that they surround. These landscapes must nurture biological diversity, first by retaining more elements of diversity within the seminatural matrix and second by easing the passage of other organisms. A less hostile matrix, one more accommodating to migrating organisms, would also drastically increase the opportunity for ecological dialogue—reciprocal movement of organisms, materials, and services—between reserved and intensively managed landscapes, to their mutual benefit. Reserved lands must be viewed as source areas for diversity but not necessarily as permanent residences for that diversity. Intensive management efforts involving both types of land will be necessary as society undertakes ecological triage, a damage-control exercise on biological diversity threatened by global climate change. As ecologists, managers, and a society, we will have to become knowledgeable, creative, innovative, overt, and holistic as we collaborate with nature to save biological diversity in the face of global warming.

## REFERENCES

Agee, J. K., and R. Flewelling. 1983. A fire cycle model based on climate for the Olympic Mountains, Washington. In *Seventh Conference on Fire and Forest Meteorology*. Boston, Mass.: American Meteorological Society.

Amaranthus, M. P., and D. A. Perry. 1987. Effect of soil transfer on ectomycorrhiza formation and the survival and growth of conifer seedlings on old, nonreforested clear-cuts. *Can. J. Forest Res.* 17:944.

Atzet, T., and D. L. Wheeler. 1984. *Preliminary Plant Associations of the Siskiyou Mountain Province*. Portland, Oreg.: USDA Forest Service Pacific Northwest Region.

Barnosky, C. W., P. M. Anderson, and P. J. Bartlein. 1987. The northwestern U.S. during deglaciation: Vegetational history and paleoclimatic implications. In *North America and Adjacent Oceans during the Last Deglaciation*, W. F. Ruddiman and H. E. Wright, Jr., eds., Geology of North America, vol. K-3, pp. 289–321. Boulder, Colo.: Geological Society of America.

Brubaker, L. B. 1986. Responses of tree populations to climatic change. *Vegetatio* 67:119.

Brubaker, L. B. 1988. Vegetation history and anticipating future vegetation change. In *Ecosystem Management for Parks and Wilderness*, J. Agee and D. Johnson, eds. Seattle: University of Washington Press.

Buchanan, T. S., and E. H. Englerth. 1940. Decay and other losses in wind-thrown timber in the Olympic Peninsula, Washington. USDA Technical Bulletin 733. Washington, D.C.: U.S. Department of Agriculture.

Dale, V. H., and J. F. Franklin. 1989. Potential effects of climate change on stand development in the Pacific Northwest. *Can. J. Forest Res.* 19:1581.

Dale, V. H., and M. Hemstrom. 1984. CLIMACS: A computer model of forest stand development for western Oregon and Washington. USDA Forest Service Research Paper PNW-327. Portland, Oreg.: Pacific Northwest Research Station.

Dale, V. H., M. Hemstrom, and J. Franklin. 1986. Modeling the long-term effects of disturbances on forest succession, Olympic Peninsula, Washington. *Can. J. Forest Res.* 16:56.

Dale, V. H., R. H. Gardner, D. L. DeAngelis, C. C. Eagar, and J. W. Webb. 1991. Elevation-mediated effects of the balsam woolly adelgid on the southern Appalachian spruce-fir forests. *Can. J. Forest Res.* 21:1639.

Davis, M. B. 1986. Climatic instability, time lags, and

community disequilibrium. In *Community Ecology*, J. Diamond and T. J. Case, eds., pp. 269–284. New York: Harper and Row.

Delcourt, P. A., and H. R. Delcourt. 1987. Long-Term Forest Dynamics of the Temperate Zone. New York: Springer-Verlag.

Franklin, J. F. 1988. Pacific Northwest forests. In *North American Terrestrial Vegetation*, M. Barbour and D. Billings, eds., pp. 103–130. New York: Cambridge University Press.

Franklin, J. F., and C. T. Dyrness. 1973. Natural vegetation of Oregon and Washington. USDA Forest Service General Technical Report PNW-8. Portland, Oreg.: Pacific Northwest Forest and Range Experiment Station.

Franklin, J. F., and R.T.T. Forman. 1987. Creating landscape patterns by forest cutting: Ecological consequences and principles. *Landscape Ecol.* 1:5.

Franklin, J. F., and R. G. Mitchell. 1967. Successional status of subalpine fir in the Cascade Range. USDA Forest Service Research Paper PNW-46. Portland, Oreg.: Pacific Northwest Forest and Range Experiment Station.

Franklin, J. F., and R. H. Waring. 1981. Distinctive features of the northwestern coniferous forest: Development, structure, and function. In *Forests: Fresh Perspectives from Ecosystem Analysis*, R. H. Waring, ed., pp. 59–86. Corvallis: Oregon State University Press.

Franklin, J. F., W. H. Moir, M. A. Hemstrom, S. E. Greene, and B. G. Smith. 1988. The forest communities of Mount Rainier National Park. USDI National Park Service Scientific Monograph Series 19.

Franklin, J. F., D. A. Perry, T. D. Schowalter, M. E. Harmon, A. McKee, and T. A. Spies. 1989. Importance of ecological diversity in maintaining long-term site productivity. In *Maintaining the Long-Term Productivity of Pacific Northwest Forest Ecosystems*, D. A. Perry et al., eds., pp. 82–97. Portland, Oreg.: Timber Press.

Gholz, H. L. 1982. Environmental limits on aboveground net primary production, leaf area, and biomass in vegetation zones of the Pacific Northwest. *Ecology* 63:469.

Greene, S. G., and M. Klopsch. 1985. Soil and air temperatures for different habitats in Mount Rainier National Park. USDA Forest Service Research Paper PNW-342. Portland, Oreg.: Pacific Northwest Research Station.

Gregory, S. V., F. J. Swanson, W. A. McKee, and K. W. Cummins. 1991. An ecosystem perspective of riparian zones. *Bioscience*: in press.

Grier, C. C., and S. Running. 1977. Leaf area of mature northwestern coniferous forests: Relation to site water balance. *Ecology* 58:893.

Hansen, J., I. Fung, A. Lacis, S. Lebedeff, D. Rind, R. Ruedy, G. Russell, and P. Stone. 1988. Prediction of near-term climate evolution: What can we tell decision-makers now? In *Preparing for Climate Change*, Proceedings of the first North American conference on preparing for climate change: A cooperative approach, J. C. Topping, Jr., ed., pp. 35–47. Washington, D.C.: Government Institutes.

Harr, R. D. 1981. Some characteristics and consequences of snowmelt during rainfall in western Oregon. *J. Hydrol.* 543:277.

Hemstrom, M. 1989. Alternative timber harvest patterns for landscape diversity. *COPE Report* 3(1): 8. (College of Forestry, Oregon State University, Corvallis.)

Hemstrom, M. A., and J. F. Franklin. 1982. Fire and other disturbances of the forests in Mount Rainier National Park. *J. Quatern. Res.* 18:32.

Henderson, J. A., and D. Peter. 1981. Preliminary plant associations and habitat types of the Quinault Ranger District, Olympic National Forest. Portland, Oreg.: USDA Forest Service Pacific Northwest Region.

Lattin, J. D., and P. Oman. 1983. Where are the exotic insect threats? In *Exotic Plant Pests and North American Agriculture*, C. Wilson and C. Graham, eds., pp. 93–137. New York: Academic Press.

Leverenz, J. W., and D. J. Lev. 1987. Effects of carbon dioxide–induced climate changes on the natural ranges of six major commercial tree species in the western United States. In *The Greenhouse Effect, Climate Change, and U.S. Forests*, W. E. Shands and J. S. Hoffman, eds. Washington, D.C.: Conservation Foundation.

Lynott, R. E., and O. P. Cramer. 1966. Detailed analysis of the 1962 Columbus Day windstorm in Oregon and Washington. *Monthly Weather Rev.* 94: 105.

Mather, J. R. 1974. *Climatology: Fundamentals and Applications*. New York: McGraw-Hill.

Mattson, W. J., and R. A. Haack. 1987. The role of drought in outbreaks of plant-eating insects. *Bioscience* 37:110.

Mitchell, R. G. 1966. Infestation characteristics of the balsam woolly aphid in the Pacific Northwest. USDA Forest Service Research Paper PNW-35. Portland, Oreg.: Pacific Northwest Research Station.

Morrison, P., and F. J. Swanson. 1990. Fire history and pattern in a Cascade Range landscape. USDA Forest Service Research Paper PNW-GTR-254-77.

Portland, Oreg.: Pacific Northwest Research Station.

Perry, D. A. 1988. Landscape pattern and forest pests. *Northwest Envir. J.* 4:213.

Perry, D. A., R. Molina, and M. P. Amaranthus. 1987. Mycorrhizae, mycorrhizospheres, and reforestation: Current knowledge and research needs. *Can. J. Forest Res.* 17:929.

Reeves, G. H., F. H. Everest, and J. D. Hall. 1987. Interactions between the redside shiner (*Richardsonius balteatus*) and the steelhead trout (*Salmo gairdneri*) in western Oregon: The influence of water temperature. *Canadian Journal of Fisheries and Aquatic Sciences* 44:1603.

Ruth, R. H., and A. S. Harris. 1979. Management of western hemlock–Sitka spruce forests for timber production. USDA Forest Service General Technical Report PNW-88. Portland, Oreg.: Pacific Northwest Research Station.

Ruth, R. H., and R. A. Yoder. 1953. Reducing wind damage in the forests of the Oregon Coast Range. USDA Forest Service, Pacific Northwest Forest and Range Experiment Station Research Paper 7. Portland, Oreg.: Pacific Northwest Research Station.

Schowalter, T. D. 1988. Forest pest management: A synopsis. *Northwest Envir. J.* 4:313.

Schowalter, T. D., R. N. Coulson, and D. A. Crossley. 1981. Role of southern pine beetle and fire in maintenance of structure and function of the southeastern coniferous forest. *Envir. Entomol.* 10:721.

Schowalter, T. D., W. W. Hargrove, and D. A. Crossley, Jr. 1986. Herbivory in forested ecosystems. *Ann. Rev. Entomol.* 31:177.

Shugart, H. H. 1984. *A Theory of Forest Dynamics.* New York: Springer-Verlag.

Tsukada, M. 1982. *Pseudotsuga menziesii* (Mirb.) Franco: Its pollen dispersal and late Quaternary history in the Pacific Northwest. *Jap. J. Ecol.* 32:159.

Waring, R. H., and J. F. Franklin. 1979. Evergreen coniferous forests of the Pacific Northwest. *Science* 204:1380.

Woodman, J. N. 1987. Potential impact of carbon dioxide–induced climate changes on management of Douglas-fir and western hemlock. In *The Greenhouse Effect, Climate Change, and U.S. Forests,* W. E. Shands and J. S. Hoffman, eds. Washington, D.C.: Conservation Foundation.

Zobel, D. B., A. McKee, G. M. Hawk, and C. T. Dyrness. 1976. Relationships of environment to composition, structure, and diversity of forest communities of the central western Cascades of Oregon. *Ecol. Monogr.* 46:135.

CHAPTER TWENTY

# Effects of Climate Change on Mediterranean-Type Ecosystems in California and Baja California

WALTER E. WESTMAN AND
GEORGE P. MALANSON

## I. INTRODUCTION

Certain changes in vegetation in the Mediterranean-climate region of California and Baja California would be expected to result from predicted regional changes in precipitation and temperature under the influence of doubled carbon dioxide levels. Such changes and their implications for conservation policy are examined here at three levels of organization: community types at the regional level, species composition at the stand level, and ecophysiological changes at the individual level.

A Mediterranean-type climate of cool, wet winters and warm, dry summers currently occurs along the Pacific coast of North America from the Canada–United States border to northern Baja California, Mexico (lat. 50°–30° N; UNESCO-FAO 1963). Within the California portion of this range, the Mediterranean-type climate extends from the coast to middle elevations (2000 m) of the Sierra Nevada (long. 121°–117° W; Donley et al. 1979); in Baja California, the climate type similarly extends from the west coast to the upper elevations of the central ranges (long. 116°–114° W). At the uppermost elevation of the Sierras, the climate is cold temperate, with average temperatures of the coldest month below freezing. On the continental side of these ranges, the climate grades rapidly into semiarid and arid types. The Mediterranean-climate region from the Rogue River basin in southern Oregon to the margin of the Vizcaino desert near El Rosario in northern Baja California, and east to the Sierra Nevada-Cascade axis, excluding deserts, constitutes the California floristic province (Jepson 1925, Howell 1957; fig. 20.1).

The California floristic province (324,000 km²) currently contains a relatively high number of endemic species (48% of the 4452 native species of vascular plants). The richness of the region is attributed to the survival

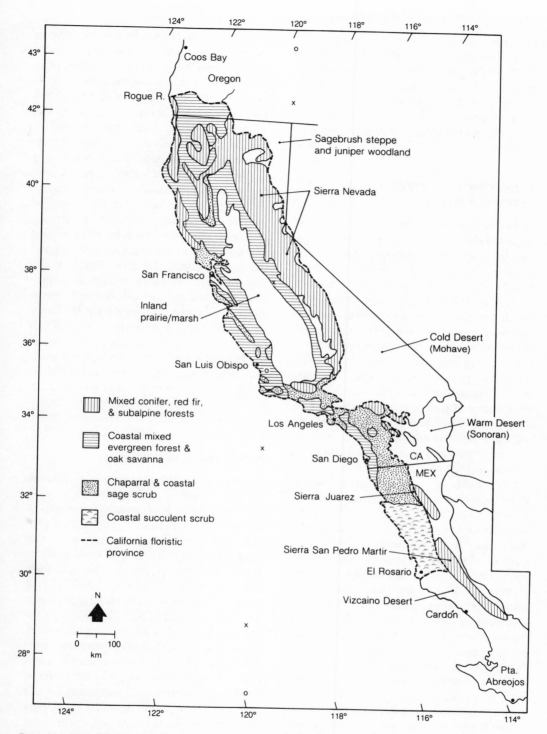

*Figure* 20.1. Simplified potential vegetation map of the California floristic province and adjacent arid lands. The centerpoint of the grid cells for global climate models are shown as × for GFDL and ○ for GISS along the 120° meridian. Sources: Hammond Ambassador World Atlas 1988, Kuchler 1977, Raven and Axelrod 1978, Westman 1983a,b.

of Tertiary relics in the equable coastal climate, shielded from continental extremes by the Coast Ranges and the Sierras; to the topographic and substrate diversity of the region; and to speciation promoted by recurrent climatic fluctuations (cool-moist, warm-dry) since the Middle Pliocene (Raven and Axelrod 1978).

## II. ANTICIPATED CHANGES AT THE REGIONAL LEVEL

At the regional level, a first approximation of expected changes in vegetation composition can be made by considering the climatic bounds within which current vegetation types occur, in relation to expected changes in climate. Although biotic interactions play significant roles in regulating relative abundances of species, the variability in composition of a vegetation type arises largely from the differing response of each component species to physical factors of the environment. Consequently, as climate changes, the abundances of each species will fluctuate in accord with their individual tolerances, resulting in biotic assemblages that differ both in composition and relative abundance from present conditions.

It is possible to postulate general changes in the location of vegetation types under new climatic conditions based on current vegetation-climate correlations, but important caveats apply. A certain inertia to species migration over large distances will inhibit colonization of climatically suitable sites (see chapter 22). This is especially true for the many tropical forest species with heavy seeds exhibiting no dormancy, for which dispersal distances are small and generation times long. Dispersal by colonization of outlier patches is also made more difficult in a landscape as topographically and geologically complex as California and northern Baja California; urban and agricultural land uses pose further barriers to migration for some species (see chapter 2). As new species assemblages arise, definitions of existing vegetation types based on their floristic composition and rela-

tive abundances are unlikely to fit many of the new or transitional types. To the extent that the assemblages are new, they are likely to be depauperate, lacking in certain pollinators, dispersal agents, or other critical-link species. As a result, species composition and relative abundance would be expected to fluctuate more widely with climatic variation; continued rapid climatic change would only aggravate this tendency. Further, any dependence of species survival on precipitation and temperature is only part of the story, as survival depends on a multitude of factors.

Indeed, the environmental changes that will influence future vegetation go beyond the climatic effects of carbon dioxide alone. The emission of other radiatively important trace gases and particulates could change climatic predictions significantly (Ramanathan et al. 1985). Carbon dioxide itself can change plant growth rates through increases in carbon fixation and water use efficiency (e.g., Mortensen 1987). Climatically induced changes in vegetation will alter wildfire characteristics in an interactive way. Climate change will also influence levels of air pollution, herbivory, and pathogenicity.

### A. Regional Climate Changes

One approach currently used to consider the possible effects of elevated carbon dioxide on future climate is to consider the output of models of global climate. Figure 20.2 presents predicted changes in climate under doubled atmospheric carbon dioxide levels derived from two general circulation models of global climate. Grid cells for the Goddard Institute for Space Sciences (GISS) model span 10° of longitude and 7.83° of latitude; cells for the Geophysical Fluid Dynamics Laboratory (GFDL) model span 7.5° of longitude and 4.44° of latitude. The data are presented as long-term monthly means for grid cells centered on the 120° meridian. Centerpoints for the grid cells are shown in figure 20.1. The data were supplied by the Data Support Section, National Center for Atmospheric Research, in Boulder, Colorado, for this chapter.

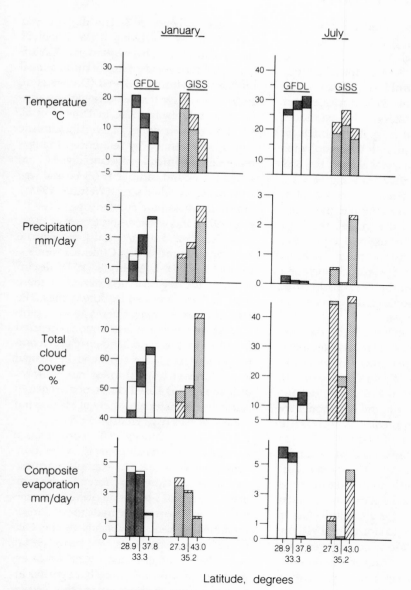

Figure 20.2. Levels of climate parameters predicted for two carbon dioxide levels by two global climate models (GFDL, GISS) in the grid cell centered on longitude 120° and the latitudes shown. $CO_2$, ppm: GFDL, 300 □ and 600 ▓; GISS, 315 ▤ and 630 ▨. The nearest Pacific coastal cities intersected by each latitude are: Cardón, Mexico (lat. 28.9°), San Diego, California (33.3°), San Francisco (37.8°), Punta Abreojos, Mexico (27.3°), San Luis Obispo, California (35.2°), and Coos Bay, Oregon (43.0°). Output from Data Support Section, National Center for Atmospheric Research.

The two global models differ in a number of respects apart from grid size and location (cf. GISS: Hansen et al. 1983; GFDL: Manabe and Wetherald 1987). For example, the models use different control levels of carbon dioxide (300 ppm, GFDL; 315 ppm GISS) and different solar constants; only the GISS model incorporates a diurnal cycle. Both models assume unrealistically smooth topography for the earth's surface. At a global level, the GISS model predicts a surface warming of 4.2°C with doubled carbon dioxide, versus 4.0°C for the GFDL model. In addition to the inherent simplifications in the model, the large size of grid cells makes regional application of the data difficult. Over northern California and southern Oregon, the grid cells include a substantial portion of the Great Basin; over southern California and Baja, the cells include substantial areas of ocean.

Despite the model limitations, the two models predict rather similar initial temperatures and temperature changes for January in the California region (4°–6.5°C rise). For July, both models predict a 2°–5°C rise, but the absolute temperatures in the GISS model are lower. Both models predict an increase in January precipitation of about 0.3–3.9 cm from southern California northward but differ in sign of change (+ or −) in northern Baja California. July rainfall is predicted to increase by 0.06–0.6 cm throughout the region under both models, except for a small decline in rainfall in the central California grid cell under the GISS model. Cloud cover is generally expected to increase (except in winter in Baja). Composite evaporation predictions vary between the two models, except for predicted summer increases in Baja and winter increases in northern California-Oregon. At present, climate modelers have more confidence in projections for temperature than for other parameters.

## B. Regional Vegetation Changes

The climatic ranges of major types of vegetation that would be present today in the absence of human disturbance (potential vegetation) in California and Baja California are presented in figure 20.3. The diagram was constructed by overlaying A. W. Kuchler's (1977) map of potential vegetation in California on California-wide maps of mean annual precipitation during 1931–60 (Donley et al. 1979) and average temperature of the warmest month (Leighly 1938, in Donley et al. 1979), to observe zones of overlap. Climatic ranges were refined by reference to ranges reported for particular vegetation types: cold desert (Vasek and Barbour 1977), coastal sage and coastal succulent scrub (Westman 1983a), mixed and subalpine conifer forest (Rundel et al. 1977), and warm desert (Axelrod 1979). Thorn scrub and dry tropical forest climate boundaries found in Baja California were estimated from D. I. Axelrod's (1979) discussions of the regional occurrence of these types and are indicated by dotted lines. The resulting climatic ranges per type were plotted. Where overlap in distributions occurred (such as in the woodland-chaparral complex), the types were separated by dashed lines. The wet tropical forest range shown in figure 20.3 is based on its occurrence in warm, wet climates elsewhere in Mexico and the neotropics (e.g., Hudson 1958).

Such a two-dimensional representation such inevitably oversimplifies the vegetation-climate relationships. For example, warmest-month temperature was chosen for the ordinate of figure 20.3 based on previous studies indicating its power as a predictor of floristic variation for xeric shrublands in the California floristic province (Westman 1981a). Nevertheless, other climatic variables or expressions of seasonality may have greater influence on the distribution of other vegetation types. P. J. Richerson and K.-L. Lum (1980), using a regression model, found that total precipitation had the strongest ability to predict floristic richness in California, accounting for 52% of variance in richness among sites. The standard deviation of mean monthly temperatures and of mean January temperatures account for an additional 5% and 4% of variance, respectively. Topographic heterogeneity (as measured by standard deviations of elevation in a region) ac-

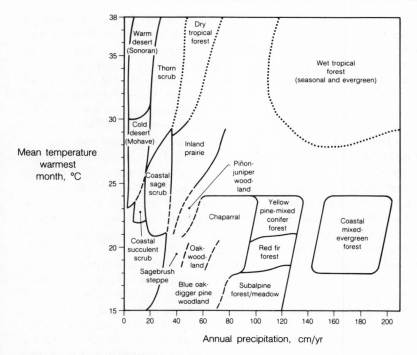

Figure 20.3. Current ranges of major potential vegetation types in California and Mexico in relation to annual precipitation and mean temperature of the warmest month. Dashed lines indicate overlap of vegetation types. Dotted lines indicate estimation from nonmap sources.

counted for an additional 19% of variance. Other authors have explored the utility of various climatic parameters in predicting vegetation occurrence in the western United States or North America (e.g., Major 1967, Sowell 1985, Stephenson 1988). Despite this complexity, we have chosen to use here relatively simple measures of precipitation and temperature to facilitate comparison with the more reliable output parameters from current global climate models.

An average mean summer temperature rise of 2°–5°C and an increase in January rainfall of as much as 4 cm, as currently predicted under carbon dioxide doubling for California, would imply a shift in ecotonal boundaries and centroids of distribution of the vegetation types in the state, but not a total displacement of any type from its current geographic range. A possible exception is subalpine forest-meadow, which would face significant invasion pressure from the upward elevational advance of red fir (Abies magnifica) forest and yellow pine (Pinus ponderosa, P. jeffreyi)–mixed conifer forests in the Sierra Nevada. Since subalpine forest-meadow vegetation already occupies the highest elevations and most temperate climate in the California floristic province, it would have little suitable habitat to retreat to under a warm-wet trend. More generally speaking, the predicted climatic changes would favor the expansion of vegetation types into the types found toward the lower left corner of figure 20.3. This would imply an expansion of chaparral at the expense of southern oak woodland (Quercus agrifolia) and blue oak–digger pine woodland (Quercus douglassii, Pinus sabiniana) and the expansion of inland prairie and sage, and eventually thorn scrub, elements into coastal sage and coastal succulent scrub vegetation in southern California and Baja California.

The climate of the California floristic province currently consists of the moderately cool-wet coastal and montane Mediterranean-climate and temperate zones and the warm-dry desert climate of the Great Basin, Mohave, and Sonoran deserts. Notably absent is a warm-wet zone, as evidenced by the absence of current California vegetation from the upper right quadrant of figure 20.3. Interpretation of the fossil record for the California floristic province suggests that alternating cool-wet and warm-dry periods have dominated the climatic history of the region. Thus northern Arcto-Tertiary elements provide more than half the genera and species of the province (Raven and Axelrod 1978). These species, typical of cool-wet climates, make up the bulk of the present conifer and mixed-evergreen forests of the province. About one fourth of the genera and one third of the species are from southern Madro-Tertiary elements. These species, typical of warmer, drier climates, consist largely of sclerophyllous taxa found in current southern oak woodland, chaparral, and sage scrub communities. The Madro-Tertiary elements were present as early as 50 million years ago but spread with an expanding dry climate (Raven and Axelrod 1978). By the Miocene (26 million years ago), much of interior southern California was inhabited by a broadleaf evergreen woodland (live oak, laurel, madrone) with a sclerophyllous shrub understory comprising many current chaparral genera. The flora also contained numerous thorn scrub taxa from higher summer-rain areas of the southwestern United States and Mexico—*Acacia, Bursera, Caesalpinia, Ficus*, and others (Axelrod 1956, 1968).

The latter is significant because subsequent climatic changes in the province have seen periods of either increased aridity (4 million–7 million years ago) or increased cold and reduced summer rain (2.5 million–4 million years ago) largely inhospitable to these warm-wet elements. The Pleistocene (starting 2.5 million years ago) saw an accentuation of that trend, as cold-wet glacial periods alternated with warm-dry ones, the last of which (Xerothermic period) led to the expansion of chaparral and desert elements at the expense of woodland (Raven and Axelrod 1978). Axelrod (1978) considers the current Mediterranean climate to have emerged in the mid-Quaternary (approximately 1 million years ago) and the origin of coastal sage scrub to date from the end of the last glacial period, 12,000 years ago. Since that time, a more equable winter-wet, summer-dry climate has prevailed in the province. If the climate becomes progressively warmer and wetter in both summer and winter in the next few centuries, the current desert and shrubland-grassland areas should see the incursion of subtropical thorn scrub and dry tropical forest elements from Mexico and the U.S. Southwest that have not been abundant in the region since the Miocene.

The climate range of inland prairie—a mixture of grassland, dry shrubland, and riparian woodland that occupied the Central Valley of California before agricultural development and can also be found in inland foothills—shown in figure 20.3 suggests that these floristic elements could expand into current areas of woodland and shrubland in coastal southern California as the climate initially becomes warmer and wetter. Such a tendency would be accentuated because many of the species are annuals or short-lived perennials with wind-dispersed seeds and short generation times. Since much of the Central Valley and dry coastal plains (e.g., the Riverside–San Bernardino region) are now home to exotic annuals from the Mediterranean region and elsewhere, it is likely that these exotic species will play a pioneering role in colonization of areas made bare by the death of shrubs less tolerant of the new climatic conditions. The possibility for major expansion of exotic annuals in the southern coastal region, especially when climatic change is combined with ancillary fire and air pollution effects, has other bases for support, discussed below.

## III. FLORISTIC CHANGES: THE CASE OF XERIC COASTAL SHRUBLANDS

### A. Shrub Formations and Associations

Four main coastal shrubland formations occur in the California floristic province along a moisture gradient from wet to dry: chaparral, northern coastal scrub, coastal sage scrub, and coastal succulent scrub. The evergreen, sclerophyllous chaparral shrubland occupies the areas of greatest and most sustained moisture availability in the coastal region. Northern coastal scrub, a mixed evergreen and drought-deciduous shrubland, occurs where low precipitation is compensated for by substantial fog (mostly in northern California). Coastal sage scrub, a drought-deciduous, mesophyllous shrubland, tends to occupy habitats with still shorter periods of effective moisture availability and higher summer evapotranspiration stress (mostly in southern California). Finally, toward the hotter, drier areas with some summer rainfall in northern Baja California (and some California Islands), the drought-deciduous shrublands show increased representation of thorny succulents and are termed coastal succulent scrub (Westman 1983a,b).

Extensive regional surveys of the latter two xeric types in the past ten years have established that the coastal sage scrub is composed of three distinct coastal floristic associations—Diablan, from San Francisco to Santa Barbara; Venturan, from San Luis Obispo to Orange County; and Diegan, from Orange County to about 100 km south of Ensenada in Baja California. There is also one inland association—Riversidian—in interior Los Angeles, Riverside, San Bernardino, and San Diego counties (fig. 1 in Westman 1983a; see also Kirkpatrick and Hutchinson 1977, Axelrod 1978). The coastal succulent scrub is composed of two floristic associations—Martirian, from Ensenada to San Quintin; and Vizcainan, from San Quintin to the Vizcaino Desert, south of El Rosario; a variant of these types occurs on the California Channel Islands (fig. 1 in Westman 1983a).

Based on data from weather stations near 92 sites of coastal sage and coastal succulent scrub (Westman 1983a), the mean temperature and rainfall for coldest and warmest months (1950–70) are plotted for each vege-

tation association in figure 20.4. In addition, the climatic conditions expected under doubled carbon dioxide are shown for the present location of each association. For the coldest month, the projected climate changes would imply a northward movement of coastal succulent scrub into Diegan and Riversidian regions and increased influence of inland prairie elements farther north. For the warmest month, which prior study suggests is more likely to control floristic changes (Westman 1981a), the changes are less extreme, but a definite northward shifting of types is implied, with coastal succulent scrub moving north along the coast of Baja and into the Riverside valley, and Riversidian sage migrating to the coastal Venturan region.

The Riversidian sage scrub region is already extensively disturbed by a combination of agricultural development, grazing, and significant ozone air pollution (mean annual concentration of 0.13 ppm, standard error 0.013; Westman 1983a). The average cover of exotic species in the Riversidian sage scrub is 36%, vs. 3%–12% in coastal types (Westman 1983a). There are several reasons for this. A major contributing factor is the reduced tendency of Riversidian shrubs to resprout following fire, permitting a longer postfire period for colonization of bare ground by annuals and short-lived perennials, both native and exotic (Westman and O'Leary 1986, O'Leary and Westman 1988). Additionally, intermittent grazing and the proximity of cultivated fields increase the seed source of exotic herbs; grazing can further inhibit resprout growth. Beyond those factors, there is some evidence from fumigation (Preston 1986) and field studies (Westman 1979, 1983a, 1990a; O'Leary and Westman 1988) that ozone will aggravate the spread of exotics. The sage shrubs respond to chronic ozone injury by additional aboveground growth, depleting root stores, and inhibiting the ability to resprout after fire (Preston 1986, O'Leary and Westman 1988, Westman 1990a). In addition, at least one exotic annual grass of importance in the Riversidian region, Bromus rubens, has been shown to have developed a pollution-

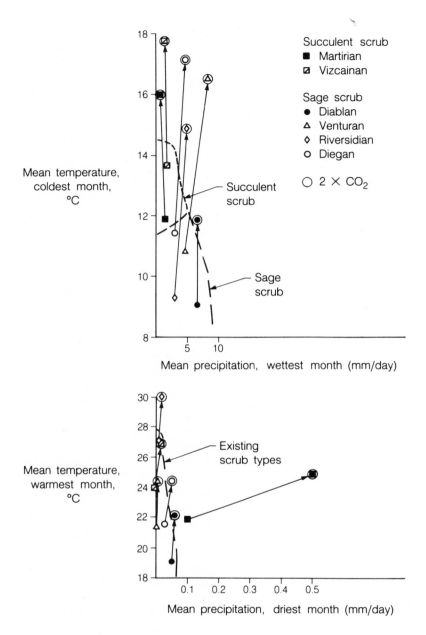

Figure 20.4. Projected changes in climate regime of four floristic associations of coastal sage scrub (Diablan, 13 sites; Venturan, 34; Riversidian, 19; Diegan, 12) and two associations of coastal succulent scrub (Martirian, 8; Vizcainan, 6), using GFDL climate model projections for a doubled carbon dioxide scenario. The arrow base shows the current climate found in the geographic region occupied by a particular association. The arrow head shows the future climate predicted to occur in that geographic region. Initial climate data from Westman 1983a.

resistant strain (Westman et al. 1985, Preston 1986). Thus, by reducing postfire resprouting and by enhancing competitive abilities of tolerant strains, ozone aids expansion of exotic cover in Riversidian sage.

Although currently the proportion of exotic species in the Riversidian sage is relatively low (8% vs. 16% for coastal associations) and species richness (both natives and exotics) is high (132 herb species vs. 59–139 for coastal types), an expansion of exotic herb cover within the Riversidian sage could ultimately reduce native herb species richness. Because herbs are the major source of floristic richness in the coastal sage scrub flora, a decrease in native herbs would decrease native diversity overall (Westman 1981a).

The analysis of change in vegetation associations also demands consideration of the interactive effects of climate, fire, and air pollution on the competitive interactions between individual species. The next section examines such dynamics for Venturan coastal sage scrub, using simulation modeling.

## B. Modeling Studies on Venturan Sage Scrub

A computer simulation model of the growth and reproduction of five dominant shrub species of Venturan coastal sage scrub was used to examine interactive effects (FINICS; Malanson 1984a, 1985a). The model uses Leslie matrices to follow the population dynamics and growth of each species. Competitive relations are modeled by assuming that species do not grow once total leaf cover has reached 90%. At this cover level, in the absence of fire, changes in composition occur only by death of individuals and colonization of resultant gaps by seedlings. The stand composition is thus determined by competition for the available space, based on the ability of each species to reestablish leaf cover by resprouting or seed reproduction after each fire. The model was parameterized using field data from the central and western Santa Monica Mountains in coastal Los Angeles and Ventura counties.

The effects of changing moisture availability were incorporated into the model using a direct gradient analysis approach (Malanson and Westman 1991; cf. Whittaker 1967). Species response to precipitation was derived from field-based data on cover values along a synthetic moisture gradient constructed from an index incorporating data on mean annual precipitation, aspect, slope, and soil moisture-holding capacity (Westman 1981a). Predicted changes in precipitation at the site were obtained by applying factors of change from the appropriate grid cell of the GISS or GFDL global climate model to the relevant weather station data at ten sites in the western Santa Monicas. The GFDL model predicts a change in mean annual precipitation from 330 mm to 560 mm; the GISS model predicts a change to 400 mm. The proportional change in cover between control and doubled carbon dioxide levels was then calculated for each species and averaged for the ten sites. The mean proportional change in cover was used to adjust equations of growth and levels of resprouting and seedling establishment in the FINICS model (Malanson and Westman 1991).

To examine the interactive effects of climate change and air pollution on growth, we considered the scenario in which 0.1 ppm ozone was present for 8 hours a day, 5 days a week, for 10 weeks during the peak growing season. This represents a 20%–30% increase over current ozone levels in the region (Westman 1990a); these sustained levels are not likely until later in the summer, when growth has all but ceased. The increase in winter temperatures could be expected to increase ozone levels in the region, because of increased rates of photooxidation and more frequent inversions (Westman 1979). Changes in leaf cover for each species were estimated from observed reductions in leaf mass after 10 weeks of fumigation with ozone in open-topped chambers using the exposure regime above (Preston 1986).

Fuel loads could increase or decrease under predicted climate changes in the region. We report here the case in which, because of the moderating influence of the ocean along the immediate coast (e.g., central and western Santa Monica Mountains), increases in winter precipitation result in net increases in shrub growth, despite regional increases in temperature and evaporative stress. With increases in shrub growth, herb understory cover would decline. To simulate these changes, we changed the fuel load used to calculate fire intensity (Malanson and West-

man 1991). In the FINICS model, fire intensity, as measured by total heat release, is calculated by a reparameterized fire model (FIREMOD; Albini 1976; for initial reparameterization to sage scrub, see Westman et al. 1981). The fire intensity in turn affects postfire resprouting vigor of shrubs in FINICS (Malanson and Westman 1991; see also Malanson 1984a).

The effects of altered moisture availability, ozone exposure, and fire intensity were tested separately and in combination. Each case was iterated in 5-year steps at fixed fire-return intervals of 10, 20, 30, and 40 years for a 200-year period.

One way to summarize results is to examine the ease of alteration of the sage scrub ecosystem under chronic stress, that is, how strongly the ecosystem responds to each modeled change in fire interval and the other important variables (Westman 1978, 1985a). A measure of the malleability of these systems is the extent to which the leaf cover values of each of the five modeled shrub species in the

Figure 20.5. Malleability of species composition and relative abundance of Venturan coastal sage scrub to altered climatic conditions under doubled carbon dioxide (GFDL or GISS global climate models). Malleability is expressed as the percentage similarity between the stressed and control system after 200 years of model run. Increased carbon fixation from altered climate shunted wholly to aboveground growth (a), seedling establishment (b), resprouting (c), or distributed between growth, seedling establishment, and resprouting (d). Fire intensity higher, but climate conditions at control level (e). Distributed growth from altered climate (GISS) and 0.1 ppm ambient ozone (f). Distributed growth from altered climate (GISS), 0.1 ppm ambient ozone, and high fire intensity (g).

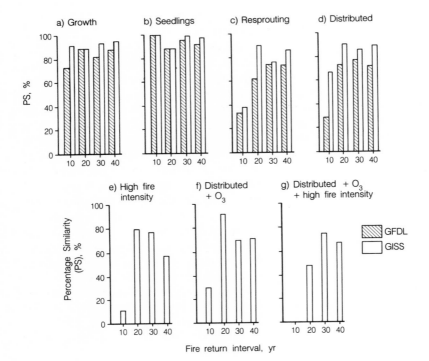

stand with changed environmental variables differ from the control after a simulation run of 200 years. Such a comparison is made in figure 20.5 by plotting the percentage similarity of species composition and relative abundance under doubled carbon dioxide conditions to the control conditions after 200 years, under various assumptions about carbon allocation. The control conditions are taken as current levels of carbon dioxide, no ozone, and fire intensities resulting from the existing stand composition (cf. Malanson 1984a). The percentage similarity index used is that of J. Czekanowski (1909; see, e.g., Westman 1985a). Runs under control conditions are shown in figure 20.6.

Since we do not know how the relative growth of various plant parts will change under the altered climatic conditions, several possibilities were modeled. In part a of figure

Figure 20.6. Changes in percent cover of five species of coastal sage scrub under FINICS model simulation, control conditions. Runs use fire intervals of 10 (a), 20 (b), 30 (c), and 40 (d) years.

20.5, all carbon fixation resulting from increased precipitation is shunted to aboveground growth. The result is a modest change in composition (PS, 73%–95% of the control). Shunting new carbon allocation to seedling establishment (part b) results in the least change from control conditions, largely because of the low incidence of reestablishment by seedling vs. resprout in these systems. Indeed, with a fire return interval as short as 10 years, no compositional change occurs because of the inability of seedlings to establish dominant positions during that interval. When new carbon allocation is shunted to resprouts (part c), more significant compositional changes occur, especially with a short fire return interval. This is because weak resprouters are now able to compete more vigorously with inherently strong resprouters, whose populations were already resprouting with nearly 100% frequency. Results are similar when carbon allocation is distributed among growth (50%), seedling establishment (25%), and resprouting (25%),

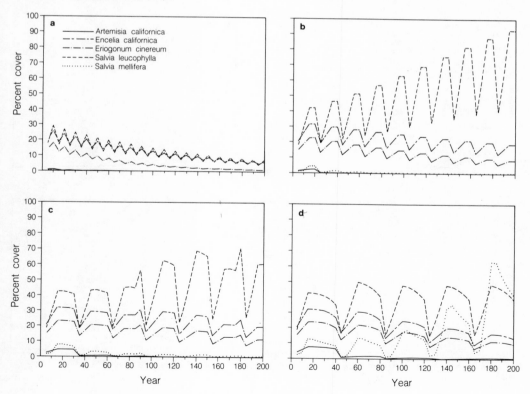

as shown in part d. These results reflect the important role of resprouting in the dynamics of this modeled system (viz. Malanson and Westman 1985).

Increasing the fire intensity without changing other model features (part e) results in significant changes in composition and relative abundance, particularly at 10-year and 40-year fire return intervals (PS, 11%, 57%, respectively). This is because of the death of root crowns under higher fire intensity (modeled as decreased resprouting), resulting in shifts in competitive advantage away from strong resprouters. When growth decrements due to 0.1 ppm ozone for 10 weeks are added to the model conditions, assuming distribution of carbon allocation among growth, seedling establishment, and resprouting for the GISS run (part f), significant changes in composition again occur, particularly at the 10-year fire return interval (PS, 30%).

When both ozone and fire intensity are added to the effects of altered precipitation

(part g), all five modeled species are eliminated at the 10-year fire return interval. At longer fire return intervals, changes are not as dramatic (PS, 48%–75%). The detailed changes in species abundance are plotted in figure 20.7. It is apparent that the maintenance of mixed-species shrub dominance in this modeled system would require a long fire return interval. During the half-century preceding 1980, when a fire exclusion policy was in force for the Santa Monica Mountains, fire return interval in coastal sage scrub averaged 20 years (Westman 1982). Since 1980 prescribed burning has begun in coastal sage, with a planned return interval of about 10 years. As noted by G. P. Malanson (1985b), such a short fire return interval would eventually impoverish these ecosystems of the

Figure 20.7. Changes in percent cover of five species of coastal sage scrub under FINICS model simulation, assuming distributed growth from altered climate (GISS), 0.1 ppm ambient ozone, and high fire intensity. Runs use fire intervals of 10 (a), 20 (b), 30 (c), and 40 (d) years.

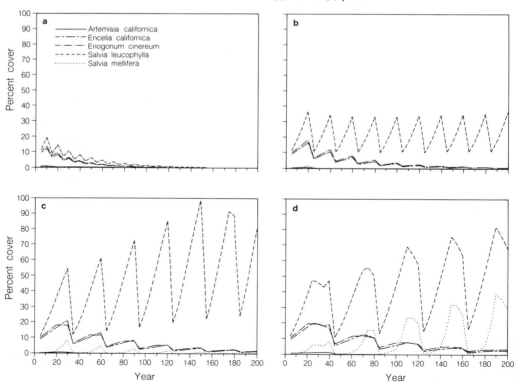

weaker resprouters, based on modeling results.

In the presence of the increased growth, ozone, and fire intensity resulting from predicted climate changes, the model results of figure 20.7 imply that practices planned to reduce fire hazard, including relatively frequent prescribed burning, will hasten the change in species composition on the site under changing climatic conditions. The changes in richness and composition of dominant shrubs have larger implications for stand composition, since changes in the dominance of coastal sage by *Salvia leucophylla* vs. *Salvia mellifera* are typically accompanied by a change in as many as 44 associated native herb species, at least some of which may be under biotic, as opposed to physical habitat, control (Westman 1983a, Malanson 1984b). Model results also imply that more effective controls on emission of oxides of nitrogen and reactive hydrocarbons could mitigate the effect of prescribed burning on biodiversity.

Of the various carbon allocation scenarios, the distributed-growth case is probably the most realistic. In such a case, the increased precipitation predicted by the GFDL model relative to the GISS model resulted in increased departure of coastal sage composition from the control, as would be expected. In the GFDL run, the ecosystem is increasingly dominated by *S. leucophylla*, and *S. mellifera* is extirpated at fire return intervals of 30 years or less. Because of the constraints of the FINICS model, any gaps created by the death of one species can be taken up only by another of the five species. In reality, these gaps may be colonized by species even better adapted to the changing climate than any sage species, thereby initiating a change in vegetation type not reflected in these model runs.

## IV. CHANGES AT THE LEVEL OF PLANT SPECIES: EARLY WARNING SIGNS

As plants are subjected to increasingly suboptimal environmental conditions, sublethal changes in growth and development should become apparent. In the last several years,

the five dominant coastal sage shrubs modeled in FINICS have also been studied from an ecophysiological viewpoint, seeking to elucidate changes that occur as a result of exposure to ozone and sulfur dioxide both in fumigation chambers (Westman et al. 1985, Westman 1990a, Preston 1986, Weeks 1987) and in the field (Westman 1985b, 1990a). The field study site has been the Santa Monica Mountains, a coastal range that extends east from the coast 68 km toward urban Los Angeles. Ozone increases from a mean annual level of 0.064 ppm in the west to 0.087–0.114 ppm in the east. There is no significant precipitation gradient (34.4 cm/yr in the west, 33.2–35.7 cm/yr in the east), but temperature regimes are markedly more extreme in the east, away from the moderating influence of the coast (mean annual temperature 19.8°C in the west, 24.7°C in the east). Both the climate and pollution figures are based on nearby lowland stations and will differ in detail from levels in the mountains, though the east-west gradient is likely to remain.

The Santa Monica Mountains therefore represent a simultaneous gradient from west to east of increasing mean temperature and ozone levels analogous to the changes expected in the region under doubled carbon dioxide levels. In past studies several changes to the coastal sage community in the Santa Monicas have been observed along this gradient. At a community level, the sage stands become less extensive and increasingly mixed with chaparral and grassland components toward the east (Westman and Price 1988). Visible leaf injury symptoms of ozone become more intense (Westman 1985b), and total nitrogen, chlorophyll-a, fiber and lignin, and a broad range of cations become more concentrated in leaves toward the east ($P<0.10$; Westman 1990a). An increase in total nitrogen in sage foliage was also observed in fumigation trials with ozone (Westman 1990a). A rise in nitrate levels under ozone fumigation has been observed in other broadleaved and conifer tree species (Krause 1988). Fiber, lignin, and pigments increased in sage species under ozone treatment, but not significantly (Westman 1990a).

These nutrient changes may be associated with differences in growth behavior along the gradient. In analyzing leaf nutrient changes along the field gradient, W. E. Westman (1990a) noted that the increased concentration of nitrogen and chlorophyll could be due to premature senescence of foliage (induced either by ozone or the earlier onset of drought stress), leading to an increased representation of younger, actively growing leaves on shrubs toward the east. At the same time, even the younger leaves could be exhibiting some thickening of secondary walls to counter increased water stress. Fibers in the central vascular bundles of S. mellifera did exhibit thicker walls at the eastern end of the gradient (Westman 1990a).

To test the hypothesis that nutrient changes were related to premature senescence of foliage, we collected early- and late-season leaves (dolichoblasts and brachyblasts, respectively; Westman 1981b) of S. mellifera for 1986 and early-season leaves for 1987, and leaves of Malosma laurina from earlier and more recent growth flushes, from the same 13 sites along the east-west gradient that Westman (1990a) used. We analyzed 25 leaves of each type and age class from each species and site separately for nutrients and correlated the results with site position along the east-west gradient.

Results for some of the nutrients are shown in table 20.1. In S. mellifera, current-year early-season leaves (1987 dolichoblasts) show a significant rise in total nitrogen toward the east, and early-season leaves from 1986 show a significant rise in lignin. All early-season leaves of this species show increases in nitrogen and lignin, but increases are greatest toward the east. Pigment content also rises in leaves of this species toward the east, though only chlorophyll-b in early 1987 leaves shows a significant trend ($P<0.05$). Trends of change in Malosma are not significant. These results are consistent with the hypothesis that increased temperatures or increased ozone levels, or both, can cause enrichment of early-season leaves of S. mellifera with nitrogen and lignin. A comparison of these results with ozone fumigation results suggests that ozone may cause the rise in nitrogen, and water stress the rise in lignin, but further study is needed to confirm these findings.

In addition, some changes in leaf morphology and stem elongation are observable in S. mellifera along the east-west gradient. Early-season leaves become significantly larger toward the east. This is consistent with observations of increased width of early-season leaves in this species under 0.1 ppm ozone in fumigation chambers (Preston 1986). The mean internodal distance between the topmost five leaf whorls also significantly lengthens toward the east, but that change does not appear to be attributable to ozone effects, since Preston (1986) found that internodal length was significantly reduced under 0.1 and 0.4 ppm ozone, though not under 0.2 ppm ozone. The mechanistic explanation for the observed changes will clearly require further research, but the occurrence of significant morphological and nutrient changes along the field gradient suggests that field analogues may be used to generate hypotheses about early changes in species growth under climate change.

## V. IMPLICATIONS FOR CONSERVATION POLICY

Figure 20.8 provides one scenario of possible changes in Venturan coastal sage scrub in the Los Angeles Basin away from the immediate coast (eastern Santa Monicas and points inland), under elevated carbon dioxide. The scenario assumes GFDL predictions of higher evapotranspiration stress in summer. Unlike the scenario at the immediate coast, this assumes that the higher evapotranspiration will now have the predominant influence on shrub growth, shortening the effective growing season and reducing net fuel accumulation. Such will occur if the effect of a shortened growing season outweighs any stimulation of growth by increased winter rains.

The shift of Venturan to Riversidian sage shown in figure 20.8 is supported by figure 20.4. Reduced fuel loads and fire intensities, increased prevalence of postfire reproduc-

Table 20.1. Correlation coefficient (r) of nutrient concentrations of early-season (dolichoblasts) or late-season (brachyblasts) leaves of Salvia mellifera and leaves of Malosma laurina from two growing periods with increasingly westward position in the Santa Monica Mountains. *P<0.05.

| | Salvia mellifera | | | Malosma laurina | |
| | 1987 | 1986 | | | |
| | early | early | late | Young | Old |
|---|---|---|---|---|---|
| Total nitrogen | −0.74* | −0.61 | −0.46 | 0.00 | −0.03 |
| Lignin | −0.23 | −0.72* | 0.31 | 0.06 | −0.04 |
| Chlorophyll a | −0.39 | −0.60 | −0.13 | 0.26 | 0.38 |
| Chlorophyll b | −0.66* | −0.51 | −0.23 | 0.03 | 0.25 |
| Chlorophyll a/b | 0.71* | 0.51 | 0.52 | 0.16 | 0.18 |
| Carotenoids | −0.50 | −0.58 | −0.17 | 0.47 | 0.36 |
| Fiber | 0.22 | −0.21 | 0.10 | 0.05 | −0.10 |
| No. of sites | 8 | 8 | 13 | 13 | 13 |

tion by seed, and increased cover of exotic annuals in Riversidian sage is supported by field studies (Westman et al. 1981, Westman and O'Leary 1986, O'Leary and Westman 1988). As noted earlier, agricultural and grazing influences in the Riverside–San Bernardino Basin also contribute to increased cover of exotics. The favoring of weak resprouters (Artemisia californica, Salvia mellifera) under low fire frequencies is supported by FINICS runs (fig. 20.6). The morphological and biochemical changes are based on the gradient studies in the Santa Monicas (table 20.1). Although ignition rate might be lower as a result of higher summer rainfall (fig. 20.2), the increased growth of annuals in early stages of postfire succession might also increase ignition rate, by extending the period of dry fuel. Thus figure 20.8 outlines only one of several possible scenarios (for additional scenarios, see Malanson and Westman 1991).

As ecosystems exhibit increased death of intolerant individuals, aggressive annual and short-term perennial invaders will likely colonize the gaps. Many of these species are exotic, from the Mediterranean Basin proper. Of 975 exotic species in the California flora (Munz and Keck 1959, Munz 1968), P. H. Raven and Axelrod (1978) consider 674 species to be naturalized; all but 11 of these are in the California floristic province. It is the current policy of both the National Park Service and the State Resources Agency in California to remove exotic species that have arrived since European settlement (Leopold et al. 1963) and replace them with native ones where feasible. Leaving aside the question of the feasibility (and cost) of removing most of the widespread, rapidly dispersing species from extensive portions of a landscape, there remains the important question of the desirability of this policy in light of anticipated climate changes.

With the widespread disruption of ecosystems that will be the inevitable result of a period of rapid climatic change, many species that were formerly not adapted to an area will invade. Some will be from the same floristic province, others not. This invasion under changing environmental conditions will emphasize one of the major weaknesses of the implementation of the present exotic species policy, namely, the futility of trying to freeze species composition at its condition 200 years ago, in light of continually changing environmental conditions. Climate change will force public park agencies to reconsider the value of exotic species and recognize that in some cases the functions they serve may be too valuable to sacrifice in the costly attempt to restore a historic vignette that is no longer self-sustaining (Westman 1990b).

Underlying the current arguments for preserving native biodiversity against the inva-

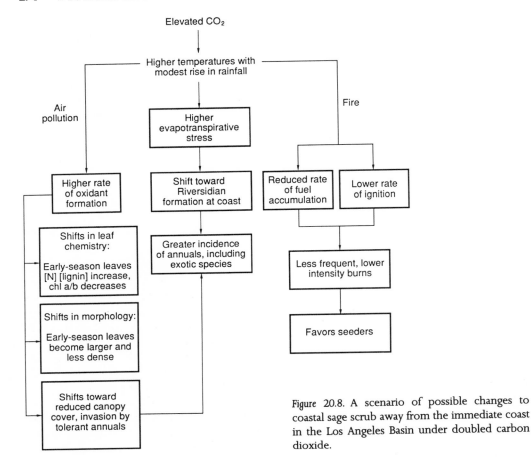

Figure 20.8. A scenario of possible changes to coastal sage scrub away from the immediate coast in the Los Angeles Basin under doubled carbon dioxide.

sion by introduced species is often a flawed understanding of biogeographic history, as well. Proponents of exotic-species removal frequently argue that these species have not coevolved with the native community and therefore do not belong. Yet the vast majority of currently coexisting species in California have not coevolved over periods long enough to account for their speciation. In fact, since the last glaciation some 12,000 years ago, species from Arctic and Madro-Tethyan origins have continued to re-sort themselves into new assemblages in California (Raven and Axelrod 1978).

Beyond the issue of exotic species, there is the problem of retaining representative samples of major species assemblages through the current array of public parks in California and Baja California. Given the uncertainty surrounding future vegetation patterns, more

general guidelines for park planning would appear to apply. Connecting corridors between existing protected areas would aid in migration and hence movement of species to habitats more suited to them under changing climate. Expanding the area of parkland in areas currently underrepresented geographically (e.g., northern Baja California coast, sagebrush steppe of northwest California, Central Valley, Mohave and Sonoran deserts) will aid in ensuring the capture of new assemblages in regions that may currently be of less interest biotically.

Because of the uncertainty that will continue to attend the effects of climate change for some time, introducing increased flexibility into the manner of acquisition and management of parklands would be helpful. The purchase of development rights to private land by public agencies may be a useful tool to hold land for possible future addition to park status, without incurring associated management costs in the short term. Transferable development credits have been used since 1970 to regulate land use within five miles of the coast in Los Angeles under the California Coastal Zone Management Act. The California Coastal Commission serves as a clearing house for purchasing, banking, selling, and trading transferable development credits (Westman 1985a:72). Such a system may serve as a useful model for maintaining options for future reserve status in a cost-efficient way in the approaching era of climatic uncertainty.

## REFERENCES

Albini, F. A. 1976. *Computer-based models of wildland fire behavior: A user's manual.* Ogden, Utah: U.S. Forest Service Intermtn. For. Range Experiment Station.

Axelrod, D. I. 1956. Mio-Pliocene floras from west-central Nevada. *Univ. Calif. Publ. Geol. Sci.* 33:1.

Axelrod, D. I. 1968. Tertiary floras and topographic history of the Snake River Basin, Idaho. *Geol. Soc. Am. Bull.* 79:713.

Axelrod, D. I. 1978. The origin of coastal sage vegetation, Alta and Baja California. *Am. J. Bot.* 65:1117.

Axelrod, D. I. 1979. Age and origin of Sonoran desert vegetation. *Calif. Acad. Sci. Occ. Papers* 132:1.

Czekanowski, J. 1909. Zur differential Diagnose der Neandertalgruppe. *Korrespbl. dt. Ges. Anthrop.* 40:44.

Donley, M. W., S. Allan, P. Caro, and C. Patton. 1979. *Atlas of California.* Culver City, Calif.: Pacific Book Center.

*Hammond Ambassador World Atlas.* 1988. Maplewood, N.J.: Hammond.

Hansen, J. E., G. Russell, D. Rind, P. Stone, A. Lacis, S. Lebedeff, R. Ruedy, and L. Travis. 1983. Efficient three-dimensional global models for climate studies: Models I and II. *Monthly Weather Rev.* 111:609.

Howell, J. T. 1957. The California flora and its province. *Leafl. West. Bot.* 8:133.

Hudson, G. D., ed. 1958. *World Atlas.* Chicago: Encyclopedia Brittanica.

Jepson, W. L. 1925. *A Manual of the Flowering Plants of California.* Berkeley: University of California Press.

Kirkpatrick, J. B., and C. F. Hutchinson. 1977. The community composition of Californian coastal sage scrub. *Vegetatio* 35:21.

Krause, G.H.M. 1988. Ozone-induced nitrate formation in needles and leaves of *Picea abies, Fagus sylvatica* and *Quercus robur. Envir. Poll.* 52:117.

Kuchler, A. W. 1977. The map of the natural vegetation of California. In *Terrestrial Vegetation of California,* M. G. Barbour and J. Major, eds., pp. 909–938. New York: Wiley-Interscience.

Leighly, J. 1938. The extremes of the annual temperature march. *Univ. Calif. Publ. Geog.* 6:191.

Leopold, A. S., S. A. Cain, C. M. Cottam, I. N. Gabrielson, and T. L. Kimball. 1963. *Wildlife Management in the National Parks.* Washington, D.C.: National Park Service.

Major, J. 1967. Potential evapotranspiration and plant distribution in western states with emphasis on California. In *Ground Level Climatology,* R. H. Shaw, ed., pp. 93–126. Washington, D.C.: American Association for the Advancement of Science.

Malanson, G. P. 1984a. Linked Leslie matrices for the simulation of succession. *Ecol. Model.* 21:13.

Malanson, G. P. 1984b. Fire history and patterns of Venturan subassociations of Californian coastal sage scrub. *Vegetatio* 57:121.

Malanson, G. P. 1985a. Fire management in coastal sage scrub, southern California, U.S.A. *Envir. Conser.* 12:141.

Malanson, G. P. 1985b. Simulation of competition between alternative shrub life history strategies through recurrent fires. *Ecol. Model.* 27:271.

Malanson, G. P., and W. E. Westman. 1985. Post-fire succession in Californian coastal sage scrub: The role of continual basal sprouting. *Am. Midl. Nat.* 113:309.

Malanson, G. P., and W. E. Westman. 1991. Modeling interactive effects of climate change, air pollution, and fire on a California shrubland. *Clim. Change:* in press.

Manabe, S., and R. T. Wetherald. 1987. Large-scale changes in soil wetness induced by an increase in carbon dioxide. *J. Atmos. Sci.* 44:1211.

Mortensen, L. M. 1987. Review: $CO_2$ enrichment in greenhouses. Crop responses. *Scientia Horticulturae* 33:1.

Munz, P. A. 1968. *Supplement to a California Flora.* Berkeley: University of California Press.

Munz, P. A., and D. D. Keck. 1959. *A California Flora.* Berkeley: University of California Press.

O'Leary, J. F., and W. E. Westman. 1988. Regional disturbance effects on herb succession patterns in coastal sage scrub. *J. Biogeogr.* 15:775.

Preston, K. P. 1986. Ozone and sulfur dioxide effects on Californian coastal sage scrub species. Ph.D. diss., University of California, Los Angeles.

Ramanathan, V., R. J. Cicerone, H. B. Singh, and J. T.

Kiehl. 1985. Trace gas trends and their potential role in climate change. *J. Geo. Res.* 90(D3):5547.

Raven, P. H., and D. I. Axelrod. 1978. Origin and relationships of the California flora. *Univ. Calif. Publ. Bot.* 17:1.

Richerson, P. J., and K.-L. Lum. 1980. Patterns of plant species diversity in California: Relation to weather and topography. *Am. Naturalist* 116:504.

Rundel, P. W., D. J. Parsons, and D. T. Gordon. 1977. Montane and subalpine vegetation of the Sierra Nevada and Cascade Ranges. In *Terrestrial Vegetation of California*, M. G. Barbour and J. Major, eds., pp. 559–600. New York: Wiley-Interscience.

Sowell, J. B. 1985. A predictive model relating North American plant formations and climate. *Vegetatio* 60:103.

Stephenson, N. L. 1988. Climatic control of vegetation distribution: The role of the water balance with examples from North America and Sequoia National Park, California. Ph.D. diss., Cornell University, Ithaca.

UNESCO-FAO. 1963. *Bioclimatic Map of the Mediterranean Zone*. Arid Zone Res. 21. Paris: UNESCO.

Vasek, F. C., and M. G. Barbour. 1977. Mojave desert scrub vegetation. In *Terrestrial Vegetation of California*, M. G. Barbour and J. Major, eds., pp. 835–868. New York: Wiley-Interscience.

Weeks, L. B. 1987. Effects of ozone and sulfur dioxide on the belowground growth of Californian coastal sage scrub. M.A. thesis, University of California, Los Angeles.

Westman, W. E. 1978. Measuring the inertia and resilience of ecosystems. *Bioscience* 28:705.

Westman, W. E. 1979. Oxidant effects on Californian coastal sage scrub. *Science* 205:1001.

Westman, W. E. 1981a. Factors influencing the distribution of species of Californian coastal sage scrub. *Ecology* 62:439.

Westman, W. E. 1981b. Seasonal dimorphism of foliage in Californian coastal sage scrub. *Oecologia* 51:385.

Westman, W. E. 1982. Coastal sage scrub succession. *Proc. Intl. Symp. Dynamics and Management of Mediterranean-type Ecosystems*. USDA Forest Service Pacific S.W. For. Range Experiment Station General Technical Report PSW-58.

Westman, W. E. 1983a. Xeric Mediterranean-type shrubland associations of Alta and Baja California and the community-continuum debate. *Vegetatio* 52:3.

Westman, W. E. 1983b. Island biogeography: Studies on the xeric shrublands of the inner Channel Islands, California. *J. Biogeogr.* 10:97.

Westman, W. E. 1985a. *Ecology, Impact Assessment, and Environmental Planning.* New York: Wiley-Interscience.

Westman, W. E. 1985b. Air pollution injury to coastal sage scrub in the Santa Monica Mountains, southern California. *Water, Air, Soil Poll.* 26:19.

Westman, W. E. 1990a. Detecting early signs of regional air pollution injury to coastal sage scrub. In *Biotic Impoverishment: Changes in the Structure and Function of Natural Communities under Chronic Disturbance*, G. M. Woodwell, ed. New York: Cambridge University Press.

Westman, W. E. 1990b. Park management of exotic plant species: Problems and issues. *Conser. Biol.* 4:251.

Westman, W. E., and J. O'Leary. 1986. Measures of resilience: The response of coastal sage scrub to fire. *Vegetatio* 65:179.

Westman, W. E., J. O'Leary, and G. P. Malanson. 1981. The effects of fire intensity, aspect, and substrate on post-fire growth of Californian coastal sage scrub. In *Components of Productivity of Mediterranean Regions: Basic and Applied Aspects*, N. S. Margaris and H. A. Mooney, eds., Tasks for Vegetation Science Series, pp. 151–179. The Hague: Junk.

Westman, W. E., K. P. Preston, and L. B. Weeks. 1985. Sulfur dioxide effects on the growth of native plants. In *Sulfur Dioxide and Vegetation: Physiology, Ecology, and Policy Issues*, W. E. Winner, H. A. Mooney, and R. B. Goldstein, eds., pp. 264–280. Stanford: Stanford University Press.

Westman, W. E., and C. V. Price. 1988. Detecting air pollution stress in southern California vegetation using LANDSAT Thematic Mapper band data. *Photo. Engr. Remote Sens.* 54:1305.

Whittaker, R. H. 1967. Gradient analysis of vegetation. *Biol. Rev.* 42:207.

This modeling work was a collaborative effort of the two authors. The remaining work reported here was conducted by W. E. Westman. Research for this article was supported by the Environmental Protection Agency in Interagency Agreement DW89933219-01-0 with the Department of Energy. New field data reported here were obtained with support of the National Aeronautics and Space Administration, Earth Science and Life Science Divisions, under grants 677-21-35-08 and 199-30-72-02. Any opinion, findings, and conclusions or recommendations expressed in this publication are those of the authors and do not necessarily reflect the views of the Environmental Protection Agency, NASA, or Department of Energy.

# CHAPTER TWENTY-ONE

# Projecting the Effects of Climate Change on Biological Diversity in Forests

DANIEL B. BOTKIN AND
ROBERT A. NISBET

## I. INTRODUCTION

In terms of human lifetimes, there is an ancient relationship between forests and people: forests provided many materials for civilization, and we have continually cleared forested landscapes as civilization has expanded. Deforestation was recognized as early as the classical Greek civilization when Plato wrote that the hills of Attica were a skeleton of their former selves (Thomas 1956). For a long time, especially since the Industrial Revolution, the direct effects of human beings on forests obscured slower, natural changes. For example, George Perkins Marsh, the first writer of the modern industrial era to point out the effects of civilization on environment, believed that forests remained constant except when suffering from human influence (Marsh 1864). Today we have achieved a much longer perspective on the history of vegetation and the history of climate. We know that both have changed over the millennia and that vegetation has changed in response to climatic change. This is made apparent by the analysis of pollen records, as illustrated in chapters 5 and 22 and elsewhere (see Davis 1983, COHMAP 1988). Now, faced with a potentially rapid and novel climatic warming of our own doing, we need to evaluate what these changes will mean to forests and to the biological diversity they contain. Change in climate that affects vegetation has been the norm, not the exception (Botkin and Sobel 1975, Botkin 1990). It is not change per se that concerns us but the rate of change and the novel qualities of the change that might take place, combined with the occurrence of these changes on a highly altered landscape, one in which forests exist primarily as broken, segregated patches so that natural migration paths for seeds have been largely fragmented.

Biological diversity of forests must be at the center of our concern. The commercial utility of forests depends not simply on the total production of wood but also on the kind of trees that produce that wood. Forests provide habitats for many kinds of organisms, not only of species of vegetation but also of animals, fungi, and bacteria; the kind and number of such habitats depend on the total number of species of trees, on which species are dominant, and on the complex spatial and temporal patterns that occur in forests.

Forests may affect the biosphere. For example, forests may affect climate; trees and forest soils are a major site of storage of carbon, and the release or uptake of carbon dioxide from the atmosphere can affect the rate of climate change. Forests also influence climate locally by altering the surface energy balance through reflection of sunlight, evaporation of water, and upward percolation of water in the soil. Forests create the microclimate necessary for maintenance of animal species habitat. The rate at which these processes—carbon dioxide uptake, water evaporation, and radiative energy exchange—occur depends, however, on what species are present and the relative abundance of the species. Thus, biological diversity of forests may influence climate, and knowledge about forest biological diversity may be important for an understanding of global climate change. In all aspects, as sources of commercial products, habitats for many kinds of organisms, and influences on the rest of the biosphere, the diversity of forests plays an important role.

Because we have cleared so much of the forests and altered so much of what remains, most forests exist as patches and remnants. The few extensive continuous tracts that remain lie primarily in the boreal forests of the north, in the most remote high mountains, and in some of the tropical regions. The effect of climate change on biological diversity in forests will differ with the size of the remnant forests and with our specific concerns. We would expect that a species with highly specific habitat requirements, requirements that are presently limited in geographic distribution, would be more greatly affected than species with broader habitat tolerances and with the potential to migrate within a continuously forested area. This chapter contrasts the responses of small and large forested areas, both in their biological diversity and the capacity for the preservation of specific species.

There are three methods open to us for assessing these changes: present measurements, our knowledge of the past, and our ability to make projections using ecological theory. Present measurements are of limited utility, because, with rare exceptions, we have not measured forests for long enough to detect changes in forest diversity due to climate change, distinguished from other kinds of change. Indeed, some ecologists even argue that current forests are little different from those of presettlement times (Russell 1983); that such a point of view can be argued suggests that we have been unable to detect significant changes in forests in response to climate or other factors for three centuries. Even where such measurements exist, the causes of changes can be confused. Measurements that show correlations between climate change and forest growth have also been attributed to the effects of acid rain and other forms of pollution. Observed current changes in forests might be responses to direct human effects on forests that took place decades or centuries ago and have been obscured by time, such as selective logging in the past.

Even where there might have been sufficient time to observe changes, such as at the limits of the present distribution of a tree species where responses to climate change may be most sensitive, we have not established proper programs to monitor changes over time, either to provide us with information leading up to the present or to provide those measurements in the future. Such monitoring programs would establish a proper array of permanent plots and repeated measurements at useful intervals.

The past, especially the information provided by fossil pollen and its correlation with the history of climate, provides important information about the long-term changes in the distribution of trees in response to past climate. Pollen records, however, are not sufficient in themselves. These records can provide insight into the rate of seed migration across the landscape and into the kinds of changes in species composition that may occur within a forest, but not the rate at which forest composition will change. Furthermore, the relationship between pollen deposits and climate is a statistical correlation and does not lead to a cause-and-effect understanding of processes, limiting the extent to which the information can be used. For example, vegetation responds to combined patterns in rainfall and temperature. If future climates involved novel combinations of rainfall and temperature patterns, the future distribution of vegetation may not be simply estimated from past climate-vegetation conditions. This is especially important because climate models suggest that future climates may develop novel relationships between temperature and rainfall, and it is not clear how useful past statistical correlations will be as a basis for extrapolation into the future.

The only method available to provide insights into the rates of changes in forest composition, especially in response to the novel aspects of the expected climate change, is through theoretical but data-based models. Unfortunately, such data-based models are relatively rare in ecology because the field has suffered during the twentieth century from an emphasis on theory unconnected to observations and from observations unconnected to theory, with the result that in many areas major generalizations have not been tested by systematic data collection. Furthermore, the twentieth century's ecology has been based on a kind of population dynamics that did not consider the relationship between organisms and their environment. Often that theory was contradicted by facts, but ecologists continued to use and believe the theory (Botkin 1990, 1982). The logistic and Lotka-Voltera equations are prime examples of where real-world observations demonstrated that populations often refused to

follow the theoretically predicted patterns (Botkin et al. 1973). To avoid this trap, when dealing with the effects of climate change on biological diversity, it is essential that we develop a theory that is connected to observations, is open to tests of validation and to calibration, and links changes in populations to environmental conditions.

## II. A BRIEF INTRODUCTION TO THE FOREST MODEL

In 1970, Botkin, J. F. Janak, and J. R. Wallis began to develop a computer model of forest growth known as JABOWA, which met two criteria stated above: it was open to test and calibration—directly linked to and developed from a basis in observation—and involved the relationships between tree growth and regeneration and environmental conditions (Botkin et al. 1970). Since their first report in 1970, the model has been shown to be accurate and realistic and has been applied to forests around the world (Botkin et al. 1970, 1972; Prentice 1986; Leemans and Prentice 1987). Other forest models that are now often referred to as gap-phase models are directly derived from earlier versions of JABOWA. This original approach and body of algorithms have been so useful that most forest growth models continue to be variants of JABOWA models (Shugart 1984). Various additions have been made by other investigators to extend the JABOWA approach to include other dimensions of the ecosystem. For example, J. Pastor and W. M. Post (1985, 1986) have added a module to analyze decomposition and to improve the incorporation of effects of nutrient cycling into the model. There are presently more than fifteen versions used by nearly forty investigators, who have applied versions of the models to a variety of forests around the world (West et al. 1981, Smith et al. 1981, Weinstein and Shugart 1983, Shugart 1988). Two have been used to investigate some aspects of $CO_2$-induced climate change (Pastor and Post 1988, Solomon and West 1987), while the original model has been used to consider some aspects of the direct effect of

$CO_2$ increase on forest growth (Botkin et al. 1973).

The current version of the JABOWA model, JABOWA-II, incorporates a number of advances including: a more complete method of handling the relationships between water and tree growth, making possible a separation of floodplain communities from bog and other wetland communities (Botkin and Levitan 1977); a more specific relationship between nitrogen concentration in the soil and tree growth (Aber et al. 1978, 1979); a more realistic treatment of growth and reproductive rates among species; and an improved soil-water balance calculation. The more accurate consideration of soil-water relations makes possible a better distinction between forests of wetlands, well-drained and well-watered sites, and dry sites. JABOWA-II also incorporates forty species of trees, which are all the major native trees found in the northern hardwoods and boreal forests of eastern and midwestern North America and all the major tree species found north of Connecticut since the end of the last major period of glaciation.[1] A precursor to JABOWA-II was used to investigate the effects of climate change since the end of the last ice age on forests of New England (Davis and Botkin 1985).

It is important to understand that this model is developed from a basis in natural

---

1. The model includes 34 species that grow in the Great Lakes states: sugar maple (Acer saccharum), yellow birch (Betula alleghaniensis), white ash (Fraxinus americana), mountain maple (Acer spicatum), striped maple (Acer pennsylvanicum), pin cherry (Prunus pennsylvanica), choke cherry (Prunus virginiana), balsam fir (Abies balsamea), white birch (Betula papyrifera), mountain ash (Sorbus americana), red maple (Acer rubrum), scarlet oak (Quercus coccinea), hornbeam (Carpinus spp.), green alder (Alnus crispa), speckled alder (Alnus incana), black ash (Fraxinus nigra), butternut (Juglans cinerea), white spruce (Picea glauca), black spruce (Picea mariana), jack pine (Pinus banksiana), red pine (Pinus resinosa), white pine (Pinus strobus), trembling aspen (Populus tremuloides), white oak (Quercus alba), northern red oak (Quercus rubra), white cedar (Thuja occidentalis), hemlock (Tsuga canadensis), silver maple (Acer saccharinum), tamarack (Larix laricina), basswood (Tilia americana), bigtooth aspen (Populus grandidentata), balsam poplar (Populus balsamifera), black cherry (Prunus serotina), and eastern red cedar (Juniperus virginiana).

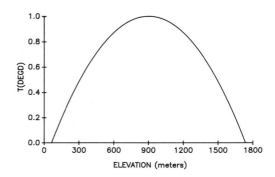

Figure 21.1. The growth response function, T(DEGD), for balsam fir, graphed against elevation. T(DEGD) is an indicator of how much growth occurs in response to total heat available during the growing season.

history observations, in plant physiology and physiological ecology, and in plant morphology. We used actual measurements of, for example, the rate of photosynthesis as the basis for an equation in the model. Figure 21.1 shows the model's projected growth of balsam fir at increasing elevations in the mountains of New England. Peak values of T(DEGD), an indicator of growth in response to heat available during the growing season, occur at intermediate levels of heat. When too little or too much heat is available during the growing season, T(DEGD), and therefore growth of balsam fir, decreases. Growth first rises and then falls as available heat increases, because rising termperature increases metabolic rate until a point is reached when too much heat begins to inhibit biochemical reactions. The T(DEGD) growth response curve is an assumption (not a prediction) of the JABOWA model based on hundreds of publications showing similar dome-shaped responses to available heat for all biochemical reactions and all cells and organisms. For balsam fir, too much heat is present when elevation approaches sea level (0 m of elevation) in northern New England, and therefore T(DEGD) approaches zero. This growth response curve for balsam fir is shown for the mountains of New England and assumes that no factor except temperature limits growth. The graphed values of T(DEGD) are calculated using the following formula, where $DEGD_{max}$ is the heat energy available for growth at the southern end of the species' range and $DEGD_{min}$ is the energy available at the northern end:

$$\frac{(DEGD_{max} - DEGD)(DEGD - DEGD_{min})}{(DEGD_{max} - DEGD_{min})^2}$$

DEGD, a critical variable in calculating T(DEGD), is the total growing degree days at a particular altitude, which is a measure of total heat energy available for growth during the growing season. DEGD is calculated by subtracting an arbitrary temperature baseline of 4°C (approximately 40°F) from the average temperature during each day. This yields a difference for each day. Adding up all the differences for all the days in the growing season gives an annual growing degree day value. For example, for a two-day period in which the first day has an average temperature of 14°C and the second day has an average of 19°C, the total number of growing degree days is $(14 - 4) + (19 - 4)$, or 25. Days when the average temperature does not exceed 4°C do not contribute to the total growing degree days and provide too little energy to affect growth substantially. Chapter 8 uses a similar measure of total available heat called the day-degree total, which differs from growing degree day in that it takes as its baseline 0°C instead of 4°C but otherwise is essentially the same. Different workers use different baselines, depending on what baseline data are available or what life form they are studying.

Moving downslope from a New England mountaintop, average temperatures increase approximately 1°C per 300 meters, the amount of heat available during the growing season increases, and its measure, DEGD, increases. As DEGD increases, the growth function T(DEGD) initially increases, as increasing amounts of heat favor increased growth. Farther downslope, DEGD increases to the point that there is more heat available during the

growing season than balsam fir is adapted to, and T(DEGD) and associated growth decline. A great number of physiological studies over the years have shown that many metabolic functions change in just this way with changing temperature, increasing from zero growth at extremes and rising to a single maximum rate. The exact temperature at which growth becomes zero varies with species. In an informal way, one can verify this relationship by observation on a hike up Mount Washington in New Hampshire or other mountains in the northeastern United States. If the extreme values are reasonably correct, balsam fir should not be found below 50 m or above 1700 m and should be most abundant at approximately 900 m, assuming soils and other environmental factors are the same.

In the model many factors other than temperature, including light intensity, operate simultaneously to produce a complexity that is difficult to perceive in an informal way but that can be made clear graphically. Figure 21.2 compares the growth of 5-cm-diameter individuals of different species at different temperatures (degree days), in full sun and in one tenth of full sunlight. The amount of growth differs greatly with the available light. For example, pin cherry grows rapidly in full sunlight but slowly in one tenth of full sunlight. Sugar maple, a shade-tolerant species, grows more slowly than pin cherry in full sunlight but much more rapidly in one tenth full sunlight.

Species also differ greatly in which temperatures are best for growth. Spruce, fir, and white birch, major boreal forest species, grow under colder conditions than sugar maple, beech, and yellow birch, which are characteristic of warmer conditions and members of the northern hardwoods forest. Some species, such as pin cherry, have a wide temperature tolerance, and others, such as fir, tolerate a comparatively narrow range.

The details and rationale of this model have been published elsewhere. Those interested can refer to Botkin et al. 1972a, 1972b; Aber et al. 1978; Botkin and Levitan 1977; Davis and Botkin 1985. The most recent modifications are available with a manual for a

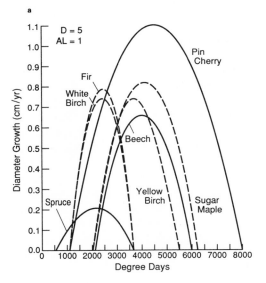

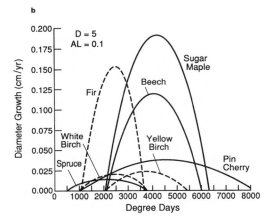

Briefly, the model simulates growth of individual trees on small plots (in the applications shown here, 100 m²), which are small enough so that a single large tree can shade and therefore influence all the other trees on the plot. At the heart of the model is a fundamental growth equation, which expresses simply the idea that the growth of a tree is directly proportional to the total amount of its leaf area and is decreased by the amount of nonphotosynthetic living tissue that must be supported. This fundamental equation represents the growth of a tree when all environmental requirements are at an optimum for that species and growth therefore is limited only by genetic potential. This optimal growth is then modified in the model in light of factors, such as competition, that decrease environmental requirements below optimum. Species-specific parameters determine the maximum size that a tree can reach and the maximum rate of growth possible for each size.

Each species is defined by a maximum height, maximum diameter, maximum age, and the maximum and minimum growing degree days under which it can grow; by allometric relationships (those relating tree diameter or diameter and height to leaf mass and total biomass); and by factors that concern water relations, including the relationship between tree growth and soil moisture-holding capacity, percentage rock in the soil, depth of the soil, as well as depth to the water table (following Botkin and Levitan 1977, Davis and Botkin 1985). Finally, in our current version, there is a relationship between tree growth and soil nitrogen (following Aber et al. 1978, 1979).

The environment is defined by mean monthly temperature, mean monthly rainfall, sunlight, soil depth, soil moisture-holding capacity, percentage of soil that is rock, depth to the water table, and soil nitrogen content. The actual status of each environmental factor reduces the growth below the maximum possible in the fundamental equation. Trees compete for light, with taller trees shading smaller ones and trees with more leaves for a given diameter producing a

Figure 21.2. The growth responses, function T(DEGD), of several tree species to temperature at different light intensities and for different sizes of trees: diameter growth of trees of 5 cm in diameter under full sunlight (a) and one-tenth full sunlight (b). D, diameter. Al, available light. These curves are assumptions of the forest model, not predictions by the model.

microcomputer version of the model, called JABOWA-II (c), which can be obtained from the Santa Barbara Institute for Environmental Studies, Box 3765, Santa Barbara, CA 93105. A complete description of JABOWA-II will be available soon in Botkin (in press).

denser shade. The growth of a tree increases as the available light increases, and there are two light response curves, one for shade-tolerant species and one for shade-intolerant species. Shade-tolerant species grow well in low light, but their growth reaches a maximum at a comparatively low light level. Shade-intolerant species grow more slowly than the tolerant species at low light levels, but intolerant species can use much more light (higher saturation level), and their growth rate increases in bright light beyond that of the shade-tolerant species. The light response equations were derived from studies in the scientific literature on the photosynthesis of trees. In the work reported here, the model was used to consider the responses of a forest to changes in temperature and rainfall.

## III. EFFECTS OF CLIMATE CHANGE ON KIRTLAND'S WARBLER HABITAT AND THE BOUNDARY WATERS CANOE AREA

The process of climate change takes place today on a landscape greatly altered by human action. Forests that once spread over great areas, interrupted only by major topographic features such as lakes and rivers, have been largely reduced to small, isolated patches. The problems that might arise for biological diversity in forests as a result of rapid climate change are exacerbated by this condition, especially because the natural migration routes for seeds have been eliminated. The previous means of forest adjustment to major climate change, the slow spread of seeds through the landscape, has become impossible over much of the earth. Small isolated animal and plant populations dependent on forested habitats are at particular risk of local extinction if the climate changes as rapidly as forecast by the current global climate models. The climate may change so greatly that entire remnant patches may no longer be able to support the species of trees that grow there now. As suggested by R. L. Peters and D. S. Darling (1985), parks and preserves may no longer provide habitat for

those species they were established to protect.

These problems can be divided into two spatial classes: events within small nature preserves and other areas with discrete external boundaries, set up for the conservation of specific species or ecological communities; and areas that are not artificially isolated but form a large, continuous pattern on the landscape. This chapter considers an example of each: a small nature preserve for the Kirtland's warbler near Grayling, Michigan, and the Boundary Waters Canoe Area (BWCA), a million-acre recreational and wilderness area in northern Minnesota. In this way we attempt to provide specific quantitative projections about the effects and the time of those effects suggested by Peters and Darling (1985).

### A. How Climate Projections Were Used with the Forest Model

Weather records were prepared for control (normal) conditions and treatments (weather as modified by the output from a specific climatic dynamic model). During control runs (climate under current greenhouse-gas concentrations) the JABOWA-II model used actual 30-year weather records (1951 to 1980) from the weather station nearest a point of interest. For projections of 100 years, the 30-year record was repeated. This provided a future climate like the recent past in terms of means and variation. For the Kirtland's warbler habitat, we used weather records from nearby Grayling, Michigan. To model forest growth in the BWCA, where there are no weather stations, we used records from Virginia, Minnesota, just outside the BWCA to the southwest.

To model climatic warming resulting from an increase in atmospheric concentrations of greenhouse gases, we obtained the climatic projections from the NASA Goddard global climate model, called the GISS model (Hansen et al. 1983, 1988b). Projections were obtained for the normal climate and two different warming scenarios, transient A and transient B. The normal projection was obtained by repeating the 1951–1980 weather records into the future. To generate the warming climates, ratios of warming and normal conditions from the climate model were calculated for each mean monthly temperature and mean monthly rainfall for each year. Each actual mean monthly value from the chosen weather station was multiplied by the corresponding ratios to generate "treatment" weather patterns. For example, given January mean temperatures of (1) 283°K for observed January temperature, (2) 263°K for steady-state 1980 climate model projection, and (3) 293°K for projected double-$CO_2$ climate model projection, the warming ratio is 293/263 and the observed temperature (283°) is multiplied by this ratio. The ratios for projected mean monthly temperature and mean monthly rainfall were calculated by R. Jenne of NCAR and provided to us in computer format.

Each warming scenario begins in 1980 and continues into the next century to a point at which the concentration of $CO_2$ in the atmosphere reaches twice the 1980 concentration (Hansen et al. 1988a). In scenario A, $CO_2$ concentration is doubled by 2070; the onset of warming occurs rapidly, but the changes in later decades take place more slowly than before. In scenario B, $CO_2$ concentration is doubled by 2040 but the onset of warming is delayed, so warming occurs rapidly as 2040 is approached.

The forest model is stochastic in that birth and death are treated as probabilities. The chance of reproduction and the chance of death are functions of environmental conditions. The chance of death is also a function of tree growth. This allows consideration of the mean and variance and the variability of forests over time. Experiments of 50 replicates with identical initial conditions and weather patterns over time were used to project means, standard errors, and confidence intervals. Initial conditions are given in table 21.1.

## B. Projecting the Effects of Rapid Climatic Warming on Kirtland's Warbler Habitat

Kirtland's warbler (Dendroica kirtlandii) is an endangered species whose highly specialized habitat requirements place it at considerable risk from a rapid climatic change.[2] The information on abundance of this species is among the best known for birds. It was the first songbird subject to a complete population census, which was done in 1951 (Byelich 1985). At that time the population was approximately 1000 birds, but the population declined rapidly to approximately 400 in 1971, causing concern among conservationists. It became clear that the decline was the result of the decrease in its nesting habitat, the young jack pine (Pinus banksiana) woodlands, which were in turn declining because of the suppression of forest fires (Line 1964, Mayfield 1969). The warbler builds its nests on the ground and, apparently because the nests must remain dry, typically only on dead tree branches still attached to a tree at ground level, and only on coarse sandy soil that drains away rain water rapidly. Although jack pine occurs throughout a large part of the boreal forests of North America, the Kirtland's warbler nests only in jack pine stands that occur on a single kind of soil, a coarse sandy soil called Grayling sand, which is found only in the lower peninsula of Michigan. In these woods, the warblers nest only in stands of trees between 5 and 20 feet high (about 6 to 21 years old); only jack pine of these ages have the dead branches at ground level necessary for nesting. Woodlands of this kind are used by the warbler only if they are larger than 80 acres, the average minimum territorial size of a male.

Jack pine is a fire-dependent species. Intolerant of shade, its seedlings cannot survive under the shade of adults, and regeneration therefore takes place only in openings. The cones open only after they have been heated by a fire, and the trees produce an abundance of dead branches, which provide fuel for wildfires. Fires are generally large enough so that regeneration depends on seeds from the mature trees within the burned area, effectively isolating forest patches after burns from

2. The following natural history of Kirtland's warbler is from Byelich 1985.

Table 21.1. Initial forest conditions for 400-year-old growth.

| Figure | Species | Stems ± 95% CI (per 100m²) | Basal Area (cm²/m² ± 95% CI) |
|--------|---------|---------------------------|------------------------------|
| 21.3 | no trees (clearing) | | |
| 21.5 | Balsam fir | 8.2 ± 0.2 | 2761 ± 47 |
| | Sugar maple | 4.0 ± 0.5 | 217 ± 7 |
| 21.6 | White cedar | 0.9 ± 0.03 | 2685 ± 91 |
| 21.7 | White birch | 12.9 ± 0.2 | 3369 ± 50 |
| | Sugar maple | 4.0 ± 0.03 | 79 ± 1 |
| 21.8 | Trembling aspen | 1.0 ± 0.1 | 2074 ± 70 |

outside seed sources. The warblers prefer woods that grow after fire to woods that grow following other kinds of clearing. The density of nesting warblers in a burned and regrown stand is about twice that of stands that have regrown following some other kind of disturbance (Byelich 1985).

At the present time about 4000 to 5000 suitable acres exist for breeding of the warbler. The species is protected by the 1973 Endangered Species Act, and a Kirtland's warbler recovery plan was prepared in 1976 and updated in 1985 (Byelich 1985). Thus management policies that involve money and time have been established, but they were devised before the new information from the global climate models became available and therefore do not consider the possibility of major, rapid climate change.

We have made projections for the warbler's habitat using 30-year weather records from Grayling, Michigan, and, lacking a direct field study of the physical and chemical characteristics of Grayling sands, using some reasonable assumptions about this soil. We have assumed that the soil has a moisture-holding capacity of 50 mm of water per meter depth of soil, a capacity of a very coarse sand. We have assumed that the soil is relatively fertile for a jack pine stand, which places it at poor (but not extremely poor) in nitrogen content. The soil is shallow in terms of the rooting depths and biological activity within it. Having relatively fertile soil for jack pine, the habitat is shared with other species, mainly white pine and red maple.

Projections for stands burned every 30 years beginning in 1980 are shown in figure 21.3. According to these projections, jack pine would persist and remain dominant if the 1951–1980 climate at Grayling, Michigan, were to be repeated into the future, but the species could not persist if the climate warms according to the projections of the GISS transient A. Most striking is the rapidity with which the pine declines. A stand burned in 1980 would not grow back to a woodland dominated by jack pine; according to the projections, for such a stand the growth of jack pine should be, in 1988, significantly less than would have been found in a stand burned in 1951 and measured in 1959. Since a stand seven years old is typically in use by the warbler, the projections of the model might be open to test today. Under the projected greenhouse-warming climate, white pine and red maple replace jack pine as dominants, but as the climate continues to warm even those species decline so that by year 2070 the projected habitat is a treeless plain. This area may perhaps assume the appearance of the stump barrens now found in the upper peninsula of Michigan, which were once occupied by white pine but since the logging and burning of the nineteenth and early twentieth centuries have been unable to support trees. Lichens, grasses, bracken fern, and small shrubs now dominate the barrens.

If that projection were to come true, Kirtland's warbler would not be able to persist in its current habitat area. If this species will nest only in jack pine forests that occur on the Grayling soils of Michigan, then the species may become extinct. The warbler might sur-

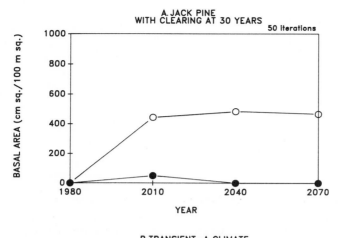

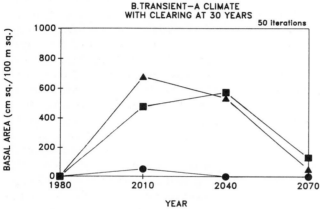

*Figure 21.3.* Projected tree growth in Kirtland's warbler habitat. Sites are on Grayling sand soil in the lower peninsula of Michigan and are burned every 30 years. A. Jack pine under 1951–1980 weather observed at Grayling, Michigan (O——O), and under the transition to doubled $CO_2$ projected by GISS (●——●). B. White pine (▲——▲), red maple (■——■), and jack pine (●——●) under the transition conditions. The soil for this hypothetical site is shallow (0.4 m from the surface to the base of biological use), has a moisture-holding capacity of 50 mm per meter depth, and a low nitrogen content. Initial conditions are a clearcut (no trees).

vive as a species if it nests northward where jack pine now grows over a wide area, but attempts to establish such breeding areas have yet to be successful.

Clearly, these projections have direct local management implications. If the projections are correct, current management actions may be to no avail. The management plans should then be reevaluated, perhaps with greater emphasis placed on learning why the warbler chooses such specific nesting sites and what might be done, considering the behavior of this species, to create suitable nesting areas to the north. A key management action would be an attempt in the near future to found a colony of Kirtland's warblers in a jack pine area of Canada, preferably in an isolated stand so that if the results are undesirable, inadvertent and undesired spread of the warbler could be controlled.

There are limitations to the projections given here. Before embarking on a major change in management, it is important to do the following: carry out field studies of the soils to determine physical and chemical characteristics; obtain measurements of the trees in actual warbler nesting areas to use as initial conditions for further simulations; carry out more simulation trials to test the sensitivity of the results to the accuracy of the parameters that define jack pine and to the equations that determine tree growth, regeneration, and mortality in the model. These tasks have not been done because no funds have been available for them.

The results have several general implications for Kirtland's warbler. Isolated woodlands even as large as 5000 acres, which exist near the edge of the range of the dominant tree species, may undergo rapid changes as a result of the greenhouse effect on climate, changes that might be observable today. Mammals and birds that depend on such woodlands are at risk and must seek similar habitats to the north.

## C. Projecting the Effects of Rapid Climatic Warming on the Boundary Waters Canoe Area

The example of the jack pine woodlands shows the implications of rapid climate change for a comparatively small, isolated remnant forest and for a species within it requiring specific habitat conditions that exist only locally. For that species, changes in its distribution rapid enough to keep up with climate change do not seem likely. But what would occur in areas large enough to permit migration of the species of interest within a continual stretch of woodland? One would expect that seeds could migrate within such an area and that the effects of climatic warming might be less severe than in the Kirtland's warbler habitat. For this example we have considered the response of major species of trees in the Boundary Waters Canoe Area (BWCA), a million-acre designated wilderness in the Superior National Forest, which provides additional contiguous woodlands, as does the Quetico Provincial Park to the north in Canada.

For the BWCA, weather records were obtained from Virginia, Minnesota, the nearest weather station with a long-term record. We examined forest growth both from clearings (initial conditions of no trees) and of old-age (400-year) forest stands for four soil types: deep relatively dry soil, deep relatively wet soil, shallow wetland soil, and shallow dry soil. These provide a broad range of forest conditions, from old-age cedar bogs and old-age balsam fir stands to regrowth of aspen on thin sandy soils.

For the BWCA, the projected transitional climatic regimes lead to surprisingly large changes in the forests (fig. 21.4), and these changes are surprisingly rapid, considering the natural rate of change in boreal and northern hardwood forests, but the character of the changes depends on soil moisture conditions. The GISS transient-A climate projects that a major change in forest composition would occur in the BWCA by year 2010, and transient B projects a major change by 2040. Although this is slower than the projections for jack pine, it is a rapid change compared with the natural rate of change in these forests. A 400-year-old stand dominated by balsam fir in 1980 and growing on a deep, fertile, moist soil would lose two thirds of its balsam fir basal area by 2010, replacing it with sugar maple as the dominant species (fig. 21.5). The climatic change actually improves the soil conditions for total tree growth. The soil had been too wet for optimal tree growth, but the warming and drying of the soil converts the area into a well-watered but well-drained site by 2010. On this site the total biomass nearly triples that which occurs under the 1951–1980 weather.

In contrast, in a true bog on a shallow wet soil, a forest undisturbed for 400 years and dominated by white cedar would decline in the next century to a treeless, shrub-dominated bog with total tree biomass less than 0.1 kg/m$^2$ (fig. 21.6). On a deep, drier, but fertile sandy upland soil, a forest undisturbed for 400 years is dominated by white birch in 1980, but that species declines to about 10% of its 1980 level in about 40 years, replaced by a sugar maple forest (fig. 21.7). This projection is consistent with that in

Before

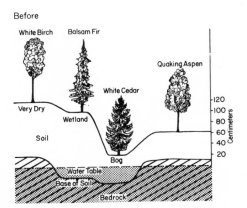

After

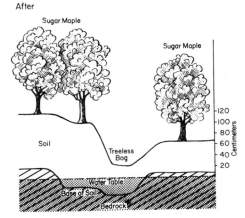

Figure 21.4 (above). Some major changes projected in forest composition in the Boundary Waters Canoe Area. Before: Conditions achieved under the 1951–1980 weather from Virginia, Minnesota. After: Growth under climatic conditions projected for this weather station when the $CO_2$ concentration of the atmosphere doubles, based on the GISS global climate model.

Figure 21.5 (below). Projected growth of balsam fir and sugar maple in the BWCA, beginning with a 400-year-old stand in 1980. Normal climate conditions, using 1951–1980 weather records from Virginia, Minnesota (O——O). GISS doubled-$CO_2$ climate (●——●). Transient A, one of the transitional scenarios projected by the GISS model from 1980 climate to doubled-$CO_2$ conditions (△——△). Transient B (▲——▲). The soil for this hypothetical site is deep (1.0 m), wet (depth to the water table, 0.8 m), moderately sandy, with a moisture-holding capacity of 250 mm per meter depth, and moderately fertile, assuming 250 kg nitrogen yearly above baseline.

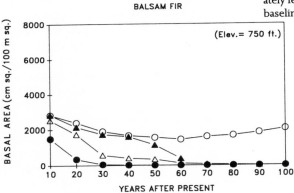

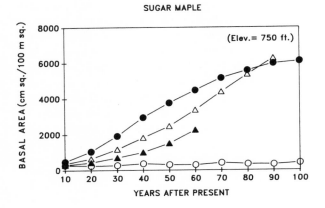

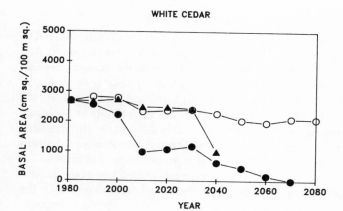

**WHITE CEDAR**

*Figure* 21.6 (*above*). Projected growth of northern white cedar in the BWCA, beginning with a 400-year-old stand in 1980. Normal climate (O——O). Transient-A climate (●——●). Transient-B climate (▲——▲). The soil for this hypothetical site is shallow (0.5 m), wet (depth to the water table, 0.2 m), moderately sandy, with a moisture-holding capacity of 250 mm per meter depth, and moderately fertile, assuming 250 kg nitrogen yearly above baseline.

*Figure* 21.7 (*below*). Projected growth of upland stands of white birch and sugar maple in the BWCA, beginning with old-growth stands in 1980. Normal climate (O——O). Transient-A climate (●——●). Transient-B climate (▲——▲). The soil for this hypothetical site is deep (1.0 m), wet (depth to the water table, 1.2 m), moderately sandy, with a moisture-holding capacity of 250 mm per meter depth, and moderately fertile, assuming 250 kg nitrogen yearly above baseline. White birch dominates under current climate; sugar maple dominates under altered climate.

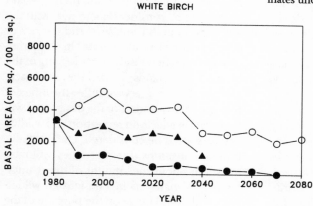

**WHITE BIRCH**

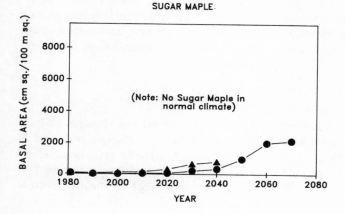

**SUGAR MAPLE**

(Note: No Sugar Maple in normal climate)

chapter 22, based on pollen analysis, for a generally southern die-off but northern expansion of sugar maple under the GISS-A scenario. Total biomass accumulation declines to about half after 90 years. On shallow, dry upland soil that begins as a clearing in 1980, trembling aspen dominates the forest under current climatic conditions but is replaced by a sugar maple forest under the transient climates (fig. 21.8).

The rapidity with which statistically significant changes occur is striking. According to these projections, the southern portion of the BWCA and the adjacent Superior National Forest may change drastically, from boreal evergreen forest to northern hardwoods, and this change may be readily observable sometime between 2010 and 2040. At least in the southern sections, the aesthetic quality of the BWCA would change dramatically from boreal forest to northern hardwoods forest. Outside the protected wilderness, within the Superior National Forest where commercial logging is important, there would be a major change in the dominant species, from boreal forest species useful for paper, pulp, and construction timber to northern hardwood species used primarily for furniture. Although the transition is rapid in terms of natural rates of change in forests, from the human viewpoint the forests would be neither aesthetically pleasing nor commercially useful for decades, a period during which the boreal forest species would be declining and the northern hardwood species not yet grown to maturity. Even when fully developed, the resulting forest might be unlike any forest of today. Sugar maple would be much more dominant than in present northern hardwood forests, and as a result the biological diversity would be less than in present northern hardwood forests.

Another problem that might arise would be continued climate change after 2070. If $CO_2$ concentration continued to increase, the climate would continue to warm. It is conceivable that even large contiguous areas such as the BWCA might continue to undergo such rapid climatic change that seedlings could not

reach maturity before the climate changed to conditions unfavorable for their growth. As a result, large areas might remain treeless, dominated instead by shorter-lived annual and perennial herbs and shrubs, as was projected for the jack pine stands in Michigan, which might only temporarily support large trees of any kind and then decline to barrens dominated by grasses, ferns, and shrubs.

To increase our confidence in the projections, we need actual measurements for the initial conditions, including measurements of actual soil and stand conditions. It is also important to carry out simulations for the northern portion of the BWCA, and for the Canadian Quetico Park to the north, to determine whether any of the designated wilderness might retain boreal characteristics. It is important to test the robustness of the results, of the accuracy in the estimation of parameters used in the model, and to set up stands to be monitored over time to validate the projections. If the results stand up to these extensions of the work, then it would be prudent to consider how to deal with the transition in both the BWCA and the rest of the Superior National Forest. In the latter, managers might consider early logging of the boreal forest stands, before they die back, and planting or otherwise aiding the influx of northern hardwoods. For the designated wilderness, the effects on recreation and wildlife habitat should be considered. Undoubtedly, if these transitions take place they will raise questions about the amount of direct interference by managers in a designated wilderness, which will hinge on the purpose of the wilderness: is it to provide an area that appears to be a presettlement boreal forest, or is it to provide a preserve undisturbed by any human influence, regardless of the consequences.

## IV. CONCLUSIONS

Forests of Minnesota and Michigan may undergo rapid, dramatic changes as a result of climatic change due to the greenhouse effect. Habitat for endangered species may be lost, and species requiring these habitats, such as

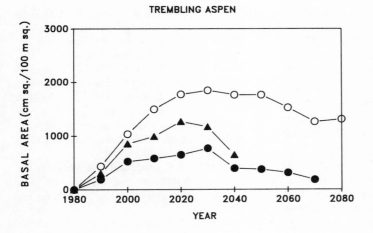

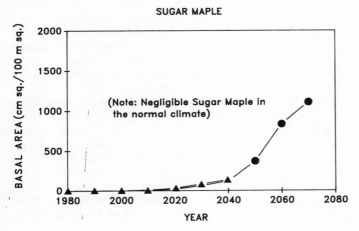

Figure 21.8. Projected growth of trembling aspen and sugar maple in the BWCA, beginning with a clearcut on a shallow, dry upland soil in 1980. Normal climate (O——O). Transient-A climate (●——●). Transient-B climate (▲——▲). The soil for this hypothetical site is shallow (0.5 m), wet (depth to the water table, 0.6 m), moderately sandy, with a moisture-holding capacity of 250 mm per meter depth, and moderately fertile, assuming 250 kg nitrogen yearly above baseline. Aspen dominates under current climate; sugar maple dominates under altered climate.

Kirtland's warbler, may become extinct unless new management policies are devised and unless the factors limiting the geographic range of species are better understood. Even in forests large enough to permit seed migration over considerable distances, dramatic changes in forest composition are projected. These changes are so rapid that there may be transitional decades when dying and poorly growing trees dominate the landscape, under which would be growing seedlings and saplings of species now found to the south. We must consider our management options at three scales: the local scale of the Kirtland warbler habitat and the Boundary Waters Canoe Area, the regional scale of national policy, and the scale of the biosphere. We can begin by consideration of the local effects, as discussed in this chapter for the warbler habitat and the BWCA. If the projected changes are untenable and no mitigating actions seem possible, we may seek global approaches that reduce the production of greenhouse gases or lead to the greater absorption of these gases by terrestrial vegetation or marine ecosystems.

The prospect of future global climate change challenges us to reconsider fundamental aspects of our conservation ethic. We may no longer be satisfied to maintain the environmental status quo, since the natural environment for most areas will change drastically during the next 50 years. We may be forced to reconsider the meaning of "wilderness" and to resolve what it is we intend to conserve when we designate such areas. Letting nature take its course may no longer be a satisfactory option. Conservationists, as well as commercial-area managers, must design strategies that account for global climate change. For example, efforts might be redirected to acquire and manage areas that will soon develop into habitats for future preservation efforts. In these efforts, we should integrate forest growth modeling into the initial planning process to analyze candidate sites.

The projections suggest severe and undesirable changes, which could occur so soon that there may be little time to wait before making policy decisions. An essential first step is to increase our confidence in the projections through a series of additional simulations. If they support the results given here, prompt action may be necessary if we are to conserve our northern forests and their habitats for wildlife.

## REFERENCES

Aber, J. D., D. B. Botkin, and J. M. Melillo. 1978. Predicting the effects of different harvesting regimes on forest floor dynamics in northern hardwoods. *Can. J. Forest Res.* 8:306.

Aber, J. D., D. B. Botkin, and J. M. Melillo. 1979. Predicting the effects of different harvesting regimes on productivity and yield in northern hardwoods. *Can. J. Forest Res.* 9:10.

Botkin, D. B. 1982. Can there be a theory of global ecology? *J. Theor. Biol.* 96:95.

Botkin, D. B. 1990. *Discordant Harmonies: A New Ecology for the 21st Century.* New York: Oxford University Press.

Botkin, D. B. In press. *The JABOWA Book: The Ecology of Forests, Theory and Practice.* New York: Oxford University Press.

Botkin, D. B., and R. E. Levitan. 1977. Wolves, moose and trees: An age specific trophic-level model of Isle Royale National Park. IBM Research Report in Life-Sciences RC 6834.

Botkin, D. B., and M. J. Sobel. 1975. Stability in time-varying ecosystems. *Am. Naturalist* 109:625.

Botkin, D. B., J. F. Janak, and J. R. Wallis. 1970. A simulator for northeastern forest growth: A contribution of the Hubbard Brook Ecosystem Study and IBM Research. IBM Research Report 3188. Yorktown Heights, N.Y.

Botkin, D. B., J. F. Janak, and J. R. Wallis. 1972a. Rationale, limitations and assumptions of a northeast forest growth simulator. *IBM J. Res. and Dev.* 16:101.

Botkin, D. B., J. F. Janak, and J. R. Wallis. 1972b. Some ecological consequences of a computer model of forest growth. *J. Ecol.* 60:849.

Botkin, D. B., J. F. Janak, and J. R. Wallis. 1973. Estimating the effects of carbon fertilization on forest composition by ecosystem simulation. In *Carbon and the Biosphere*, G. M. Woodwell and E. V. Pecan, eds., Brookhaven National Laboratory Symposium 24, pp. 328–344. Oak Ridge, Tenn.: Technical Information Center, USAEC.

Byelich, J., M. E. DeCapita, G. W. Irvine, R. E. Radkey, N. I. Johnson, W. R. Jones, H. Mayfield, and W. J. Mahalak, eds. 1985. *Kirtland's Warbler Recovery Plan.* Washington, D.C.: Dept. of Interior, Fish and Wildlife Service.

COHMAP. 1988. Climatic changes of the last 18,000 years: Observations and model simulations. *Science* 241:1043.

Davis, M. N. 1983. Holocene vegetational history of the eastern United States. In *Late Quaternary Environments of the United States.* Vol. 2, *The Holocene*, H. E. Wright, Jr., ed., pp. 166–181. Minneapolis: University of Minnesota Press.

Davis, M. D., and D. B. Botkin. 1985. Sensitivity of the cool-temperate forests and their fossil pollen to rapid climatic change. *Quatern. Res.* 23:327.

Hansen, J., G. Russell, D. Rind, P. Stone, A. Lacis, S. Lebedeff, R. Ruedy, and L. Travis. 1983. Efficient three-dimensional global models for climate studies: Models I and II. *Monthly Weather Rev.* 111:609.

Hansen, J., I. Fung, A. Lacis, S. Lebedeff, D. Rind, R. Ruedy, and G. Russell. 1988a. Prediction of near-term climate evolution: What can we tell decision-makers now? In *Preparing for Climate Change*, Proceedings of the first North American conference on preparing for climate change: A cooperative approach, J. C. Topping, Jr., ed. Washington, D.C.: Government Institutes.

Hansen, J., I. Fung, A. Lacis, D. Rind, S. Lebedeff, R. Ruedy, and G. Russell. 1988b. Global climate changes as forecast by Goddard Institute for Space Studies three-dimensional model. *J. Geo. Res.* 93:9341.

Leemans, R., and I. C. Prentice. 1987. Description and simulation of tree-layer composition and size distributions in a primaeval *Picea-Pinus* forest. *Vegetatio* 69:147.

Line, L. 1964. The bird worth a forest fire. *Audubon* 66:371.

Marsh, G. P. 1864. *Man and Nature.* 1967 reprint edited by D. Lowenthal. Cambridge, Mass.: Belknap Press.

Mayfield, H. 1969. *The Kirtland's Warbler.* Bloomfield Hills, Mich.: Cranbrook Institute of Science.

Pastor, J., and W. M. Post. 1985. Development of a linked forest productivity–soil process model. Oak Ridge Natl. Lab Rep. ORNL/TM-9519.

Pastor, J., and W. M. Post. 1986. Influence of climate, soils moisture, and succession on forest carbon and nitrogen cycles. *Biogeochemistry* 2:3.

Pastor, J., and W. M. Post. 1988. Response of northern forests to $CO_2$-induced climate change. *Nature* 334:55.

Peters, R. L., and D. S. Darling. 1985. The greenhouse effect and nature reserves. *Bioscience* 35:707.

Prentice, I. C. 1986. The design of a forest succession model. In *Forest Dynamics Research in Western and Central Europe,* J. Fanta, ed., pp. 253–256. Wageningen: Center for Agricultural Publishing and Documentation.

Russell, E.W.B. 1983. Indian-Set Fires in Northeastern United States. *Ecology* 64:78.

Shugart, H. H. 1984. *A Theory of Forest Dynamics.* New York: Springer-Verlag.

Shugart, H. H. 1988. *International Forest Modeling Newsletter.* Charlottesville: University of Virginia.

Smith, T. M., H. H. Shugart, and D. C. West. 1981. The use of forest simulation models to integrate timber harvest and nongame bird habitat management. *46th North American Wildlife and Natural Resources Conf.* Washington, D.C.: Wildlife Management Institute.

Solomon, A. M., and D. C. West. 1987. In *The Greenhouse Effect, Climate Change, and U.S. Forests,* W. E. Shands and J. S. Hoffman, eds., pp. 189–217. Washington, D.C.: The Conservation Foundation. Climate projections from J.F.B. Mitchell, *Q. J. R. Meteor. Soc.* 109(1983):113.

Thomas, W. L., ed. 1956. *Man's Role in Changing the Face of the Earth.* Chicago: University of Chicago Press.

Weinstein, D. A., and H. H. Shugart. 1983. Ecological modeling of landscape dynamics. In *Disturbance and Ecosystems,* H. Mooney and M. Godron, eds., pp. 29–45. New York: Springer-Verlag.

West, D. C., H. H. Shugart, and D. B. Botkin, eds. 1981. *Forest Succession: Concepts and Applications.* New York: Springer-Verlag.

This work was supported in part by Environmental Protection Agency grant CR-814595-0-10, National Science Foundation grant DEB-80-12159, 1980–83, and a grant from the Andrew J. Mellon Foundation. Results do not necessarily reflect views of any of the granting organizations, and no official endorsement should be inferred from this publication.

# PART V

# Implications for Conservation

# CHAPTER TWENTY-TWO

# Changes in Geographical Range Resulting from Greenhouse Warming: Effects on Biodiversity in Forests

MARGARET B. DAVIS AND
CATHERINE ZABINSKI

## I. INTRODUCTION

Changes in the geographical distributions of plant and animal species in response to future greenhouse warming threaten to reduce biotic diversity. Many plants and animals are now confined to fragmented habitats that are protected within reserves. Although these small populations of plants and animals are already vulnerable to chance fluctuations in climate, a persistent warming trend will place them at increased risk.

The risk posed by $CO_2$-induced warming depends on the distances that regions of suitable climate are displaced northward and on the rate of displacement. If the geographical limits of a species' range move only a few kilometers northward, reserves at the southern edge of the range are the only ones affected. But if areas of suitable climate are displaced hundreds of kilometers within a few decades, there may be no reserves left that have a climate suitable for continued survival (Peters and Darling 1985). If the change occurs too rapidly for colonization of newly available regions in the north, population sizes may fall to critical levels, and extinction will occur (fig. 22.1).

This chapter presents a quantitative estimate of the range displacement expected as the result of doubled $CO_2$ for four widespread tree species that commonly grow together in northern hardwood forests. These species are not endangered but were chosen for analysis because their present distribution is well known, and there is a body of literature on their physiology and response to climate. Further, an extensive fossil record reveals their response to climate in the past and their ability to colonize new areas. Because they are dominant canopy trees in the forest, their continued presence in the forest community affects many animals and understory plants, among them rare species for which reserves have been established.

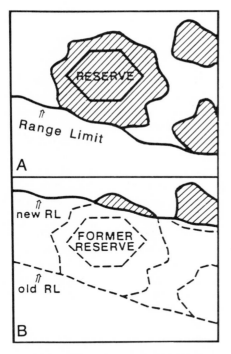

Figure 22.1. Effect of a change in geographical distribution brought on by greenhouse warming on a species population in a reserve. Hachured area is habitat where the species of interest survives. RL, range limit. A: Present reserve and species distribution. B: Future reserve and species distribution. Modified slightly from Peters and Darling 1985; copyright © 1985 by the American Institute of Biological Sciences.

After predicting those geographical displacements, we discuss the probable time course of changes in forest reserves, focusing primarily on reserves containing American beech (*Fagus grandifolia*). Four locations are considered: reserves near southern range limits, reserves near the center of species' ranges, reserves in the region of overlap between present and future ranges where the trees can be expected to survive, and reserves in the new regions of potential habitat that will be opening up to the north. To project a time course for response at these locations, we have had to assume a time course for changes of climate. We are taking a fairly conservative view by assuming that climatic conditions projected for doubled $CO_2$ by general

circulation models will be reached at A.D. 2090 (Jaeger 1988). The actual timing of warming will of course depend on the rate of production of greenhouse gases and the sensitivity of the earth system. The discussion of effects and their timing results in a series of suggestions for reserve management to meet the challenge posed by greenhouse warming.

## II. COMPARISONS OF PRESENT GEOGRAPHICAL RANGES AND FUTURE POTENTIAL RANGES

Present geographical ranges of four important forest trees—eastern hemlock (*Tsuga canadensis*), yellow birch (*Betula alleghaniensis*), beech (*Fagus grandifolia*), and sugar maple (*Acer saccharum*)—were compared with maps of climatic variables to determine the threshold values of temperature and moisture that correspond to each species' range limits. These values comprise the boundaries of an area of suitable climate where each species is able to grow. We then used the predictions of general circulation models for climate at equilibrium with doubled $CO_2$ to predict future locations of suitable climate areas for each species. We have not included the direct effects of increased levels of $CO_2$; for some plants elevated $CO_2$ seems to compensate for high temperature and water stress (Gates 1983), but experiments have not yet been done for these four species.

The relation between climate and plant distribution is often complex (Denton and Barnes 1987, Woodward 1986), but because of the high correlation between measures of climate variables and because of the limited range of parameters that the climate models specify, we selected the simplest variables that correspond most closely with range distributions: mean January temperature, mean July temperature, and annual precipitation. The northern distribution limit for many trees of the eastern deciduous forest occurs where winter minimum temperatures drop as low as −40°C and also where the mean January temperature is −15°C. Although the winter minimum is probably a more meaningful variable, since xylem freezes at −41°C for three of the four species (Sakai and Weiser 1973, George et al. 1974), the coincidence of the northern distribution limits with mean January temperature suggests that mean temperature is correlated with temperature extremes. The 25°C mean July isotherm coincides closely with the southern distribution of hemlock and yellow birch, as do cumulative measures of days with temperatures above 40°F (Botkin et al. 1970).

In cases where there is not an isopleth that coincides with the species' geographic limit, tabulations were made of the values of all three variables (January and July mean temperatures, and precipitation) within the modern geographic range. Future distribution was estimated by comparing the climate model's output at a gridpoint with current climate data within the species' distribution. A gridpoint was assumed to be within the species' potential distribution if the combination of predicted January and July mean temperature and the annual precipitation values are identical to a combination of those three variables within the current range distribution. We cannot preclude the possibility that new combinations of precipitation and temperature might also permit growth, but in practice this procedure was used mainly along the western and southern limit where moisture stress is likely to continue to be the limiting variable.

Two rather different models were used to generate two climate scenarios, a conservative and a more radical prediction of effects on species' ranges. The GISS model (Hansen et al. 1983) predicts elevated temperatures throughout eastern North America with slightly increased moisture in the Great Plains region. The GFDL model (Manabe and Wetherald 1987) predicts elevated temperatures and decreased soil moisture. The two future potential range maps that result provide a range of possible outcomes that summarize our best information on climate change and the species' response to that change.

The locations of boundaries on these maps are approximations. The degree of certainty is constrained both by the limits of accuracy of the global climate models and the simplifying assumptions we have made in the construction of our predictions. Because the global climate models make different simplifying assumptions, they differ from one another in the magnitude of temperature rise and the precipitation change they predict for the future. The models generalize geography, leaving out features such as the Great Lakes that would have an effect on weather patterns. The gridpoints at which predictions have been made are as much as 500 km apart—see figure 22.2, where the gridpoints for GISS and GFDL are indicated. We made a linear interpolation of climate values between the gridpoints, ignoring topographical features such as the Appalachian Mountains, which are too detailed to show on maps at this scale.

Climate scenarios from both models will result in large displacements of geographical ranges—greater than 500 km for all four species. The GISS model output translates to a loss of sugar maple from a region 200–600 km wide along the southern edge of the range, contraction eastward about 100 km in Minnesota, and expansion northward 800 km (fig. 22.2a). A combination of summer temperatures and low levels of precipitation will limit sugar maple in both the south and the west. The new northern limit will be restricted not by cold winter temperatures, but by cool summer temperatures.

With GFDL output, we predict that sugar maple will die out throughout its range except in Maine, eastern Quebec, and Nova Scotia (fig. 22.2b). Sugar maple will be limited by the combination of low precipitation and high summer temperatures in the south, by low precipitation in the western edge of its range, and by cold winter temperatures in the north. New potential habitat would extend 500 km northward in Quebec and continue in a narrow tongue 400 km west of James Bay.

Similar patterns are seen for yellow birch and for hemlock (figs. 22.3 and 22.4), except that hemlock's range is more restricted in the

SUGAR MAPLE

A                     B

*Figure 22.2.* Present geographical range of sugar maple (horizontal lines) and potentially suitable range under doubled $CO_2$ (vertical lines). Cross-hatching indicates the region of overlap. A: Predictions using climate scenario derived from the GISS general circulation model. B: Predictions using climate scenario derived from the GFDL model. Gridpoints are sites of climatic data output for each model.

*Figure 22.3.* Present geographical range of yellow birch (horizontal lines) and potentially suitable range under doubled $CO_2$ (vertical lines). A: GISS scenario. B: GFDL scenario.

west, a pattern reflective of its lower tolerance for warm, dry conditions. Once again, under the GISS scenario the boundaries of the potential ranges are limited by low summer temperatures in the north and by a combination of low precipitation and high summer temperatures in the west and the south. Using the GFDL scenario, the limits are determined by cold winter temperature in the north, low precipitation in the west, and high summer temperature in the south.

Beech is the most severely affected (fig. 22.5), becoming rare or dying out even under the GISS scenario over a region 1500 km wide in the eastern United States (except for high

YELLOW BIRCH

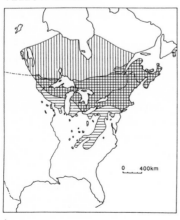 

A                     B

HEMLOCK

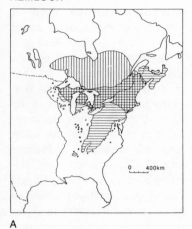 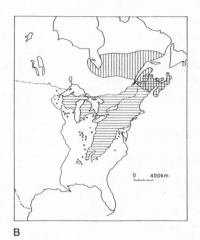

A                                              B

Figure 22.4. Present geographical range of hemlock (horizontal lines) and potentially suitable range under doubled $CO_2$ (vertical lines). A: GISS scenario. B: GFDL scenario.

Figure 22.5. Present geographical range of beech (horizontal lines) and potentially suitable range under doubled $CO_2$ (vertical lines). A: GISS scenario. B: GFDL scenario. Numerals indicate hypothetical locations of reserves that are (1) near the southern limits of the present range, (2) near the center of the present range, (3) in the region of overlap, and (4) in the region of potentially suitable range. The four types of reserve are in similar locations regardless of the climatic scenario considered.

elevations), while new habitat of much smaller area would open up in Ontario and Quebec. Under the GFDL scenario, beech would continue to grow only in northernmost Maine and Nova Scotia, while a larger but mostly noncontiguous habitat would open up in Quebec and northern Ontario. The patterns produced by either climate model are primarily the result of a shortage of precipitation. Even though beech is currently present throughout the southeastern United States, which implies that it might tolerate climatic warming in the northern section of its range, the predicted annual

BEECH

A                                              B

precipitation under either scenario would be limiting in the west and in the south, given the increase in temperature. The northern limit would be determined by low precipitation and cool summer temperatures under the GISS scenario, and low winter temperature under the GFDL scenario.

## III. TIME COURSE OF CHANGE

### A. Reserves near Southern Range Limits

Trees growing near their southern range limits will be the most severely affected by climatic warming and will be affected earliest. Within a few decades seedlings of species near range limits will disappear from the forest, while seedlings of species that tolerate higher temperatures will become more abundant. Growth rates of trees, also sensitive to climate (Fritts 1976), will decline. Nevertheless, most canopy trees will continue to live for several decades longer (Davis and Botkin 1985), because adult plants and animals can tolerate a wider range of climatic conditions than earlier life stages (Neilson and Wullstein 1983, Brubaker 1986; see also chapter 12). Many species of perennial plants can resprout after damage from frost or drought, maintaining a presence in plant communities under climatic conditions where flowering and seedling establishment can occur only in exceptional years (Davis 1984). Well-documented examples are from populations along northern range limits: *Picea mariana*, which reproduces by layering, has persisted for centuries without sexual reproduction in groves north of the forest limit in the Canadian Arctic (Payette 1983), and *Tilia cordata*, which can reproduce as sprouts, grows in the Lake District of England although the climate is too cold for successful seed set. Fossil pollen indicates that *Tilia* spread into the region during an interval of warmer climate 5000 years ago (Pigott and Huntley 1978).

As warming proceeds, higher temperatures and decreased moisture will eventually affect the trees directly. Thresholds that will cause physiological failure are not known for the four species we have studied, making it difficult to predict the precise time course for tree death. Greenhouse experiments are of limited relevance for mature trees growing in forest stands. There are a few field reports of tree deaths following unusual weather events; for example, hemlock trees in Wisconsin died during the drought years of the 1930s (Secrest et al. 1941). Long-term monitoring is needed to obtain additional information. For example, a drought in Panama associated with El Niño caused high, differential mortality of stems between 1980 and 1983 in a mapped rain forest plot (Hubbell and Foster 1991). Increased incidence of disease and insect outbreaks on weakened trees is to be expected. The immediate cause of death of hemlock trees in Wisconsin during the 1930s, for example, was insect attack; the attacks proved fatal to trees that had already lost up to 90% of their roots because of moisture stress (Secrest et al. 1941).

In old-growth northern hardwood forests under today's climate, disturbance by wind or fire is a common cause of death of old and large trees. Once trees have reached the canopy of Great Lakes hardwood forests, they live on average 150–175 years before removal by wind (Frelich 1986). In the Great Smoky Mountains, where disturbances occur more frequently, trees average 83 years in the canopy (Runkle 1982). Climatic changes associated with greenhouse warming will affect disturbance regimes (chapter 19), increasing the numbers of tropical hurricanes; dry years will occur more often, increasing the frequency of drought-associated wildfires (Schroeder and Buck 1970, Haines and Sando 1969). Increased attack from insects and pathogens on trees that are stressed by climate will also shorten the average lifespan of forest trees, accelerating the removal of older trees. Trees that persist as long as 50 years might in any case experience direct physiological stress from high temperatures, as the general circulation models we are using suggest that 50 years from now (given $CO_2$ doubling by A.D. 2090) summer temperatures in the southeastern United States might be 1° to 4°C warmer than today.

In forested reserves in the Southeast (site 1 in fig. 22.5; examples are the Big Thicket in southeast Texas, the Apalachicola Bluff and Ravines Preserve in Florida, or the Joyce Kilmer Reserve in North Carolina), beech can be expected to show an early response to changed climate through decreased seed production, decreased frequency of good seed years, and reduced growth rates. The effect may be immediate, because even at present there is some indication of infrequent seed production by beech near its western limit, where mast years may be less frequent than elsewhere within the range (Ward 1961, cf. Woods and Davis 1988). Northern populations, such as those in the Jung Beech-Hemlock Forest in Wisconsin, however, form root sprouts, which might permit continued reproduction within forest stands even after seed production ceases (Ward 1961, Held 1983). Southwestern populations, which will be subject to the earliest climatic stress, appear to sprout less frequently (Ward 1961, Held 1983). Beech is easily damaged by fire, and it competes poorly in forests subjected to frequent disturbance by windstorm (Henry and Swan 1974). Droughts, exceptionally warm summers, or winters too warm to maintain dormancy will stress trees and increase susceptibility to pathogens and insect attack. With increased disturbance and the temperature scenarios we are using here, persistence of beech for longer than 50 years in reserves in the Southeast seems unlikely.

To compound the difficulties of management, it may become difficult to protect forest reserves. Moribund beech forests will have little aesthetic appeal and may also be viewed as fire hazards or centers for disease that endanger surrounding vegetation. In addition, the forests will become increasingly isolated, losing buffer habitats. Commercial interests can be expected to harvest timber on surrounding land on an accelerated schedule, to replace trees that are no longer growing rapidly with seedlings of species more likely to thrive under warmer climate.

## B. Reserves near the Centers of Species Ranges

Reserves located farther north, near the center of the beech range (site 2 in fig. 22.5), will show the same effects of climatic change, but on a schedule delayed by four or five decades. Again, as soon as climatic thresholds associated with the range limit are exceeded, individual trees will display physiological stress, growing slowly and failing to flower and fruit. Actual loss of local populations may occur much later as the last adults succumb, perhaps 40 or 50 years after reproduction begins to decline. The Appalachian Mountain chain makes up the center of the range of beech; the variety of habitat and the rugged topography that inhibits fire and wind disturbances, as well as logging, may enable beech trees to persist longer than elsewhere. The highest elevations in the Smoky Mountains National Park, for example, will provide a refuge even after doubled-$CO_2$ climate has been reached. With a lapse rate for July temperature of about 2°C per 1000 m (Botkin et al. 1970) the summits of the mountains will remain suitable for beech, even under the more extreme GFDL scenario. Beech grows as high as 1800 m today on south-facing slopes in the central Appalachians (Fowells 1965); presumably these trees will grow more vigorously under the warmer climate (providing rainfall is sufficient) and new stands will become established on north-facing slopes and in ravines and valleys. In the New England mountains (White Mountain National Forest) beech grows today as high as 750 m (Bormann et al. 1970); here it is expected to move upslope, perhaps as high as the present-day treeline. The problem of ecotypic differentiation discussed below is less serious in mountainous regions, where distances are short between elevations with quite different climate regimes. Presumably birds will disperse seed from low elevation trees to higher elevations.

## C. Reserves in Regions where Survival Is Expected

Survival of populations in the region of overlap between present-day and predicted ranges (site 3 in fig 22.5) depends on the degree of ecotypic specialization. If local individuals can tolerate the full range of climate that characterizes the geographic range of the entire species, a change in Maine to climate similar to Georgia will have little effect on local trees. But if the local population is highly specialized and adapted to the present-day climate of Maine, the changes in climate will exceed the tolerance limits of individual trees and cause tree death, despite the similarity of the changed climate to conditions presently tolerated by southern populations of the same tree species.

Beech is reported to have three well-differentiated subspecies (Camp 1950): white beech, found mainly at low elevations in the south and west; red beech, which grows at low elevations in the Appalachians and at sea level northward to Nova Scotia; and gray beech, which grows in association with spruce and fir at high elevations in the Appalachians and Adirondacks, northward to the border of the coniferous forest in Nova Scotia and westward to the Great Lakes region. On morphological criteria, Camp recognized that these three types were sharply delineated in some regions and mixed in others. If the populations are as distinct physiologically as he believed they are genetically, it seems unlikely that northern gray beech would be able to tolerate future climatic conditions that resemble those presently characterizing the range of white beech. If so, the high-elevation populations in the Appalachians will die out unless replaced by genotypes from lower elevations, and northern populations will die out unless replaced by white beech from the south. Seed or seedlings from populations located several hundreds of kilometers to the south will have to be introduced artificially into forests to replace ecotypes less well adapted to the changed climatic conditions.

Tree species differ in the pattern of genetic variability. Those with variable local populations may contain genotypes that can tolerate the new conditions—and that can increase within the population as the climate warms.

One hundred years is only one or at most two generations of trees—too short a time for the many generations necessary to accomplish evolutionary adaptation. It is possible, however, that significant change in allelic frequencies could occur within certain populations in response to selection by the changing climate. In theory, genetic material from southern populations could also move northward by natural means, but the rates of diffusion of genes through a population are probably not much faster than rates of range extension and therefore too slow to have much impact in the coming century.

Additional studies of geographical variation are sorely needed, especially for climatic sensitivity of characteristics of the tree that enhance its chance for survival (such as fecundity and ability to sprout). For example, there is evidence for the existence of two physiological races of hemlock in Wisconsin: seedlings grown from seed taken from warmer, drier sites showed higher water use efficiency (Eickmeier et al. 1975). Photoperiod responses are a particular problem, as the day length at particular latitudes and the climate there represent unique combinations. Consequently, many forest trees display latitudinal variations in photoperiod responses (Vaartaja 1959, Kramer and Kozlowski 1979).

Provenance studies investigate geographical variation in physiological response. In provenance trials, seed from many geographical sources is planted in uniform habitats, and growth is monitored. To test for variation in the responses of full-grown trees, provenance studies will need to be continued for many decades. To date, most provenance studies concern commercially valuable trees that are grown in plantations; despite the economic importance of sugar maple and yellow birch, only one ongoing provenance study (for sugar maple) is listed in a recently published compendium (Guries et al. 1981). This single orchard is only 18 years old, hardly time enough to assess the performance of trees that generally spend a century in the forest canopy. Many more such experiments are needed for a much greater variety of trees if we are to predict their chances of survival in the regions of range overlap where future growth seems most likely. These studies should be started now, as many years of observation are needed to predict the behavior of canopy forest trees in natural stands.

## D. Future Reserves North of Present Species Limits

How rapidly can populations of trees extend their ranges to take advantage of the new habitats opening up to the north? Possible rates of population expansion are better known for trees than for other plant species because past changes in range are documented by the fossil record. Fossil pollen in sediments records the northward diffusion of trees as climate warmed beginning about 15,000 years ago at the opening of the present interglacial interval. The hundreds of sedimentary pollen records available from both eastern North America and Europe document population diffusion at rates varying between 10 and 45 km per century (Davis 1981, Huntley and Birks 1983, Davis and Jacobson 1985, Gaudreau and Webb 1985, Webb 1988; see also chapter 5). The fastest rate recorded, 200 km per century, is for spruce moving into northwest Canada 9000 years ago (Ritchie and MacDonald 1986). At this time the climate of that region was warming rapidly, and prevailing winds from the southeast may have facilitated seed dispersal.

Beech diffusion northward has been studied in particular detail (Davis 1981, 1987; Davis et al. 1986; Woods and Davis 1989; S. L. Webb 1986, 1987; Bennett 1985, 1988; T. Webb III 1986). The pollen record in lake sediments suggests that the population frontier advanced at rates that averaged around 20 km per century, being more rapid along the eastern seaboard and slower in the Great Lakes region. On a continental scale and on a time scale of millennia, the rates of dispersal and expansion reflected the rate at which the climate ameliorated (T. Webb III 1986).

The dispersal of beech nuts by jays has been studied by W. C. Johnson and C. S. Ad-

kisson (1985). Jays carry nuts several kilometers and bury them in caches in wooded areas, thus dispersing seed to favorable habitats. Given the long generation time for beech, 40 years before seed production (Fowells 1965), dispersal by jays is just adequate to explain range extensions at rates of 20 km per century. S. L. Webb (1986) argues that passenger pigeons could have facilitated dispersal, especially across large geographical barriers. As the passenger pigeon is now extinct, jays will be the principal natural agent for future dispersal northward.

Even the most rapid dispersal rates known from the past, however, are not rapid enough to track greenhouse warming. The rate of future climatic warming depends on future rates of production of greenhouse gases, and on the sensitivity of the climate system (Jaeger 1988). The scenarios we are using suggest warming at rates that move the beech limit northward 700–900 km within the next 100 years (fig. 22.5). To track this shift in range, beech would need to diffuse northward 40 times more rapidly than the rates recorded in the past. Ignoring the fossil record and considering the population biology of beech, it is obvious that even if seeds were to reach the far north and colonies were to become established, a number of years—40 for beech (Fowells 1965)—must elapse before trees are sufficiently mature to produce seed. Under natural conditions, invading beech trees could not possibly populate a large region within a century.

Soils, light, microclimate, and plant-animal interactions, all affected by community context, are critical for the establishment of seedlings of both trees and herbs within a forest stand (see chapter 24). It therefore seems likely that the kind of vegetation growing in the new areas that become climatically suitable will affect the success of colonization by species moving northward. The fossil record provides little information on this question, because community composition and climate cannot be separated when past invasions are compared in different geographical regions (Moe 1970). Modeling would be helpful here, if models can be appropriately designed to incorporate the complex community interactions and substrate properties that affect seedling dispersal, germination, and growth. The present generation of forest stand models does not attempt to simulate these processes, entering saplings rather than seeds into the simulations (Botkin et al. 1970, Shugart 1984).

## IV. RECOMMENDATIONS

Foresters will surely establish plantations of desirable tree species wherever they can grow, thus dispersing and establishing many commercially valuable trees far beyond the region to which they would be dispersed by natural means. Seeds are commonly broadcast from the air following forest fires, and this methodology could be employed wherever conditions are favorable for seedling establishment. The resulting plantations may provide habitat for some animals and plants that move northward in pace with the climate, providing they are not species-specific in their requirements for forested habitat (Salomonsen 1951, Van Devender 1984, Rogers 1980).

For the survival of many other plant and animal species, however, entire forest ecosystems, including dominant trees, understory species, and animals that affect ecosystem function, will have to be reproduced artificially on a large scale in northern locations. Methods for speeding up succession, such as manipulation of nutrients and selective thinning of early successional species, will have to be developed to create forests that resemble forests to the south. Techniques now being developed for ecosystem restoration will have to be applied on a large scale to transplant ecosystems to the new climate areas that will become available. The time frame is too short, however, to create communities resembling old-growth forest. Furthermore, the suitable climate area at doubled $CO_2$ will represent a moving target, as additional changes in greenhouse-gas concentrations continue to change the earth-climate system.

Research on physiological thresholds of forest trees growing in natural environments is needed to predict the time course of species losses in the southern regions where they are expected to die out, and more information on ecotypic variation is imperative to predict the chances for survival of local populations in the regions where changed climate will fall within the tolerance limits of the species as a whole. Predictions of the time course of changes will allow orderly planning.

To preserve biodiversity, new reserves must be established at all latitudes, since range displacements will affect species in the subtropics as well as the temperate regions; even spatial displacements half as large as those we have projected will pose significant dispersal problems. Efforts should be encouraged to bring wild plants into cultivation, paying attention to the preservation of genetic diversity. Propagules should then be moved to new reserves established expressly for this purpose. Existing reserves should be exempted from efforts to establish southern species and genotypes—the present flora and fauna should be protected from extinction as long as possible, so as to preserve genotypes adapted to warm climate.

## V. SUMMARY

Geographical ranges, or potential ranges, will shift by 500–1000 km as a result of doubled greenhouse gases in the atmosphere. Because many plant species will not be able to adjust to so large a displacement rapidly enough to track the change, they will be much reduced in geographical range and thus in population size.

Organisms near the southern limits of their geographic range will be affected first. Although adult trees may persist for four or five decades, seedlings will be more sensitive to climatic change and may disappear within a few decades. As warming continues, tolerance limits of canopy trees will also be exceeded. At this stage, forests may have increased susceptibility to disease or other forms of disturbance.

The disappearance of forest trees can have dramatic effects on understory herbs that depend on the light, nutrient, and microclimatic conditions produced by canopy dominants. Thus local loss of a tree species will affect the chances of survival of numbers of other species. To prevent wholesale extinctions, new reserves must be established in the regions of potential habitat that will open up to the north as the climate changes. Research should begin now on methods of establishing ecosystems, especially methods to speed up natural processes such as succession and soil development. Without such efforts at all latitudes, many species will become extinct because of the rapid and extensive displacements of geographical ranges that will result from greenhouse warming.

## REFERENCES

Bennett, K. D. 1985. The spread of *Fagus grandifolia* across eastern North America during the last 18,000 years. *J. Biogeogr.* 12:147.

Bennett, K. D. 1988. Holocene geographic spread and population expansion of *Fagus grandifolia* in Ontario, Canada. *J. Ecol.* 76:547.

Bormann, F. H., T. G. Siccama, G. E. Likens, and R. H. Whittaker. 1970. The Hubbard Brook Ecosystem Study: Composition and dynamics of the tree stratum. *Ecol. Monogr.* 40:373.

Botkin, D. B., J. F. Janak, and J. R. Wallis. 1970. The rationale, limitations and assumptions of a northeast forest growth simulator. *IBM Research* RC3188 (14604).

Brubaker, L. B. 1986. Responses of tree populations to climatic change. *Vegetatio* 67:119.

Camp, W. H. 1950. A biogeographic and paragenetic analysis of the American beech (*Fagus*). *Am. Phil. Soc. Year Book* 1950:166.

Davis, M. B. 1981. Quaternary history and the stability of forest communities. In *Forest Succession: Concepts and Applications*, D. C. West, H. H. Shugart, and D. B. Botkin, eds., pp. 134–153. New York: Springer-Verlag.

Davis, M. B. 1987. Invasions of forest communities during the Holocene: Beech and hemlock in the Great Lakes region. In *Colonization, Succession and Stability*, A. J. Gray, M. J. Crawley, and P. J. Edwards, eds., pp. 373–393. Oxford: Blackwell.

Davis, M. B. 1984. Climatic instability, time lags, and community disequilibrium. In *Community Ecology*,

J. Diamond and T. J. Case, eds., pp. 269–282. New York: Harper and Row.

Davis, M. B., and D. B. Botkin. 1985. Sensitivity of cool-temperate forests and their fossil pollen record to rapid temperature change. *Quatern. Res.* 23:327.

Davis, M. B., K. D. Woods, S. L. Webb, and R. P. Futyma. 1986. Dispersal versus climate: Expansion of *Fagus* and *Tsuga* into the Upper Great Lakes region. *Vegetatio* 67:93.

Davis, R. B., and G. L. Jacobson, Jr. 1985. Late glacial and early Holocene landscapes in northern New England and adjacent areas of Canada. *Quatern. Res.* 23:341.

Denton, S. R., and B. V. Barnes. 1987. Tree species distribution related to climatic patterns in Michigan. *Can. J. Forest Res.* 17:613.

Eickmeier, W., M. Adams, and D. Lester. 1975. Two physiological races of *Tsuga canadensis* in Wisconsin. *Can. J. Bot.* 53:940.

Fowells, H. A. 1965. *Silvics of Forest Trees of the United States.* Agric. Handbook 271. Washington, D.C.: Forest Service, USDA.

Frelich, L. E. 1986. Natural disturbance frequencies in the hemlock-hardwood forests of the upper Great Lakes region. Ph.D. diss., University of Wisconsin, Madison.

Fritts, H. C. 1976. *Tree Rings and Climate.* New York: Academic Press.

Gates, D. M. 1983. An overview. In $CO_2$ and Plants: *The Response of Plants to Rising Levels of Atmospheric Carbon Dioxide*, E. R. Lemon, ed., pp. 7–20. Boulder, Colo.: Westview Press.

Gaudreau, D. C., and T. Webb III. 1985. Late-Quaternary pollen stratigraphy and isochrone maps for the northeastern United States. In *Pollen Records of Late-Quaternary North America*, V. M. Bryant, Jr., and R. G. Holloway, eds., pp. 247–280. Dallas: Am. Assoc. Strat. Palyn.

George, M. F., H.M.J. Burke, H. M. Pellett, and A. G. Johnson. 1974. Low temperature exotherms and woody plant distribution. *Hort. Sci.* 9:519.

Guries, R., S. Brown, and J. Kress. 1981. A guide to forest tree collections of known source or parentage in northeastern and north central United States and adjacent Canadian provinces. College of Agriculture Life Science Publ., Res. Bul. 3142. Madison: University of Wisconsin.

Haines, D. A., and R. W. Sando. 1969. Climatic conditions preceding historically great fires in the North Central region. USDA Forest Service Research Paper NC 34.

Hansen, J., G. Russell, D. Rind, P. Stone, A. Lacis, S. Lebedeff, R. Ruedy, and L. Travis. 1983. Efficient three-dimensional global models for climate studies: Models I and II. *Monthly Weather Rev.* 3: 609.

Held, M. E. 1983. Pattern of beech regeneration in the east-central United States. *Bull. Torrey Bot. Club* 110:55.

Henry, J. D., and J.M.A. Swan. 1974. Reconstructing forest history from live and dead plant material: An approach to the study of succession in southwest New Hampshire. *Ecology* 1974:772.

Hubbell, S. P., and R. B. Foster. 1991. Short-term population dynamics of trees and shrubs in a neo-tropical forest: El Niño effects and successional change. *Ecology* (in press).

Huntley, B., and H.J.B. Birks. 1983. *Past and Present Pollen Maps for Europe 0–13,000 Years Ago.* Cambridge: Cambridge University Press.

Jaeger, J. 1988. Developing policies for responding to climatic change. World Climate Programme Impact Studies WCIP-1, WMO/TD 225.

Johnson, W. C., and C. S. Adkisson. 1985. Dispersal of beechnuts by blue jays in fragmented landscapes. *Am. Midl. Nat.* 13:319.

Kramer, P. J., and T. T. Kozlowski. 1979. *Physiology of Woody Plants.* Orlando, Fla.: Academic Press.

Manabe, S., and R. T. Wetherald. 1987. Large-scale changes in soil wetness induced by an increase in carbon dioxide. *J. Atmos. Sci.* 44:1211.

Moe, D. 1970. The post-glacial immigration of *Picea abies* into Fennoscandia. *Botaniska Notiser* 123:61.

Neilson, R. P., and L. H. Wullstein. 1983. Biogeography of two southwest American oaks in relation to atmospheric dynamics. *J. Biogeogr.* 10:275.

Payette, S. 1983. The forest tundra and present treelines of the northern Quebec-Labrador peninsula. In *Tree-line Ecology*, P. Morisset and S. Payette, eds., Collection Nordicana 47, pp. 3–23. Quebec: Université Laval.

Peters, R. L., and J.D.S. Darling. 1985. The greenhouse effect and nature reserves. *Bioscience* 35: 707.

Pigott, C. D., and J. P. Huntley. 1978. Factors controlling the distribution of *Tilia cordata* at the northern limits of its geographical range. *New Phytologist* 81:429.

Ritchie, J. C., and G. M. MacDonald. 1986. The patterns of post-glacial spread of white spruce. *J. Biogeography* 13:527.

Rogers, R. S. 1980. Hemlock stands from Wisconsin to Nova Scotia: Transitions in understory composition along a floristic gradient. *Ecology* 61: 178.

Runkle, J. R. 1982. Patterns of disturbance in some old-growth mesic forests of eastern North America. *Ecology* 63:1533.

Sakai, A., and C. J. Weiser. 1973. Freezing resistance of trees in North America with reference to tree regions. *Ecology* 54:118.

Salomonsen, F. 1951. The immigration and breeding of the field-fare (*Turdus pilaris* L.) in Greenland. *Proc. Intl. Ornith. Cong.* 10:515.

Schroeder, M. J., and C. C. Buck. 1970. *Fire Weather.* USDA Forest Service Agric. Handbook 360.

Secrest, H. C., H. J. MacAloney, and R. C. Lorenz. 1941. Causes of decadence of hemlock at the Menominee Indian Reservation, Wisconsin. *J. Forestry* 39:3.

Shugart, H. H. 1984. *A Theory of Forest Dynamics.* New York: Springer-Verlag.

Vaartaja, O. 1959. Evidence of photoperiodic ecotypes in trees. *Ecol. Monogr.* 29:91.

Van Devender, T. R. 1984. Climatic cadences and composition of Chihuahuan desert communities: The late Pleistocene packrat midden record. In *Community Ecology*, J. Diamond and T. J. Case, eds., pp. 285–299. New York: Harper and Row.

Ward, R. T. 1961. Some aspects of the regeneration habits of the American beech. *Ecology* 42:828.

Webb, S. L. 1986. Potential role of passenger pigeons and other vertebrates in rapid Holocene migrations of nut trees. *Quatern. Res.* 26:367.

Webb, S. L. 1987. Beech range extension and vegetation history: Pollen stratigraphy of two Wisconsin lakes. *Ecology* 68:1993.

Webb, T., III. 1986. Is vegetation in equilibrium with climate? How to interpret Quaternary pollen data. *Vegetatio* 67:75.

Webb, T., III. 1988. Eastern North America. In *Vegetation History*, B. Huntley and T. Webb III, eds., pp. 385–414. Dordrecht: Kluwer Publishers.

Woods, K. D., and M. B. Davis. 1989. Paleoecology of range limits: Beech in the upper peninsula of Michigan. *Ecology* 70:681.

Woodward, F. I. 1986. *Climate and Plant Distribution.* Cambridge: Cambridge University Press.

This research was supported by the Environmental Protection Agency, Cooperative Agreement CR814607-01-0, and the National Science Foundation, grant BSR-8615196. We gratefully acknowledge critical comments on various drafts of the manuscript by L. Frelich, R. R. Calcote, and E. J. Sucoff.

CHAPTER TWENTY-THREE

# Between the Devil and the Deep Blue Sea: Implications of Climate Change for Florida's Fauna

LARRY D. HARRIS AND
WENDELL P. CROPPER, JR.

## I. INTRODUCTION

Evidence suggests that sea level in the Gulf of Mexico and Caribbean Sea has been steadily rising since the last glacial period 15,000–20,000 years ago (fig. 23.1). Although the evidence is less direct, it appears that a gradual climate change is accompanying major transformations of the fauna and flora of Florida and the southeastern coastal plain (Delcourt and Delcourt 1987, Watts and Hansen 1988). These changes have occurred at a much slower rate than those now predicted to result from the rising levels of atmospheric carbon dioxide and related greenhouse gases, but they still provide the single best indication of the nature of consequences of the rapid change that is likely to be associated with the greenhouse effect. This chapter describes potential effects of rapid climate change on Florida biota and offers new conservation strategies that will minimize the consequences.

## II. FLORIDA'S POST-PLEISTOCENE HISTORY

Sea level near Florida has been rising at approximately 4 cm per century during the past few millennia. During the past century, tide gauges in Florida indicate that sea level has increased at a rate of 2.3 mm/yr (fig. 23.1), equivalent to an increase of 1.2 m over 500 years. The principal physiographic, biogeographic, and ecological effects of importance to the fauna and flora are:

- The land area of the Florida peninsula has been reduced by approximately 50%.
- The width of the peninsula has been reduced relative to the length, and therefore the peninsular shape has been greatly amplified.
- Maximum topographic relief of the state has been reduced by about 50%, and the

A

B

C

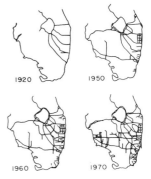

Figure 23.1. A: Empirically measured rise in sea level in south Florida over the course of the last 5000 years. From Wanless 1982. B: Florida peninsula 10,000 years ago (shaded area) and today (white area). The average rate of inland transgression of 8 meters per year for the last 10,000 years has reduced the size of the Florida peninsula by half and has accentuated the peninsular shape. From Cooke 1939. C: The increase in expanse of drainage canals in south Florida as a burgeoning human population has attempted to colonize the wetlands at the same time that the sea level is rising. From Fernald and Patton 1984.

relief over most of the peninsula has been reduced by a much greater amount.

- The low-lying circum-gulf plain that constituted the primary colonization route for the peninsular fauna during the Pleistocene has been inundated (Webb and Wilkins 1984), while the more highly excised terraces that now abut the sea have undergone dramatic vegetation change that has effectively closed the corridor to easy transverse movement by plants and animals.
- The water table lying beneath the peninsula has risen in proportion to the rising sea and has given rise to a greatly altered vegetation structure (Watts and Hansen 1988).
- Areal extent of the former Bahamian land mass has been reduced by as much as 90%, transforming it into an archipelago. West Indian islands have also been reduced in size, and the distance between the islands and the Florida peninsula has increased,

which is relevant to certain North American migrant species that overwinter in the Caribbean.

Concurrent with these physiographic and topographic changes, major transformation of the fauna and flora of the Florida peninsula and the Gulf coastal plain has also occurred. For example:

- A high proportion of the native vertebrate fauna is now extinct. In the case of mammals this totals slightly more than 50% of the species that were present at the end of the Pleistocene (Harris and Eisenberg 1989, Webb and Wilkins 1984).
- Many species that were formerly distributed throughout the Gulf coastal plain now occur only in Florida as small remnant populations isolated from their larger population centers in the southwestern United States. These include species such as gopher tortoise (*Gopherus polyphemus*), fence lizard (*Sceloporus undulatus*), scrub jay (*Aphelocoma coerulescens*), caracara (*Caracara cheriway*), and Florida panther (*Felis concolor*), many of which are rare and listed as threatened or endangered species.
- The vegetation has changed from a more arid savannah with the overstory dominated by flowering plants (angiosperms) to a coastal heathland dominated by gymnospermous pines and cypress. Because most conifers do not produce seeds and fruits evolved to attract animal dispersal agents,

and because conifer foliage generally constitutes a lower-quality browse resource, the shift from hardwood to conifer forest has probably reduced overall habitat suitability for birds and mammals.

- The extinction of mammal species from the West Indian islands and the Florida peninsula is inversely proportional to size of area and degree of proximity to the North American continent (table 23.1). Among mammals, the more vagile bats that can fly over water have not been affected nearly as much as the less vagile taxa that are restricted to the ground (Harris and Eisenberg 1989).

- Although the extinction rate of Florida mammals has been high, it has not been as severe as that of the more isolated Caribbean islands. Moreover, the peninsula has been colonized by several taxa that did not occur at the end of the Pleistocene. Both of these phenomena result directly from the interconnection of the peninsula with the North American continent.

- As the boreal conifer forest has migrated several hundred kilometers northward since the Pleistocene, the Bahamian land mass has regressed to a disjunct series of small islands, and the West Indies have become smaller and more isolated. Those species that were adapted to breed in the conifer forest but overwinter in the Caribbean have been subjected to increasingly long migrations and decreasing expanses of overwintering habitat.

The net result of these historical climate changes is to increase the vulnerability of Florida's biota to ecological stresses, including the greenhouse effect. This increased vulnerability has been greatly exaggerated by Florida's increasing human population and the associated urbanization.

## III. CONSEQUENCES OF RAPID CLIMATE CHANGE

### A. Life-Zone Modeling

Temperature and precipitation are two climatic variables whose change has profound effects on biological communities. A model that relies on these two variables to predict ecological effects offers some advantages over more complex models depending on other data sources. The Holdridge (1967) system of life-zone mapping (see figure 10.1) permits rapid delineation of the natural vegetation formations from basic temperature and precipitation data. Thus it may be useful for projecting shifts in life zones likely to result from change in these basic climatic variables.

We used the Holdridge model to generate maps of the life-zone distribution within the state of Florida under present climatic conditions to test its applicability for further use as a predictor of change (fig. 23.2). Great similarity exists between the calculated life-zone distributions and previously published vegetation maps and known zones of vegetation occurrence, giving confidence that the exercise is not trivial. Modeling was conducted primarily by Robert Dohrenwend and is reported in greater detail elsewhere (Dohrenwend and Harris 1975).

Several examples of natural vegetation zonation in Florida can be used to test the valid-

Table 23.1. Extinction patterns of nonvolant land mammals on West Indian islands. Data from Morgan and Woods 1986.

| | Size of island (1000 km²) | Remaining species | Fossil species |
|---|---|---|---|
| Cuba | 115 | 5 | 26 |
| Hispaniola | 76 | 2 | 25 |
| Jamaica | 11 | 1 | 5 |
| Puerto Rico | 13 | 0 | 6 |

*Figure* 23.2. A: The distribution of forest types in Florida at the turn of the twentieth century, long before European settlers or the consequences of increasing atmospheric concentrations of greenhouse gases could have altered the statewide vegetation distribution. From Brown 1909. B: The statewide distribution of extant life zones based on early to mid twentieth-century climatic data and the Holdridge life-zone prediction equations. C: The zone where changing climate is predicted to cause the most rapid changes in biota lies transverse across the peninsula almost exactly congruent with the Interstate 4 development corridor, which is a barrier to latitudinal migration of fauna and flora.

ity of Holdridge model predictions. The salt marshes of both the Atlantic and Gulf coasts are characterized by three distinct associations (Provost 1967), the boundaries of which seem to be sensitive to frequency of freezing temperatures. The present distribution of mangroves (*Rhizophora* and *Avicennia*) along the subtropical southern peninsula grades into *Batis maritima* and *Salicornia perennis* in the transition zone near the middle of the peninsula and cordgrass (*Spartina* spp.), saltgrass (*Distichlis* sp.), and *Juncus* salt marshes of the more northern coasts.

At the turn of the twentieth century, widespread mapping of forest conditions throughout the country included a survey of Florida. The first forest survey identified three notable forest zones, a north Florida zone, a southern Florida or Antillean zone, and a broad transitional zone in between (Brown 1909)(fig. 23.2). The author observed that "the distinguishing features of this region are

the transitional characteristics of the tree growth; that is a few northern trees have their southern limit, and some subtropical ones, their northern, here. Within this belt longleaf pine [*Pinus palustris*] grades from north to south, from large pure stands to short, scrubby, mixed stands of very open growth." These and related empirical data lend credence to the validity of the Holdridge projections of life zones that would exist under extant climatic conditions, with no human management (fig. 23.2).

Model predictions reported here include three climate change scenarios: a temperature increase of 2°C with a rainfall decrease of 200 mm, a 2°C increase with no change in rainfall, and a 2°C increase with an increase of 200 mm in rainfall (fig. 23.3). These scenarios are consistent with predictions of general circulation models (GCMs) for grid boxes that include Florida. Because of problems of scale (Ackerman and Cropper 1988) and difficulties in parameterization of key processes such as cloud formation (Schneider 1987), GCMs cannot provide precise predictions of future climate for any particular region. Still, GCMs represent the best method for integrating the complex feedbacks of earth, atmosphere, and ocean into a coherent picture of the future climate.

Under existing conditions most of the Florida panhandle falls within the warm temperate moist forest life zone and is separated from the subtropical moist forest of the southern peninsula by a broad transitional zone. Different life zones are most apparent because of the occurrence of different species, and so it is no surprise that life-zone boundaries are largely coincident with distributional limits of a large number of plant and animal species. Life-zone boundaries are therefore congruent with discontinuities between distinct vegetation associations and reflect areas of high sensitivity to slight changes in climate. It should be expected that the most rapid and obvious changes in biota will occur at the boundaries of life zones or in the transition between the two prevalent life zones that currently exist in Florida. The transition zone across Florida appears as a belt nearly 200 km wide that traverses the peninsula from the Tampa Bay area on the west to the Daytona Beach area on the east. It is here

*Figure* 23.3. Calculated distributions of the life zones of Florida based on current climate data and three climate change scenarios. Based on the Holdridge life zone prediction model, the effects of either wetter or drier conditions appear modest relative to the dramatic effect of a 2° rise in temperature. Under equilibrium conditions, a 2° temperature rise will cause a 500-km northward displacement of the existing life-zone distribution.

that the most rapid and notable responses to climate change are predicted to occur.

*Case 1:* Life-zone shifts that accompany a 2° warming above ambient temperature with no change in precipitation entail a 300-km northward migration of the various boundaries. Warm temperate moist forest migrates to the north of Florida, the transition between the warm temperate and subtropical life zones comes to dominate the panhandle, and subtropical moist forest comes to dominate the peninsula. Tropical dry forest invades the southern tip of the peninsula while a refugium of transition vegetation from subtropical moist forest and subtropical dry forest is sequestered west of Lake Okeechobee.

*Case 2:* A warming of 2°C accompanied by an annual decrease of 200 mm in precipitation again results in a 300-km northward displacement of life zones. The panhandle becomes dominated by transition vegetation between warm temperate and subtropical moist forest. Remnants of warm temperate moist forest will remain as disjunct patches on the higher elevations of the panhandle. Subtropical moist forest migrates to the northeasternmost corner of Florida, but be-

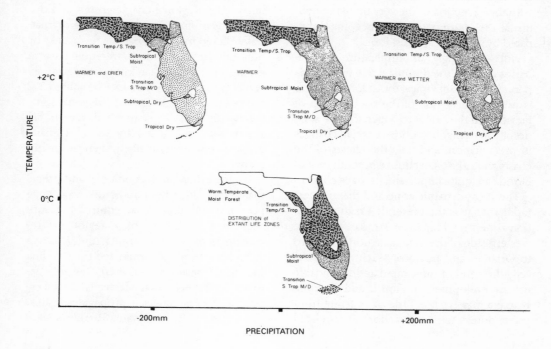

cause of decreased precipitation this life zone is greatly limited in extent. Subtropical vegetation that presently occurs only on the southernmost tip of the peninsula expands greatly throughout the peninsula. A small island of subtropical dry forest exists immediately west of Lake Okeechobee, and tropical dry forest invades the southern tip of the peninsula.

In this scenario, an additional factor comes into play. Along with the shift from moist forest to dry forest comes an attendant simplification of forest structure from that of dense, closed canopy forest to more sparse, open-canopied woodland. Thus, not only would the flora be expected to change but the structural complexity of the forest would also be reduced. This would seem to hold consequences for the survival of both floral and faunal species.

*Case 3:* The most dramatic change in biota is predicted to result when annual precipitation and temperature increase simultaneously, by 200 mm and 2°C. This climate change scenario results in the simplest of all life-zone responses. The Florida panhandle is dominated by vegetation that is transitional between warm temperate and subtropical, and with the exception of the very tip of the peninsula, the entire state would become dominated by subtropical moist forest.

Not surprisingly, all temperature-rise scenarios result in the dominance of the Florida peninsula by subtropical moist forest such as is now characteristic of the southern tip of Florida. But the extent to which this life zone dominates is highly sensitive to slight changes in precipitation, and thus the effects of the coasts and Lake Okeechobee in south central Florida are quite important, perhaps critical.

The most dynamic zone and that of most rapid change is the presently existing transition zone stretching from Tampa Bay on the western side of the peninsula northeastward to Orlando and Daytona Beach. This is exactly the band of most rapid and formidable urban development in Florida and the zone where a massive blockade of concrete turnpikes, hotels, tourist attractions, and related strip development will forever act as impediments to movement of biota from south to north.

## B. Caveats to Model Predictions

Models that include only the principal climatic variables cannot reflect the importance of many proximal factors and events that actually mediate the effects of climate change. The modeling exercises reported here project equilibrium conditions that could only occur some considerable time after altered climate has begun to affect local ecology. In the short term and at the regional level, climate translates into weather patterns, which are translated into regional ecology through the occurrence of episodic events such as storms, freezes, and droughts. Recent GCM analysis indicates that global warming may lead to increased drought, convective wind storms, coastal flooding, and hurricanes (Overpeck et al. 1990).

The degree to which these episodic events alter the regional ecology is only modestly under the influence of humans. With available technology, humans are ineffective at controlling the frequency and severity of freezing temperatures, catastrophic fires, or hurricanes. Regional characteristics and distributions of flora and fauna can be dramatically altered in the course of a few weeks by only one or two of these infrequent storm events. As any Floridian knows, the zone acceptable for citrus culture may be shifted 200 km in a single week. Thus at a regional scale, the rapidity and degree to which present life zones will be affected by severe climatic events is not substantially influenced by humans.

On the other hand, conditions and events that allow native flora and fauna to change naturally and those that control the subsequent colonization of a region by replacement biota are very much under human influence. People exterminated the Carolina parakeet (*Conuropsis carolinensis*), the only native animal species that effectively moved cypress (*Taxodium distichum*) propagules against the current of flowing water. Parakeets were no doubt important to the migration of cypress as far north as Missouri and New Jersey after the last ice age. But further northward migration by natural mechanisms will be greatly retarded by the absence of this single vertebrate dispersal agent (Harris 1988a).

A most common impact of humans on the Florida environment involves the alteration of natural drainage patterns and wetlands. Only modest hydrological alterations in the Everglades have facilitated rapid invasion and dominance of the region by a galaxy of exotic vegetation including *Melaleuca, Cassuarina, Shinus, Hydrilla,* and *Eichhornia* (Ewel 1986). These species not only dominate much of the environment but they also seriously threaten the future of the native freshwater marshes. At present, approximately one third of the plant species of the Florida peninsula are exotic to Florida (R. Wunderlin, University of South Florida, personal communication), and in the course of only three decades, the structure and composition of an entire regional flora have changed because of altered hydrology. The high intensity of effort necessary to control these exotic species in and around Everglades National Park is neither economically nor philosophically acceptable under present conditions. Thus, whereas the boundary of an existing floral and faunal association can quickly move hundreds of kilometers in a response to climate events not under the control of man, the rapidity with which a climatically adapted replacement biota can colonize in order to accommodate the new climatic conditions is very much under the influence of man.

Unlike the drama associated with temporarily barren landscapes created by freeze-killed citrus, uncontrollable wild fire, or transgression of a hurricane and the attendant tornadoes, most of the consequences resulting from rapid climate change will be subtle and indirect. It may not be possible to prove that they are causally linked to climate change at all. For example, empirical data suggest that hurricane frequency along the west coast of Florida is perhaps 20% lower during this century than the long-term aver-

age. Besides allowing construction and maintenance of human structures that would otherwise not be possible because of severe wind and flooding, it also greatly alters the frequency of severe flooding of low-lying wetlands and coastal areas. In the absence of severe flooding and of burning, natural wetlands are invaded by dryland species that not only change the structure and composition of the wetlands themselves but also affect the regional ecological and human development patterns (Marois and Ewel 1983).

A similar series of changes is likely to occur on the other topographic extreme. The highest elevations of the Florida peninsula are dominated by a native scrub community that has existed since the Pliocene. As many as fifty species of plants and animals are endemic to this community type and occur nowhere else on earth (Christman 1988). The scrub communities and the Pliocene sand dunes on which they occur have accommodated the extremes of several ice-age climate cycles and under natural circumstances they could no doubt withstand the rapid climate changes predicted to result from the greenhouse effect. But residential and agricultural development in the area have now restricted the options for adaptation of the native scrub community in a highly dynamic milieu.

## C. Accelerated Sea Level Rise

Biological systems throughout Florida are clearly vulnerable to temperature increases and precipitation changes predicted to occur in the next 100 years. Florida's biological communities are perhaps most vulnerable to an indirect effect of climate change, that is, the rising sea level. One of the most significant effects of rising sea level would be the loss of wetlands through erosion and inundation. Half of all the wetlands in the United States could be lost with a 100-cm rise in sea level (Broadus 1989). Even if sea level continues to rise only at the present rate of about 30 cm per century (Wanless 1989), the sandy coasts and mangrove ecosystems of south Florida will be destroyed by coastal erosion. The Everglades could also be subject to massive saltwater intrusion with greenhouse-induced sea level increases (Wanless 1989). Of the national parks on earth that are likely to be lost, the most notable is Everglades National Park. These climate and ocean changes must be considered in a context of a biota that is increasingly at risk from human activities.

## IV. THE PROCESS OF FAUNAL COLLAPSE AND THE CLIMATE CONNECTION

A description of processes causing the present, ongoing loss of vertebrate diversity in Florida will help clarify what negative effects can be expected from the future interaction between human pressures (notably habitat conversion and constriction) and climate change. Erosion of biodiversity includes the loss of species, loss of genetic and phenotypic diversity within species, and alteration of native communities by processes such as invasion and dominance by alien species. Together, these changes alter the biological characteristics that distinguish one native biotic complex from another. The following four primary types of ongoing biodiversity erosion suggest the most probable patterns of future change.

### A. Loss of Large, Wide-Ranging Species

There is abundant evidence that large animals are more prone to extinction than small ones. From the approximately 10,000 years of Florida's post-Pleistocene period, there is no paleontological evidence of extinction among very small mammals such as small bats and shrews (i.e., 0%), while evidence for larger mammals reveals a linearly increasing frequency of extinction as body size increased, up to the point at which mammals weighing more than 1000 kg (megamammals) experienced 100% extinction (Harris and Eisenberg 1989).

This bias against large mammals also has been evident in the historic past. Until 200 years ago Florida was occupied by eleven species of native mammals larger than 5 kg. Of those eleven, the bison (Bison bison), monk seal (Monachus tropicalis), and red wolf (Canis rufus) are locally or globally extinct; the manatee (Trichechus manatus latirostris), Florida panther (Felis concolor coryii), and key deer (Odocoileus virginianus clavium) are federally listed as endangered, and the black bear (Ursus americanus), bobcat (Lynx rufus), and otter (Lutra canadensis) are either listed by the state of Florida or by the Convention on International Trade in Endangered Species (CITES). Considering that these reductions took place in only 200 years compared with 10,000 years for the post-Pleistocene species loss, the rate of extinction is clearly much greater during the first half-millennium of settlement by Europeans. And this loss is despite concerted conservation efforts for nearly the last century.

Large animals are particularly vulnerable for several reasons. One is that relatively low population densities are coupled with a need for extensive amounts of habitat, something increasingly in short supply in Florida. Unfortunately, natural biogeographical patterns and the distribution of human settlement have left many of the large animals with their largest remnant populations sequestered in habitat near the coast. These artificial coastal endemics or near-endemics are at particular risk because the coastal zone is precisely where human occupancy and exploitation of resources is the greatest—85% of Florida's human population lives here. An indication of the immensity of human pressure is that death by vehicle collision is now the principal source of mortality for manatee, black bear, Florida panther, key deer, American crocodile (Crocodylus acutus), bald eagle (Haliaeetus leucocephalus), and even the red wolf reintroduced into a wildlife refuge in North Carolina (M. Phillips, Alligator River National Wildlife Refuge, personal communication). For slowly breeding species that are already reduced in abundance, such a new and severe source of mortality can be a major threat.

Although the U.S. Department of Interior has purchased for refuges many narrow tracts of coastal wetlands that generally lie parallel to the seashore, these acquisitions are vulnerable to rising sea level. Vulnerable species, whose remaining populations are increasingly sequestered between the rising sea and growing human populations, are typified by the Everglades mink (*Mustela vison*), Florida panther, key deer, American crocodile, and the endangered sea turtles.

## B. Loss of Endemic Coastal Zone Habitat Specialists

In addition to those once-widespread species now heavily dependent on coastal zone habitat, there are many vulnerable species and subspecies that evolved to live only in highly specialized coastal habitats. They include several species of beach mice (*Peromyscus* spp.), the Cape Sable sparrow (*Ammospiza maritima mirabilis*), and the closely related dusky seaside sparrow (*Ammospiza maritima nigrescens*). The dusky seaside sparrow occurred in saltmarsh habitats between the 3-m and 5-m contour lines in a few hundred square kilometers of eastern Florida and was perhaps North America's most habitat-restricted vertebrate species (Kale 1977, Baker 1978). Since Florida was discovered by the Spanish nearly 500 years ago, the sea level has risen more than one meter, and the narrow zone of suitable habitat has presumably migrated inland a considerable distance during this time. In this century, as dense human populations began to encroach, the bird's habitat became progressively reduced and fragmented, and despite management and captive-breeding efforts, the last full-blooded dusky seaside sparrow died in captivity in 1987, and what had been classified as an identifiable species until 1973 became extinct (new knowledge that the dusky was not yet genetically differentiated from other east-coast seaside sparrows is not central to the issue of extinction).

## C. Loss of Variability and Viability within Sequestered Coastal Populations

Genetic problems due to small population size will increase as rising sea level, climate-caused changes in habitat, and anthropogenic habitat conversion continue. This will be a particular problem for large animals because they typically have low population densities and need to range widely.

The Florida panther is a species already showing signs of genetic problems related to these factors. Historically, the panther was distributed throughout the southeastern United States from Texas and Arkansas to South Carolina and Florida. Human settlement has limited distribution of the remaining few dozen panthers to the southern tip of Florida, south of Lake Okeechobee (USFWS 1987), a peninsula that is narrowing as the post-Pleistocene sea level rises. Resultant isolation and inbreeding is the most probable explanation for the observed congenital malformation and nonviability of sperm, which in adult male panthers is about 95 percent (USFWS 1987).

Both habitat quality and panther condition appear better on the higher, richer lands to the north of the Everglades and the Big Cypress National Preserve (Maehr 1989). But without land-use designations and incentives that assure sustained occurrence of wildlife on these predominantly private lands, no secure retreat from the advancing sea exists (Maehr 1989).

## D. Loss of Neotropical Migrants

Neotropical migrants—birds that breed in North America but overwinter in the tropics—travel long distances on migration and are vulnerable to both climate-caused and anthropogenic disturbances at either end of the migratory route (Hamel 1986). Particularly at risk is a small group of warblers, including Bachman's (*Vermivora bachmanii*) and Kirtland's (*Dendroica kirtlandii*), that evolved to overwinter in Cuba, the Bahamas, or other West Indian islands. As climate warmed during the Pleistocene, their summer habitat shifted toward the north, undoubtedly increasing necessary migration distances. At the same time, rising sea level and climate change shrank land masses and expanses of overwintering habitat in the Caribbean and further isolated the overwintering habitats.

Recent destruction of Caribbean habitat has exacerbated the situation, and many (perhaps all) of the North American migrant bird species restricted to Caribbean overwintering sites are now listed as threatened or endangered species. Bachman's warbler has not been recorded in the last decade and is now presumably extinct. Further warming and rise in sea level will raise havoc with remaining species.

The monarch butterfly (*Danans plexippus*) is another species whose survival may hinge on the present climate regime. The eastern U.S. population annually migrates northward in millions from its overwintering colonies in a few disjunct, remnant patches of high mountain cloud forest, dominated by the fir *Abies religiosa*, in the central highlands of Mexico (Calvert et al. 1989). The Mexican colonies are found only in a small area of about 4000 km² in area, between 2928 and 3400 m altitude, where the temperature is cold enough to induce reproductive torpor but not so cold as to kill the unprotected insects. It must be warm enough to allow the clusters to stay together but not stimulate excessive activity that uses their energy reserves, and it must be moist enough to prevent dessication and forest fires but not so wet as to harm the individuals or colonies (Calvert and Brower 1986). Survival of this species and the phenomenon of mass migration depends on a highly restricted combination of climate and habitat conditions that occurs only in one small location. Amplified and accelerated climate change that alters temperature, precipitation, and the distribution of Oyamel fir forests will most certainly threaten the eastern North American monarch population.

## V. A NEW CONSERVATION PARADIGM FOR THE TWENTY-FIRST CENTURY

Although the conservation movement has a long history and deep roots in Florida, with a focus on preservation of south Florida wading birds (Pearson 1937, Doughty 1975), the traditional approaches and magnitude of conservation efforts have not been sufficient.

Despite the establishment of Everglades National Park and the continuing addition of contiguous lands to the north that now total nearly a million hectares, the numbers of wading birds in the Everglades region have dropped dramatically, from between about two million birds a century ago to less than 5% of that number today (see Harris 1988a). Much of this drop is recent—"The numbers of nesting attempts [in 1986–87] represented a drop of at least an order of magnitude compared with nesting numbers during the period 1940–1960, and in some cases with surveys during the 1970s" (Frederick and Collopy 1988). At least eleven native birds, including the Everglades kite (*Rostrhamus sociabilis*, now named snail kite), no longer reproduce within Everglades National Park (table 23.2).

If we look at those and similar losses from other parks and preserves around the world (e.g., Newmark 1987), it is apparent that the nineteenth-century solution of simply protecting a small percentage of land by designating national parks will not adequately conserve viable populations of wildlife during the twentieth and twenty-first centuries. Species such as the panther that require large blocks of habitat are having a hard time surviving in areas of the size that people are willing to protect. Under conditions of climatic change, survival will be even more difficult.

Large mammals such as panthers and bears are not the only animals that must move throughout the landscape daily and seasonally; in the face of changing climate and rising sea levels, entire faunas will need to shift across the landscape in search of climatically suitable habitat. It is clearly not feasible, however, to purchase the entire landscape for public ownership and wildlife protection. This forces conservationists to turn to the vast majority of habitat that is unprotected and typically used by people to extract a variety of resources. Are there ways that present methods of using nonreserve land can realistically be adapted to meet the economic and recreational needs of human society

Table 23.2. Native species of birds and mammals that no longer occur in Everglades National Park (in consultation with W. B. Robertson).

| | |
|---|---|
| Native AmerIndian | Eastern bluebird |
| Monk seal | American kestrel |
| Red wolf | Red-cockaded woodpecker |
| Passenger pigeon | White-breasted nuthatch |
| Carolina parakeet | Brown-headed nuthatch |
| Ivory-billed woodpecker | Summer tanager |
| Wood duck | Snail kite* |

*Snail kites may sometimes occur in the park, but they do not currently reproduce successfully there.

while becoming more sensitive to the requirements of wild species now and when the climate changes?

We believe that an integrated group of policies can substantially increase the contributions that nonreserve lands make to the maintenance of biological diversity. The aim is to create graduated zones of land use where fundamentally different conservation philosophies, approaches, laws, and agencies can be integrated in complementary patterns. Specifically, in areas immediately surrounding natural preserves, we can increase the amount of buffering, land-use control, and management flexibility. Land-use control in a buffer area, for example, will be needed to designate protected migration corridors that link reserves. This will be especially critical for animal survival as climate changes.

On a wider scale, society must encourage uses of nonreserve lands that preserve specific elements of natural habitat and serve specific functions in the overall system. Uses including sustainable forestry, livestocking, or hunting and fishing give people (and economic interests) an incentive to maintain the habitat that produces the resources in a relatively undespoiled condition. At the same time, it is necessary to recognize the tension between encouraging some use, so as to provide the users with incentives to protect the resources, and the tendency of economically profitable uses to intensify until they threaten the resource.

## VI. SUGGESTED LAND-USE POLICIES FOR DEALING WITH RAPID CLIMATE CHANGE

### A. Landscape Linkages

Landscape linkages are tracts of land that lie between existing parks and reserves and that are being acquired or leased to ensure the opportunity for organisms to move from one isolated reserve to another. Although wildlife movement easements would be adequate under many circumstances, the best course of action is commonly fee-simple purchase of the land. Several of these strategic acquisitions are now being implemented (Harris 1988b, Harris and Gallagher 1989). Pinhook Swamp is one such area that constitutes an 8000-ha tract lying between the Okefenokee National Wildlife Refuge in Georgia and the Osceola National Forest of north Florida. Although water and wide-ranging animal species were known to traverse the area freely from one preserve to the other, the area remained in private ownership until 1988. Much of the tract has now been purchased by the Nature Conservancy for transfer into the U.S. National Forest System. The combined interagency, interstate national forest and wildlife refuge will total nearly 400 km². More important, it will span a north-south distance of 100 km and greatly enhance the opportunity for native biological communities to migrate in times of rapid climate change.

## B. Wildlife Dispersal Corridors

Wildlife dispersal corridors are identified and protected specifically to facilitate the movement of native plants and animals across the human-dominated landscape. These linear strips of habitat are commonly associated with riparian forests, recreational greenways, restored railroad rights-of-way, or easements for utilities such as power lines and pipelines. From the climate change perspective, the provision of breeding habitat is not so important as the provision for movement through a human-dominated landscape without undue risk of mortality. For example, restoration and conservation of the north-south Wekiva and Oklawaha riparian corridors and the east-west cross-Florida barge canal lands will be critical to the maintenance of viable populations of hundreds of species that are increasingly sequestered in the Ocala National Forest of north-central Florida.

Movement corridors can not be limited to terrestrial environments, because numerous large vertebrates such as manatees, marine turtles, and dozens of fish species depend on seasonal migrations for survival and reproduction. Heavy boat traffic, locks, dams, overly shallow waters, and chemical pollution constitute effective barriers to movement and must be modified accordingly.

## C. Streamside Buffer Zones

Streamside buffer zones that prevent intensive human development to the water's edge not only protect the integrity of the aquatic system but also allow it to change its course and characteristics in response to changing sea level, consequent changes in water table, and altered flooding regimes. Again, entire aquatic and riparian faunas and floras must be allowed the opportunity to move.

## D. Reserve Buffer Zones and Restoration Areas

Reserve buffer zones and restoration areas are being acquired and implemented. In south Florida, a former agricultural and mineral mining tract known as East Everglades has been purchased to buffer Everglades Na-tional Park from the effects of intense human development to the east. Encompassing about 60,000 ha, the area consists of roughly 50% of the land remaining that could be developed for agricultural uses, and its aquifer recharge function is critical to Miami's future potable water supplies. At the same time, it represents a critical headwater and flow-through area for both Taylor and Shark River sloughs, natural subsystems that feed water and nutrients to Everglades National Park and the estuarine wildlife communities of Florida Bay (Abrams et al. 1989).

To abate the specter of highrise condominiums standing on the eastern boundary of Everglades National Park, Congress has now authorized purchase of the East Everglades Area by the National Park Service. Given that resources are available for purchase, a plan for land use and management that effectively accommodates competing demands on the area represents a new step in the creation of conservation technologies for sustained human development.

## E. Highway Underpasses

Highway underpasses that permit wildlife movement beneath automobile traffic streams, fish ladders, lifts, and specially designed passages that direct aquatic organisms (e.g., migratory manatees) around obstructions and through areas of intense boat traffic are necessary new technologies to allow movement of organisms as well as entire biological assemblages. For example, a series of 35 underpasses presently being constructed beneath the east-west extension of Interstate 75 that bisects the Everglades will not only facilitate sheet flow of fresh water to the south for the mitigation of saltwater intrusion but will also allow the biological communities that are increasingly sequestered between the roadway and the sea to migrate northward in response to the rising sea levels.

## F. Fences

Fencing along roads through wetlands is necessary to prevent the formation of ecological

traps. When artifacts such as elevated roadbeds are introduced into wetland environments, the local ecology is significantly
altered. For various reasons, the elevated
roadbeds provide desirable habitat for hundreds of wildlife species, including many endangered species, luring them into dangerous situations. For example, the black body
and heat-holding properties of roadways
make them attractive to the endangered
American crocodile, and that is why vehicle
collisions constitute the number-one source
of mortality among adult crocodiles. Numerous other vertebrates ranging from Everglades mink to key deer and Florida panthers
are endangered by such ecological traps.
One 4.5-year survey of 3.25 km of highway
through a state park in north-central Florida
revealed that mortality from automobile collisions accounted for 13,000 snakes of twelve
species weighing a total of 1.3 metric tons (R.
Franz, Florida Museum of Natural History,
personal communication). Raptors such as
bald eagles feed on the resulting carrion and
fall victim to the same traffic that created the
carrion. Although the roadbeds are not lethal
in and of themselves, death resulting from
collision with vehicles now threatens the
continued existence of several species. Such
problems will be amplified during times of
rising sea levels and water tables. The combination of strategic road alignment, movement underpasses, fencing, and other new
technologies will be necessary to prevent total elimination of mobile fauna (except birds
and bats) in landscapes bisected by roads and
traffic streams.

## G. Rotated Boundary Orientation of Coastal Preserves

A rotated boundary orientation of coastal
preserves will be necessary to mitigate the
effects of inundation and total loss of coastal
wetlands that currently run parallel with the
shore. Boundaries of the narrow coastal
strips of preserve and public access beaches
should ideally to be rotated 90° to ensure that
some wetland will exist in public ownership
no matter what the sea level. Realigning the

boundaries to follow the natural topographic
contour will be a necessary first step, but additional modifications up to and including
land ownership laws will also be necessary.
We envision that a series of riparian preserves
following the surface watercourses and penetrating toward the interior of Florida will be
necessary as sea levels continue to rise.

## H. Land Swaps

Land-swapping policies and procedures for
trading title of ownership between private
and public landholdings are increasingly necessary. As low-lying coastal and inland areas
are inundated by rising water levels and as
land use becomes increasingly regulated by
wetlands protection, private landowners will
choose ownership elsewhere if such is possible. Conversely, as these same wetlands are
identified as strategic parcels for public ownership, the mutual advantages of land trading
will be obvious. In Florida, two different
land-swap negotiations between private
landowners and the USDI, Fish and Wildlife
Service, have been executed to extend the
boundaries of the Big Cypress National Preserve, create the Florida Panther National
Wildlife Refuge, and expand the Suwannee
River National Wildlife Refuge. Government
surplus property in higher and drier areas
of the United States will assume increasing
value to the private sector.

## I. Debt Swaps

Debt swapping has gained popularity as a
strategy for reducing the national debt of developing countries while allowing benefac-
tors to initiate public land acquisition for the
purposes of nature conservation. Until now,
most applications of this technique have involved international negotiations between
lending banks and conservation organizations in the developed countries and debtor
nations in the developing world.

The massive debt incurred by the savings
and loan industry during the late 1980s in the
United States opens many new possibilities,
however. Savings and loan institutions that
have become financially insolvent and those

that require federal assistance to maintain solvency have relinquished title to thousands of mortgaged properties that may hold strategic value in the realignment of natural-area boundaries. Rather than flooding the real estate market and simply reselling its repossessed properties to the highest bidders, the congressionally created Resolution Trust Corporation should first consult the Department of Interior and other agencies knowledgeable about the utility of these now government surplus properties.

## J. Mitigation Land Banks

Mitigation land banks consist of areas that responsible conservation organizations and agencies have identified as valuable to the conservation of natural systems. Land developers and others who are obliged to provide mitigation for the loss of wetlands or other critical habitats or species may purchase land in the designated mitigation land bank. If these designated areas are chosen with particular reference to the changing climatic conditions, the effects of rising sea levels and shifting biological communities can be significantly abated.

## VII. SUMMARY

During the next 100 years atmospheric trace gases such as $CO_2$, methane, and $N_2O$ will continue to increase as a result of human activities. The increasing concentration of these and other greenhouse gases has the potential to produce climate change of unprecedented speed and magnitude. Estimates from general circulation models of the atmosphere indicate that the global mean temperature may increase by 2°–5°C, with a doubling of atmospheric $CO_2$. Climate effects on local and regional areas may include changes in the frequency and duration of precipitation and alterations of the probability of extreme climatic events. Associated increases in sea level could have devastating results in coastal regions. Biological communities are clearly vulnerable to climate change of the magnitude predicted for the green-

house century. When coupled with an independent crisis in biodiversity due to other human activities, climate change is an issue that must be considered in any conservation planning.

The history of Florida over the last 10,000 years demonstrates the sensitivity of biological communities to changing climate and sea level. The rise in sea level has decreased the Florida peninsula in size and topographic relief, and there have been major changes in vegetation type during the past 5000 years. Holdridge life-zone modeling of Florida indicates a zone of high sensitivity to a temperature increase of 2°C and to annual precipitation changes of 200 mm. Calculations of equilibrium conditions indicate a minimum movement of life-zone boundaries 300 km north and a dramatic transition of most of Florida to subtropical moist forest with a 2°C increase in temperature and a 200 mm increase in precipitation. Although climate change will appear slow from a human perspective, episodic climatic events such as hurricanes, freezes, and droughts will play a major role in the rapid restructuring of biological communities.

The potential effects of climate change in Florida must be considered in the context of species and communities already at great risk from the rapidly expanding human population. During the past 200 years several native mammal species have become extinct, and the remaining large mammals are highly threatened. It is not likely that a successful strategy to abate the climate change can be developed, but it is essential that conservation management develop new landscape-level strategies that take into account the heterogeneous nature of the environment over both space and time. Ecologists and conservationists now have a considerable empirical and theoretical underpinning for understanding and mitigating the biological consequences of climate change and human activities.

What is missing from the conservation effort is an adequate implementation of a conservation technology. We are still operating on nineteenth-century conservation paradigms, even when we know that they don't work. Conservation strategy must be more flexible than the simple approach of setting aside preserved natural islands surrounded by the rising tide of unlimited human activities. A landscape management approach, including zones surrounding preserves, will be critical to mitigating climate change effects. Elements of this approach may include wildlife movement corridors, streamside buffer zones, and realignments of reserve boundaries to follow the natural topographic contours.

## REFERENCES

Abrams, K., M. Marvin, H. Gladwin, and B. Brumback. 1989. The east Everglades planning case study. Final report to U.S. Environmental Protection Agency, Atlanta, Georgia. Joint case for environmental and urban problems, Fla. Intl. Univ., North Miami.

Ackerman, T. P., and W. P. Cropper, Jr. 1988. Scaling global climate projections to local biological assessments. *Environment* 30(5):31.

Baker, J. 1978. Dusky seaside sparrow. In *Rare and Endangered Biota of Florida*. Vol. 2, Birds, H. Kale II, ed., pp. 16–19. Gainesville, Fla.: University Presses of Florida.

Broadus, J. M. 1989. Impacts of future sea level rise. In *Global Change and Our Common Future*, R. S. DeFries and T. F. Malone, eds. Washington, D.C.: National Academy of Sciences.

Brown, N. C. 1909. Preliminary examination of the forest conditions of Florida. Special report on behalf of the newly formed Forest Service and Roosevelt's 1908 Governors Conference. Manuscript, University of Florida library.

Calvert, W. H., and L. P. Brower. 1986. The location of monarch butterfly (*Danaus plexippus* L.) overwintering colonies in Mexico in relation to topography and climate. *J. Lepid. Soc.* 40:164.

Calvert, W. H., S. B. Malcom, J. I. Glendinning, L. P. Brower, M. P. Zalucki, T. Van Hook, J. B. Anderson, and L. C. Snook. 1989. Conservation biology of monarch butterfly overwintering sites in Mexico. *Vida Sylvestre Neotropical* 2:38.

Christman, S. 1988. Endemism and Florida's interior sand pine scrub. Final report submitted to Florida Game and Fresh Water Fish Commission, Tallahassee, Fla.

Cooke, C. W. 1939. Scenery of Florida interpreted by a geologist. *Geol. Bull.* 17. Florida Geol. Survey, Tallahassee.

Delcourt, P. A., and H. R. Delcourt. 1987. Longterm forest dynamics of the temperate zone: A case study of late Quaternary forests in eastern North America. New York: Springer-Verlag.

Dohrenwend, R. E., and L. D. Harris. 1975. A climatic change impact analysis of peninsular Florida life zones. In *Impacts of Climatic Change on the Biosphere*, part 2, *Climatic Impact*, pp. 5/107–5/122. Washington, D.C.: Climate Impact Assessment Program, U.S. Dept. Transportation.

Doughty, R. 1975. *Feather Fashions and Bird Preservation.* Berkeley: University of California Press.

Ewel, J. J. 1986. Invasibility: Lessons from south Florida. In *Ecology of Biological Invasions of North America and Hawaii*, H. A. Mooney and J. A. Drake, eds., pp. 214–230. New York: Springer-Verlag.

Fernald, E. A., and D. J. Patton. 1984. *Water Resources Atlas of Florida.* Tallahassee: Florida State University Press.

Frederick, P., and M. Collopy. 1988. Reproductive ecology of wading birds in relation to water conditions in the Florida Everglades. Final report submitted to the U.S. Army Corps of Engineers, Jacksonville, Fla. University of Florida, Dept. of Wildlife and Range Sciences.

Hamel, P. 1986. *Bachman's Warbler: A Species in Peril.* Washington, D.C.: Smithsonian Institution Press.

Harris, L. D. 1988a. The nature of cumulative impacts on biotic diversity of vertebrates in wetlands. *Environ. Mgmt.* 12:675.

Harris, L. D. 1988b. Landscape linkages: The dispersal corridor approach to wildlife conservation. *Trans. N. Am. Wildlife and Nat. Res. Conf.* 53:595.

Harris, L. D., and J. Eisenberg. 1989. Enhanced linkages: Necessary steps for success in conservation of faunal diversity. In *Conservation for the 21st Century*, M. Pearl and D. Western, eds., pp. 166–181. New York: Oxford University Press.

Harris, L. D., and P. Gallagher. 1989. New initiatives for wildlife conservation, the need for movement corridors. In *In Defense of Wildlife: Preserving Communities and Corridors*, G. Mackintosh, ed., pp. 11–34. Washington, D.C.: Defenders of Wildlife.

Holdridge, L. R. 1967. *Life Zone Ecology.* San Jose, Costa Rica: Tropical Science Center.

Kale, H. 1977. Dusky seaside sparrow. *Fla. Naturalist* 50:16.

Maehr, D. 1989. The Florida panther and private lands. *Conservation Biology* 4:1.

Marois, K. C., and K. C. Ewel. 1983. Natural and

management-related variation in cypress domes. *For. Sci.* 29:627.

Morgan, G. S., and C. A. Woods. 1986. Extinction and the zoogeography of West Indian land mammals. *Biological J. Linnean Society* 28:167.

Newmark, W. 1987. A land-bridge perspective on mammalian extinctions in western North American parks. *Nature* 325:430.

Overpeck, J. T., D. Rind, and R. Goldberg. 1990. Climate-induced changes in forest disturbance and vegetation. *Nature* 343:51.

Pearson, T. 1937. *Adventures in Bird Protection.* New York: Appleton-Century.

Provost, M. 1967. Managing impounded salt marsh for mosquito control and estuarine resource conservation. In *Proc. Marsh and Estuarine Management Symposium,* J. Newsom, ed., pp. 163–171. Baton Rouge: Louisiana State University.

Schneider, H. 1987. Climate modeling. *Scientific American* 256(5):72.

USFWS. 1987. Florida panther (*Felis concolor coryi*) recovery plan. Prepared for the U.S. Fish and Wildlife Service by the Florida Panther Interagency Committee, Atlanta, Georgia.

Wanless, H. R. 1982. Sea level is rising. So what? *J. Sed. Petrology* 52:1051.

Wanless, H. R. 1989. The inundation of our coastlines. *Sea Frontiers* 35(5):264.

Watts, W. A., and B.C.S. Hansen. 1988. Environments of Florida in the late Wisconsin and Holocene. In *Wet Site Archaeology,* B. A. Purdy, ed., pp. 307–323. Caldwell, N.J.: Telford Press.

Webb, S., and K. Wilkins. 1984. Historical biogeography of Florida Pleistocene mammals. In *Contributions in Quaternary Vertebrate Paleontology,* H. Genoways and M. Dawsen, eds. Carnegie Museum Special Publication 8. Pittsburgh: Carnegie Museum.

The work reported here was supported by contracts and grants to L. D. Harris from the U.S. National Park Service, the U.S. Man and the Biosphere Program, and the U.S. Department of Transportation (CIAP).

# CHAPTER TWENTY-FOUR

# The Nature and Consequences of Indirect Linkages Between Climate Change and Biological Diversity

JOHN HARTE, MARGARET TORN,
AND DEBORAH JENSEN

## I. INTRODUCTION

Because of the complex interactions that link the earth's atmosphere, soils, waters, and biota, the impending greenhouse warming can affect biodiversity in many ways. Most research to date has focused on the direct responses of organisms to an altered climate, particularly physiological responses to changes in temperature and precipitation. Generally overlooked, however, are the many indirect ways in which climate change can affect biodiversity. First, climate shapes the physical environment—affecting such processes as fire, flooding, and soil formation—which in turn affects living organisms. Second, the impacts of climatic change on a plant, animal, or microbial population will have ripple effects within ecosystems, through interactions among species. Third, climatic change will exacerbate other anthropogenic stresses, such as acid deposition and deforestation, and increase species vulnerability to these stresses. We show here that these indirect linkages have the potential to exert a major influence on biological diversity, that they are not adequately addressed by current research strategies, and that they pose a serious challenge to the way in which we currently formulate conservation policies.

Analysis of the biotic consequences of the greenhouse effect is in its infancy, but already a prevailing forecasting strategy has emerged. A look at this strategy and its shortcomings helps illustrate the importance of the distinction made here between direct and indirect linkages. At the risk of oversimplification, the strategy can be represented by four steps:

1. Use general circulation models (GCMs) to estimate future temperatures and precipitation levels under various assumptions about the rate of increase of greenhouse gases in the atmosphere.

2. Correlate the present-day distribution of plants and animals with these same climate variables. Both field-derived correlations and physiological information based on laboratory studies can be used to determine the tolerances of biota for ranges of climatic conditions.

3. Determine the ability of species to persist in their present location given a climatic change, and determine where the biota would have to move to if they were to remain in a suitable climate. This type of analysis typically focuses on changes in growth and death rates caused by altered thermal and moisture conditions. Both correlations between climate and species distributions (e.g., chapter 22) and forest succession models (e.g., chapters 11 and 21) are used for this purpose. Estimates of the direct consequences of $CO_2$ enrichment on plant growth are sometimes included in these analyses.

4. Assess the conservation implications of the previous step. For example, determine the land use of areas projected to have a climate suitable for a particular species or community. If such an area is already a vast agricultural region, or if it will become one as crop production shifts geographically in response to climate change, it will be of little benefit to that species or community. If a suitable new habitat does exist, determine what is required to permit biota to relocate there, such as a connecting habitat corridor or transplantation of trees and soil.

This strategy assumes that the factors governing the distribution of plants and animals are the direct climate variables whose values the GCMs predict. Few argue that temperature and precipitation, alone, characterize the suitability of habitat for biota; the premise is, rather, that these climate variables are useful surrogates for the complex factors that do characterize habitability (such as stream flow, the chemical composition of soil, air, and water, and a host of other geospheric conditions). The strategy also treats superficially the interspecific couplings within an ecosystem, presuming in some applications that species respond individually and in isolation

to climate change and in others that assemblages of species will change as an intact unit. Finally, it presumes that other anthropogenic stresses do not interact with the stress of climate warming. In short, the prevailing strategy neglects indirect effects.

Support for this forecasting strategy has been drawn from an analysis of the association between paleoclimate and vegetation type since the last ice age. For example, Thompson Webb (chapter 5) analyzes the correlation between climate variables reconstructed from GCM studies and tree distributions deduced from pollen data over a 15,000-year period. He shows that the dominant species in the forests of North America moved northward so that they were always found within regions of relatively constant climate. Given the radical differences between past climate change and projected greenhouse warming, the success of Webb's research does not validate the use of the four-point strategy for predicting future conditions. In particular, the rate of paleoclimate change was roughly two orders of magnitude slower than that anticipated during the coming greenhouse warming (chapter 4). Mineral weathering and soil formation, along with other indirect geospheric processes that influence habitability, could well have kept pace with climate change over 15,000 years, whereas they might not under the much more rapidly changing climate of the next 100 years. The paleoclimatic studies provide evidence that the impending climate warming will greatly stress ecosystems, but they offer little assurance that the idealized strategy above will be useful in forecasting ecological changes over the coming period of climate warming.

We are by no means the first to try to broaden the scope of inquiry in climate effects research. Some of the papers presented at this conference refer to indirect linkages (e.g., chapter 20). Moreover, M. Oppenheimer (1986) has surveyed the possible effects of climate change on atmospheric concentrations of phytotoxins. But to our knowledge, the missing links have not been discussed systematically before, nor have their implications for research and conservation policy been considered.

Figure 24.1 places the links among the climate, geosphere, anthropogenic stresses, and biota in context. The term "geosphere" refers to the atmosphere, hydrosphere, and lithosphere. After discussing geosphere-mediated processes, species interactions, and linkages to other anthropogenic stresses, this chapter explores some implications of these linkages for research on climate, geospheric processes, and ecology and for conservation policy. We refer to feedback processes (see fig. 24.1) only peripherally here, but D. Lashof (1989) analyzes some of the most important of them.

## II. OVERVIEW OF INDIRECT GEOSPHERIC LINKAGES

Many biologically important geospheric processes and conditions will change as a result of the greenhouse effect. Table 24.1 lists many important geospheric linkages according to the environmental medium in which the processes occur. Here we discuss a few examples in sufficient detail to convey their potential importance for biological diversity and to permit generalizations about research needs and conservation-policy implications.

### A. Coastal Upwelling

The climate and biota of each ocean's eastern boundary current region, such as coastal California and Peru, are strongly influenced by nearshore upwelling. These regions are characterized by highly productive fisheries, diverse and abundant bird populations, and vegetation communities associated with frequent coastal fog. Upwelling occurs when wind-driven ocean currents are deflected offshore by the Coriolis effect. The diverging surface water is then replaced with water from lower layers of the ocean by a process called Ekman pumping. The differential heating of the planet caused by the greenhouse effect is expected to alter global air pressure patterns and therefore wind patterns. We

## Greenhouse-Biota Linkages

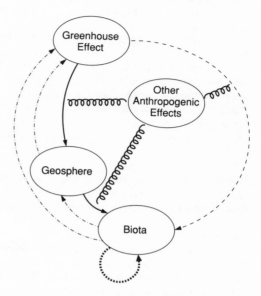

Figure 24.1. Linkages between the greenhouse effect and biota: direct processes (- - -), geosphere-mediated indirect processes (——), interspecific connections (·····), feedback processes (-·-·-), linkages between greenhouse warming and other anthropogenic stresses (ᖆᖆᖆ).

predict that this will lead to changes in coastal upwelling.

Changes in either the magnitude of upwelling or the timing of the upwelling season may have major effects on marine and terrestrial biodiversity. Examples of the many linkages between upwelling and biodiversity are shown in figure 24.2. Water upwelled from the deeper ocean is cold and nutrient-rich, which is why regions of upwelling are known for highly productive fisheries. Study of the El Niño phenomenon shows that a reduction in nutrient inputs decreases primary productivity, leading to mass starvation of anchovy and other fishes. If the greenhouse effect weakens upwelling, interrupted nutrient flows are expected, but the effects of increased upwelling are less well understood.

Reproduction and larval recruitment are particularly sensitive to the timing of upwelling because the physical advection of surface

*Table 24.1.* Major geospheric processes linking the greenhouse effect to biodiversity. Climatic variables that control the individual processes are cloud evaporation (CE), evapotranspiration (ET), humidity (H), lapse rate (LR), precipitation, frequency or amount (P), temperature (T), large-scale tropospheric circulation (TC), and wind (W).

*Atmosphere*
Photochemical and catalytic formation of atmospheric pollutants (T, H)
Transport and deposition of pollutants (TC, P)
Concentration of atmospheric pollutants (LR, CE)
Transport of biologically essential trace substances from ocean to land (TC)
Fire frequency and intensity (T, P, W, ET, H)
Dispersal of pollen, chemical cues, seeds, and larvae (TC, W, T)

*Soil*
Nutrient availability, mineral weathering, organic content (T, P, ET)
Leaching, gullying, erosion (P, W)
Biotoxicant formation and accumulation, acid neutralization (T, P, ET)

*Fresh water*
Timing of snow melt, i.e., dry-season stream flow; ephemerality of ponds; areal extent of wetlands (T, P, ET)
Nutrient cycling; pollutant formation and toxicity (T, P, ET)
Concentrations of nutrients, salinity, and pollutants (P, ET)
Lake stratification and circulation (W, T)
Siltation and sedimentation, resulting from erosion (P, W)

*Marine*
Sea level and extent of intertidal, shore, and estuarine habitat (T)
Upwelling of cold, nutrient-rich water; consequent fog (W, TC)
Salinity of polar waters and continuity of ice floes (T, P, ET, W)
Turbidity and nutrient status of coastal waters affected by river inflow (P, W)

water in the Ekman layer carries larvae far offshore, outside the nursery where productivity and food supplies are high (Roughgarden et al. 1988). For example, larval survivorship of hake (Pacific whiting) is decreased if upwelling is unusually early or strong (Bailey 1981). Reductions in hake numbers could have serious consequences for certain California sea lion populations. The diet of the Farallone Marine Sanctuary sea lion population is 80% hake in the spring months before the sea lions pup (Ainley et al. 1982), though it is more diverse at other times of the year. This is one example of the potential influence of upwelling on several levels of a food web.

Terrestrial ecosystems also are influenced by upwelling. The cool ocean surface cools coastal air and gives San Francisco its famous summer air conditioning, advection fog. Fog permits drought-intolerant plants such as redwood seedlings to survive without summer rain (Veirs 1982). By reducing solar radiation and temperature, fog also slows the formation of the constituents of photochemical smog such as ozone—a major stress on vegetation in the San Bernardino Mountains and other parklands surrounding Los Angeles. Fire, too, is influenced by fog; if upwelling persists later in the fall, it will reduce the fire potential in coastal California.

We have estimated upwelling along the west coast of the United States using a model based on the simple equations that describe the physical mechanisms of upwelling caused by alongshore wind stress. GCM output was used for control and doubled-$CO_2$ climates. A preliminary analysis (Torn, unpublished) suggests that the upwelling season may peak later along the coast and that annual upwelling magnitude will increase slightly in some areas and decrease slightly in others. It is premature, however, to predict possible ecological effects confidently; improved wind data from GCMs are needed.

Global circulation model predictions of the wind fields that cause upwelling are generally inadequate for modeling upwelling. Running the model with control (current $CO_2$) GCM output of monthly winds gave upwelling indices that were 10–20 times lower than the historical values determined by J. E. Mason and A. Bakun (Bakun 1973, Mason and Bakun 1986). One reason for these low estimates is that the GCM monthly winds are averaged as vectors. Figure 24.3 shows how this can lead to systematically low predictions. GCM outputs of wind drag (proportional to wind squared at each time step) can be used in upwelling calculations to avoid this problem. Unfortunately, the drag components may overestimate upwelling under conditions of high wind variability. A second reason for the low estimates is that surface roughness and wind speed vary significantly within the GCM coastal grids, which include both land and sea-surface topography. An improvement to the roughness

averaging may await models of finer spatial resolution.

Figure 24.4 shows several different mechanisms by which wind influences marine ecosystems. Note that for each, different kinds of GCM output are needed. For example, wind drag can be used for research on upwelling and currents but not for turbulence; monthly means are probably adequate for modeling currents, but finer time scales are needed for the other processes. Getting the most appropriate GCM output (to use as input) depends on knowing the properties of the mechanism or process being modeled.

## B. Wildfire

Fire can play a diverse role in ecosystem dynamics (fig. 24.5). In semiarid regions like the western United States, wildfires may be the most important indirect, geospheric link between climate change and biodiversity, influencing the distribution and composition of plant communities. Changes in the timing, location, and severity of fire will undoubtedly abet or inhibit vegetation shifts induced by direct effects of climate change on, for example, seedling survival (chapters 19 and 20).

Fire severity is primarily a function of the four climate variables shown in figure 24.5 and, secondarily, of vegetation type. Fire rec-

*Figure 24.2.* Linkages between upwelling and biota.

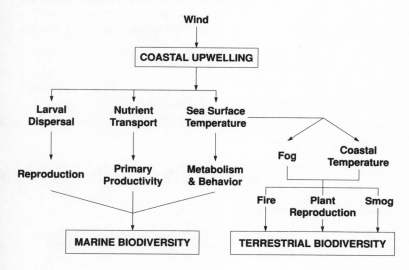

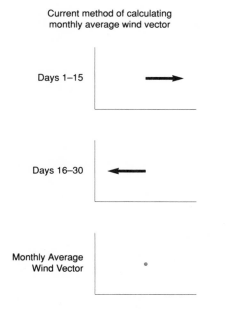

Current method of calculating
monthly average wind vector

Days 1–15

Days 16–30

Monthly Average
Wind Vector

Figure 24.3. The error that can occur in ecosystem models if monthly mean winds are averaged as vectors. If the wind blows east half the month and west the other half, the net vector wind speed is zero, suggesting upwelling is also zero, even though there may have been significant wind-driven upwelling during the month.

Figure 24.4. Some mechanisms by which wind influences marine ecosystems.

ords suggest that the climate changes forecast for the coming decades will alter wildfire patterns. Even the relatively small climatic variation of the past millennium was accompanied by such changes. For example, J. S. Clark (1988) found that fire frequencies in Minnesota were significantly lower during the Little Ice Age than they were during warmer periods over the last 750 years. Although the greenhouse effect will result in higher annual-averaged temperatures nearly everywhere, some regions may get drier and others wetter. Regions that become both warmer and drier will likely experience more burning, although changes in wind speed and direction will also affect fire behavior.

To predict changes in acreage burned by wildfire, J. S. Fried and M. S. Torn (1990) have developed a computer-based model that incorporates discrete weather and fire events, fire suppression efforts, and site-specific information on vegetation and slope. GCM output and daily weather records are used to generate climate scenarios. Their study site in the Sierra foothills includes four of the main vegetation types found in California: coniferous forest, oak woodland, chaparral, and grassland. Wildfires are suppressed to protect the large numbers of houses and other buildings in the area. The model predicts that the annual acreage burned by wildfire will increase as a result of climate change. An initial analysis of a doubled-$CO_2$ climate predicts that twice as many fires in grasslands will escape initial suppression efforts and become

| | OCEANIC PROCESSES | | |
| | CURRENTS | UPWELLING | TURBULENCE |
| --- | --- | --- | --- |
| DEPENDENCE ON WIND SPEED | Square $(w^2)$ | Square $(w^2)$ | Cube $(w^3)$ |
| DEPENDENT ON WIND DIRECTION | Yes | Only alongshore component | No |
| TIME SCALE OF VARIABILITY | Seasons | Weeks | Hours |
| SPATIAL SCALE | Oceanic | Coastal | Local |

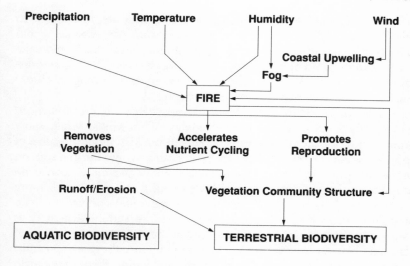

Figure 24.5 (*above*). Linkages between climatic variables and fire and between fire and biota.

Figure 24.6 (*below*). Two ways in which a 15% increase in monthly rainfall might be distributed over time: the conventional assumption (top), in which additional rainfall is distributed in proportion to existing rainfall; and an alternative assumption (bottom), in which the number of rainy days increases. The conventional method would predict more-severe fire danger than would the equally plausible alternative scenario.

campaign fires. A less severe response is expected in the chaparral, oak, and conifer habitats because of the higher capacity of those vegetation types to retain moisture after a rainfall.

As with upwelling, forecasting of fire frequency and intensity is hindered by mismatches between GCM output and the data requirements of fire models. Discrete fire models call for daily weather values, but GCM output is most readily available (and is only considered reliable) as monthly means. Fire potential, particularly in grasslands and chaparral, depends on the time since the last rainfall, a figure not available from current GCM output. As figure 24.6 illustrates, different assumptions about the way in which the predicted change in a monthly mean of rainfall will be distributed yield different predictions of fire potential. In addition, the vector-averaging of GCM wind speeds, as described above for upwelling, generates underestimates of the burning height and the rate of spread of fires.

## C. Timing of Snowmelt

Many montane ecosystems derive most of their water from snowmelt. In the Sierra Nevada of California and in the Colorado Rock-

**Increasing the Rainfall 15%**
**Conventional Scaling Method (e.g., EPA)**

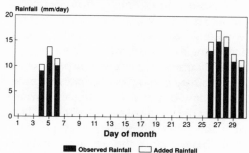

**Alternative Distribution**

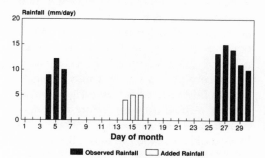

ies, for example, 80%–90% of annual precipitation falls as snow. The volume of winter snowpack is of obvious importance to ecosystems as well as to downstream water users, but less obvious is the critical role of the timing of snowmelt. A change in climate can cause snow to melt earlier or later in the year and can cause it to melt faster or slower.

Under climate warming, snow is expected to melt earlier than it does now (Gleick 1988). In regions where summer rain provides little of the available moisture for plants and animals, an earlier melt means a longer and therefore drier summer, particularly if summers are also hotter with an increase in evaporation rates. This can stress biota in several ways. Late summer dryness can stress seedlings. Reduced runoff and increased evaporation can cause ponds that under the present climate are wet throughout the year to dry up each year in late summer. The conversion of permanent water bodies into ephemeral ones will greatly affect the survival of numerous populations of waterfowl, fish, and amphibia, as well as aquatic plants, insects, and plankton.

A change in the timing of snowmelt can affect biota in springtime, when montane plants and animals emerge from dormancy or hibernation and migrants return from wintering areas. Some organisms respond to increasing day length, while others respond to the disappearance of the snowpack, which exposes bare ground and liberates a considerable flow of water. Greenhouse warming will alter the timing of snowmelt but will not alter the seasonally shifting daily photoperiod. Hence, biota that respond more to snowmelt cues may find themselves out of synchrony with others that respond to photoperiod cues. For example, broad-tailed hummingbirds begin their northern migration in response to photoperiod (W. Calder, personal communication); however, many of the subalpine plants that they pollinate, such as Delphinium nelsonii, an early blooming species, respond to temperature and snowmelt. If snowmelt were earlier, the hummingbirds could arrive too late to be effective pol-

linators. Available data are inadequate to assess the effect such asynchronies could have on community structure in montane ecosystems. J. P. Myers and R. T. Lester discuss several examples involving migratory species in chapter 15.

Montane ecosystems are also influenced by the rapidity of snowmelt. When snow melts rapidly in the high mountains, less of the meltwater can soak into the thin soils on the mountain slopes and more flows overground. Downhill flow is also much more rapid, because water flows orders of magnitude faster over the surface than it does through soil. If climatic change forces a rapid snowmelt, in addition to the greater potential for downstream flooding, the pulse of meltwater flowing through streams and rivers will be of shorter duration, potentially reducing the duration of algal blooms and altering other aspects of aquatic communities.

The length of time that soil and meltwater are in contact determines the extent of chemical reactions between them, thus influencing the chemical composition of both water and soil. If a rapid melt leaves soils moist for a shorter period of time, mineralization rates of nutrients would likely decrease, reducing soil fertility. In areas with acidic precipitation, rapid runoff may mean that less of the acidic meltwater will be buffered by soil reactions such as carbonate dissolution or cation exchange. As a result, the springtime acid pulse will be more acidic, threatening fish, amphibia, and other aquatic biota. Recent research by Harte and Hoffman (1989) suggests that Ambystoma tigrinum populations in the Rocky Mountains are declining because of the spring-snowmelt acid pulse. An earlier, more acidic pulse is an example of a synergy between two seemingly unrelated environmental stresses—climate warming and acid precipitation. (See also chapter 25.)

A rough estimate of the change in the timing of snowmelt in the Sierra Nevada suggests that under doubled-$CO_2$ climate conditions, snowmelt might be one or two months earlier than the current 100-year average (Gleick 1988). In other words, there may be significant changes in the phenology of snowmelt, with potentially widespread ecological repercussions. To improve model predictions of whether snow melts more rapidly or more slowly, it would be useful to combine climate model output with snowmelt models such as J. Dozier's (1980).

There are two sources of uncertainty in using GCM data to forecast the effect of greenhouse warming on the timing of snowmelt in a specific region, such as the Sierra Nevada. First, there is considerable uncertainty in the regional predictions of GCMs; predictions of changes in precipitation are particularly uncertain, but even regional temperature predictions vary from one model to another and are not considered as reliable as globally averaged temperature change. Second, each GCM temperature or precipitation datapoint is an average value for a grid much larger in area than the high mountain region where most of the snowpack accumulates.

## D. Soil

Changes in soil structure and chemistry wrought by the greenhouse effect are of particular importance to plant growth and community structure. Changes in soil temperature and moisture are likely to exert significant influence on organic matter content and the availability of nutrients, such as nitrogen and phosphorus, to plants and soil microorganisms. Changes in storm patterns can influence leaching and erosion rates. On a longer time scale, climate-induced changes in mineral weathering rates may influence the chemistry and physical character of soil.

Correlation studies have been an important source of information on the relationship between climate and soil chemistry. H. Jenny (1930), who pioneered this approach, determined the nitrogen content of soils along temperature and moisture gradients in North America. Lashof (1989) estimated the change in global stocks of soil carbon implied by the greenhouse effect, assuming that presently observed correlations of soil carbon with temperature and precipitation persist as the climate changes. Correlation studies, however, may be poor predictors of how soil conditions will track shifting climate contours, because the time required for soil carbon to reach a new equilibrium is long compared with the time scale of the impending climate change. Using microcosms, W. D. Billings et al. (1982, 1984; chapter 18) showed that in tundra ecosystems $CO_2$ levels are not likely to affect vegetation directly but that increased soil temperature and decreased moisture will increase soil carbon turnover and possibly increase nitrogen availability.

As these four examples show, indirect geospheric responses to climate change may be at least as important to biodiversity as direct biological responses.

## III. SPECIES INTERACTIONS

Geospheric processes and conditions influence the environmental parameters and disturbance that limit the distribution of individual species and determine community composition. But species distributions and abundances are also determined by other species. The responses of individual species to climate change will change relative species abundances, create novel species associations, and alter patterns of species interactions such as disease, predation, competition, and mutualism. These, in turn, will induce further indirect effects on biodiversity.

Wherever sharp transitions in species abundance are the result of species interactions, a climate-induced change in the relative advantage of one species could alter the community composition. For example, the avian malaria vector in Hawaii is found only at elevations below 600 m. Susceptible bird species are found only above this range (Warner 1968). Global warming might increase the elevational range of the mosquito vector, further restricting the distribution of susceptible bird species. While birds vulnerable to malaria might not be pushed off the top of the mountain by climate change directly, the increased elevational range of the mosquito, resulting from a warming, might cause local bird-population extinctions.

Individual species can play a large, or keystone, role in the distribution and abundance of other species in the community. The presence of a predator may allow several prey species to coexist by holding the population of each in check (Paine 1966). Similarly, some plants provide food that is critical during low resource periods (Terborgh and Winter 1980). Removal or introduction of such keystone species by climate warming would have significant repercussions elsewhere in the ecosystem.

Mutualisms also influence the ability of species to move to and survive in new locales. Numerous plant species depend on mutualistic relationships with mycorrhizal fungi or other soil microorganisms for nitrogen, phosphorus, and other nutrients. The plant may be able to disperse its seeds to new areas of suitable climate, but if the fungi cannot also disperse to and survive in the new locations, the plant will fail to establish and grow. Animal-pollinated plants and animal-dispersed plants may also suffer if their mutualists respond differently to the new climate.

Many plant species alter local nutrient availability, microclimate, local disturbance patterns, and other aspects of their environment. In doing so, they may facilitate or inhibit succession (Connell and Slayter 1977). Where climate change alters the ability of individual species to regenerate, for example, by affecting competitive advantages, reproduction, or the frequency of local disturbance, the successional patterns will change. Much of what is known about the role of individual species in succession is based on the observed effects of introduced species. For example, *Myrica faya*, an introduced, nitrogen-fixing plant in Hawaii, is able to invade the nitrogen-poor soils of young lava flows. It forms single-species stands and alters nutrient cycles, thereby changing the successional pattern in areas recovering from volcanic eruptions (Vitousek 1986).

Novel species combinations are expected as species migrate to new locations in response to direct or geosphere-mediated effects of climate change. From studies of in-

troduced species, we know that the introduction of some species has few notable consequences, but other species cause dramatic changes. For example, introduced predators and large herbivores have caused extinctions, particularly on islands.

This brief discussion suggests that understanding the consequences of the greenhouse effect on biodiversity requires an investigation of the interspecific connections within biotic communities. Although we cannot yet forecast the magnitude of the changes that will result from these connections, numerous examples indicate that they can be quite large.

## IV. LINKAGES WITH OTHER ANTHROPOGENIC STRESSES

The magnitude and consequences of climate change cannot be viewed in isolation from other anthropogenic global stresses. Table 24.2 lists many of the major interactions of greenhouse warming with deforestation, acid deposition, tropospheric pollution, and stratospheric ozone depletion. Interestingly, global stresses often interact synergistically, meaning the combined effect of the two stresses is more harmful than the sum of the separate effects of the stresses. The most widely recognized class of such synergies is physiological in nature: when plants and animals are weakened by one stress, they often become more vulnerable to other stresses. Hence, the stress on biota from climate change will often be exacerbated synergistically by stresses from other anthropogenic sources. In addition to increasing species vulnerability, the linkages with climate change also affect the severity of the stresses, themselves. For example, there are interactions that alter the magnitude of climate change, the concentration of air pollutants, the acidity of lakes, the extent of deforestation, and the magnitude of ozone depletion. These interactions, like those affecting vulnerability, are often synergistic.

Deforestation accelerates greenhouse warming by transferring stocks of carbon from veg-

*Table 24.2.* Linkages between the greenhouse effect and other anthropogenic stresses.

*Deforestation*
Poleward relocation of populations barred by deforested areas
Transfer of carbon stocks from vegetation to atmosphere, increasing greenhouse effect
Altered flux to atmosphere of greenhouse gases such as methane and nitrous oxide from deforested areas
Decreased transpiration and altered albedo contributing to climate change
Increased pressure on forested land for agricultural use

*Acid Deposition and Tropospheric Pollutants*
Rate of oxidation reactions in troposphere altered by climate change
Pollutants and climate change synergistically stress biota
Reduced acid-anion uptake from soil by climate-stressed vegetation, increasing acidity of soil and runoff
Pests damage vegetation more because pollutants reduce plants' resistance and milder winters fail to check
    pest populations
Earlier snowmelt leads to longer, drier summer, augmenting pollutant stress on forests and ephemeral
    aquatic ecosystems
Change in the rate of snowmelt affects ability of soil to neutralize acid snowmelt
Change in fossil fuel consumption and therefore pollution emission

*Stratospheric Ozone Decline*
UV radiation and climate change synergistically stress biota
Greenhouse warming leads to cooler stratosphere, affecting rate of ozone depletion
Greenhouse warming alters microbial production of nitrous oxide, affecting natural ozone sink

etation and soil to the atmosphere (chapter 3). This is an additive, not synergistic, process, but synergistic interactions between large-scale deforestation and climate warming can occur as well. One such synergy arises because deforestation leaves in its wake vast tracts of land that are inhospitable to many plants and animals, creating geographical barriers to the relocation of plants and animals. To the extent that plants and animals will respond to climate change by migrating, the barriers that are created by deforestation (and other types of habitat destruction) will inhibit range shifts, resulting in more damage to populations than would result from deforestation plus climate warming, individually (chapter 2). By the same token, climate change will affect the recovery of deforested areas left fallow.

Just as climate change may influence the acidity of soil and runoff and the sensitivity of biota to acidification, there are many other interactions by which greenhouse warming may affect the severity and consequences of tropospheric air pollution (table 24.2). Op-

penheimer (1986) reviews several of these, emphasizing potential effects of greenhouse warming on the formation of phytotoxins, such as ozone, in the troposphere.

Climatic change may result in multiple stresses on an ecosystem, increasing plant or animal vulnerability to pests and pathogens. For example, the susceptibility of vegetation to insect pests is often enhanced if the vegetation is stressed by air pollutants (Smith 1981)—concentrations of which may be greater in a warmer atmosphere. In addition, drought induced by climate change would also increase plant vulnerability to pests and pathogens. Cold winters exert a brake on some pest populations, thereby reducing the severity of pest outbreaks. Milder winters resulting from greenhouse warming will be less effective in checking the growth of these pest populations. Thus the stress on plants from climate change may include both increased pest outbreaks and increased vulnerability to pests. Other human activities exacerbate pest infestations because agricultural and managed-forest monocultures encourage outbreaks.

Societal responses to climate change may also create a link to other anthropogenic stresses. In a greenhouse climate, people may choose to clear areas of boreal forest for agriculture, thereby increasing the hazards and feedbacks triggered by deforestation. Fossil fuel use may change in response to an increased demand for air conditioning and a decreased demand for heating, resulting in increases or decreases in air pollution and acid rain intensities.

Stratospheric ozone depletion threatens biological diversity because the resulting increase in UV radiation inhibits plant growth and may damage aquatic larvae. The greenhouse effect is predicted to cool the stratosphere, which could slow the rates of ozone-destroying reactions, leading to slower loss of ozone. Cooler temperatures could also lead to a decrease in the concentrations of chemicals that protect ozone, leading to accelerated ozone destruction. In particular, stratospheric nitrogen (usually as nitric acid) slows ozone loss by inactivating the chlorine catalyst that destroys ozone. Nitrogen, however, is sequestered by stratospheric ice particles, so in a cooler stratosphere this inhibitory mechanism may be weaker and the rate of ozone loss may increase. Just such a mechanism is postulated to explain the stratospheric ozone hole observed over the Antarctic in recent austral winters (McElroy and Salawitch 1989).

The greenhouse effect may also influence stratospheric ozone by altering the net global flux of nitrous oxide to the stratosphere. Nitrous oxide in the stratosphere breaks down into nitric oxide, which initiates a catalytic cycle that consumes ozone. The bacterial formation of nitrous oxide in water and soil, which is the primary source of atmospheric nitrous oxide, is influenced by temperature and soil moisture. To predict whether climate change will increase or decrease global nitrous oxide production requires more tenable regional climate predictions and a greater understanding of the dependence of nitrous oxide production on environmental conditions. Together with the potential for both synergistic and antagonistic effects of stratospheric cooling, many questions concerning the net effect of climate change on ozone depletion are unresolved.

## V. RESEARCH NEEDS

As shown in the previous three sections, a complex web of processes mediates between climate and the well-being of life on earth. Despite the current difficulties in quantifying their effects, it is prudent to assume that the indirect effects of climate change on biodiversity are as important as the direct effects. Our confidence in predictions of the biological consequences of the greenhouse warming derived from combining GCM output with information on the climatic associations of plants or animals is confounded by geospheric linkages mediating between biota and climate change, by interspecific connections, and by linkages with other anthropogenic stresses.

Given the number and complexity of the indirect processes by which the greenhouse warming could influence biodiversity (tables 24.1 and 24.2 list a representative sample but do not include all the surprises nature undoubtedly has in store for us), how can we hope to make reliable forecasts? The current journal literature in ecology and conservation biology does not provide as much help as we would like. A survey of all the articles appearing in recent volumes of some leading journals illustrates the Pluto syndrome that afflicts ecology today. Much research in ecology might just as well have been carried out on the planet Pluto—climate and the chemical and physical properties of the soil, the water, and the atmosphere of Earth were irrelevant to the research. Only 22% of the articles mention at least one of those factors, and even then the reference is generally made as part of a site description and plays no role in the interpretation of results. Not one of the papers surveyed mentions the interactions among the factors. (The 285 articles surveyed were from four consecutive issues of *Ecology, American Naturalist, Oecologia,* and *Conservation Biology* during 1987–88.)

In light of the importance of indirect effects, we discuss elements of a research strategy that could improve our understanding of the potential effects of greenhouse warming on biodiversity. Although a few of the elements presented here are focused on understanding the geospheric processes that link climate and biota, most are applicable to research containing any indirect linkage.

## A. Improving the Resolution of Climate Forecasts

We have seen that today's GCMs cannot give us the information required to forecast effects of greenhouse warming on biodiversity. How can this situation be improved? The inadequacy of the existing models originates, in part, with their coarse spatial resolution. Habitats such as alpine peaks, where considerable population extinction could occur during climate warming, sit like needles in the haystack of a GCM zone; hence, zonally averaged climate predictions are not likely to be applicable to the alpine regions. There is a fundamental difficulty with trying to make GCMs so fine-grained that they could describe ecologically distinct habitats—computer capacity would be so vastly oversaturated that even the next generation of computers is not likely to have the capability. One approach to this dilemma of spatial scale is to use GCM zonal output as a constraint on a local or regional climate model that describes the specific habitat. Models for this purpose could range from quite simple, one-dimensional energy balance models to three-dimensional models with parameterized air motions and realistic vegetation type and topography. For this approach to be rigorous, the small region modeled must not exert a significant feedback on the larger GCM zone; this will have to be demonstrated in each particular case.

A second limitation of existing GCM output is its coarse temporal resolution. As shown in the discussion of the climate data required for fire and upwelling models, monthly averaged data are often inadequate for assessing the consequences of indirect geospheric linkages. GCMs do yield output at 60- or 90-minute intervals, but GCM experts express much less confidence in those predictions, and even in the daily or weekly averages, than they do in the monthly averages. Fortunately, the actual time series of precipitation or temperature is often not necessary for impact assessment; needed are such statistical characteristics as the average time between storms or the frequency of years in which two weeks, say, will pass between precipitation events in summer. One possible way to improve the reliability of GCM predictions of such statistical measures is to derive them from longer GCM runs.

To ecologists working in specific habitats, knowledge of monthly or even daily averaged temperature, precipitation, or wind is often of less concern than knowledge that the type of weather pattern is expected to change. GCM output now is couched in terms such as "a 10% increase in September precipitation in Southern California." If this were supplemented with the information that the additional rain represents an earlier occurrence of a northerly front, or alternatively that subtropical hurricanes are more likely in the region, then a greater understanding of the meaning of the GCM numbers would be possible and the information would be of more use to ecologists.

## B. Acquiring Useful Baseline Data

Baseline data are essential for making a convincing case that greenhouse warming has actually affected a population. A potential pitfall lurks here: while the impending warming is rapid on the time scale of paleoclimate change, it is slow on the time scale of scientific innovation. The pace of technological change in scientific measurement techniques is incredibly rapid and threatens to render obsolete the existing techniques and methods for baseline measurement. In research on acid precipitation this has already happened: older data on lake acidification often cannot be compared with new data because the technology and methods for pH measurement have vastly improved, and changed, in the past few decades. As new techniques and strategies for measurement come along, they must be calibrated with the old, to allow use of long time series of measurements and not just those data acquired since the latest breakthrough in measurement methods. No one should expect that scientific instrumentation fifty years from now will resemble that used today in any of the disciplines that bear on climate impact assessment.

## C. Conducting Field Manipulations

Mathematical models will be of value in understanding how some geophysical processes respond to climate change and, more rarely, how biota respond to altered geophysical conditions, but confidence in those models will be limited unless they can predict correctly the results of controlled experiments on climate warming. Laboratory microcosms are unlikely to be of much value except for determining physiological rate constants for use in models, because of unavoidable problems of spatial scale in laboratory systems (Harte and Jassby 1978). Field manipulations, in which sizable plots are subjected to combinations of warming, precipitation alteration, and $CO_2$ increase, can provide a wealth of data for forecasting ecological consequences of the greenhouse effect and for validation of models. Information obtained from field manipulations cannot easily be generalized to other sites, but here correlative information can be of great use. Suppose that a field manipulation of temperature shows that a biological or geospheric condition, like soil nitrogen content, depends on temperature in a specific way. If that same dependence is seen in field measurements along a wide variety of existing temperature gradients, one's confidence in the generality of the relationship is greatly enhanced.

## D. Using Correlations and Mathematical Models Effectively

Some information on indirect effects of climate change on biodiversity can be derived from historical and paleoclimate records. The example of wildfire illustrates some of the drawbacks of this analogue approach, however. A documented, historical relationship between wildfire and climate could reflect either the fire regime's change as a direct result of climatic change or the fire regime's response to climate-induced vegetation changes. Similar ambiguities arise for most of the geospheric linkages discussed here.

To make more effective use of correlative data from the past, as well as from current baseline and proposed manipulative experiments, mechanistic models are needed. Mechanistic models, as the name suggests, attempt to describe observed behavior in terms of underlying processes; these processes might be physiological if the object of study is an organism, or they might by thermodynamic if the object of study is a physical system. Mechanistic models contrast with correlative models, which attempt only to characterize relations among observed behaviors but not to describe the underlying processes that govern the behavior. Mechanistic models can range from the simple to the complex, depending on how many processes are included. There is merit in including many processes in a model, thereby making the model more realistic. But often the desire for realism runs up against the lack of quantitative information about the underlying processes. In those circumstances, modelers must either revert to simpler models or include in their complex models constants that can be determined only by fitting the predictions of the model to experimental data. Complex models with fitted parameters are sometimes called simulation models.

We emphasize here the benefits of the combined use of correlations and simple mechanistic models because of the enormous value of this strategy in the resolution of past environmental dilemmas. The debate that raged during the 1960s over the cause of accelerated lake eutrophication illustrates this. Complex simulation models were paraded forth by advocates of various explanations, but it was not until two developments occurred that an unambiguous resolution of the cause of, and solution to, the problem was possible. The first was the demonstration that for the majority of lakes in the world, phosphorus loading correlates remarkably well with midsummer lake-water chlorophyll (Vollenweider and Dillon 1974). Simple mechanistic modeling suggested a correlation, while empirical studies confirmed it and pinned down the numerical value of the slope of the regression line. This work was startlingly successful and is even today used as a theoretical tool for managing lake-water quality. The second development was a simple box model of the Great Lakes used to describe the consequences of a reduction in phosphorus loading to those lakes (Reckhow and Chapra 1983). When the model prediction for phosphorus loading to each lake was combined with the widely applicable relation between phosphorus loading and trophic condition, the model predicted correctly the state of eutrophication of each lake. The predicted benefits from affordable reductions in phosphorus loading were sufficiently large and the argument that the predictions were valid was so convincing that cleanup soon began. Follow-up studies documented the accuracy of the model's predictions of improvement in the trophic status of the Great Lakes following reduction of phosphorus loading.

The alternative to this simple mechanistic approach is to use large, highly parameterized simulation models. These do have the potential to include a greater number of the components and processes in the system under study, but they are difficult to validate (Cosby et al. 1986). Our reading of the track record of biogeochemical modeling over the past several decades suggests that simple mechanistic models have been of the most practical value. This can be seen by contrasting simulation models of watershed acidification (e.g., Chen et al. 1983) with simpler and more mechanistic approaches (e.g., Cosby et al. 1985, Kirchner 1990). Models work best for predicting change when the important underlying mechanisms are well understood; for geospheric processes, for example, the model is pinned down by the laws of physics and chemistry. But that ideal is rare, and most often some empirical input is required to reduce the ambiguity in the model. In those situations, correlative information, such as the relation between phosphorus loading and midsummer lake-water chlorophyll, is needed to make the model a powerful predictive tool. We submit that the same combination—simple mechanistic models, augmented with correlative empirical information—will provide the most incisive approach to modeling effects of climate warming on geophysical processes.

Simple mechanistic models also provide insights into the type of climate model output appropriate for a particular ecosystem impact analysis. A sensitivity analysis of a model is often needed to determine the features of the input data to which the results are most sensitive. Such analyses are far easier to carry out with simple mechanistic models than with complex simulation models. The upwelling discussion illustrates the kind of insight provided by an understanding of the mechanisms by which climate (wind in this case) influences the ecosystem of concern (fig. 24.4). The spatial and temporal resolution of wind data necessary for assessing the effect of climate change on marine ecosystems depends on the specific mechanisms linking climate change to biodiversity.

## VI. IMPLICATIONS FOR CONSERVATION POLICY

Climate change, generally, and the presence of indirect effects, particularly, present new conservation challenges and exacerbate current conservation problems. The indirect effects of climatic change will influence ecosystems in ways that could not be anticipated by looking at direct effects alone. Forecasting either the magnitude or the direction of indirect effects on populations is more difficult and has more uncertainties than forecasting direct effects. The uncertainties in forecasting biological responses are further increased because society has the means to minimize or accelerate the greenhouse effect by changing, for example, rates of fossil fuel combustion and emissions of chlorinated hydrocarbons (for example, CFCs, which are potent greenhouse gases). The unknown societal response combined with the scientific uncertainties make it impossible to predict the climate at a specific location 20, 40, or 60 years in the future.

Actions to protect biodiversity in the face of past threats have included regulating hunting and fishing, passing laws to protect particular species or habitats (e.g., Endangered Species Act, Migratory Bird Act, wetlands protection laws), setting aside land in parks and preserves, and constraining human land use to allow wild species to coexist with people (e.g., open space in urban areas, multiple use of managed national forests). In light of uncertainties about the future, what elements of these approaches and what novel approaches can be combined to design a policy to protect biodiversity in a world of changing climate?

## A. Parks Are Not a Panacea

The worldwide system of parks and protected areas is a critical component of international conservation efforts. Ideally, parks containing representative examples of all major ecosystems would be distributed throughout every biogeographic region. Yet today's park systems operate under substantial constraints. Only a limited amount of land can be dedicated to a single, nonconsumptive use, especially in countries greatly dependent on land-intensive natural resources. Scientific understanding of how to design preserves to maintain viable populations of even a few target species is abstract and largely untested (Jensen 1987). Finally, parks are not isolated pieces of nature walled off from the rest of the world. Instead, park boundaries are lines on maps across which people, pollution, and species regularly pass (Schonewald-Cox 1988). Many park management problems, such as elephant poaching in Africa, grizzly bear attacks on livestock near

Yellowstone, or decreasing water supply in the Everglades, occur because park boundaries are permeable. Setting aside parks and preserves as safe sites for species cannot always protect diversity, even without climate change.

Nor will parks be a panacea in a world of changing climates. Even considering only direct effects, parks will provide limited protection to species. As chapter 2 shows, species' ranges may shift out of park boundaries as climate changes. Some populations in the park will go extinct and populations of other species will invade. If parks are treated as self-sufficient refuges, the regions surrounding parks may not be managed to protect biodiversity. Where human land uses eliminate or fragment native vegetation, many species will find insurmountable barriers or inhospitable land outside the parks. If the park becomes a habitat island, extinctions are more likely than invasions, but the consequences of the species reshuffling that will likely occur cannot be readily predicted.

The major problem with relying solely on parks is the failure of a key assumption on which the park approach is based. Proposals to design new parks to protect species from global warming assume implicitly that there will be a new equilibrium climate and that species will find a permanent home in some new location with a suitable environment (including climate). But it is unlikely that climate will equilibrate in the next hundred years, and it is also unlikely that the other physical and chemical factors affecting habitat suitability will equilibrate. Geospheric conditions and processes will be altered by climate change, and in many cases, such as rising sea level and soil weathering, response times will be centuries long.

Parks will and should continue to be the core of our efforts to protect biodiversity, by providing lands dedicated to species conservation. Adding more parks to our system of protected areas will increase the meager acreage dedicated to conservation, but designing a park to protect a single species or habitat or predicting which species will be protected by a park may not be possible. Given the uncertainty of future distributions of species and the relatively long time horizon for change, we should minimize species extinctions from existing threats and look for conservation tools in addition to parks.

## B. Other Solutions

The best strategy when an outcome is uncertain is to hedge your bets; try to decrease the risk of failure by taking several approaches. If parks are the only conservation strategy, the prescription is to diversify the number, location, and size of the parks in our portfolio. But why should our conservation efforts be limited to parklands? Less than 5% of the earth's land surface is in parkland, and even if this number were to double, it would be hard to hedge our bets with such small holdings. Achieving the goal of conservation of biodiversity requires additional strategies.

Most of the area of the globe is neither in parks nor under intensive development or cultivation. Unprotected rangeland, forest land, and deserts cover the majority of the earth's land surface. Although these lands are not dedicated to the conservation of biodiversity, countless species live in this semi-natural matrix (Brown 1988). The conservation of biodiversity on these lands could be promoted by a multitude of management policies directed at, for example, selecting sustainable timbering practices that preserve old growth (chapter 19), regulating recreational uses, constraining locations of roads and human dwellings, confining grazing activities to protect streams and riparian zones, and a host of other strategies of similar intent. This is not an easy task. Multiple-use management often means balancing conflicting objectives; society must decide that biodiversity is to be a strongly competing objective before this approach can succeed. Incentives for private landowners to join into the conservation task, as well as the means for them to do so, must be created. These are only a few ways to increase the options available for the protection of biodiversity. Numerous creative means of habitat conservation can be and

need to be found to enhance our system of parks and protected areas.

Another necessary component is to design a safety net for heavily exploited resources that are not in parks or that parks intrinsically cannot protect. Laws to provide baseline protection for migratory species, river organisms, and large vertebrates that move in and out of protected areas would be part of such a safety net. This is especially important and difficult for river- and wetland-dependent species. In arid regions, water is always a contested resource; climatic change may make it even more likely that water will be fought over. Under existing policies and water rights laws, water conservation alone will not necessarily provide additional water to fish and wildlife. Unless protections such as minimum flow requirements are designed for fish and other aquatic species, it is unlikely they will fare well.

## C. Human Institutions: Part of the Problem, Part of the Solution

Habitat loss is often cited as the major cause of species endangerment. Will human settlement patterns also change as a result of global warming? Where will we grow food? Cut timber? Obtain water? Will wild species migrating to a new location because of global warming find that people have the same idea? Certainly one indirect effect of climate change will be changes in human institutions and societies. Although not addressed in this paper, these responses constitute indirect linkages that are just as important as geospheric processes, interspecific connections, and synergies.

Conservation policies designed without considering the role of existing institutions and societal responses to climate change are unlikely to be successful. Particularly when attempting to protect biodiversity outside parks, new policies must integrate resource law and management practices into the plans. Existing resource laws allocate rights assuming that the quantity and location of the resources are not going to change greatly. Most research indicates that natural resources will not remain distributed as they are now. The winners and the losers cannot be determined, but for those resources that have a history of being contested, people currently controlling resources like timber and water will try to protect their rights. Resources valued for their nonmarket or amenity benefits do not have such powerful advocates with vested personal interests. As natural resources are redistributed by global warming, who will argue for new parks or for including biodiversity among the multiple uses to which overcommitted public lands are put?

Global warming is a problem created by human society, and it can be solved only by that same society. Human lifestyles, settlement patterns, and institutions must be included in the problem definition as well as in the formulation of viable solutions. Energy conservation and a halting of deforestation to reduce carbon dioxide emissions must be part of the solution, as must reductions of chlorinated hydrocarbons such as CFCs. The most important component of any strategy to protect biological diversity from climatic change must be preventive measures to slow down the rate of warming. But prevention is not sufficient to protect biological diversity, because the quantities of greenhouse gases already emitted commit the planet to some degree of warming. Both prevention and protection are required. In addition, the uncertainties created by the existence of a myriad of indirect linkages between climate change and biodiversity necessitate that we hedge our bets and implement a wide variety of protective policies, for no single approach is likely to be successful.

## VII. SUMMARY

Forecasting the ecological consequences of the greenhouse effect is among the most important and challenging intellectual tasks that scientists have ever faced. In addition to its direct physiological consequences to plants and animals, climate change will trigger innumerable changes in the physical environment, interspecific connections, and synergies with other anthropogenic stresses that will shape life on earth during the coming decades. Societal response to shifting resource distribution under a warming climate will have further consequences. The challenge of formulating a viable species conservation policy is no easier nor less important than the challenge of forecasting effects. Meeting these challenges will force us to overcome counterproductive disciplinary barriers to research, will add immeasurably to our understanding of the biosphere, and will provide society with the knowledge needed to accommodate to future warming. Moreover, we must confront head on the consumptive patterns worldwide that are leading to this impending and biologically potent change in our planet's climate.

## REFERENCES

Ainley, D. G., H. R. Huber, and K. M. Bailey. 1982. Populations of California sea lions and the Pacific fishery of Central California. *Fishery Bull.* 80(2): 253.

Bailey, K. 1981. Larval transport and recruitment of Pacific hake. *Marine Ecology Progress Series* 6:1.

Bakun, A. 1973. Coastal upwelling indices, west coast of North America 1946–1971. NOAA Tech. Rep. NMFS SSRF-671, NOAA, Washington, D.C.

Billings, W. D., J. O. Luken, D. A. Mortensen, and K. M. Peterson. 1982. Arctic tundra: A source or sink for atmospheric carbon dioxide? *Oecologia* 53:7.

Billings, W.D., K. M. Peterson, J. O. Luken, and D. A. Mortensen. 1984. Interaction of increasing atmospheric carbon dioxide and soil nitrogen on the carbon balance of tundra microcosms. *Oecologia* 65:26.

Brown, J. 1988. Paper presented at the 1988 AIBS meetings in Davis, Calif. Cited by Leslie Roberts in Hard choices ahead on biodiversity. *Science* 241:1759.

Chen, C., S. Gherini, R. Hudson, and J. Dean. 1983. The integrated lake-watershed acidification study. Vol. 1: Model principles and application procedures. Final report, Sept. 1983. Lafayette, Calif.: Tetra Tech.

Clark, J. S. 1988. Effect of climate change on fire regimes in northwestern Minnesota. *Nature* 334: 233.

Connell, J. H., and R. O. Slayter. 1977. Mechanisms of succession in natural communities and their role in community stability and organization. *Am. Naturalist* 111:1119.

Cosby, B. S., R. F. Wright, G. M. Hornberger, and J. N. Galloway. 1985. Modeling the effects of acid deposition: Assessment of a lumped parameter model of soil water and streamwater chemistry. *Water Resources Res.* 21:51.

Cosby, B. S., G. M. Hornberger, E. B. Rastetter, J. N. Galloway, and R. F. Wright. 1986. Estimating catchment water quality response to acid deposition using mathematical models of soil ion exchange processes. *Geoderma* 38:77.

Dozier, J. 1980. A clear-sky spectral solar radiation model for snow-covered mountainous terrain. *Water Resources Res.* 16:709.

Fried, J.S., and M. S. Torn. 1990. Analyzing localized climate impacts with the Changed Climate Fire Modeling System. *Nat. Resource Model.* 4:229.

Gleick, P. 1988. Regional hydrologic consequences of increases in atmospheric $CO_2$ and other trace gases. *Clim. Change* 10:137.

Harte, J., and A. Jassby. 1978. Energy technologies and natural environments: The search for compatibility. *Ann. Rev. Energy* 3:101.

Harte, J., and E. Hoffman. 1989. Possible effects of acidic deposition on a Rocky Mountain population of the tiger salamander *Ambystoma tigrinum*. *Conser. Biol.* 3(2):149.

Jenny, H. 1930. A study on the influence of climate upon the nitrogen and organic matter content of the soil. Univ. Missouri Agric. Experiment Station Res. Bull. 152, Columbia, Missouri.

Jensen, D. B. 1987. Concepts of preserve design: What have we learned? In *Conservation and Management of Rare and Endangered Plants*, T. S. Elias, ed., pp. 595–603. Sacramento: California Native Plant Society.

Kirchner, J. W. 1990. A strategy for predicting watershed acidification. Ph.D. thesis, Energy and Resources Group, University of California, Berkeley.

Lashof, D. 1989. The dynamic greenhouse: Feedback processes that may influence future concentrations of atmospheric trace gases in climatic changes. *Clim. Change* 14:213.

Mason, J. E., and A. Bakun. 1986. Upwelling index update, U.S. West Coast, 33N–48N latitude. NOAA Tech. Mem. NOAA-TM-NMFS-SWFC-67, NOAA, Washington, D.C.

McElroy, M. B., and R. J. Salawitch. 1989. Changing composition of the global stratosphere. *Science* 243:763.

Oppenheimer, M. 1986. *The Atmosphere and the Future of the Biosphere: Points of Interactive Disturbance.* New York: Environmental Defense Fund.

Paine, R. T. 1966. Food web complexity and species diversity. *Am. Naturalist* 100:65.

Reckhow, K., and S. Chapra. 1983. *Engineering Approaches for Lake Management.* Stoneham, Mass.: Butterworth.

Roughgarden, J., S. Gaines, and H. Possingham. 1988. Recruitment dynamics in complex life cycles. *Science* 241:1460.

Schonewald-Cox, C. M. 1988. Boundaries in the protection of nature reserves. *Bioscience* 38:480.

Smith, W. 1981. *Air Pollution and Forests.* New York: Springer-Verlag.

Terborgh, J., and B. Winter. 1980. Some causes of extinction. In *Conservation Biology*, M. E. Soulé and B. A. Wilcox, eds., pp. 119–134. Sunderland, Mass.: Sinauer Assoc.

Veirs, S. D. Jr. 1982. Coast redwood forest: Stand dynamics, successional status, and the role of fire. In *Proceedings of Symposium: Forest Succession and Stand Development in the Northwest*, J. E. Means, ed., pp. 119–141. Corvallis: Forest Research Laboratory, Oregon State University.

Vitousek, P. M. 1986. Biological invasions and ecosystem properties: Can a species make a difference? In *Ecology of Biological Invasions in North America and Hawaii*, H. A. Mooney and J. A. Drake, eds., pp. 163–178. New York: Springer-Verlag.

Vollenweider, R. A., and P. J. Dillon. 1974. *The Application of the Phosphorous Loading Concept to Eutrophication Research.* Ottawa: National Research Council, Assoc. Comm. Sci. Criteria Envir. Qual.

Warner, R. E. 1968. The role of introduced diseases in the extinction of endemic Hawaiian avifauna. *Condor* 70:101.

# CHAPTER TWENTY-FIVE

# Synergisms: Joint Effects of Climate Change and Other Forms of Habitat Destruction

NORMAN MYERS

## I. INTRODUCTION

On top of the myriad assaults that wildlife species are subjected to by human activities, we must now add another: the greenhouse effect with its associated changes in climate. This looks set to cause broad disruption and destruction of species' habitats (Peters 1988, Peters and Darling 1985). It may even wreak such havoc that climate change could eventually come to rank as the number-one threat to species, causing more extinctions than any other single factor.

Yet even this dismal scenario, baldly stated, may underestimate the ultimate scale of the extinction spasm that appears probable in the wake of the greenhouse effect. We must also factor in the synergistic interactions between habitat destruction of conventional type and this new climate-derived form of habitat destruction. When the two processes operate in a manner that allows each to reinforce the other's effects, the result could well be a powerful compounding of impacts.

It is this theme, the mutual amplification of effects from two discrete ecological processes working in concert, that is the subject of this chapter. After taking a quick look at the phenomenon of synergism, the chapter will consider how it applies to the present situation, in which species and biotas are being overtaken by two separate sets of threats: the conventional destruction of habitats (tropical deforestation, spread of deserts, pollution, etc.) and the new form of habitat destruction (climatic change). The potential manifestation of synergistically aggravated threats is illustrated by examples of tropical forests, tropical coral reefs, wetlands, and Mediterranean-type zones.

Synergisms occur when two or more ecological processes interact in such a manner that the product of their interactions is not merely additive but multiplicative (Ehrlich and Roughgarden 1987; Myers 1987a, 1989; Odum 1971). For instance, a biota's tolerance of one stress tends to be lower when other stresses are in operation at the same time. A plant that experiences depleted sunlight and reduced photosynthesis is unduly prone to the adverse effects of cold weather and thereby suffers more in the cold than a plant enjoying normal growth and full vigor (Graham and Patterson 1982, Levitt 1980). A similar amplified effect operates the other way round as well. The compounding interactions of a synergistic relationship can be so powerful that the result may be an order of magnitude greater than the simple sum of the components (Lange et al. 1981, Pretcht et al. 1973).

Despite their obvious importance, we know all too little about synergisms. Ecologists cannot even identify and define their main manifestations in nature, let alone document their more frequent effects. To the extent that we can discern some of the possible synergistic mechanisms at work in the impending extinction spasm, the better we shall start to understand some potential patterns and processes as the spasm works itself out—and the better we shall be able to anticipate and even prevent some of them.

An example of synergism is the susceptibility to disease of plants under high-temperature or reduced-water stresses. The plants need additional energy and thus acceptable levels of temperature and moisture to cope with stresses, especially to repair any damage the stresses cause (Larcher 1980). Conversely, plants that are diseased or otherwise damaged are less able to cope with the onset of unduly elevated temperatures or reductions in moisture. This applies particularly to plants that have been weakened through other types of environmental insults, such as UV-B radiation and chemical pollutants (Worrest and Caldwell 1986). In addition, plants could well become subject to pandemic diseases such as might occur through the ecological disruptions of a greenhouse-affected world (chapter 16), and pathogen-carrying insects may become more

numerous as higher UV-B sensitivity affects birds and other insect predators.

A greenhouse-affected world, then, will feature more than a general warming and some changes in rainfall. In many regions, droughts will become more common, more protracted, and more severe, while in other regions rainfall may become both heavier and more concentrated, causing environmental damage such as soil erosion (IPCC 1990). Certain sectors of the world will feature more heat waves than before, and they will be longer and hotter, too. In many coastal zones there will be increased sea storms (including typhoons, hurricanes, and cyclones), along with storm surges (DeSylva 1986). When we consider all these disruptive phenomena together and allow for the mutual amplification at work between just some of them, we see a massive scope for exceptional numbers of synergistic interactions, with exceptionally pronounced effects. (See chapter 24 for additional discussion.)

The upshot is that we should be careful not to consider various categories of environmental assaults in isolation from each other. Rather, we should consider how their interactive and cumulative effects can exacerbate each other. In view of the multidimensional and largely simultaneous effects of ecological interactions in a greenhouse-affected world, a merely linear account of effects will surely underestimate the eventual outcome overall.

Of course there can be cases of synergistic effects that prove to be beneficial rather than harmful. But judging from what we know about synergisms, it is realistic to suppose that mutually compounding responses will mainly turn out to be harmful.

This chapter considers some of the amplificatory ramifications that will likely ensue during the decades ahead. It is essentially exploratory in character and scope, describing some of the more significant forms of synergisms that we can anticipate as climate change interacts with the longstanding kinds of habitat destruction. In light of our scant understanding of synergisms, I do not attempt a systematic analysis of such processes at work with regard to mass extinction. Indeed, this should be viewed as an exercise in creative speculation: it makes no claim to present definitive arguments or to adduce conclusive evidence. On the contrary, its modest aim is to stimulate thinking in an emerging subject area that is critical to conservation biology.

## II. ILLUSTRATIVE EXAMPLE: AMAZONIA

Even if Amazonia's ecosystem suffers no climatic disruptions, the present pattern of deforestation would lead to a sizable fallout of species (Simberloff 1986). If deforestation continues at mid-1980s rates until the year 2000 and then halts completely, we should anticipate a loss of about 15% of plant species. If Amazonia's forest cover is ultimately reduced to those areas now set aside as parks and reserves, we should anticipate that 66% of plant species will eventually disappear, together with almost 69% of bird species and similar proportions of other major categories of animal species.

But this pioneering analysis does not take into account the greenhouse effect. Climatic dislocations could prove to be critical for the survival of many of Amazonia's species. Although the greenhouse effect is expected to produce more rainfall along the equator, it could result in less rainfall in tropical areas away from the equator, an outcome postulated through disruption of Amazonia's hydrological systems (Salati and Vose 1984, Shukla 1990) and through shifts in albedo regimes (Henderson-Sellers and Gornitz 1984, Shuttleworth 1988). In short, the greenhouse effect could severely intensify the habitat-loss repercussions already cited. Of 26 centers of endemism identified in South America, the great bulk are in Amazonia (Prance 1982), and 9 are in the southern border zone of Amazonia.

Moreover, reduced rainfall in southern Amazonia could lead to positive feedback processes with still more repercussions, possibly working in synergistic accord to generate compounding impacts. A decline in pre-

cipitation would lead to water stresses for plants, which in turn would lead to a more xerophytic vegetation with *cerrado*-type scrubland. That much is plain. Not so apparent is that such changes could start to feed on themselves, insofar as a further reduction in evapotranspiration could cause an accelerated shift toward dryland vegetation (Dickinson and Virji 1987)—with all that would mean for rainforest species. Even partial deforestation of Amazonia could initiate an irreversible trend toward greater aridity and more-seasonal precipitation, leading to a collapse of surviving forest ecosystems, even ecosystems in the best-protected parks. In addition, the drying-out process could foster more wildfires in remaining forest tracts, which in turn would likely accentuate the speed of collapse (chapter 10 discusses changes in flowering and fruiting caused by climate shifts, both of which affect animal pollinators and seed disseminators). The convective patterns of rainfall that characterize the humid tropics could also have amplifying effects. With greater warmth from the greenhouse effect, this rainfall will tend to become more intense and concentrated, possibly leading to greater flooding and soil erosion while leaving less moisture in the soil and thus giving rise to more stressful dry seasons (Dickinson and Virji 1987).

In addition to southern Amazonia with its rich centers of endemism, a number of other tropical forest areas are unusually rich in endemics and unusually threatened with destruction (Myers 1988a and 1990). They are apparently at severe risk from climatic vagaries in the wake of the greenhouse effect, notably gross irregularities of rainfall (Emanuel 1987, Myers 1988b, Woodwell 1986, Zhao and Kellogg 1988). These hot spots include the Atlantic-coastal forest of Brazil, the southwestern Ivory Coast, East African montane forests, eastern Madagascar, southwestern Sri Lanka, the eastern Himalayas, the Philippines, and New Caledonia. All these forest areas are in seasonal-rainfall zones, where vegetation is already vulnerable to water stress at unusually dry times of the year. All are likely, even with the best onsite management, to be biotically impoverished as a result of the rainfall irregularities entrained by the greenhouse effect. Worse, if their biotas are severely degraded by slash-and-burn cultivation, among other human activities, even before the full force of the greenhouse effect is felt, their species communities will be all the less capable of withstanding climatic upsets.

## III. NONLINEAR CHANGES

Certain ecological changes, synergistically driven, may not be linear and gradual in their eventual outcome but rather nonlinear and sudden in occurrence. To be specific, certain forest ecosystems may undergo creeping degradation for a time, manifesting a slow, cumulative decline and chronic stress, before finally and suddenly exhibiting stress of an altogether more severe type (chapter 22; Duinker et al. 1991, Emanuel et al. 1985, Shands and Hoffman 1987). As F. H. Bormann (1985) has remarked (see also Zhao and Sun 1986), albeit with respect to acid rain (much the same applies to forest ecosystems undergoing different stresses), "Often enough, a forest ecosystem can successfully buffer stress for long periods, with the results that biotic regulation is scarcely affected and ecosystem changes are little perceptible. Yet even while the ecosystem remains apparently unchanged and healthy, its inherent complexity may mask symptoms of damage. As the buffering capacity is depleted, so the ecosystem moves nearer to the limits of its resilience—and thereafter toward a potential collapse to a state of lower productivity and markedly less biotic regulation of energy flow and biochemical cycles."

These jump events, plus related phenomena such as environmental threshold effects, breakpoints, and other ecological discontinuities, are especially pertinent to the question of synergisms stemming from climate changes. Even our most advanced climate models tend to discount, by virtue of their very structure, the possibility of certain kinds of synergistic interactions in the real world, leading to threshold-type repercussions. Far from knowing how to incorporate such complex interactions into our models, we scarcely know how to identify and define them as yet. As W. S. Broecker (1987) has recently reminded us, the earth's past climate, as revealed by ice cores, ocean sediment, and bog mucks, has often responded in a far from smooth, gradual, and hence predictable manner. On the contrary, it has often reacted in the form of sharp jump events, "which involve large-scale reorganization of Earth's ecosystems." Suppose such jump effects were to occur, perhaps in multiple forms, in the near future. What, for instance, if the Gulf Stream were to be significantly disrupted, even diverted southward, rather than flowing northeastward to warm northwestern Europe?

## IV. TROPICAL CORAL REEFS

The second-richest biome after tropical forests, tropical coral reefs cover some 600,000 km² (depending on how one measures their areal extent), accounting for only 0.17% of the marine realm. Apart from their own coral species and reef-related communities, they support large numbers of open-ocean species, perhaps 65% or more of all marine fish species, which depend on coral-reef ecosystems for some part of their life cycles. Already subject to many environmental insults at human hands (Wells 1988), these reefs may ultimately prove yet more susceptible to the adverse effects of greenhouse warming than the hot-spot areas in tropical forests.

Tropical marine organisms exist much closer to their intrinsic physiological limits for temperature and for oxygen than species of temperate-zone seas (Longhurst and Pauly 1987). For this reason, among others, coral-reef communities (also those of mangroves) are more ecologically fragile than might be supposed (Goreau 1990, Johannes and Hatcher 1986). For instance, release of heated wastewater is more stressful in the tropics than in temperate zones on the grounds that tropical marine organisms live at environmental temperatures closer to their upper thermal limits. We know, moreover, that abnormal warming episodes in the past have caused localized mass fatalities; witness the 1983 El Niño phenomenon. El Niño brought on a heat increase of 3°–4°C for just six months, yet it was enough to inflict widespread and significant damage on tropical coral reefs (Glynn 1984, Goreau 1990).

Tropical organisms also live closer to their lower oxygen limits. Not only are their metabolic rates higher, but also dissolved oxygen concentrations in tropical seas are lower (seawater saturated with air contains 35% less oxygen at 30°C than at 8°C). Any environmental perturbation that lowers the oxygen concentration, such as sewage pollution, is likely to exert a greater adverse effect on tropical biotas (Pastorok and Bilyard 1985, Smith et al. 1981). Similar considerations apply to certain other forms of pollution, notably oil pollution. In the Caribbean, with some of the richest coral reefs anywhere (at least 30,000 species), whole communities are at risk through oil spillage. One sixth of the world's oil is produced in or shipped through the Caribbean, and supertankers plus offshore oil rigs inject more than 100 million barrels of oil into the sea each year (Rodriguez 1981). As far as we can discern from changes in species abundance and diversity and from the contrast between rich and depauperate reefs, extensive pollution can decrease species richness by at least an order of magnitude (Johannes and Hatcher 1986).

Another form of pollution is siltation of soil and other materials washed off from inland watersheds (Cortes and Risk 1985). This is particularly stressful for reefs already suffering from localized pollutants such as sewage, pesticides, and industrial effluents—another instance of multiplier effects (Hodgson 1988, Kuhlmann 1988). As inland watersheds of the tropics suffer growing deforestation, the siltation problem will surely grow more frequent and pronounced. It is likely to become most problematic in Southeast Asia, where deforestation is more widespread than else-

where—and where there are 20,000 islands with the greatest diversity of corals and coral reefs in the entire tropics (Gomez 1988, McManus 1988).

On top of these environmental problems for coral-reef ecosystems, there comes the additional problem of greenhouse warming. With higher sea-surface temperatures, pollutant effects in coral-reef communities will be synergistically more severe—a phenomenon that itself will impose significant extra stresses on coral reefs. But the superstressed ecosystems will then be subject to the most adverse consequence of all from greenhouse warming, sea-level rise entrained by heating of the ocean's surface layer. Corals can grow, of course, but they do not generally do so faster than a modal rate of 3 mm per year, which is only one fifth of the rate of sea-level rise anticipated; even their fastest rate is only half of what they would need to keep up with rising sea levels (Buddemeier and Smith 1988). So they are likely to die out for lack of light, possibly too for lack of warmth and oxygen. This factor in itself could be enough to eliminate many coral-reef ecosystems.

Of course certain coral species will probably respond better than others. For instance, the rapidly growing antler corals may well be favored. Conversely, they could then prove susceptible, precisely because of their branched forms, to undue damage from storms—and the greenhouse effect is expected to increase the incidence of tropical storms.

## V. WETLANDS

Also susceptible to widespread damage from rising sea levels are coastal wetlands. A 10-cm rise in sea level can cause tidal rivers to advance as much as 1 km inland; a 30-cm rise allows seawater to penetrate 30 m inland along average coastlines; a 1-m rise can inundate low-lying coastal areas almost 1 km inland; and a 2-m rise could eliminate 80% of all U.S. coastal wetlands (Hoffman 1987, Leatherman 1987). Of course the coastal wetlands will try to move inland, but they will find their way blocked by levees, highways, seawalls, housing, and other man-made structures. According to some observers (e.g., Titus 1986), the demise of coastal wetlands as a result of rising sea level could well prove to be the greatest wildlife-related impact of the greenhouse effect in the entire United States.

As for wetlands inland, the outlook is scarcely better. Including marshes, fens, peatlands, mires, swamps, bogs, sloughs, swales, and wet heaths and moors, they cover about 6% of the earth's land surface or roughly the same as remaining tropical forests. They comprise some of the richest biotic zones anywhere. But thanks largely to draining, half the world's wetlands have been lost since the year 1900, with the most rapid rates still in developed nations (Tiner 1984). Draining continues apace and is spreading now to developing nations. Thus, by the time the greenhouse effect arrives, wetlands are likely to be further degraded, fragmented, and otherwise depleted—precisely the state that will leave them all the more vulnerable to drying out in a warmed-up world (Gopal et al. 1982).

## VI. MEDITERRANEAN-TYPE ZONES

Also exceptionally rich in endemic species are Mediterranean-type zones (cool, wet winters and hot, dry summers) (Cody 1986, DiCastri and Mooney 1973). California contains 5050 native plant species (30% of them endemic) in 411,000 km²; the Cape floristic province in South Africa contains 8600 such species (73% endemic) in 77,400 km²; and southwestern Australia contains 3630 species (80% endemic) in 2340 km². Yet large numbers of their plants are rare or threatened (or already extinct): at least 1136 species in California, 1621 in South Africa, and 860 in southwestern Australia.

As the greenhouse effect overtakes these beleaguered communities, they will face not only temperature increases but also changes in rainfall (probably reduced), soil chemistry, and community composition (Busby 1988, Pittock and Nix 1986). Many of their populations will have been further fragmented be-

yond today's discontinuous states, and much genetic adaptability—a key factor in coping with environmental change—will have been lost. In these reduced circumstances, the biotas will seek to migrate toward territories with climates akin to what they are accustomed to (see, e.g., chapter 20). They will be almost entirely unsuccessful. The California biotas will find their way blocked by human settlements in Oregon and Washington, while the other two will find no land at all to receive them, only sea.

## VII. UV-B RADIATION

Besides the greenhouse effect, another atmosphere-related source of environmental disruption deserves mention. It is depletion of the ozone layer, coupled with an increase in UV-B radiation. The radiation over much of the Southern Ocean could rise 5%–20%, even as much as 50%, early next century, together with a 1.7% loss in latitudes from Florida to Pennsylvania and a 3% loss from Pennsylvania north to mid-Canada (Hoffman 1987, Woodwell 1982, Teramura et al. 1980, Worrest and Caldwell 1986). Every 1% loss of ozone allows roughly 2% more UV-B light to reach the earth's surface.

As harmful as UV-B radiation is to humans through skin cancers and cataracts, it is more broadly injurious to plants and certain categories of animals. In the case of plants, the radiation cuts back on photosynthesis; it reduces leaf size, limiting the area available for energy capture; it reduces water-use efficiency; and it adversely affects plant growth generally (Pittock 1987, Teramura 1986). Of course we can still hope that not all plants on earth will be affected; if the Montreal Protocol can be strengthened fast enough, perhaps the effects can be confined to those plant communities harmed by increased UV-B radiation through the ozone hole over Antarctica (notably, in South Africa, southern Australia, New Zealand, and southern South America—the first two being areas with exceptional concentrations of endemic species in their Mediterranean-type zones, which may well be hard hit by the greenhouse effect). Unfortunately, we can expect that the two environmental insults will serve to reinforce each other, as plants weakened by one assault become less able to withstand the second assault—an outcome that works the other way round as well (Pittock 1987). This is not to overlook the possibility that one environmental change may actually compensate for the other. Increased carbon dioxide can, through its fertilizer effect, increase photosynthesis; it can expand leaf size; and it can enhance water-use efficiency. But in the main, a likely outcome is that global warming and UV-B increase will exacerbate each other's deleterious effects (U.S. Department of Energy 1985).

Equally significant could be UV-B's effect on marine ecosystems. Increasing radiation slows the process of photosynthesis in phytoplankton: just a 5% increase can cut their lifetime by half, and a 10% increase causes them to die off almost entirely. The process weakens the basis of entire food webs in the oceans (Calkins 1982, El-Sayed 1988, Hoffman 1987). This would be all the more harmful in the rich marine ecosystem surrounding Antarctica—precisely the area where UV-B radiation may be most pronounced because of the ozone hole. Were the phytoplankton to decline, the herbivorous krill would be immediately affected—and then the many other creatures that depend on krill: squid, fish, penguins, seals, whales, and other animal life in the Southern Ocean. Worst of all, the radiation damage would be imposed on an ecosystem already deeply disrupted by greenhouse warming, projected to be most pronounced in polar regions.

## VIII. SOME MISCELLANEOUS QUESTIONS

A whole series of miscellaneous questions arises, just a few of which are raised here by way of illustration. The reader who likes to engage in that crucial scientific process known as creative speculation will quickly come up with many more questions.

First, the spread of acid rain into the trop-

ics. Although acid rain has been mainly confined thus far to North America and Europe, there are signs of its emergence in such humid-tropic areas as southern China, peninsular Malaysia, Java, south-central Thailand, northwestern India, the West African coast, southern Brazil, and northern Venezuela (Rodhe and Herrera 1988). A good number of other countries, notably the Philippines, the Ivory Coast, Colombia, and Mexico, are starting on rapid industrialization, while giving scant attention to costly pollution controls. We should not be surprised to find that as early as the end of the century, acid rain will become a potent force in degrading certain tropical ecosystems, especially those of tropical forests. It may become a problem even sooner in forest localities already close to industrial centers, for instance the southern end of the Atlantic-coastal forest in Brazil—a forest tract that is ultra-rich in endemic species and ultra-threatened by agricultural encroachments.

Second, consider the case of some important populations of "charismatic vertebrates" such as the tiger population in the Sunderbans Reserve on the coastline that faces the Bay of Bengal. This is the largest tiger population in the Indian subcontinent, and it lies precisely in an area that is highly vulnerable to rising sea level and associated sea storms and storm surges. A one-meter sea-level rise in Bangladesh would affect nearly 12% of Bangladesh's land area, including practically all the Sunderbans.

## IX. THE BIG PICTURE

The biggest miscellaneous question of all is the big picture against which we should view the greenhouse effect's implications for threatened biotas. It is this perspective that points up the most potent synergisms of the coming decades, deriving from growing human numbers with growing human expectations.

The world we have built already, a world of only 5 billion people and built with much effort, has imposed drastic costs on the earth's biodiversity (Wilson 1988). Yet within the next few decades we must anticipate a world with twice as many people, who might well be seeking three times as much food and fiber, consuming four times as much energy, and engaging in five to ten times as much economic activity if the Third World majority of humankind is to climb out of its poverty (Myers 1987b, WCED 1987). A downside scenario would envisage multitudes of impoverished peasants penetrating into every last corner of forest, woodland, savannah, grassland, and whatever other area might afford them living space, albeit at cost of wildlife habitat. Because of his numbers, nobody is more destructive of natural environments than a destitute farmer who takes his machete and matchbox, or his hoe and his hopes, into unoccupied wildland territory. His subsistence pressures could accelerate tropical deforestation or spread of deserts or soil erosion, and he could speedily reduce biotas to a state where they would be far more vulnerable to climatic dislocations.

That is the outlook for sub-Saharan Africa, especially if the region continues to languish in extreme poverty. The region contains just over 500 million people today, projected to increase through uniquely rapid growth to 680 million by the year 2000. Already 30 million people are starving and hence ready to gain their livelihood at whatever cost to natural environments. Unless there is a huge and almost instantaneous advance in the socioeconomic status of the region, these famished throngs are projected to increase to 130 million by the year 2000—expanding the proportion of starving people in the region from less than 7% to 19% (Myers 1989, Timberlake 1988).

Another dimension to the overall downside scenario lies with the activities of the developed nations. They can likewise lay waste to the earth's biotas through their appetites for tropical timber, hamburger beef, whale meat, and a host of other commodities that they consume at an unduly cheap price. They are the ones who have visited CFCs on the biosphere, plus, of course, fossil-fuel carbon loading of the atmosphere.

If, through economic incompetence and political chauvinism, we bring to pass this pessimistic scenario, it will generate a super-sized synergism as the direct (conventional) depletion of wildlife habitats interacts with the new and indirect depletion via the greenhouse effect. Degraded and destabilized ecosystems will enable climatic dislocations far greater than those expected among healthy ecosystems. Conversely, the dislocations would, through their aggravated effect, enable ecological instability to magnify the effects of global warming.

Fortunately, there is also an upbeat scenario that postulates reformed economic analysis (such as natural resource accounting) and enlightened political initiatives (for instance, a global compact embracing the atmosphere, climate, biodiversity, and other common-heritage resources—plus of course an acceptable resolution to the debt crisis and other major international inequities). It also requires a speedy end to such problems as population growth and pervasive poverty in the Third World. In short, it demands sustainable development, practiced in both the developing and developed worlds (WCED 1987). There are lots of success stories to point the way. In terms of socioeconomic advancement, population planning, rational development, energy conservation, and also biodiversity safeguards, we know what to do, and the funding resources are available to us. All we have to do is go ahead, whereupon we would encounter a super-sized synergism of a different sort, as one constructive measure buttresses and reinforces another. To gain an insight into this positive scenario, one need only envisage the capacity of population stabilization to catalyze speedy demarches in many other development fields (Myers 1991).

The most straightforward and cost-effective way out of the greenhouse problem is to stem and contain it. Apart from a sizable amount of warming that is already in the pipeline from past burning of fossil fuels and forests, we could call a halt to the great bulk of the greenhouse effect and the full rigors of its impacts. Again, we know what to do for the most part: the policy initiatives have been laid out in some detail (Keepin and Kats 1988, Mintzer 1987, Schneider 1989, WMO and U.N. Environment Programme 1988). Moreover, that would generate a synergistic spin-off of its own. Not only would it save our fellow species but it would also save us money.

The descriptions and discussions above raise clear implications for conservation policy and on-ground planning. There is not space here to go into the manifold ramifications. Suffice it to say that conservation must be based on scientific knowledge and understanding; and suffice it to add that in light of the synergistic interactions outlined here, we could well be facing a greater mass-extinction episode, arriving more rapidly, than has generally been suspected to date. Should we not consider, then, a research agenda to confront the conservation questions implied by the dynamic interactions of synergisms? To this writer's knowledge, the amount of research under way is all too little, and such research as is in train or envisaged is almost entirely uncoordinated. Herein lies a major challenge—one might say an unusually creative challenge too (even a synergistic challenge)—for conservation biologists.

## REFERENCES

Bormann, F. H. 1985. Air pollution and forests: An ecosystem perspective. *Bioscience* 35:434.

Broecker, W. S. 1987. Unpleasant surprises in the greenhouse? *Nature* 32:123.

Buddemeier, R. W., and S. V. Smith. 1988. Coral reef growth in an era of rapidly rising sea level: Predictions and suggestions for long-term research. *Coral Reefs* 7:51.

Busby, J. R. 1988. Potential implications of climate change on Australia's flora and fauna. In *Greenhouse: Planning for Climate Change*, G. I. Pearman, ed., pp. 387–398. New York: E. J. Brill.

Calkins, J. 1982. *The Role of Solar Ultraviolet Radiation in Marine Ecosystems*. New York: Plenum Press.

Cody, M. L. 1986. Diversity, rarity, and conservation in Mediterranean-climate regions. In *Conservation Biology: The Science of Scarcity and Diversity*, M. E. Soulé, ed., Sunderland, Mass.: Sinauer Associates.

Cortes, J. N., and M. J. Risk. 1985. Reef under silta-

tion stress: Cahuita, Costa Rica. *Bull. Marine Sci.* 36:339.

DeSylva, D. 1986. Increased storms and estuarine salinity and other ecological impacts of the greenhouse effect. In *Effects of Changes in Stratospheric Ozone and Global Climate*, J. G. Titus, ed., Vol. 4, *Sea Level Rise*, pp. 153–164. Washington, D.C.: Environmental Protection Agency.

DiCastri, F., and H. A. Mooney. 1973. *Mediterranean-Type Ecosystems: Origin and Structure*. New York: Springer-Verlag.

Dickinson, R. E., and H. Virji. 1987. Climate change in the humid tropics, especially Amazonia, over the last twenty thousand years. In *The Geophysiology of Amazonia: Vegetation and Climate Interactions*, R. E. Dickinson, ed., pp. 91–105. New York: John Wiley and Sons.

Duinker, P. N., M. Y. Antonovsky, and A. M. Soloman. 1991. *Impacts of Changes in Climate and Atmospheric Chemistry on Northern Forest Ecosystems and Their Boundaries: Research Directions*. IIASA Working Paper. Laxenburg, Austria: International Institute for Applied Systems Analysis.

Ehrlich, P. R., and J. Roughgarden. 1987. *The Science of Ecology*. New York: Macmillan.

El-Sayed, S. Z. 1988. Fragile life under the ozone hole. *Natural History* 97:72.

Emanuel, K. A. 1987. The dependence of hurricane intensity on climate. *Nature* 326:483.

Emanuel, W. R., H. H. Shugart, and M. P. Stevenson. 1985. Climatic change and the broad-scale distribution of terrestrial ecosystem complexes. In *The Sensitivity of Natural Ecosystems and Agriculture to Climate Change*, M. L. Parry, ed., pp. 29–43. Laxenburg, Austria: International Institute for Applied Systems Analysis.

Glynn, P. W. 1984. Widespread coral mortality and the 1982–83 El Niño warming effect. *Envir. Conser.* 11:133.

Gomez, E. D. 1988. Overview of environmental problems in the East Asian seas region. *Ambio* 17:166.

Gopal, B., R. E. Turner, and R. G. Wetzel. 1982. *Wetlands Ecology and Management*. Jaipur, India: International Scientific Publishers.

Goreau, T. J. 1990. Coral bleaching in Jamaica. *Nature* 343:417.

Graham, D., and B. D. Patterson. 1982. Responses of plants to low or freezing temperatures: Proteins, metabolism and acclamation. *Ann. Rev. Plant Physl.* 33:347.

Henderson-Sellers, A., and V. Gornitz. 1984. Possible climatic impacts of land cover transformations, with particular emphasis on tropical deforestation. *Clim. Change* 6:231.

Hodgson, G. 1988. *Sedimentation Damage to Corals Due to Coastal Logging*. Ph. D. dissertation, Department of Zoology, University of Hawaii, Honolulu.

Hoffman, J. S., ed. 1987. *Assessing the Risks of Trace Gases That Can Modify the Stratosphere*. Washington, D.C.: U.S. Environmental Protection Agency.

IPCC (Intergovernmental Panel on Climate Change). 1990. *Climate Change: The IPCC Scientific Assessment*. Cambridge: Cambridge University Press.

Johannes, R. E., and B. G. Hatcher. 1986. Shallow tropical marine environments. In *Conservation Biology: Science of Scarcity and Diversity*, M.E. Soule, ed., pp. 371–382. Sunderland, Mass.: Sinauer Associates.

Keepin, B., and G. Kats. 1988. Greenhouse warming: Comparative analysis of nuclear and efficiency abatement strategies. *Energy Policy* 16:538.

Kuhlmann, D.H.H. 1988. The sensitivity of coral reefs to environmental pollution. *Ambio* 17:13.

Lange, O. L., P. S. Noble, C. B. Osmond, and H. Ziegler, eds. 1981. *Encyclopedia of Plant Physiology*. New York: Springer-Verlag.

Larcher, W. 1980. *Physiological Plant Ecology*. New York: Springer-Verlag.

Leatherman, S. P. 1987. *Impact of the Greenhouse Effect on Coastal Environments: Marshlands and Low-Lying Population Areas*. College Park: Laboratory for Coastal Research, University of Maryland.

Levitt, J. 1980. *Responses of Plants to Environmental Stresses*. New York: Academic Press.

Longhurst, A. R., and D. Pauly. 1987. *Ecology of Tropical Oceans*. New York: Academic Press.

McManus, J. W. 1988. Coral reefs of the ASEAN region: Status and management. *Ambio* 17:189.

Mintzer, I. M. 1987. *A Matter of Degrees: The Potential for Controlling the Greenhouse Effect*. Washington, D.C.: World Resources Institute.

Myers, N. 1987a. The extinction spasm impending: Synergisms at work. *Conser. Biol.* 1:14.

Myers, N. 1987b. Emergent aspects of environment: A creative challenge. *Environmentalist* 7:163.

Myers, N. 1988a. Threatened biotas: "Hotspots" in tropical forests. *Environmentalist* 8:1.

Myers, N. 1988b. Tropical deforestation and climate change. *Envir. Conser.* 15:293.

Myers, N. 1989. Population growth, environmental decline and security issues in sub-Saharan Africa. In *Ecology and Politics: Environmental Stress and Security in Africa*, A. Hjort and M. A. Mohamed Salih, eds., pp. 211–231. Uppsala: Scandinavian Institute of African Studies.

Myers, N. 1990. The biodiversity challenge: Expanded hot-spots analysis. *Environmentalist* 10:1.

Myers, N. 1991. *Population and the Environment: Issues, Prospects and Policies*. New York: United Nations Population Fund.

Odum, E. G. 1971. *Fundamentals of Ecology*, 3rd ed. Philadelphia: Saunders Publishers.

Pastorok, R. A., and G. R. Bilyard. 1985. Effects of sewage pollution on coral-reef communities. *Marine Ecology—Progress Series* 21:175.

Peters, R. L., and J.D.S. Darling. 1985. The greenhouse effect and nature reserves. *Bioscience* 35:707.

Peters, R. L. 1988. The effect of global climatic change on natural communities. In *Biodiversity*, E. O. Wilson, ed., pp. 450–461. Washington, D.C.: National Academy Press.

Pittock, A. B. 1987. Forests beyond 2000: Effects of atmospheric change. *Aust. Forestry* 50:205.

Pittock, A. B., and H. A. Nix. 1986. The effect of changing climate on Australian biomass production: A preliminary study. *Clim. Change* 8:243.

Prance, G. T., ed. 1982. *Biological Diversification in the Tropics*. New York: Columbia University Press.

Pretcht, H., J. Christopherson, H. Hensel, and W. Larcher, eds. 1973. *Temperature and Light*. New York: Springer-Verlag.

Rodhe, H., and R. Herrera. 1988. *Acidification in Tropical Countries*. Chichester: John Wiley and Sons.

Rodriguez, A. 1981. Marine and coastal environmental stress in the wider Caribbean region. *Ambio* 10:283.

Salati, E., and P. B. Vose. 1984. Amazon Basin: A system in equilibrium. *Science* 225:129.

Schneider, S. H. 1989. *Global Warming*. San Francisco: Sierra Club Books.

Shands, W. E., and J. S. Hoffman, eds. 1987. *The Greenhouse Effect, Climate Change and U.S. Forests*. Washington, D.C.: The Conservation Foundation.

Shukla, J., C. Nobre, and P. Sellers. 1990. Amazon deforestation and climate change. *Science* 247:1322.

Shuttleworth, W. J. 1988. Evaporation from Amazonian rain forest. *Proc. Roy. Soc. Lond.* B 233:321.

Simberloff, D. 1986. Are we on the verge of a mass extinction in tropical rain forests? In *Dynamics of Extinction*, D.K. Elliott, ed., pp. 165–180. New York: John Wiley and Sons.

Smith, S. V., et al. 1981. Kaneohe sewage diversion experiment: Perspectives on ecosystem responses to nutritional perturbation. *Pacific Sci.* 35:279.

Teramura, A. H. 1986. Overview of our current state of knowledge of UV-B effects on plants. In *Effects of Changes in Stratospheric Ozone and Global Climate*, vol. 1, *Overview*, J.G. Titus, ed., pp. 165–173. Washington, D.C.: U.S. Environmental Protection Agency.

Teramura, A. H., R. H. Biggs, and S. Kossuth. 1980. Interaction between ultraviolet B and photosynthetically active radiation on net photosynthesis, dark respiration, and transpiration. *Plant Physl.* 65:483.

Timberlake, L. 1988. *Africa in Crisis: The Cause, the Cures of Environmental Bankruptcy*. London: Earthscan Publications.

Tiner, T. W. 1984. *Wetlands of the United States: Current Status and Recent Trends*. Washington, D.C.: U.S. Fish and Wildlife Service.

Titus, J. G., ed. 1986. *Effects of Changes in Stratospheric Ozone and Global Climate*, vol. 4: *Sea Level Rise*. Washington, D.C.: U.S. Environmental Protection Agency.

U.S. Department of Energy. 1985. *Direct Effects of Increasing Carbon Dioxide on Vegetation*. Washington, D.C.: Carbon Dioxide Office, U.S. Department of Energy.

Wells, S. M., ed. 1988. *Coral Reefs of the World*. Cambridge: World Conservation Monitoring Centre.

Wilson, E. O. 1988. *Biodiversity*. Washington, D.C.: National Academy Press.

Woodwell, G. M. 1982. The biotic effects of ionizing radiation. *Ambio* 11:142.

Woodwell, G. M. 1986. Forests and climate: Surprises in store. *Oceanus* 29:71.

WCED (World Commission on Environment and Development). 1987. *Our Common Future*. New York: Oxford University Press.

WMO (World Meteorological Organization) and U.N. Environment Programme. 1988. Developing policies for responding to climatic change (report of the Villach and Bellagio workshops, late 1987), *World Climate Program Impact Studies*. Geneva and Nairobi: WMO and U.N. Environment Programme.

Worrest, R. C., and M. M. Caldwell, eds. 1986. *Stratospheric Ozone Reduction, Solar Ultraviolet Radiation and Plant Life*. New York: Springer-Verlag.

Zhao, D., and B. Sun. 1986. Air pollution and acid rain in China. *Ambio* 15:2.

Zhao, Z.-C., and W. Kellogg. 1988. Sensitivity of soil moisture to doubling of carbon dioxide in climate model experiments: Part II, the Asian monsoon region. *J. Clim. Appl. Meteor.* 1:367.

# CHAPTER TWENTY-SIX

# Effects of Climate Change on Biological Diversity in Western North America: Species Losses and Mechanisms

DENNIS D. MURPHY AND
STUART B. WEISS

## I. INTRODUCTION

Predictions of the effects of climate change on regional biological diversity are necessarily imprecise. Nonetheless, we must attempt to make projections of the fate of biological systems as the earth warms. Conservation biologists are concerned with the effects of global warming on biodiversity, from extinctions of single populations of highly habitat-specific endangered plants to the extirpation of entire species, communities, and ecosystems. This chapter explores two situations in which reasonably accurate projections do appear possible. First, at a regional scale, we explore the potential effect of global warming on entire montane biotas of mammals, birds, butterflies, and plants in the Great Basin of North America. Second, on a local scale, we examine the potential effects of global warming on the fate of the Bay checkerspot, *Euphydryas editha bayensis*, a butterfly of native grasslands in the San Francisco Bay area listed as a threatened species by the U.S. Fish and Wildlife Service.

## II. CLIMATE CHANGE AND REGIONAL BIODIVERSITY: THE GREAT BASIN

R. L. Peters and J. D. Darling (1985) discuss the potential consequences of global warming on nature reserves, emphasizing both latitudinal and elevational shifts in climate conditions. As climate warms, belts of vegetation on mountain range slopes move upward in elevation. Some high-elevation species may get squeezed out when climate zones exceed the highest elevation in a range or when continuous habitat is reduced to fragments on mountain summits. Since less habitat area exists at higher elevations, extinction probabilities increase for species forced to ascend to track appropriate thermal regimes and resources.

Great Basin mountain ranges are ideally suited for projecting changes in species diversity under such circumstances. The Great Basin of western North America is a region of interior drainages between the Rocky Mountains and the Sierra Nevada. Its arid landscape is characterized by basin and range topography, which has resulted from numerous north-south trending uplifts. These mountain ranges, which rise a thousand or more meters above valley floors, constitute archipelagoes of boreal habitat (fig. 26.1). That boreal habitat consists of riparian plant associations along stream corridors, piñon-juniper forests, subalpine forests, and alpine zones at the highest elevations. The lower boundary of piñon-juniper forest in this region is approximately 2300 m (West et al. 1978), and alpine zones extend from approximately 2750 m to the highest peaks at 3000–4000 m (Billings 1978). Climate projections concur

Figure 26.1. Projected changes in boreal habitat areas in east-central Nevada under a 3°C warming scenario. The outer line is the 2300-m contour representing the current lower limit of boreal habitat. The black area represents the area above 2800 m, the lower limit of boreal habitat after warming. Note the large area reductions and fragmented distribution of boreal habitat after the regional warming. The Ruby–East Humboldt ranges are to the north, and the Grant–Quinn Canyon ranges are to the south.

that this region is likely to become both warmer and drier if global warming occurs (see chapter 4).

Boreal habitat was practically continuous in the Great Basin during the Pleistocene. Mountain ranges supported glaciers, basins contained pluvial lakes, and the lower limit of alpine vegetation was approximately 1850 m at glacial maximum (Billings 1978). The distributions of many present-day boreal species were more extensive then. Fossils of the montane vole *Phenacomys intermedius*, for example, are known from the Toquima Range of central Nevada, far to the southeast of the current distribution of the species. In the same mountain range, fossils of the pika, *Ochotona princeps*, have been recovered more than 1000 m lower in elevation than the current minimum elevation of the species (Grayson 1983). Postglacial warming similar to that projected in the immediate future resulted in a 1°C rise in temperature from 7500 to 4500 years before the present (Wells 1983). Piñon-juniper forest and timberline apparently shifted upslope by 100–150 m in the White Mountains of the western Great Basin during that period (Lamarche and Mooney 1967).

A number of studies have established lists of boreal species for major mountain ranges

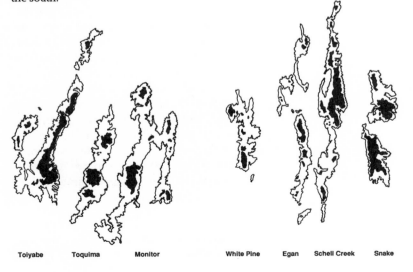

Tolyabe       Toquima       Monitor            White Pine    Egan    Schell Creek    Snake

Table 26.1. Species-area parameters for Great Basin mountain ranges. $S = CA^z$, where $S$ is species number, $A$ is area, and $C$ and $z$ are parameters fitted by a regression on log-transformed data (see fig. 26.2). Higher $z$-values indicate a steeper species-area slope; $C$ reflects the species pool number and other region-specific phenomena. The $r^2$ values indicate the amount of variation explained by area. Sources: (1) Brown 1978, (2) Wilcox et al. 1986, (3) Harper et al. 1978.

| | Species-Area Curves for Taxa | | |
|---|---|---|---|
| | $z$ | $C$ | $r^2$ |
| Mammals | 0.34 | 0.79 | 0.68(1) |
| Birds | 0.16 | 2.16 | 0.49(1) |
| Butterflies | 0.15 | 23.11 | 0.39(2) |
| Bf1 (sedentary) | 0.27 | 2.01 | 0.43 |
| Bf2 (semivagile) | 0.15 | 12.50 | 0.26 |
| Bf3 (vagile) | 0.06 | 9.53 | 0.06 |
| All plants | 0.31 | 49.02 | 0.84(3) |
| Annuals | 0.30 | 5.07 | 0.46 |
| Perennial grasses | 0.36 | 5.21 | 0.73 |
| Perennial forbs | 0.33 | 24.78 | 0.84 |
| Shrubs | 0.24 | 8.06 | 0.70 |
| Trees | 0.11 | 5.86 | 0.12 |

in the Great Basin. Coverage includes mammals (Brown 1971, 1978), birds (Johnson 1978, Behle 1978, Brown 1978), plants (Harper et al. 1978), and butterflies (Murphy and Wilcox 1986, Wilcox et al. 1986, Austin and Murphy 1987). Each of these studies has noted that the distribution of boreal habitat in the Great Basin is analogous to the distribution of habitats across oceanic islands. Several authors have therefore employed island biogeographic theory to explain patterns of species diversity (see MacArthur and Wilson 1967). The essential features of island biogeography pertinent to these studies are: larger islands have more species than smaller islands, islands farther from mainland sources have fewer species than islands of similar size closer to the mainland, and these differences can be explained by opposing rates of colonization and extinction of species.

The results of Great Basin biogeographic studies share several conclusions. All boreal mammal, bird, and butterfly species found in the Great Basin are widely distributed in ei-

ther one or both of the mainland areas. The major source of less widely distributed species appears to be the Rocky Mountain "mainland" to the east, rather than the Sierra Nevada to the west. Distance from those mainland areas, however, is a poor predictor of species number in most taxa. Furthermore, the flora and fauna of the Great Basin are depauperate compared with those of the Rocky Mountains and Sierra Nevada.

Almost all taxa that have been studied previously exhibit significant increases in species number with increased boreal habitat area (table 26.1, fig. 26.2). Indeed, area explains much of the variance in species numbers for two taxa in particular, 68% in mammals and 84% in plants (see $r^2$ values in table 26.1), although area explains somewhat less variance in species numbers among ranges for birds and butterflies, at 49% and 39%, respectively. The slopes of the species-area curves also differ among taxa. Species numbers of small mammals decrease relatively rapidly with decreasing area (see $z$ values in table 26.1). Mammal extinctions apparently

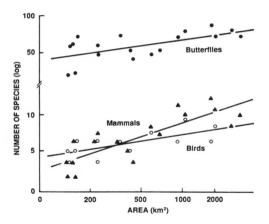

*Figure* 26.2. Species-area curves of mammals, birds, and butterflies on Great Basin mountain ranges. The data are graphed on a log-log scale. Note the differences in the slopes of the regression lines and the large number of butterfly species compared with mammals and resident birds.

have not been offset by recolonizations, because nonvolant boreal mammals are largely unable to cross arid lowlands between ranges (Brown 1971, 1978). Mountain ranges lose proportionally fewer species of birds and butterflies with diminishing area, apparently because some recolonization can take place after extinction in these vagile taxa (Brown 1978, Wilcox et al. 1986). The importance of vagility is underscored when butterflies are divided into vagility classes. Sedentary butterflies are more sensitive to decreased area than moderately vagile butterflies. Wide-ranging and migratory butterfly species, in contrast, do not show significant species-area relationships (Wilcox et al. 1986).

The use of area as the fundamental variable in this study should not imply that reduction in habitat area alone is the direct causal factor in extinctions. Many important habitat features contribute to or covary with area of mountain ranges, including elevation of the highest peak, diversity of habitats, and extent of perennial flowing water. Indeed, a habitat diversity score for birds, based on the number of coniferous tree species and the number of major habitat types, has proven to function better than area as a predictor of

resident bird diversity (Johnson 1978, Behle 1978, Brown 1978).

Species-area relationships, however, do allow us to generate first-cut estimates of species losses from Great Basin mountain ranges after regional warming of 3°C. We make four assumptions in using species-area relationships for these projections:

1. Boreal habitat zones will ascend in elevation 500 m for every 3°C rise in average temperature.

2. The lower limit of boreal habitat is currently at 2300 m, and its extent can be estimated by the area delineated by the 2300-m contour; thus, the change in boreal habitat area following a regional warming of 3°C can be estimated by the area above the 2800-m contour.

3. The same processes that led to current species-area relationships among taxa will continue to operate; therefore, the new number of species in a range will be governed by the same species-area relationship presently found in the region.

4. Estimates can predict the magnitude of species losses but cannot tell us which species will become extinct.

We are able to estimate whether 10% of present resident species may become extinct in a given mountain range, as opposed to, say, 50% or 80%. Although we would like to differentiate between circumstances in which 30% or 35% of species would be lost, that level of resolution is beyond the predictive capabilities of this model (particularly for ranges supporting already depauperate mammal and resident bird faunas). Some interpretations of species-area statistics have been criticized (e.g., Conner and McCoy 1979), but we believe that the relationships between area of Great Basin mountain ranges and species number in the taxa examined are robust enough to provide a basis for our predictions.

Those caveats stated, we predict that all mountain ranges will suffer reductions and fragmentation of boreal habitat with regional warming of 3°C (fig. 26.1). Reductions range from 66% to 90%, with a mean of about 80%

(table 26.2). Though a number of central Great Basin ranges currently support more than 900 km² of habitat, no range would support more than 400 km² of boreal habitat after 3°C warming.

Significant numbers of extinctions are predicted in all three animal groups for all mountain ranges. The average loss of mammal species is 44%. Resident birds appear less affected. Butterflies suffer an average decrease of 23% of species per range in this set of mountain ranges. Species losses among sedentary and moderately vagile butterflies are higher, at approximately 30%. The large ranges (Toiyabe-Shoshone, Snake, Schell Creek–Egan, Ruby–East Humboldt, and Monitor-Toquima) have postgreenhouse species numbers comparable to current species numbers in the smaller ranges (Grant–Quinn Canyon and White Pine).

The southern Snake Range was recently designated Great Basin National Park, hence it warrants special note. Previous surveys that included the southern Snake Range established species-area relationships for plants (Harper et al. 1978). We applied these data to

the projected reduction of boreal habitat with regional warming in Great Basin National Park (fig. 26.3, table 26.3). Substantial losses of plant species are projected, from 305 currently (326 predicted from the regression equation using fifteen Great Basin ranges) to 254 under the 3° warming scenario. Thus, approximately 17% of plant species could be lost from the newest addition to the national park system. A breakdown of plant groups shows that the burden of species losses is not shared equally. Perennial grasses and forbs show the greatest species losses, about 30% (note the high $z$ values in table 26.1), while approximately 17% of shrub species are lost. Great Basin National Park appears to be somewhat buffered against species losses from regional warming, because of its relatively higher proportion of high-elevation habitat to low-elevation habitat, less linear configuration than other ranges, as well as the great height of Wheeler Peak, at nearly 4000 m (fig. 26.1, fig. 26.3).

All in all, these numbers suggest that species losses at the scale of individual mountain ranges will be on the order of 20%–50% after

Table 26.2. Current and projected boreal habitat areas and associated species numbers for animal taxa in Great Basin mountain ranges. Numbers in parentheses are projected species numbers under a 3°C regional warming that raises the lower limit of boreal habitat from 2300 m to 2800 m. Mammals include neither bats nor large game species (see Brown 1971, 1978). Birds are year-round residents only (see Brown 1978). Bf, butterflies; Bf1, sedentary butterflies; Bf2, moderately vagile butterflies.

| Range | Area (km²) Above | | | Mammals | Birds | Bf | Bf1 | Bf2 |
|---|---|---|---|---|---|---|---|---|
| | 2300m | 2800m | 3100m | | | | | |
| Snake | 982 | 338 | 125 | 10(6) | 9(6) | 74(54) | 12(10) | 45(30) |
| White Pine | 630 | 65 | 5 | 7(3) | — | 52(42) | 9(6) | 34(23) |
| Schell Creek–Egan | 2048 | 335 | 83 | 8(6) | — | 77(54) | 14(10) | 48(30) |
| Monitor-Toquima | 3009 | 377 | 73 | 10(6) | — | 63(55) | 15(10) | 34(30) |
| Toiyabe-Shoshone | 1685 | 372 | 78 | 13(6) | 6(6) | 84(55) | 21(10) | 43(30) |
| Ruby–East Humboldt | 968 | 342 | 80 | 12(6) | 6(6) | 67(55) | 17(10) | 39(30) |
| Grant–Quinn Canyon | 400 | 53 | 5 | 5(3) | 5(4) | 52(41) | 7(6) | 29(23) |
| Average species loss (%) | | | | 44% | | 23% | 32% | 27% |

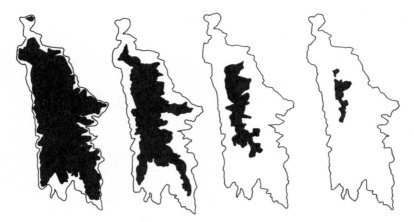

*Figure* 26.3. Boreal habitat area reductions in Great Basin National Park in the southern Snake Range. The outer line is the 2300-m (7500-feet) contour, and successive black areas from left to right are 2450 m (8000 ft), 2750 m (9000 ft), 3150 m (10,000 ft), and 3450 m (11,000 ft).

a 3°C regional warming, but we cannot say which species will be lost without a more detailed look at the basic biology and habitat requirements of the species. The conjecture that alpine and subalpine zone species will be the likely early victims, since those zones may practically disappear from many mountain ranges, may not necessarily prove correct. Some populations of high-elevation species may actually increase over the short term, should growing seasons extend in length (Ehrlich et al. 1980, Murphy and Weiss

1988c). Although these first-cut approximations tell us that substantial numbers of extinctions are likely to occur, they shed little light on either the time scales or mechanisms of individual species extinctions. In an attempt to identify the factors that could cause local extinction in a 3°C global warming scenario, we review here the role of climate in butterfly population dynamics.

## III. GLOBAL CLIMATE CHANGE AND THE BAY CHECKERSPOT BUTTERFLY

We now focus on butterflies for a number of reasons. The butterfly extinctions projected above are numerous compared with other animal taxa. Most species require one or at most a few plant species as larval hosts and

*Table* 26.3. Plant species losses from Great Basin National Park. The first column is the recorded number of plant species, estimated from graphs in Harper et al. 1978. Numbers in parentheses are species numbers estimated by the species-area curves. Other columns are predictions of species numbers after regional warming.

|                 | 2300 m     | 2800 m     | 3100 m    |
|-----------------|------------|------------|-----------|
| Area            | 448 km$^2$ | 204 km$^2$ | 83 km$^2$ |
| All plants      | 305(326)   | 254        | 192       |
| Annuals         | 20(32)     | 25         | 19        |
| Perennial grass | 55(47)     | 35         | 26        |
| Perennial forb  | 200(186)   | 143        | 106       |
| Shrubs          | 35(35)     | 29         | 23        |
| Trees           | 15(11)     | 10         | 10        |

require an array of adult nectar sources. This dependence on plants usually makes butterflies highly habitat specific. For that reason and others, butterflies as a group are sensitive to thermal conditions and often show immediate dramatic responses to climate fluctuations.

The close relationships of butterflies and their host plants, coupled with the steep species-area curves exhibited by plants (table 26.1), suggest that many butterfly host-plant species may become extinct as habitat area diminishes. Yet, butterfly populations often go extinct well before their plant resources disappear completely, as conditions on both macroclimatic and microclimatic scales become unsuitable.

Many of the complex interactions between macroclimate, microclimate, and population persistence in butterflies are vividly illustrated by long-term studies of checkerspot butterflies in coastal California (see Ehrlich et al. 1975; Ehrlich and Murphy 1981, 1987; Murphy and Weiss 1988a for reviews). Here we discuss how this one species reacts to weather patterns and how habitat topography buffers the effects of weather extremes. These observations form the basis for tentative projections of the fates of butterflies under different local climate change scenarios.

The Bay checkerspot is restricted to patches of serpentine soil–based grassland in the San Francisco Bay area (fig. 26.4). Its recent addition to the federal threatened species list reflects local losses of populations to land development and drought (U.S. Federal Register, Sept. 19, 1987). The serpentine soil–based grasslands inhabited by the butterfly can be viewed as another set of island patches: suitable native grassland habitat in a matrix of unsuitable habitat dominated by alien grassland species.

The life cycle of the Bay checkerspot butterfly is closely tied to the winter-spring growing season of California's Mediterranean climate. Larvae emerge from summer diapause following rains in late fall and early winter. The black larvae feed on the annual forb *Plantago erecta* and spend much of their

Figure 26.4. Distribution of serpentine soil–based grasslands in the San Francisco Bay area. The Morgan Hill habitat area is the large patch at the bottom right.

time basking in direct sun to speed metabolism and growth (Weiss et al. 1988). Larvae pupate for several weeks, and adults emerge from late February to early May. Females lay egg masses at the base of *Plantago* and less frequently on the secondary hosts, *Orthocarpus densiflorus* and *O. purpurascens*. Eggs hatch after approximately two weeks, and larvae must reach fourth-instar size to enter dry season diapause.

The single most important factor in population size changes in this butterfly is the survivorship rate of larvae as they enter diapause (Singer 1972). Larvae require 4–5 weeks to develop from egg to fourth instar. When winter rains give way to summer drought, the annual host plants of the larvae senesce rapidly. Once *Plantago* has senesced, larvae may transfer to *Orthocarpus*, which remains edible later in spring. The temporal phase relationship between adult flight and host-plant senescence and the density of *Orthocarpus* largely determine subsequent population size, hence the likelihood of population persistence.

This phase relationship is affected by macroclimatic conditions during the winter-spring growing season, suggesting that population persistence may be tied to future changes in regional climate. Sunny winter days speed larval growth and pupal development, and spring rains extend the host-plant growing season. Patterns of rainfall vary widely between years, and extremes of drought and deluge can cause severe population reductions. During droughts, host plants senesce early, and the dry sunny conditions that speed larval growth cannot compensate for early plant senescence (Singer and Ehrlich 1979, Ehrlich et al. 1980). The 1975–77 drought in California caused severe declines in population sizes and numerous extinctions in the southern range of the butterfly in Santa Clara County, including all inhabited patches save the largest in the region (fig. 26.4). The 1987–90 drought period may produce similar results. At the opposite climatic extreme, the wet El Niño weather sequence in 1981–83 also caused population declines because larval and pupal development was slowed markedly by extended rainy weather, and delayed host-plant senescence did not compensate (Dobkin et al. 1987). Weather extremes, in other words, put Bay checkerspot populations at great risk.

These responses to weather are influenced in each population by the topography of the habitat patch (Singer 1972; Singer and Ehrlich 1979; Weiss et al. 1987, 1988; Murphy and Weiss 1988b). Topographic diversity produces differential solar exposures across habitats, resulting in a complex local mosaic of topographically induced microclimates, or topoclimates (Geiger 1965). Topoclimates can be quantified using calculated clear-sky insolation based on the geometry of sun, earth, and local slope. The shape of insolation curves for different slope exposures is directly reflected in larval development rates and host-plant phenology (fig. 26.5). Larvae on warm south-facing slopes gain an early season growth advantage over larvae on north-facing slopes because south-facing slopes receive significantly more insolation at the winter solstice (at the beginning of postdiapause larval development). Since larval weight gains are directly proportional to insolation, larvae on the warmest habitat slopes

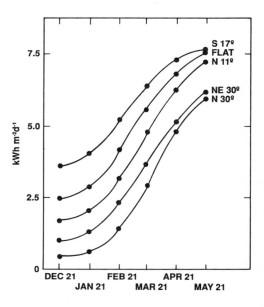

Figure 26.5. Clear-sky insolation curves for selected slopes at latitude 37.5° N.

may emerge as adults one month earlier or more on average than larvae on the coolest slopes (fig. 26.6). Host plants and adult nectar sources on the warmest slopes grow, flower, and senesce many weeks sooner than those on the coolest slopes. The phase relationship of butterflies and host plants, therefore, varies in space as topoclimates interact with macroclimate (Weiss et al. 1987, 1988; Murphy and Weiss 1988b).

Not surprisingly, spatial patterns of prediapause larval survivorship change from year to year, reflecting macroclimatic conditions and the distribution of the previous generation of larvae. Severe droughts tend to shift the larval distribution proportionally toward cooler slopes, which exacerbates the effects of subsequent drought years because few adults are able to emerge early enough for their offspring to find green, edible host plants (Singer and Ehrlich 1979, Ehrlich et al. 1980, Murphy and Weiss 1988b). A shift to warmer slopes during wetter years can buffer a population against subsequent droughts by facilitating earlier adult emergence.

This understanding of the roles of macroclimate and topoclimate in the persistence of Bay checkerspot butterfly populations is crucial to our prediction of the effects of regional warming. More appropriately, we ask whether the Bay checkerspot can survive in its existing habitats, particularly in the large Morgan Hill habitat (fig. 26.4). Because of its size (more than 1500 ha, nearly an order of magnitude larger than any other habitat) and its diverse topography, Morgan Hill supports the population that is most likely to survive a major regional change in climate. In this discussion, the Morgan Hill population may be viewed as existing within the intersection of a suitable habitat space and suitable climate space. The suitable habitat space may be defined as topographically diverse, serpentine soil–based grassland that supports the necessary larval host plants and adult nectar sources. The suitable climate space presently occurs across much of inland San Mateo and Santa Clara counties and the surrounding Inner Coast Range of central California.

The Morgan Hill population, however, is in the portion of the distribution that now receives the lowest rainfall because of the rain shadow of the nearby Santa Cruz Mountains. Many populations inhabiting smaller habitat patches adjacent to Morgan Hill were lost during the 1975–77 California drought, while several populations inhabiting smaller patches in wetter San Mateo County survived. For this reason, we believe that optimal mean rainfall for the butterfly may be somewhat greater than that presently occurring at Morgan Hill; hence a macroclimatic shift that results in a moderate increase in rainfall could well enhance the persistence of this species. Conversely, even a slight decrease in mean seasonal rainfall there could result in substantial population declines over time.

To predict population persistence in an era of climate change, we have developed a simulation model that accounts for the complex interactions among macroclimate, microclimate, and population dynamics of the Bay checkerspot butterfly. We iterate empirical relationships between insolation and growth derived from field studies across an array of slopes over daily weather records for San Jose, near Morgan Hill. The model can predict adult emergence dates from a wide array of topoclimates. We present two examples of that model output (fig. 26.7)—the results of simulation runs for 1976–77, a severe drought year (30% of average annual rainfall), and 1982–83, a very wet year (250% of average).

Gross differences in larval developmental phenologies between these two extreme years include the initiation date of larval feeding, early January in 1976–77 and early December in 1982–83. The larval and pupal periods are much shorter in the drought year (fig. 26.7). Development in much wetter 1982–83 is delayed by extended rainfall in February and March. The mean emergence date (of the four chosen slopes) in 1976–77 is April 1, with a spread of 37 days from the south-facing 11° slope to the north-facing 25° slope. These figures compare with a mean emergence date of April 7 for 1982–83 and a spread of 52 days. The butterflies that emerged early in the drought year were likely

Figure 26.6. Growth of postdiapause larvae on four different slope exposures at Morgan Hill in 1987–88. Note the spread of about one month between the larvae on the warmest slope and the larvae on the coolest slope.

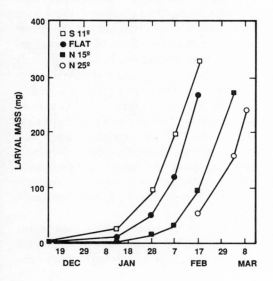

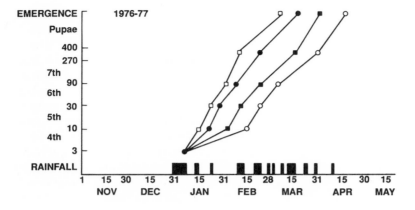

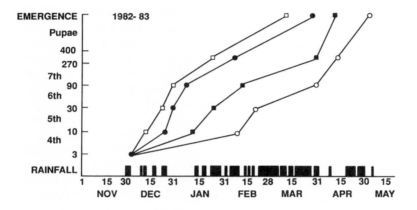

*Figure* 26.7. Simulation runs of Bay checkerspot postdiapause development for 1976–77 (a drought year) and 1982–83 (a very wet year). The vertical axis shows developmental stages, instar number on the outside, and molt weights directly adjacent to the axis. Rainfall days are indicated by black bars along the horizontal axis. Slopes: South-facing, 11° (□); flat (●); north-facing, 17° (■); north-facing, 25° (○).

to be the only ones that reproduced successfully, while the early emerging adults in the deluge year were subject to weeks of rain in March and were unable to fly.

Insights gleaned from these and other model runs can be applied to various global warming scenarios for central California. California is located within an unusual climate zone. Temperature increases during an average global warming may be less there than in

other regions, and precipitation may increase with global warming because of the proximity of the Pacific Ocean (Kellogg and Schware 1981, Wigley et al. 1981; chapter 20). However, predicting the effects of these changes on the Bay checkerspot butterfly depends on a knowledge of climate specifically during the November-April rainy season at a level of detail as yet unachieved in climate models. Indeed, during overall global warming, coastal California could evidence seasonal cooling. To cover all bases, we tentatively explore four broad cases of local climate change: warmer-drier, warmer-wetter, colder-drier, and colder-wetter conditions.

A warmer and drier climate, similar to that in coastal southern California, would increase the chances of severe drought years, like

1975–76, 1976–77, and 1987–88, which undeniably are detrimental to populations of the Bay checkerspot (Ehrlich et al. 1980, Murphy and Weiss 1988a). Under such circumstances, populations decline when larvae evidence a poor phase relationship with host plant availability. Sequential dry years cause larval populations to become concentrated on cooler slopes (that is, they undergo thermal retreat). Over a sequence of dry years a habitatwide shift of surviving larvae from south-facing slopes to flat areas, flat to north exposures, and north to steeper north exposures occurs, with the result that the warmest exposures cease to contribute significantly to the carrying capacity of the habitat. Populations may become extinct in habitats lacking topoclimatic refugia, particularly in habitats in the driest portions of the distributional range. Emergence in surviving populations is then delayed the following year because most larvae are on cooler slopes, leaving those individuals vulnerable to a subsequent deleterious phase relationship. Most important, an overall net loss of suitable habitat within presently occupied habitats is probable.

A colder and drier climate would be even worse for the Bay checkerspot than a warmer and drier climate. Larval development is slowed by cold weather, even during sunny days. Plant senescence still would occur early because of reduced soil moisture in spring. This combination of extended larval and pupal development with a shortened host-plant growing season is particularly disruptive to the phase relationship between butterflies and their host plants.

Warmer and wetter conditions could result in increases in both sizes and numbers of populations of the Bay checkerspot butterfly. Warm winters in California have long periods of sunny days with daytime highs that produce perfect conditions for rapid development of larvae and pupae. Moisture in such years comes from multiday soaking rains of subtropical origin. Larval and pupal development, therefore, can be relatively rapid, and soil moisture remains high in spring because

the ground is saturated late in the season. A good example of such a year is 1985–86, which saw a substantial increase in size of the Morgan Hill population (Murphy and Weiss 1988b). High survivorship because of a favorable phase relationship between larvae and host plants left the majority of larvae residing on warm and moderate slopes—a thermal advance. A sequence of such years could well result in population explosions, reestablishment of peripheral satellite populations, and even defoliation of host plants on slopes heavily populated by larvae. Under such circumstances a population could be a victim of its own success and defoliate the slopes that offer optimal topoclimatic conditions, creating a check on further population growth.

Colder and wetter conditions would make the local climate more like that farther north on the West Coast. Two examples of cold, wet winters are 1981–82 and 1982–83 (El Niño years). This period otherwise produced warm conditions worldwide, pointing to the difficulties of making projections for specific regions (Murphy and Weiss 1988c). Postdiapause larvae at Morgan Hill were largely restricted to cool slopes by 1984 because consecutive spring deluges in 1982 and 1983 delayed larval growth and pupal development more than they extended the host-plant growing season, forcing populations into sequential thermal retreats to cooler slopes. Marked population declines were observed in a number of habitats, although no population extinction events were recorded.

These projections are based on overall trends in climate. Three out of the four scenarios appear unfavorable to the Bay checkerspot, but unpredictable yearly rainfall patterns profoundly affect the impacts of such scenarios on the butterfly. The key factor in determining population persistence is the relative frequency of years causing population declines versus the frequency of years resulting in population increases. Model runs for a variety of years show disparate responses in years with similar rainfall totals, indicating that the timing of rainfall, rather than seasonal totals, is the primary determinant of population fluctuations (Murphy et al. 1990). Furthermore, population responses are dependent on the recent history of the population as manifested by the larval distribution along the topoclimatic gradient.

These projections address only the immediate responses of butterfly populations to climate change. Variations in local temperature and rainfall regimes often result in changes in the composition of plant communities. Such changes can drastically reduce the availability of necessary larval host-plant and nectar resources and can reduce solar exposure at the soil surface. For example, serpentine outcrops in more mesic areas of California tend to be dominated by a chaparral community, with grasslands restricted to the driest, warmest sites. Indeed, the higher elevations at the southern end of the Morgan Hill serpentine outcrop primarily support chaparral species. More moisture could encourage the spread of chaparral species, particularly on the most mesic grassland slopes where isolated shrubs already exist. These north and northeast exposures currently provide a last refuge for the Bay checkerspot in the case of severe multiyear droughts (which are still possible in a wetter climate).

## IV. APPLICATIONS OF BAY CHECKERSPOT BUTTERFLY STUDIES TO PROJECTED EXTINCTIONS IN THE GREAT BASIN

How do the Bay checkerspot butterfly studies apply to the projected extinctions in the Great Basin? First, they suggest mechanisms by which climatic factors determine the persistence of populations of butterflies and other herbivorous insects. Our view of the Bay checkerspot butterfly tracking appropriate thermal and moisture regimes from certain slope exposures to others in its grassland habitat assists us in understanding how Great Basin species may track appropriate topoclimates and resources. The process in boreal habitat in the Great Basin can be represented by laterally ascending stripes across mountain slopes: species move both upward and

to cooler, more mesic exposures. Common proximate causes of natural local extinctions of butterfly populations are extreme weather events, particularly season-long drought or deluge and late or early freezes (e.g., Ehrlich et al. 1972). Our studies of the Bay checkerspot indicate that the diversity of topoclimates within a habitat space provides a buffer against such weather patterns. Diverse topoclimates offer potential local refugia for populations faced by erratic and extreme macroclimate conditions. The chance of population extinction, therefore, is reduced in landscapes of high topographic diversity. This observation implies that mountainous regions, such as the Great Basin, may prove to be well buffered against species losses associated with regional warming compared with areas with less relief.

Local topographic diversity, however, can contribute only so much to fending off butterfly extinctions during a prolonged regional warming trend. As demonstrated by our Great Basin habitat projections, some plant species will undergo radical distributional shifts, and others may be unable to disperse rapidly enough to track shifting climate spaces, both circumstances with potentially fatal results for populations of butterflies and other species dependent on them. Many plant distributions will narrow as plants become restricted to higher elevations or to slopes of cooler exposure. Certain topoclimates that now act as refugia in stressful years may become the only available habitat. Conversely, species adapted to xeric conditions may expand their ranges.

All of the factors producing topoclimatic variations apply to Great Basin mountain ranges and provide a framework in which to assess changes on a local scale. The elevational gradient in temperature may be viewed as being fractioned into topoclimates on different slopes. Because they receive more rain, the western sides of these mountain ranges are generally more mesic than the eastern sides. North- and east-facing slopes tend to be more mesic than adjacent south- and west-facing slopes because of lower insolation.

Some of these factors have been discussed as they affect species diversity in piñon-juniper forests (West et al. 1978), and numerous other situation-specific examples exist in the literature on mountain ecology. For example, relatively low elevation areas at the base of Great Basin glacial cirques often support alpine species because cold air drainage from higher elevations increases the frequency of frosts (Billings 1978). But, with few exceptions, as species ascend in elevation to follow ascending habitat spaces, they will find that the extents of cool and moist habitats contract.

Even this rather pessimistic view of dwindling habitat spaces is actually optimistic, because it presupposes the ability of organisms to track moving climate spaces and resources. In cases of vagile species like birds or readily dispersed plants, the optimistic view is perhaps justified. But, for the vast majority of species—species that are sedentary, not easily dispersed, highly host-specific, or dependent on other species that have limited vagility—the outlook is grim. Many global warming models suggest a dauntingly short time scale for substantial change, which would not permit populations to make major range adjustments or to evolve in situ to increasing temperature and aridity.

Furthermore, an array of insidious processes is likely to accompany regional warming. For instance, a major mechanism shaping ecosystems in western North America is fire. Lightning fires are a common feature in this largely arid region. One can envision a scenario where parched Great Basin piñon-juniper forests on lower slopes ignite and eliminate mature trees over large areas. Climatic conditions in subsequent years might not allow seedling establishment or survival (see chapter 19 for a more extended discussion of forest regeneration). Many populations of perennial plants depend on intermittent good years for reproduction. Should such years decrease in number, occurring at intervals longer than the lifespans of individual plants or seed bank survival times, then regional extinction could become substantially more likely.

Although we have only begun to address the potential effects of global warming on biological diversity, this chapter and others in this volume echo a common theme—species backed into ecological corners, population disappearances, and disrupted ecosystems. Predicted local and regional climate scenarios promise at the very least a different natural world. The speed at which climate may change assures that this different world will be one that is biologically less rich and less stable than our present one. It will be a world less able to absorb and ameliorate our inevitable future mistakes in land and resource use, a natural world that will, in many ways, seem highly unpredictable and terribly unfamiliar.

## REFERENCES

Austin, G. T., and D. D. Murphy. 1987. Zoogeography of Great Basin butterflies: Patterns of distribution and differentiation. *Great Basin Naturalist* 47:186.

Behle, W. H. 1978. Avian biogeography of the Great Basin and Intermountain Region. *Great Basin Naturalist Memoirs* 2:55.

Billings, W. D. 1978. Alpine phytogeography across the Great Basin. *Great Basin Naturalist Memoirs* 2:105.

Brown, J. H. 1971. Mammals on mountaintops: Nonequilibrium insular biogeography. *Am. Naturalist* 105:467.

Brown, J. H. 1978. The theory of insular biogeography and the distribution of boreal birds and mammals. *Great Basin Naturalist Memoirs* 2:209.

Conner, E. F., and E. D. McCoy. 1979. The statistics and biology of the species-area relationship. *Am. Naturalist* 113:791.

Dobkin, D. S., I. Olivieri, and P. R. Ehrlich. 1987. Rainfall and the interaction of microclimate with larval resources in the population dynamics of checkerspot butterflies (*Euphydryas editha*) inhabiting serpentine grasslands. *Oecologia* 71:161.

Ehrlich, P. R., and D. D. Murphy. 1981. The population biology of checkerspot butterflies (*Euphydryas*): A review. *Biol. Zentral.* 100:613.

Ehrlich, P. R., and D. D. Murphy. 1987. Conservation lessons from long-term studies of checkerspot butterflies. *Conser. Biol.* 22:122.

Ehrlich, P. R., D. E. Breedlove, P. F. Brussard, and M. A. Sharp. 1972. Weather and the "regulation" of subalpine butterfly populations. *Ecology* 53:243.

Ehrlich, P. R., R. R. White, M. C. Singer, S. W. Mc-Kechnie, and L. E. Gilbert. 1975. Checkerspot butterflies: A historical perspective. *Science* 188: 221.

Ehrlich, P. R., D. D. Murphy, M. C. Singer, C. B. Sherwood, R. R. White, and I. L. Brown. 1980. Extinction, reduction, stability, and increase: The responses of checkerspot butterfly populations to the California drought. *Oecologia* 46:101.

Geiger, R. 1965. *The Climate near the Ground.* Cambridge: Harvard University Press.

Grayson, D. K. 1983. Paleontology of Gatecliff Shelter: Small mammals. In *The Archeology of Monitor Valley,* D. H. Thomas, J. O. Davis, D. K. Grayson, W. N. Melhorn, T. Thomas, and D. Trexler, 2: *Gatecliff Shelter.* Anthropological Papers of the American Museum of Natural History 59.

Harper, K. T., D. C. Freeman, W. K. Ostler, and L. G. Klikoff. 1978. The flora of Great Basin mountain ranges: Diversity, sources, and dispersal ecology. *Great Basin Naturalist Memoirs* 2:81.

Johnson, N. K. 1978. Patterns of avian geography and speciation in the Intermountain region. *Great Basin Naturalist Memoirs* 2:137.

Kellogg, W. W., and R. Schware. 1981. *Climate Change and Society: Consequences of Increasing Atmospheric Carbon Dioxide.* Boulder, Colo.: Westview Press.

LaMarche, V. C., and H. A. Mooney. 1967. Altithermal timberline advance in western United States. *Nature* 213:980.

MacArthur, R. H., and E. O. Wilson. 1967. *The Theory of Island Biogeography.* Princeton: Princeton University Press.

Murphy, D. D., K. E. Freas, and S. B. Weiss. 1990. An "environment-metapopulation" approach to population viability analysis for a threatened invertebrate. *Conser. Biol.* 4:41.

Murphy, D. D., and S. B. Weiss. 1988a. Ecological studies and the conservation of the Bay checkerspot butterfly, *Euphydryas editha bayensis.* Biol. Conser. 46:183.

Murphy, D. D., and S. B. Weiss. 1988b. A long-term monitoring plan for a threatened butterfly. Conser. Biol. 2:367.

Murphy, D. D., and S. B. Weiss. 1988c. Drought, deluge, and endangered species. *End. Spe. Update* 5(6):6.

Murphy, D. D., and B. A. Wilcox. 1986. Butterfly diversity in natural forest fragments: A test of the validity of vertebrate-based management. In *Modeling Habitat Relationships of Terrestrial Vertebrates,* J. Verner, M. L. Morrison, C. J. Ralph, and R. H. Barret, eds., pp. 287–292. Madison: University of Wisconsin Press.

Peters, R. L., and J. D. Darling. 1985. The greenhouse effect and nature reserves. *Bioscience* 35: 707.

Singer, M. C. 1972. Complex components of habitat suitability within a butterfly colony. *Science* 173:75.

Singer, M. C., and P. R. Ehrlich. 1979. Population dynamics of the checkerspot butterfly *Euphydryas editha.* Fortschr. Zool. 25:53.

Weiss, S. B., R. R. White, D. D. Murphy, and P. R. Ehrlich. 1987. Growth and dispersal of larvae of the checkerspot butterfly, *Euphydryas editha.* Oikos 50:161.

Weiss, S. B., D. D. Murphy, and R. R. White. 1988. Sun, slope, and butterflies: Topographic determinants of habitat quality for *Euphydryas editha bayensis. Ecology* 69:1386.

Wells, P. V. 1983. Paleobiogeography of montane islands in the Great Basin since the last galciopluvial. *Ecol. Monogr.* 53:341.

West, N. E., R. J. Tausch, K. H. Rea, and P. T. Tueller. 1978. Phytogeographical variation within juniper-pinyon woodlands of the Great Basin. *Great Basin Naturalist Memoirs* 2:119.

Wigley, T.M.L., P. D. Jones, and P. M. Kelley. 1981. Scenario for a warm, high carbon dioxide world. *Nature* 283:17.

Wilcox, B. A., D. D. Murphy, P. R. Ehrlich, and G. T. Austin. 1986. Insular biogeography of the montane butterfly faunas in the Great Basin: Comparisons with birds and mammals. *Oecologia* 69: 188.

# Contributors

Vera Alexander
Institute of Marine Science
University of Alaska
Fairbanks, Alaska

W. Dwight Billings
Botany Department
Biological Sciences
Duke University
Durham, N.C.

Daniel Botkin
Department of Biological Sciences
University of California, Santa Barbara
Santa Barbara, Calif.

Arthur J. Bulger, Jr.
Department of Environmental Sciences
University of Virginia
Charlottesville, Va.

Robin Carper
Department of Ecology and
Evolutionary Biology
Princeton University
Princeton, N.J.

Wendell P. Cropper, Jr.
Department of Forestry
University of Florida
Gainesville, Fla.

Virginia H. Dale
Environmental Sciences Division
Oak Ridge National Laboratory
Oak Ridge, Tenn.

Margaret B. Davis
Department of Ecology
and Behavioral Ecology
University of Minnesota
Minneapolis, Minn.

William R. Dawson
Museum of Zoology
University of Michigan
Ann Arbor, Mich.

Andrew Dobson
Department of Ecology and
Evolutionary Biology
Princeton University
Princeton, N.J.

William K. Ferrell
Professor Emeritus
College of Forestry
Oregon State University
Corvallis, Oreg.

Jerry F. Franklin
USDA Forest Service
Pacific Northwest Research Station
and College of Forest Resources
University of Washington
Seattle, Wash.

Peter H. Gleick
Pacific Institute for Studies in Development,
Environment, and Security
Berkeley, Calif.

Russell W. Graham
Quaternary Studies Center
Illinois State Museum
Springfield, Ill.

Stanley V. Gregory
Department of Fish and Wildlife
Oregon State University
Corvallis, Oreg.

Larry D. Harris
Department of Wildlife and Range Sciences
University of Florida
Gainesville, Fla.

John Harte
Energy and Resources Group
University of California
Berkeley, Calif.

Mark E. Harmon
College of Forestry
Oregon State University
Corvallis, Oreg.

Gary S. Hartshorn
World Wildlife Fund
Washington, D.C.

Bruce P. Hayden
Department of Environmental Sciences
University of Virginia
Charlottesville, Va.

Deborah Jensen
Energy and Resources Group
University of California
Berkeley, Calif.

David Larsen
College of Forest Resources
University of Washington
Seattle, Wash.

John D. Lattin
Department of Entomology
Oregon State University
Corvallis, Oreg.

Robert T. Lester
W. Alton Jones Foundation
Charlottesville, Va.

Thomas E. Lovejoy
Smithsonian Institution
Washington, D.C.

George P. Malanson
Department of Geography
University of Iowa
Iowa City, Iowa

M. Geraldine McCormick-Ray
Department of Environmental Sciences
University of Virginia
Charlottesville, Va.

Arthur McKee
College of Forestry
Oregon State University
Corvallis, Oreg.

Joseph E. Means
USDA Forest Service
Pacific Northwest Research Station
Corvallis, Oreg.

Linda Mearns
National Center for Atmospheric Research
Boulder, Colo.

Dennis D. Murphy
Center for Conservation Biology
Department of Biology
Stanford University
Stanford, Calif.

J. P. Myers
W. Alton Jones Foundation
Charlottesville, Va.

Norman Myers
Headington, Oxford
United Kingdom

Robert A. Nisbet
Department of Biological Sciences
and Environmental Studies Program
University of California
Santa Barbara, Calif.

David A. Perry
College of Forestry
Oregon State University
Corvallis, Oreg.

Robert L. Peters
W. Alton Jones Foundation
Charlottesville, Va.
and Global Change Program
Conservation International
Washington, D.C.

Kim Moreau Peterson
Department of Biological Science
Clemson University
Clemson, S.C.

G. Carleton Ray
Department of Environmental Sciences
University of Virginia
Charlottesville, Va.

Daniel I. Rubenstein
Biology Department
Princeton University
Princeton, N.J.

Stephen H. Schneider
Interdisciplinary Climate Systems
National Center for Atmospheric Research
Boulder, Colo.

Timothy D. Schowalter
Department of Entomology
Oregon State University
Corvallis, Oreg.

Herman H. Shugart
Department of Environmental Sciences
University of Virginia
Charlottesville, Va.

Thomas M. Smith
Department of Environmental Sciences
University of Virginia
Charlottesville, Va.

Michael E. Soulé
Department of Environmental Studies
University of California
Santa Cruz, Calif.

Thomas A. Spies
USDA Forest Service
Pacific Northwest Research Station
Corvallis, Oreg.

Frederick J. Swanson
USDA Forest Service
Pacific Northwest Research Station
Corvallis, Oreg.

Margaret Torn
Energy and Resources Group
University of California
Berkeley, Calif.

C. Richard Tracy
Department of Biology and Program for
Ecological Studies
Colorado State University
Fort Collins, Colo.

Thompson Webb III
Department of Geological Sciences
Brown University
Providence, R.I.

Stuart B. Weiss
Center for Conservation Biology
Department of Biology
Stanford University
Stanford, Calif.

Walter E. Westman
Lawrence Berkeley Laboratory
Applied Science Division
University of California
Berkeley, Calif.

Walter G. Whitford
Department of Biology
New Mexico State University
Las Cruces, N.Mex.

F. Ian Woodward
Department of Animal and Plant Sciences
University of Sheffield
Sheffield, Yorkshire
United Kingdom

George M. Woodwell
Woods Hole Research Center
Woods Hole, Mass.

Catherine Zabinski
Department of Ecology and
Behavioral Ecology
University of Minnesota
Minneapolis, Minn.

# Index